Frank Rösch

Nuclear- and Radiochemistry

Also of Interest

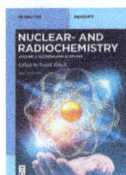

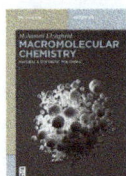

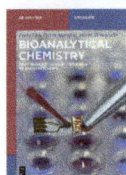

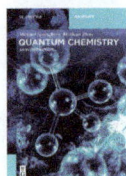

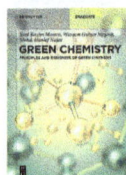

Frank Rösch

Nuclear- and Radiochemistry

Volume 1: Introduction

2nd, extended edition

DE GRUYTER

Author
Prof. Dr. Frank Rösch
Institute of Nuclear Chemistry
Johannes Gutenberg University Mainz
Fritz-Strassmann-Weg 2
55128 Mainz, Germany
E-mail: frank.roesch@uni-mainz.de

ISBN 978-3-11-074271-8
e-ISBN (PDF) 978-3-11-074272-5
e-ISBN (EPUB) 978-3-11-074280-0

Library of Congress Control Number: 2022938840

Bibliographic information published by the Deutsche Nationalbibliothek
The Deutsche Nationalbibliothek lists this publication in the Deutsche Nationalbibliografie;
detailed bibliographic data are available on the Internet at http://dnb.dnb.de.

© 2022 Walter de Gruyter GmbH, Berlin/Boston
Cover image: agsandrew/iStock/Getty Images Plus
Typesetting: Integra Software Services Pvt. Ltd.
Printing and binding: CPI books GmbH, Leck

www.degruyter.com

Preface

Except for the element hydrogen with its most common stable isotope ^{1}H, the nuclei of all atoms consist of mixtures of protons and neutrons.[1] For certain well-defined nucleon compositions (i.e. protons + neutrons), nuclei are stable – they exist forever. This holds true for one or more nuclei of the isotopes of almost all chemical elements (except technetium, $Z = 43$) ranging from hydrogen ($Z = 1$) to bismuth ($Z = 83$). Yet, there are less than 300 stable nuclei altogether.

In contrast, all the known elements beyond bismuth (i.e. from $Z = 84$ to $Z = 119$) and many of the lighter elements (i.e. from $Z = 1$ to $Z = 83$) have isotopes comprising nuclei of nucleon compositions not suitable to stability. This means that there are more than 3000 unstable nuclei that must transform into new nuclei of more stable nucleon compositions. This textbook is about the many atomic nuclei and their characteristic states that form the atoms composing our physical world: stable nuclei (K), nuclei with excited nuclear levels ($^\circ$K), unstable nuclei (*K), nuclei at ground state (^{g}K), or metastable states (^{m}K).[2]

Radiochemical transformations proceed in an exothermic way. The velocity of a transformation is expressed by a transformation constant and/or half-life.

The transformation itself is accompanied by the release of various kinds of emissions. It is the velocities and emissions of these transformations that stand for the phenomenon of "radioactivity".

The understanding of basic nuclear and radiochemical processes is a prerequisite to exploring the potential of radionuclides and radioactivity as a source of energy, as tools in fundamental research and analytics, for environmental purposes, for industrial applications, in medicine, etc. For example, nuclear fission is one of the most important sources of electricity, while the beta- and alpha-decay processes are the essence of several molecular imaging processes adopted in diagnostic nuclear medicine and for patient treatments, respectively. These and several other topics are addressed in the second volume of this textbook.

The first volume of this textbook introduces the basic aspects of these processes. It focuses on explaining the fundamentals, rather than on specific details and mathematical treatments. Mathematics, and in particular (quantum) physical models, are referred to only when conventional physics cannot explain an important experimental observation.

Thus, this textbook serves as a guide to qualitatively understand the essence of radioactivity, considering questions such as:

---

1 H consists of a single proton.
2 K = Kern (German for "nucleus").

https://doi.org/10.1515/9783110742725-202

- What makes a nucleus stable (or unstable)?
- What is the "motivation" for an unstable nucleus to transform?
- What are the velocities of these transformations?
- What are the different primary routes of transformation?
- What is the rationale of each route?
- What are the consequences of primary transformation routes in terms of secondary pathways and post-effects?
- How is a particular route explained by both experimental evidence and theoretical models?
- What are the main properties of the emissions (or "radiation") released during a transformation?
- How are the emission parameters of certain radionuclides adopted for practical applications, making them valuable tools in research, and industrial and medical fields?
- And finally, how can some of the particularly important radionuclides be produced artificially?

The intention of this textbook is to illuminate the concepts of radioactive transformations. Excellent textbooks on nuclear and particle physics and quantum mechanics are available for more detailed discussions of many of the phenomena. For a comprehensive description of specific aspects of nuclear- and radiochemistry, the six-volume, over 3000 page "Handbook of Nuclear Chemistry", A Vértes, S Nagy, Z Klencsár, RG Lovas, F Roesch (eds.), second edition, Springer, 2011, can be recommended. Thus, this textbook may be useful for bachelor's students who are interested not only in nuclear sciences, radio- and nuclear chemistry, radiopharmaceutical chemistry, nuclear analytics, nuclear energy, but also in nuclear medicine.

This textbook may also assist professionals working in the latter field (chemists, physicists, physicians, medical-technical assistants), who handle increasingly numerous radioactive isotopes in diagnostic and therapeutic nuclear medicine.

Mainz, December 2013

Preface for the 2nd edition

Nuclear and radiochemistry is, compared to many other branches of chemistry, relatively young. Beginning with Marie Curie's identification of radium and coining of the term "radioactivity", the field of nuclear and radiochemistry is about 120 years old (or young, if you will). Within this period, significant knowledge has been gained. In the 12 years since the first edition of this textbook was published, no fundamental changes occurred which had to be considered in the present edition. This second edition gives some modified explanations and makes several corrections. I am extremely grateful to all the students and colleagues who added valuable comments.

If not indicated otherwise, data used throughout the first edition of this textbook were from the AME2012 atomic mass evaluation. This database is regularly updated, with new editions published in 2016 and 2020. The second edition of the textbook refers to the 2020 database by M Wang et al., Chinese Phys. C 45 (2021) 030003 (https://iopscience.iop.org/article/10.1088/1674-1137/abddaf/pdf). Because more and more the value of mass excess is preferred compared to mass defect, the second edition highlights this value more than in the first edition (covered in Chapter 2).

Finally, the readers' attention is directed to the second volume of this teaching book, which covers the many applications of nuclear and radiochemistry. An impressive and steadily increasing number of chemists, physicists, physicians, technologists, radiation safety personnel, and many other professionals are handling a variety of radionuclides for very different purposes. Volume 2 of the textbook is entitled NUCLEAR- AND RADIOCHEMISTRY: MODERN APPLICATIONS. Prominent scientists illustrate in 13 chapters the most relevant directions, namely (1) Radiation measurement, (2) Radiation dosimetry, (3) Elemental analysis by neutron activation, (4) Radioisotope mass spectrometry, (5) Nuclear dating, (6) Chemical speciation of radionuclides in solution, (7) Radiochemical separations, (8) Radioelements: Actinides, (9) Radioelements: Transactinides, (10) Nuclear energy, (11) Life sciences: Isotope labeling with tritium and carbon-14, (12) Life sciences: Nuclear medicine diagnosis, and (13) Life sciences: Therapy.

Vichuquen, December 2021 | Berlin, May 2022
Frank Rösch

https://doi.org/10.1515/9783110742725-203

Contents

1 The atom's structure I: Electrons and shells

Aim: The concept of atoms as the smallest, indivisible constituents of matter is introduced. It arises from more than two millennia old Greek philosophy, reaches the atom's renaissance in the nineteenth century, and reflects the dramatic improvements achieved at the beginning of the twentieth century.

Atoms are made of electrons, which exist in the shell of an atom, and a set of different particles located in the atom's nucleus. To understand the chemical properties of an atom, the electron shell structure – i.e. the number, characteristics, and transitions of electrons in various shells – is essential. To a large extent, this understanding needs a significant reflection on quantum physics.

Developments of concepts of ancient and modern atom theory are introduced, turning from the atomic philosophy of being the ultimate particle to concepts of the atom being a substance of various, subatomic particles. The latter ideas are illustrated as a scientific need to understand pioneering experiments in chemistry and physics, i.e. to correlate experimental evidence and subsequent theoretical explanation.

The latter is meant to serve as an introduction to (quite similar) considerations relevant to the structure of nucleons in the atomic nucleus.

1.1 The philosophy of atoms

1.1.1 The beginning: Just philosophy

The concept of an atom was approached about 24 centuries ago when ARISTOTLE (384–322 BC) suggested that all existing matter is built of four components: air, water, fire, and earth. PLATO (427–347 BC) added a fifth element, the ether, which allows for interaction and transformation between the others. The five Platonic solids are illustrated in Fig. 1.1. Interestingly, they are each composed of just one of only three single abstract geometric figures (equilateral triangle, square, and a regular pentagon), which when assembled form the tetrahedron, hexahedron, octahedron, icosahedron, and dodecahedron.

At almost the same time, the Greek philosopher LEUCIPPUS (ca. 450–370 BC) and his scholar DEMOCRITUS (460–371 BC) proposed a material world made of indivisible components. Assuming that every material could be divided into smaller parts, and these again into even smaller fragments; finally, a level of indivisible things is reached. They called them *atomos*, which is the Greek name of atoms, the indivisible. The philosophers ascribed them with defined properties, namely to be:
– very small,
– and thus invisible,

https://doi.org/10.1515/9783110742725-001

- indivisible,
- hard,
- of different forms (however without color, taste, or smell),
- moving spontaneously and continuously (in an empty space, i.e. *in vacuo* or in "ether").

FIRE	EARTH	AIR	WATER	ETHER

tetrahedron	hexahedron	octahedron	icosahedron	dodecahedron

Fig. 1.1: The five Platonic solids. Each is composed of a number of identical isosceles surfaces: tetrahedron (4 triangles), hexahedron (6 squares), octahedron (8 triangles), icosahedron (20 triangles), dodecahedron (12 pentagons).

To conceive of the ultimate building blocks[1] as having "different forms" included the elegant idea that the matter built of them finally reveals properties of the "atoms", which themselves remain invisible. Ultimately, the building blocks are not only the composites of our material world, they are responsible for its properties.

1.1.2 2000 Years later: Experimental evidence

It took more than 22 centuries until this conceptual idea was proven experimentally. Chemists like LAVOISIER (1743–1794), PROUST (1755–1826), and DALTON (1766–1844) realized that individual chemical elements combine mass-wise to form compounds according to a set of rules.

Conservation of mass (LAVOISIER): The overall mass represented by all reaction partners remains constant. For a chemical reaction, the total mass is divided among all the species involved and, in the case of complete reaction, the mass of all reaction products is equal to the mass of the reactants. If hydrogen gas and oxygen gas undergo the oxyhydrogen gas reaction, then e.g. 4 g of H_2 reacts with 32 g of O_2 (together making 36 g) to form 36 g of water. This experimental evidence is the law of conservation of mass (see Tab. 1.1).

---

1 It would become obvious only more than 2000 years later that this is not really true; it is more the number and the internal infrastructure of a well-defined, but small group of subatomic building blocks called "elementary particles" which define an "atom". These constructs indeed define the physical and chemical properties of atoms and chemical elements.

Law of definite proportions (PROUST): Species undergoing a chemical reaction combine not stochastically, but in characteristic ratios. Today, we know that this is according to the number of moles, which in turn represent the number of species. In the case where the two gases hydrogen and oxygen react to form water, hydrogen, and oxygen combine in molar ratios of 2:1. According to their individual masses, this reflects a mass ratio of 1:8 of the initial reactants. This experimental fact constitutes the law of definite proportions (see Tab. 1.1).

Tab. 1.1: The laws of conservation of mass and of definite proportions exemplified for the formation of water out of the two gases hydrogen and oxygen. The ratio between the two elements is 1:8 if masses are counted, and 2:1 if moles are considered. Using the AVOGADRO number to convert moles into numbers of atoms, it is obvious that the overall number of atoms of the reactants (H_2 and O_2) is the same for the product (H_2O), and that the number of both hydrogen and oxygen atoms starting the reaction did not change, even if the reaction product is chemically a completely different species.

	$2\,H_2$	+	$1\,O_2$	→	2	H_2O	
mass	$2 \times (2 \times 1\text{ g})$	+	$1 \times (2 \times 16\text{ g})$	=	$2 \times$	$(2 \times 1 + 16)\text{ g}$	$= (2{:}16)\text{ g}$
							mass ratio H:O $= 1{:}8$
			36 g	=	36 g		
mol	2×2 mol of H		1×2 mol of O		$2 \times$	(2 mol of H + 1 mol of O)	
							molar ratio H:O $= 2{:}1$
atoms	$4 \times 6.022 \cdot 10^{+23}$	+	$2 \times 6.022 \cdot 10^{+23}$	=	$2 \times$	$(2 \times 6.022 \cdot 10^{+23} + 1 \times 6.022 \cdot 10^{+23})$	
total			$6 \times 6.022 \cdot 10^{+23}$	=	$6 \times 6.022 \cdot 10^{+23}$		
H			$4 \times 6.022 \cdot 10^{+23}$	=	$4 \times 6.022 \cdot 10^{+23}$		
O			$2 \times 6.022 \cdot 10^{+23}$	=	$2 \times 6.022 \cdot 10^{+23}$		

These chemical observations allowed DALTON (1766–1844) to develop a "modern" theory on atoms. The fundamental postulates in 1808 were:
- chemical elements consist of extremely small particles, the atoms,
- all atoms of one specific chemical element are identical, while
- atoms of different elements are different,
- chemical reactions reflect the combination or separation of atoms,
- within these reactions, atoms do not disappear, i.e. are not destroyed or created,
- no atom of one chemical element can be transformed into an atom of a different chemical element,
- a chemical compound is a combination of characteristic atoms of one (e.g. H_2), two (e.g. H_2O) or more chemical elements,
- a specific chemical compound contains atoms in a fixed, compound-specific mass ratio (see Tab. 1.2).

Tab. 1.2: Key messages of DALTON'S theory ("philosophy") on atoms.

Atoms	Atoms are extremely small particles composing chemical elements.
	Atoms of the same element are identical.
	Atoms of different elements are different.
Chemical compounds	A chemical compound is the combination of characteristic atoms of individual chemical elements.
	A specific chemical compound contains atoms in a fixed, compound-specific mass ratio.
Chemical reactions	A chemical reaction is the combination of atoms or the separation of atoms forming a chemical compound.
	Within these reactions, atoms do not disappear, i.e. are neither destroyed nor created.
	No atom of one chemical element is transformed into an atom of a different chemical element.

DALTON'S key point was the renaissance of the assumptions made by the Greek philosophers more than two millennia ago. Chemical elements are made of atoms, representing the smallest indivisible constituents of matter (Fig. 1.2).

1.2 The inner structure of the atom

1.2.1 The electron and the proton

The identification of electricity and electrolysis were key factors in the development to make the "classical" atom "transparent". These two effects appeared to be completely new to science and stimulated new theories on the atom. FARADAY (1791–1867) experimentally observed that electric current may disrupt chemical compounds. STONEY (1826–1911) suggested the existence of "carriers of electric charges", and that those carriers were associated with the atom. In 1891 he called them "electrons". PLÜCKER (1801–1868) tried to conduct electric current through vacuum and in 1859 observed cathode rays. THOMSON (1856–1940) systematically analyzed these rays and realized experimentally that small, electrically-charged particles – electrons – were emitted from the atoms of the cathode (see Fig. 1.3). Independent of the type of hot cathode, the electron properties were the same.

ELEMENTS

Simple

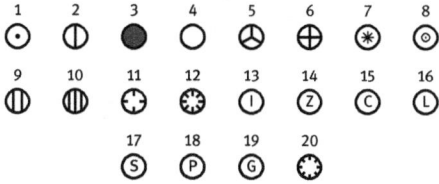

Binary

Ternary

Quaternary

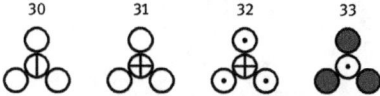

Quinquenary & Sextenary

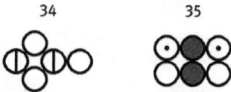

Septenary

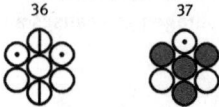

Fig. 1.2: DALTON's "New System of Chemical Philosophy" describing the elements with individual symbols and chemical compounds as individual elements interconnected in specific relationships (stoichiometry), without losing their identity.

This was the experimental evidence of the existence of the electron. THOMSON[2] even managed to measure the intensity of the electrons' deviation in external electric and magnetic fields. From these observations, he derived the first experimental value for the charge-to-mass quotient of the electron, which today is $q/m = -1.759 \cdot 10^8$ C/g.

---

2 JJ THOMSON received the Nobel Prize in Physics for this fundamental invention in 1906 "in recognition of the great merits of his theoretical and experimental investigations on the conduction of electricity by gases".

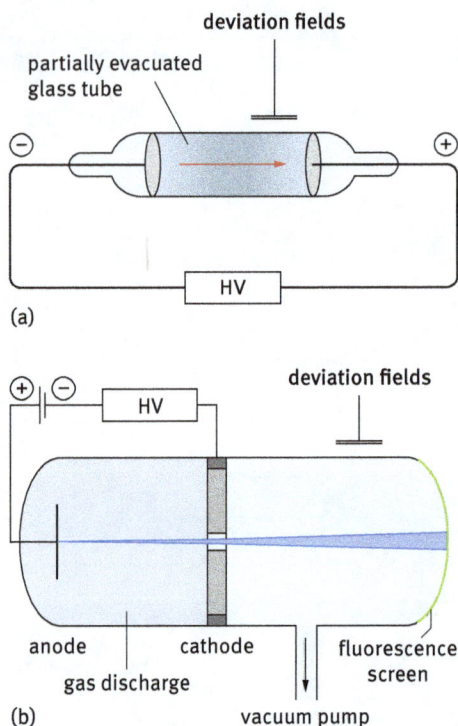

Fig. 1.3: Gas discharge tube experiments. Electric current (HV = high voltage) conducted through vacuum causes emissions – the electrons (a). Gas discharge of hydrogen gas causes a flux of "naked" protons towards a fluorescent screen (b).

The most striking point to realize is that the electron itself is a subatomic particle, thus challenging DEMOCRITUS'S and DALTON'S concept of the atom as the ultimate indivisible species.

Further experiments with similar devices, namely the gas-filled discharge tube, soon revealed another form of particle emission. In 1886 GOLDSTEIN induced gas discharge and detected positively charged components, which were emitted through a hole inside the cathode (see Fig. 1.3). Similar to JJ THOMSON'S approach, the degree of deviation of these positively charged ions of the gas (cations) by external electric and magnetic fields could be analyzed. On the basis of hydrogen emissions, W WIEN derived in 1897 a q/m value of $+ 9.58 \cdot 10^4$ C/g for this cation. He named it "proton"[3] as a tribute to the ancient Greeks; proton means the first, kind of primordial particle: another subatomic particle.

---

3 Nobel Prize in Physics, 1911, "for his discoveries regarding the laws governing the radiation of heat".

Consequently, JJ THOMSON proposed his "plum pudding" model of the atom:

1. The atom is a "substance" (this is new), in which negative charges (electrons) as well as positive charges (protons) are embedded.
2. Because the atom as a whole is neutral (this was known), the number of electrons must equal the number of protons.
3. Electrons and protons are stochastically distributed inside the atom (like raisins in an English plum pudding, or like seeds within a watermelon).

1.3 The shell and the nucleus

With the discussion of two types of subatomic particles, a new conceptual level was reached. However, further insights into the atom's structure were not possible with the classic infrastructure of chemical and physical experimental and analytical methodologies available in the nineteenth century.[4] Similar to the impact of the appearance of electricity on the progress of atomic theory, it was the phenomenon of radioactivity that opened the door to the experimental studies ahead. (Interestingly, while radioactivity required significant progress in the understanding of the inner structure of the atom, radioactivity itself was at the same time a tool to do so.)

In 1896 WC ROENTGEN first measured a new type of radiation originating from uranium ores, which he was unable to identify and therefore named X-rays.[5] Soon after, M CURIE chemically separated the main elements from uranium ore responsible for X-ray emission: radium and (later) polonium.[6] Samples of radioactive radium became a popular tool for a variety of seminal experiments – not only at M and P CURIE'S laboratory in Paris, but also elsewhere. By performing experiments similar to those conducted for the electron and the proton, it was discovered that the principle emissions were small, positively charged species.[7] The charge was twice that of the proton and the mass fourfold that of the proton. Since they were the first radioactive particle emission investigated, they were named "α-particles" after the first letter of the Greek alphabet – in particular by E RUTHERFORD.

---

4 As we know today, excursions into the atom required subatomic tools and energies much higher than those provided by classic chemistry and physics.

5 (First) Nobel Prize in Physics, 1901, "in recognition of the extraordinary services he has rendered by the discovery of the remarkable rays subsequently named after him".

6 Nobel Prize in Physics, 1903, "in recognition of the extraordinary services they (i.e. with Pierre CURIE) have rendered by their joint researches on the radiation phenomena discovered by Professor Henri BECQUEREL; and Nobel Prize in Chemistry, 1911", "in recognition of her services to the advancement of chemistry by the discovery of the elements radium and polonium, by the isolation of radium and the study of the nature and compounds of this remarkable element".

7 See Chapter 5 for the radioactive transformations originating from uranium, radium etc. and Chapter 9 for details on α-transformations.

In 1911, RUTHERFORD conducted a simple but key experiment using a radium-based α-source. He let a beam of emitted α-particles penetrate a thin foil of gold. These subatomic projectiles, emitted from the radium source with a certain kinetic energy, reached the gold atoms. While most of the α-particles had enough kinetic energy to penetrate the foil (and were measured along the line of their origin, albeit with a somewhat weaker abundance and energy than those released from the radium source, as we know today), a significant number of α-particles were recorded at detectors located along all directions possible, indicating reflected or scattered α-particles (Fig. 1.4).

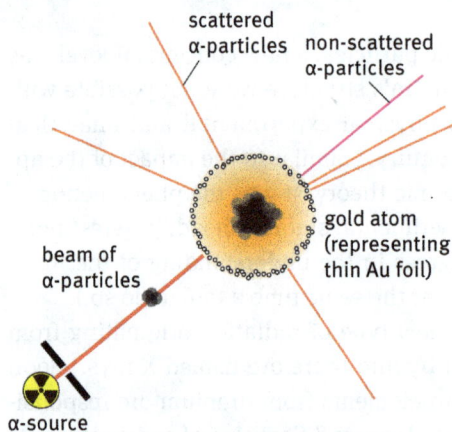

Fig. 1.4: Sketch of the RUTHERFORD experiment, analyzing α-particle[8] scattering[9] on atoms of a thin gold foil.

This observation was not in agreement with the THOMPSON model of an atom, which considered electrons and protons to be distributed homogeneously. In this case, the α-particles should have been unable to penetrate the gold atom for one of two reasons: the density of the 76 protons and the 76 electrons inside the gold atom, or because of coulombic repulsion between the 76 protons and the incoming +2 charged α-particles. Alternatively, if the α-particles could penetrate this substance, it should have occurred along their line of origin. RUTHERFORD explained the experimental result by suggesting a new "RUTHERFORD atom model". It divided the atom into two

---

8 The α-particles emitted from the source originate from the ^{226}Ra itself, but also from other radionuclides involved in ^{226}Ra transformation processes such as ^{218}Po and ^{214}Po.
9 Despite penetrating the atom and scattering (elastically or inelastically), another option would be a nuclear reaction (see Chapter 13). This, however, could not happen in RUTHERFORD'S experiment because of the kinetic energies of the α-particles relative to the high proton number of the gold nucleus.

principal components – the protons, concentrated inside a **nucleus**, and the electrons, located within a **shell.**

RUTHERFORD'S model created a sufficiently empty space between the two subatomic particles known at that time. This space was large enough for two processes: to permit α-particles entering the gold atom just between the nucleus and the outer shell to pass through (and reach the detectors behind the gold foil), and at the same time to reflect and/or scatter incoming α-particles which hit the nucleus (thus reaching detectors in a 4π geometry positioned around the gold foil).

According to the RUTHERFORD atom model,[10] each chemical element is characterized by its proton number (Z). For a neutral atom, the number of protons within the nucleus equals the number of electrons in its shells. Electrons positioned in a shell were considered to "circulate" around the nucleus. Significant energy was anticipated for the speed of this movement in order to withstand the coulomb attraction of the protons and prevent the electrons from being sucked into the oppositely-charged nucleus (Fig. 1.5).

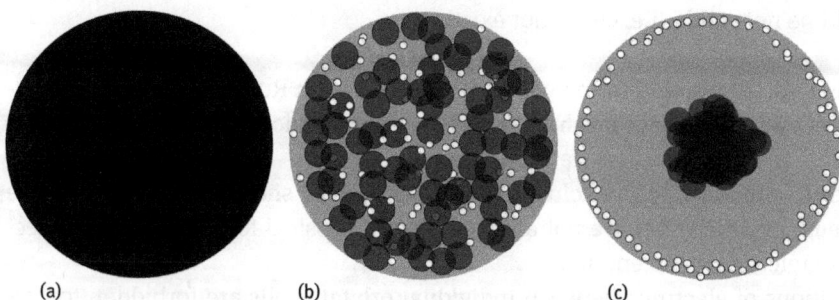

(a) (b) (c)

Fig. 1.5: Schematic representation of the successive levels of atomic "philosophy": (a) The atom as an indivisible component of matter (Democritus, Dalton). (b) Subatomic structures of electrons and protons distributed homogeneously inside the atom (Thomson). (c) Electrons and protons positioned in the shell and nucleus of the atom, respectively (Rutherford). Sizes are not to scale: the proton mass is about 2000× that of the electron mass, with which size scales accordingly. The diameter of a spherical nucleus (<10 fm) is about 1000× smaller than the diameter of the whole atom (ca. 100 pm or 1 Å). Consequently, the "empty" space within the electron shell and the nucleus is much larger than illustrated. The area of the gold atom is on the order of 10^{-20}m², that of the gold nucleus about 10^{-28}m², making an area of empty space of about 99.999999% of the atom's cross-section (see Chapter 2). About 99.99% of the atom's mass is concentrated within the little nucleus!

---

10 Nobel Prize in Chemistry, 1908, "for his investigations into the disintegration of the elements, and the chemistry of radioactive substances".

1.3.1 New model – new problems

While the new atom model could explain the experimentally observed α-scattering, it provoked fundamental questions. Three main concerns were:

1. Why should all the protons be "willing" to concentrate tightly together and stick in the little nucleus – despite the coulomb repulsion of the positively charged particles?
2. When a chemical element is identified by its number of protons (and the same number of electrons in the case of a neutral atom) – why may the chemical properties of one and the same element differ for different chemical compounds containing the same atom?
3. Why might electrons circulate in the shell? Electrically-charged moving particles are known through classic electrodynamics to lose energy. Electrons circulating in a shell around the nucleus and within the field of the nucleus must emit energy. Consequently, they are expected to approach closer and closer to the nucleus – until they are finally trapped there. The Rutherford atom thus would be not stable, i.e. could not exist!

In 1913, N Bohr responded to the contradictions between Rutherford's experimental findings and the laws of mechanics and electrodynamics by extending classical physics into quantum physics. He postulated:

1. Electrons are allowed to exclusively exist at defined shells (orbits) around the nucleus, in which they are not affected by any classical force causing the electron to release kinetic energy.
2. Transitions of electrons between individual orbital shells are forbidden to proceed continuously, but in defined intervals (quanta) only.

Although this idea was deduced somehow from the Ptolemaic conception of our planetary system – with electrons circulating around the atomic nucleus like planets around our sun – this new "Bohr atom model"[11] was not in accordance with classical physics. However, it was able to perfectly explain new experimental observations obtained for the line spectrum of hydrogen (see Fig. 1.6).

Once the hydrogen atom received energy, e.g. by elevated temperature, a set of individual photon emissions was observed, lying within the wavelength spectrum

---

11 Nobel Prize in Physics, 1922, "for his services in the investigation of the structure of atoms and of the radiation emanating from them". Remarkably, Aage Niels Bohr, son of Niels Bohr, continued research on atomic nuclei and in particular on the interaction among nucleons causing nuclear distortion. He shared the Nobel Prize in Physics, 1975, "for the discovery of the connection between collective motion and particle motion in atomic nuclei and the development of the theory of the structure of the atomic nucleus based on this connection".

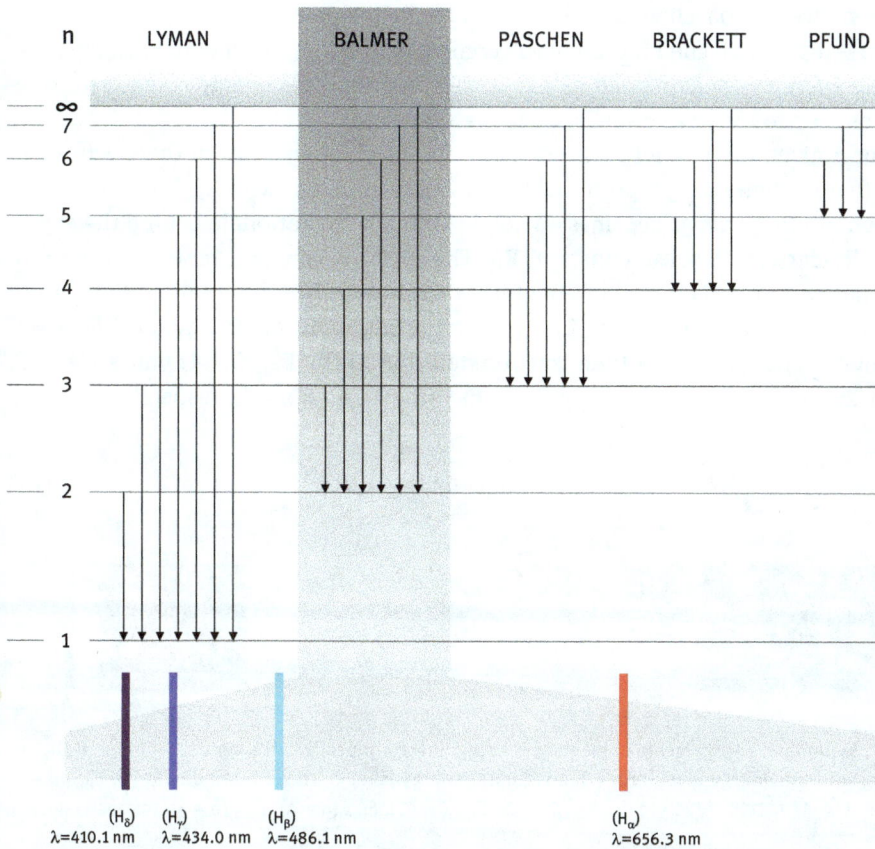

Fig. 1.6: Electron transitions of the hydrogen emission spectrum (schematic). Each emission series (Lyman, Balmer, etc.) corresponds to a different final energy level (n). Notably, emissions in the BALMER series occur at visible wavelengths, shown below in nm.

accessible to the human eye. However, there was just one electron within the neutral hydrogen atom – but sets of characteristic emissions.

1.3.2 Electron shells and electron binding energy

This observation is explained by allowing the electron to occupy several shells (n) of distinct energy. The hydrogen electron is assumed to normally exist in the energetically lowest shell ($n = 1$). When energetically stimulated, it may reach an orbital shells of higher energy ($n = 2, 3, \ldots$), i.e. further from the hydrogen nucleus. Immediately, it would fall back to a lower-energy shell.

Once a certain shell is taken as the baseline, the electron is energetically excited to reach a higher-energy shell, from where it immediately returns. The difference in energy between a baseline and higher-energy shell is emitted as electromagnetic radiation – light. Each shell represents a certain electron binding energy $E_{B(e)}$. (Electron binding energies are expressed as – $E_{B(e)}$, meaning that this energy is required to remove an electron from any shell n up to dissociation.)

For a given atom where Z is constant, the binding energies of the electrons in the different shells are proportional to $1/n^2$. For hydrogen (with Z = 1 and n = 1), the electron binding energy becomes – 13.6 eV, which is the rationale behind the experimentally-derived RYDBERG constant, R_H. The RYDBERG constant represents the proportionality and adds a unit of energy: R_H = 13.6 eV.[12] The internal relationship in energy then looks as indicated in Fig. 1.7. Starting with the energy level for n = 1, any higher energy level is then fractionated due to the $E_{B(e)}(n = 1)$ value 1/4, 1/9, 1/16, 1/25 etc. for $E_{B(e)}(n = 2)$, $E_{B(e)}(n = 3)$, $E_{B(e)}(n = 4)$, $E_{B(e)}(n = 5)$, respectively.

$$E_{B(e)}(n) = -\text{constant} \cdot \frac{1}{n^2} = R_H \cdot \frac{Z^2}{n^2} \qquad (1.1)$$

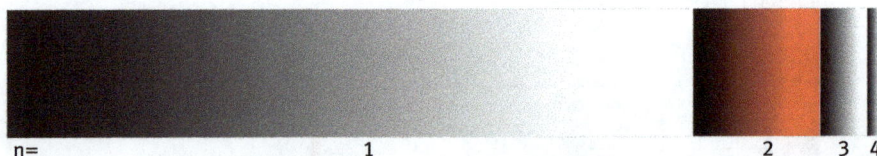

n=   1   2   3 4

Fig. 1.7: Schematic energy scale of electron binding energy in proportions according to eq. (1.1). Starting with the energy level for n = 1, any higher energy level is fractionated due to the $E_{B(e)}(n = 1)$ value as 1/4, 1/9, 1/16, 1/25 etc. for $E_{B(e)}(n = 2)$, $E_{B(e)}(n = 3)$, $E_{B(e)}(n = 4)$, $E_{B(e)}(n = 5)$, respectively.

For different atoms, electron binding energies in individual shells are very much dependent on the number of protons in the nucleus: the higher Z is, the higher $E_{B(e)}$ (n) is by $\approx(1/n)^2$. With eq. (1.1), it is straightforward to calculate the electron binding energy for any electron shell of any atom. For gold (Z = 76), the n = 1 electron binding energy is – 78 553.6 eV; for shells 2, 3, 4 and 5 it is – 19 638.4 eV, – 8 728.2 eV, – 4 909.6 eV and – 3 142.1 eV, respectively. The higher Z is and the closer the electron is to the nucleus, the higher the energy needed to remove it from its shell. For example, the K-shell electron of gold is bound almost six thousand times stronger than that of hydrogen.[13]

---

12 In atomic physics, the energy unit electron volt (eV) is preferred to joule (J): 1 eV = $1.602177 \cdot 10^{-19}$ J; 1 J = 1 kg·m² / s². Alternatively, energy might be translated into mass *via* E = mc², and the RYDBERG constant may be expressed in terms of mass, typically related to the mass of an electron or to the atomic unit.
13 For the following chapters on binding energies of protons and neutrons inside the atom nucleus, it is interesting to note this typical range of electron binding energy, namely less than 100 keV. Typically, chemical reactions proceed at eV scales: 1–7 eV between identical atoms (atom

The transition of an excited electron down to lower-lying shells is well defined by energy, according to the difference in electron binding energy $\Delta E_{B(e)}$ between the two shells. $\Delta E_{B(e)}$ is emitted as electromagnetic radiation of a certain frequency ν and wavelength λ. In 1888 J RYDBERG generalized a semi-empirical expression of the wavelength λ (and wavenumber $\bar{\nu}$) of the hydrogen emission lines. In the case of the BALMER series ($n = 2$), the reciprocal wavelength is

$$\frac{1}{\lambda} = \frac{4}{B}\left(\frac{1}{2^2} - \frac{1}{n^2}\right) \text{ for } n = 3, 4, 5, \ldots \tag{1.2}$$

$$\nu = \frac{\nu}{c} = \frac{1}{\lambda} \tag{1.2a}$$

The $1/2^2$ term corresponds to the $n = 2$ baseline (BALMER); for the other series, this would become $1/1^2$ (LYMAN), $1/3^2$ (PASCHEN), $1/4^2$ (BRACKETT), etc., with n always corresponding to the higher shell level. The lowest energy emission of the BALMER series (α-line) thus corresponds to the $E_{B(e)}(n = 3) \rightarrow E_{B(e)}(n = 2)$ transition, and according to eq. (1.2) the wavelength λ of this quantum emission is 656 nm. It represents a photon lying in the red area of the optical spectrum. The whole BALMER series is within the visible part of the electromagnetic spectrum.[14] Figure 1.8 illustrates the transitions within the different main electron shells of hydrogen, indicating the individual $\Delta E_{B(e)}(n)$ values for the BALMER series (see also Fig. 1.6).

The energy of the electromagnetic radiation is found using $E = m \cdot c^2$ and $E = h \cdot \nu$ (frequency ν, wavelength λ, and wavenumber $\bar{\nu}$). The PLANCK constant h was introduced in quantum mechanics to identify the sizes of energy quanta.[15] It is basically the proportionality constant between the energy E of a photon and its frequency ν ($E = h\nu$). Because frequency is proportional to wavelength by a proportionality factor c (the speed of light), the PLANCK constant also reflects the correlation $E = h \cdot c/\nu$.

$$\Delta E = h \cdot \nu \tag{1.3}$$

1.3.3 Electron shell occupancies

The BOHR atom model not only attributed electrons to specific "allowed" orbital shells, it also arranged the number of electrons to fit into individual shells. A crude

---

binding), 3–15 eV for ion binding, 1–8 eV for metal binding and 0.01–0.2 eV for VAN DER WAALS interactions. The binding energies of nucleons will be within an MeV scale.
14 The line spectrum of hydrogen is a "brilliant" tool in astrophysics, for example. Detecting the α-line and the deviation caused by the DOPPLER effect derives the speed of stars moving in the universe.
15 Units of h are J·s or eV·s, i.e. $6.626 \cdot 10^{-34}$ J·s or $4.135 \cdot 10^{-15}$ eV·s.

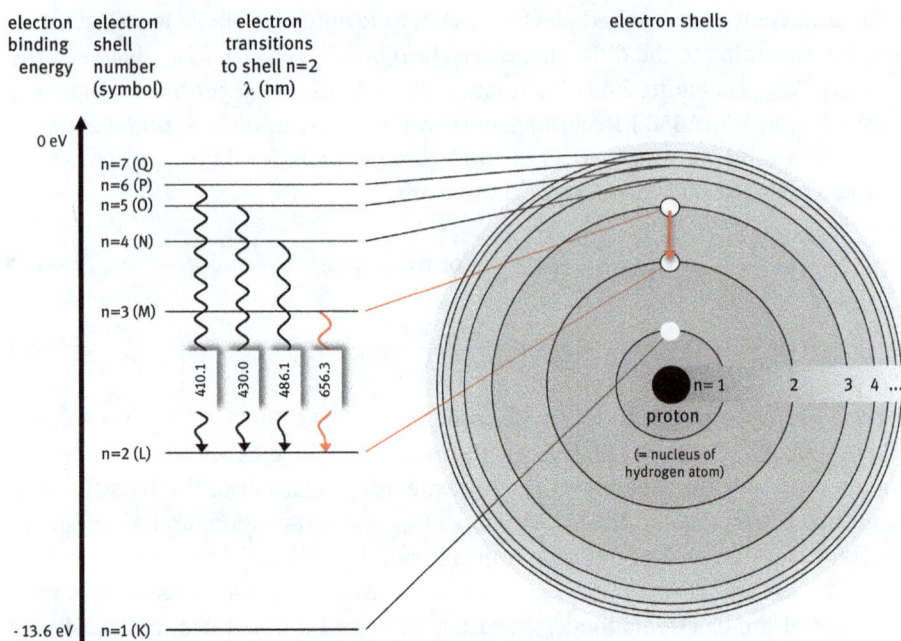

Fig. 1.8: Electronic transitions of the BALMER series. Four emission wavelengths are shown, including the α-line at 656.3 nm in red. The binding energy of the "normal" $n = 1$ electron is 13.6 eV.

viewpoint is to believe that shells close to the nucleus are "smaller" than those more distant from the nucleus. "Larger" shells can accept more electrons. Figure 1.9 illustrates the filling of electrons into the first three shells of an atom. Shell $n = 1$ is considered "closed" when it contains 2 electrons, shell $n = 2$ accepts a maximum of 8 electrons, shell $n = 3$ accepts 18, etc.

1.3.4 Electron shell structure and the Periodic Table of Elements

Over the centuries, a variety of individual chemical elements have been identified. Their physical and chemical properties were measured according to the analytical technologies available. This created a huge dataset of element-specific chemical and physical parameters. It appeared that some elements behaved chemically quite similarly and could thus be considered group-wise. Chemists like MENDELEEV (in 1896) realized that there should be a system to classify and group the elements accordingly. An initial criterion was to line up the elements following their increasing mass, starting with the lightest one, hydrogen, with number $Z = 1$. The other criterion was to consider the chemical similarities of some elements and to put them into different groups.

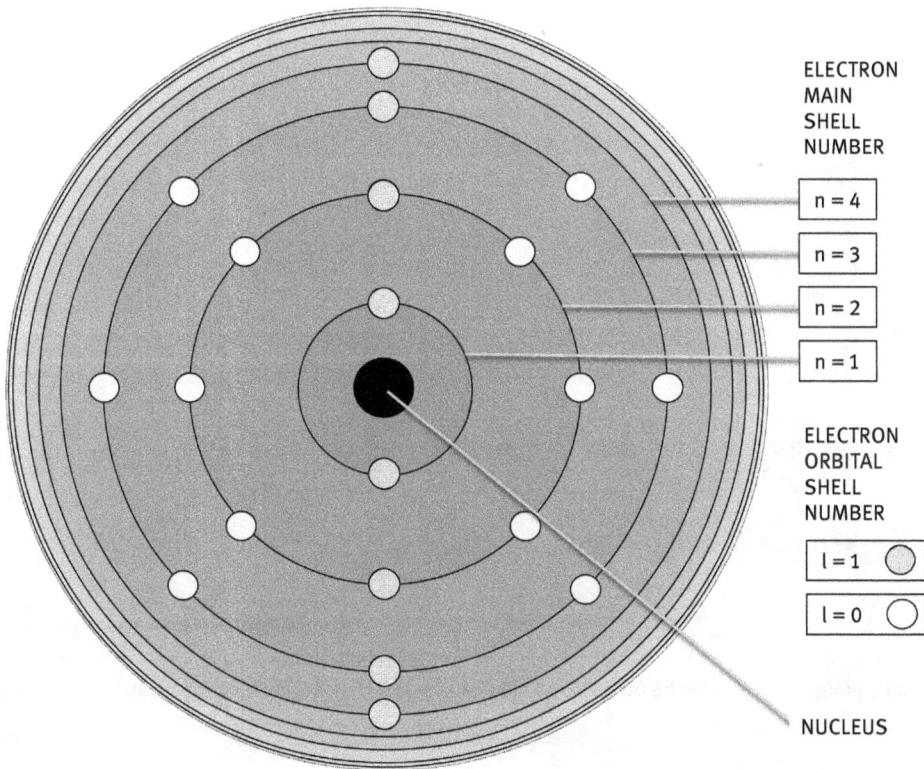

ELECTRON
MAIN
SHELL
NUMBER

n = 4

n = 3

n = 2

n = 1

ELECTRON
ORBITAL
SHELL
NUMBER

l = 1

l = 0

NUCLEUS

Fig. 1.9: An atom of the element calcium in its neutral atomic state containing 20 protons and 20 electrons. All electrons are distributed within the various main shells n according to the maximum shell occupancy numbers and electron binding energies (qualitatively).

These thoughts led to the concept of a "Periodic Table of Elements" (PTE) (see Fig. 1.10). With the later identification of the atom's electronic structure (electrons arranged in different shells of the atom), it becomes obvious that the earlier semi-empiric structure of the PTE was perfectly explained by properties of the atomic electrons of the various chemical elements. The seven horizontal "periods" reflect the number of main electron shells ($n = 1$ to 7). The filling of each period from left to right corresponds to the number of electrons allowed in each main shell. Theoretically, there is a maximum of two electrons for the first period, eight for the second period, 18 for the third, 32 for the fourth, etc.

Legend:
- ☐ s-Elements
- ☐ p-Elements
- ☐ d-Elements
- ☐ f-Elements
- Fr All isotopes: radioactive
- ☘ stable isotope not known

	1 Ia	2 IIa	3 IIIa
1	1 **H** Hydrogen 1.01		
2	3 **Li** Lithium 6.94	4 **Be** Beryllium 9.01	
3	11 **Na** Sodium 22.99	12 **Mg** Magnesium 24.31	
4	19 **K** Potassium 39.10	20 **Ca** Calcium 40.08	21 **Sc** Scandium 44.96
5	37 **Rb** Rubidium 85.47	38 **Sr** Strontium 87.62	39 **Y** Yttrium 88.91
6	55 **Cs** Caesium 132.91	56 **Ba** Barium 137.33	57 **La** Lanthanum 138.91
7	87 **Fr** Francium (223)	88 **Ra** Radium (226)	89 **Ac** Actinium (227)

Period 6 f-block: 58 Ce Cerium 140.12 · 59 Pr Praseodymium 140.91 · 60 Nd Neodymium 144.24 · 61 Pm Promethium (145) · 62 Sm Samarium 150.36 · 63 Eu Europium 151.97 · 64 Gd Gadolinium 157.25 · 65 Tb Terbium 158.93 · 66 Dy Dysprosium 162.50 · 67 Ho Holmium 164.93 · 68 Er Erbium 167.26 · 69 Tm Thulium 168.93 · 70 Yb Ytterbium 173.04

Period 7 f-block: 90 Th Thorium 232.04 · 91 Pa Protactinium 231.04 · 92 U Uranium 238.03 · 93 Np Neptunium (237) · 94 Pu Plutonium (244) · 95 Am Americium (243) · 96 Cm Curium (247) · 97 Bk Berkelium (247) · 98 Cf Californium (251) · 99 Es Einsteinium (252) · 100 Fm Fermium (257) · 101 Md Mendelevium (258) · 102 No Nobelium (259)

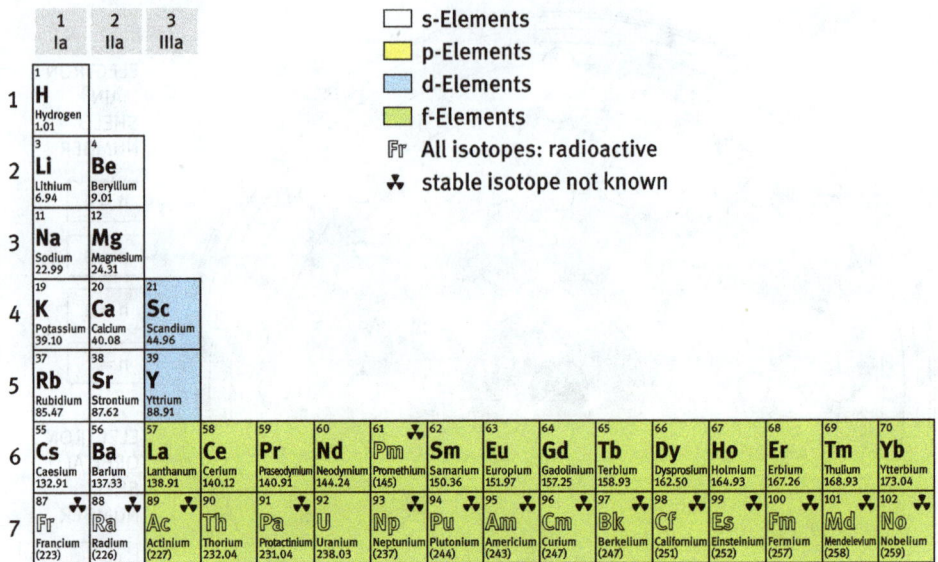

Fig. 1.10: Periodic Table of Elements (PTE). The seven horizontal "periods" reflect the main electron shells, with $n = 1$–7. The 18 vertical columns form "groups" of elements with similar chemical properties (because of similar outer electron shell occupancies). Groups 1, 2, and 13–18 are the "main" groups.[16] The value below each element's name is its molecular weight in g/mol.

1.3.5 Ionization energies

If the excited electron reaches a sufficiently high energy level to escape from the nucleus' attraction, this energy value is the ionization energy $E_{i(e)}$. In the case of the hydrogen atom, this value is 13.6 eV. (see Tab. 1.3).

Tab. 1.3: Binding and ionization energies of the hydrogen electron.

n	1	2	3	4	5	6	...	∞
$E_{B(e)}$ (eV)	−13.6	−3.4	−1.5	−0.9	−0.5	−0.4	...	0
$E_1 \to n$	0	10.2	12.1	12.7	13.1	13.2	...	13.6

---

16 Color-coded substructures of periods mirror the subshells s, p, d and f, with maximum occupancies of 2, 6, 10 and 14 electrons, respectively. This can be explained by quantum number systematics, see below.

4 IVa	5 Va	6 VIa	7 VIIa	8 VIIIa	9 VIIIa	10 VIIIa	11 Ib	12 IIb	13 IIIb	14 IVb	15 Vb	16 VIb	17 VIIb	18 0
														2 **He** Helium 4.003
									5 **B** Boron 10.81	6 **C** Carbon 12.01	7 **N** Nitrogen 14.01	8 **O** Oxygen 15.999	9 **F** Fluorine 18.998	10 **Ne** Neon 20.18
									13 **Al** Aluminium 26.98	14 **Si** Silicon 28.09	15 **P** Phosphorus 30.97	16 **S** Sulfur 32.07	17 **Cl** Chlorine 35.45	18 **Ar** Argon 39.95
22 **Ti** Titanium 47.88	23 **V** Vanadium 50.94	24 **Cr** Chromium 52.00	25 **Mn** Manganese 54.94	26 **Fe** Iron 55.85	27 **Co** Cobalt 58.93	28 **Ni** Nickel 58.70	29 **Cu** Copper 63.55	30 **Zn** Zinc 65.41	31 **Ga** Gallium 69,72	32 **Ge** Germanium 72.64	33 **As** Arsenic 74.92	34 **Se** Selenium 78.96	35 **Br** Bromine 79.904	36 **Kr** Krypton 83.80
40 **Zr** Zirconium 91.22	41 **Nb** Niobium 92.91	42 **Mo** Molybdenum 95.96	43 Tc Technetium (98)	44 **Ru** Ruthenium 101.07	45 **Rh** Rhodium 102.91	46 **Pd** Palladium 106.42	47 **Ag** Silver 107.87	48 **Cd** Cadmium 112.41	49 **In** Indium 114.82	50 **Sn** Tin 118.71	51 **Sb** Antimony 121.76	52 **Te** Tellurium 127.60	53 **I** Iodine 126.90	54 **Xe** Xenon 131.29
72 **Hf** Hafnium 178.49	73 **Ta** Tantalum 180.95	74 **W** Tungsten 183.84	75 **Re** Rhenium 186.21	76 **Os** Osmium 190.23	77 **Ir** Iridium 192.22	78 **Pt** Platinum 195.08	79 **Au** Gold 196.97	80 **Hg** Mercury 200.59	81 **Tl** Thallium 204.38	82 **Pb** Lead 207.20	83 **Bi** Bismuth 208.98	84 Po Polonium (209)	85 At Astatine (210)	86 Rn Radon (222)
104 Rf Rutherfordium (267)	105 Db Dubnium (268)	106 Sg Seaborgium (271)	107 Bh Bohrium (272)	108 Hs Hassium (270)	109 Mt Meitnerium (276)	110 Ds Damstadtium (281)	111 Rg Roentgenium (280)	112 Cn Copernicum (285)	113 Nh Nihonium (284)	114 Fl Flerovium (289)	115 Mc Moscovium (288)	116 Lv Livermorium (293)	117 Ts Tennessine (294)	118 Og Oganesson (294)

71 **Lu** Lutetium 174.97 / 103 Lr Lawrencium (262)

Fig. 1.10 (continued)

Ionization energy is defined as the energy required for an atom to release an outer-shell electron. Its value is different for each chemical element because it relates to Z and the outer electron shell energy. There are first and second ionization energies. First ionization energy refers to the first electron released from the neutral atom. Second ionization energy refers to another electron emitted in a subsequent step from the $+1$ ion of the element. Ionization energy shows a systematic trend within the periods and groups of the PTE (see Fig. 1.11): within a group, ionization energy decreases with increasing proton number; within a period, ionization energy increases.

1.4 Quantum mechanics

Within the BOHR atom model, not only does each electron have a characteristic energy as defined by its shell, but each shell was also postulated to accept a specific number of electrons. Moreover, although multiple electrons can occupy the same main shell, each electron of an atom is still considered to be unique. Of the 76 electrons located in various shells of the gold atom, all are different in at least one parameter!

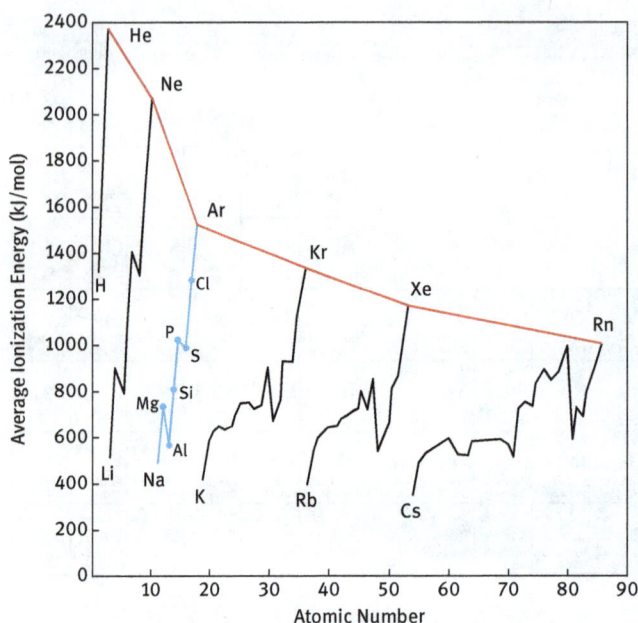

Fig. 1.11: First ionization energy of the first six periods of chemical elements. Within a group (e.g. the noble gases, red line), ionization energies decrease with increasing proton number: He $(Z = 2)$ $\rightarrow$ Ne $(Z = 10)$ $\rightarrow$ Ar $(Z = 18)$ $\rightarrow$ Kr $(Z = 36)$ $\rightarrow$ Xe $(Z = 54)$ $\rightarrow$ Rn $(Z = 86)$. Within a period, ionization energies increase; e.g. for the $n = 3$ (L) period (blue line): Na $(Z = 11)$ $\rightarrow$ Mg $(Z = 12)$ $\rightarrow$ Al $(Z = 13)$ $\rightarrow$ Si $(Z = 14)$ $\rightarrow$ P $(Z = 15)$ $\rightarrow$ S $(Z = 16)$ $\rightarrow$ Cl $(Z = 17)$ $\rightarrow$ Ar $(Z = 18)$.

1.4.1 Quantum numbers

The nomenclature applied is "quantum numbers". It bridges the energy levels of the BOHR atom model with quantum mechanics introduced by SCHRÖDINGER et al. and molecular orbital theory proposed by HUND and MULLIKEN. Accordingly, there is a set of four quantum numbers which characterize each electron, like a fingerprint. Three (n, l, m) are dedicated to the orbital parameters of the electron's existence in terms of energy and orbital angular momentum, and the fourth one (s) to the intrinsic spin of the electron.

The principal (or <u>main</u>) <u>quantum number</u> is n. It defines the different main shell the electrons exist. It quantifies the maximum number of electrons to fit into each main shell. The energy of these shells, however, depends on the individual atom. There are basically seven principal quantum numbers with $n = 1, 2, 3, 4, 5, 6,$ 7, also symbolized by Latin capital letters K, L, M, N, O, P, Q. Artificial elements of $Z > 118$ will start filling electrons in the $n = 8$ shell.

Shells may be structured into subshells. All shells except the K-shell $(n = 1)$ involve multiple subshells. The subshells are indicated by the <u>orbital quantum</u> <u>number</u> l.

The value of the electron's orbital angular momentum reflects individual orbits of electrons within a given main shell. Thus, many electrons may "circulate" within one and the same main shell, but do not overlap, as they populate individual orbital spaces.

The orbital angular momentum correlates with the principal quantum number. The higher the principal quantum number n, the more subshells may be populated within that shell. The maximum number of subshells is $l = n - 1$, i.e. $0 \leq l \leq (n - 1)$. Consequently, there is $l = 0$ for $n = 1$ (because $n - 1 = 0$). In contrast, there are four subshells for $n = 4$, namely $l = 0$, $l = 1$, $l = 2$, $l = 3$. Those orbital orbitals are denoted as s, p, d, and f.[17]

Quantum mechanics not only introduced specific energetically "allowed" orbitals for the circulating electrons, it also characterized electrons in more detail. The key point was to consider electrons not as localizable particles like planets, which can be correlated at any point in time to a specific position. Instead, their existence in space and time was proposed to follow probabilities only, defining states of high and low probability. Because only discrete values may exist in quantum mechanics (as indicated by the PLANCK constant,[18,19]), the "real" orbital angular momentum L is identified through the relation

$$L^2 = \frac{\hbar^2}{l(l+1)} \tag{1.4}$$

$$\hbar = \frac{h}{2\pi} \tag{1.4a}$$

Figure 1.12 illustrates the different features of electrons with different principal and angular quantum numbers existing in space along an x-y coordinate system.[20]

All s-electrons ($l = 0$ orbitals) exist in sphere-shaped orbitals around the nucleus. The distribution within that sphere is homogeneous, i.e. there is a certain probability for the s-electron to be in the center of the coordinate system.[21] Because

---

17 This quantum physical idea can predict "real chemistry", i.e. the chemical behavior of elements. The PTE shows a specific block of "p-elements" containing elements with the outer electrons located in the p-shell of the main shells $n > 1$; these are the main PTE groups III through VIII. There are also "d-elements" and "f-elements", i.e. the lanthanides and actinides. Again, experimental evidence is beautifully correlated with theoretical concepts on the atom's electronic structure.

18 The symbol ℏ stands for the "reduced" PLANCK constant, where ℏ = h/2π. The reduced PLANCK constant is preferred in cases where radial dependencies are relevant, such as angular frequency, solid angle, etc. (see Fig. 1.20).

19 Nobel Prize in Physics to PLANCK, 1918, "in recognition of the services he rendered to the advancement of Physics by his discovery of energy quanta".

20 This is mathematically derived for the simplest case – the one electron system of hydrogen.

21 One of the main features to understand nuclear transformations is the principal difference between probabilities of existence of electrons of s orbital quantum number compared to other orbital quantum numbers. As depicted in Fig. 1.12, s orbital electrons do have a probability (although

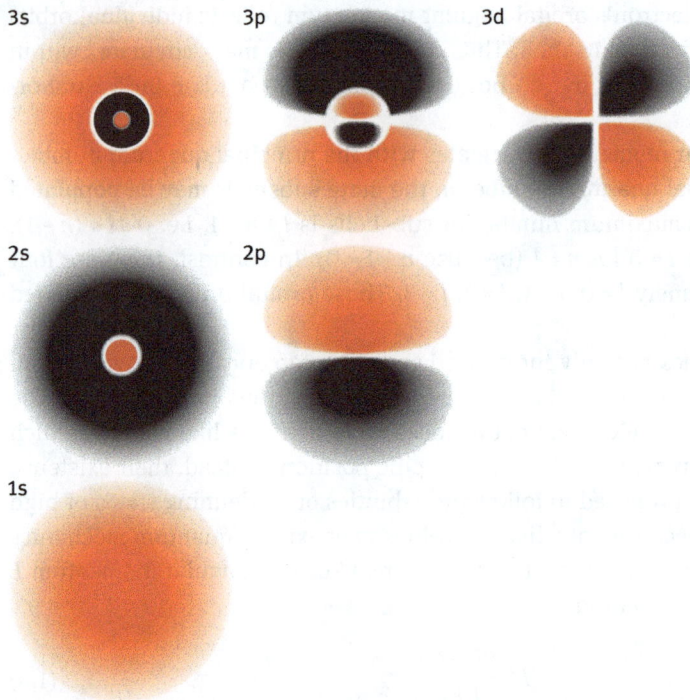

Fig. 1.12: Two-dimensional illustration of the orbital distribution of s, p, and d electrons for $n = 1$–3, imaged as densities around the atom's nucleus. For $l = 1$ (1s, 2s, 3s, etc. electrons), the density distributions are spherical. The sphere's radius increases slightly with n, following the relationship in Fig. 1.7 and according to eq. (1.1). The change in size is more dramatic as Z increases, when electrons become more and more attracted to the nucleus. For instance, the 1s-electron gets closer to the nucleus if the chemical element has a larger proton number and electrons are filled in 2s (i.e. starting from lithium) or 3s (i.e. starting from sodium) orbitals. The same is true for p-orbitals of $n = 2$ and $n = 3$ shells (2p, 3p), etc.

the electron binding energy differs between shells (see Tab. 1.3), the radial distance between each orbital and the center of the coordinate system increases with n. For example, the s-electrons of the $n = 2$ shell (L) are further from the nucleus than $n = 1$ (K) electrons.

---

extremely small) to exist close to the center of the Cartesian coordinate system (i.e. the nucleus); shells with different orbital quantum numbers do not. Despite the extremely large dimensions of electron shells compared to the atomic nucleus, s shell electrons are "allowed" to exist close to and even within the atomic nucleus. Of course, these probabilities drop dramatically when increasing the principal quantum number from $n = 1$ to $n > 1$. At minimum, each atom will have K-shell s-electrons, and important nuclear transformations of radioactive nuclides (such as the electron capture branch of the β process or secondary processes like inner conversion) rely on this feature of s orbital electrons (see Chapters 8 and 11).

From $l > 0$ onward, the electron density distribution is no longer spherical. The orbital distribution of p-electrons ($l = 1$) is dumbbell shaped and the d-electrons ($l = 2$) exist in orbitals of other shapes (see Fig. 1.12). Thus, a three-dimensional system is needed. Figure 1.13 illustrates the different electron features of p-orbital electrons existing in space along an x-y-z coordinate system.[22] The general profile is dumbbell shaped. The two symmetrical orbitals on the left and the right represent so-called bonding or nonbonding parts. Within a Cartesian system, there are three different options for a dumbbell: a p-electron of identical characteristics may either occupy space along the x-axis, the y-axis, or the z-axis (p_x, p_y, p_z). In the center of the coordinate system, i.e. at $x = y = z = 0$, the probability to exist is zero.

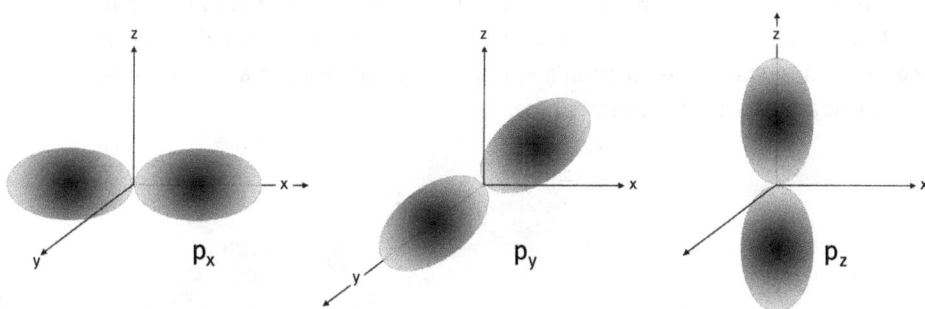

Fig. 1.13: The three-dimensional dumbbell-shaped orbitals of p-electrons along x-y-z axes of a Cartesian coordinate system.[23]

For $l = 0$, there was one s orbital. For $l = 1$, there was one p, but with three individual profiles: p_x, p_y, p_z. The number of those subshells increases with increasing values of l. Mathematically, it is expressed by $- l, -(l - 1), \ldots, 0, \ldots, +(l - 1), + l$ (see Tab. 1.4). For $l = 2$, for example, there are 5 possible states, namely $- 2, - 1, 0, + 1$ and $+ 2$.

Each given l-orbital thus contains a set of electrons described by a new parameter: their <u>magnetic quantum number</u> m. It may be thought of as if electrons within a certain subshell may form groups of "clouds". In fact, it reflects the projection of the orbital angular momentum of the electron within a magnetic field, referred to as an axis of the polar coordination system. (Negatively charged electrons moving around a positively charged center are influenced by an electromagnetic field.) The final orbital angular momentum thus becomes $L_z = \hbar m_l$.

---

22 This again is mathematically derived for the simplest case – the one-electron system of hydrogen.
23 This organization of orbital distribution of the 3p-electron states is not an abstract concept. It is reflected in real chemistry, where the p_x, p_y, p_z states make a huge difference in terms of forming chemical bonds and thus forming chemical compounds.

Tab. 1.4: The four quantum numbers making each electron of an atom unique.

Quantum number	Symbol	Values		Names	Example
Main (principal)	n	$n \geq 1$	$1, 2, 3 \ldots$	K, L, M, . . .	$n = 3$
Orbital angular	l	$0 \leq l \leq (n-1)$	$0, \ldots, n-1$	s, p, d, . . .	$l = 0, 1, 2 \ldots$
Magnetic	m	$-l \leq m \leq +l$	$-l, \ldots, 0, \ldots +l$	for p: p_x, p_y, p_z	$m = -2, -1, 0, 1, 2$
Spin	s	$\downarrow, \uparrow$	$-\frac{1}{2}, +\frac{1}{2}$		$s = -\frac{1}{2}, +\frac{1}{2}$

Finally, there is a quantum number reflecting the angular spin momentum, s, of an electron. Like the earth's movement around its axis lasts one day (see Fig. 1.14), one may understand the electron's spin angular momentum as a spinning process around an axis of the electron's sphere.

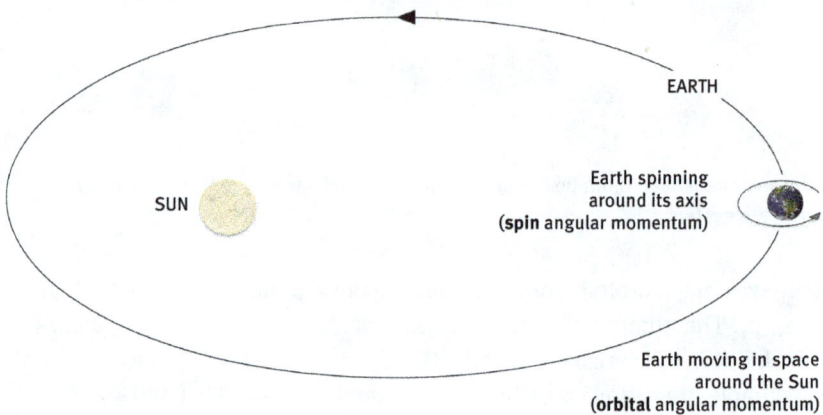

Fig. 1.14: Conception of the two different angular moments of the planet earth circulating around the sun. The earth's movement in space, taking one year, would mirror the orbital angular momentum of an electron around an atom's nucleus – reflected by the quantum number l. Similarly, the earth's own movement around its axis, lasting one day, may help to understand the spin angular momentum of the electron – denoted by the spin quantum number s.

However, the momentum is either in one direction or the other – there are no further options. The corresponding spin quantum number s thus is either $+\frac{1}{2}$ or $-\frac{1}{2}$.[24] It reflects an intrinsic parameter of the electron. For example, for a given set of n, l, m

---

24 Note that there is actually no "positive" or "negative" direction of the angular spin.

there are always two options for an electron, which differ in angular spin of $+\frac{1}{2}$ or $-\frac{1}{2}$. This is denoted as HUND'S rule. The value of this spin quantum number s is:

$$S^2 = \hbar^2 \cdot s(s+1) \tag{1.5}$$

Now each electron is fully identified by $(n, l, m, +\frac{1}{2})$ and $(n, m, l, -\frac{1}{2})$. Each suborbital, defined by (n, l, m), can contain a maximum of two electrons of opposite spin, and the general notation for how magnetic quantum number and spin are linked is $m_s = \pm\frac{1}{2}$. Finally, each electron is supposed to be unique. This is known as the PAULI exclusion principle:[25] in any quantum mechanical system (such as electrons in the atom's shell, nucleons in the atom's nucleus), two particles of identical quantum parameters should not exist.

1.4.2 Electron configurations

Finally, electron configurations are named according to a set of quantum numbers, expressed by a number (main shell / main quantum number) followed by a Latin letter (subshell / orbital quantum number) and a final number as the superscript (indicating the number of electrons per subshell). In many cases, only the composition of the last (outer) shell (or subshell) is indicated, assuming that all the lower (inner) shells are filled completely. For instance, $1s^1$ indicates the first main electron shell (n = 1), its s-orbital character (l = 0), and that this orbit contains one electron only. Chemically speaking, this is the electron configuration of a neutral hydrogen atom. In comparison, $3p^2$ says that there are two electrons located in the third principal electron shell (n = 3) and that both are of p-orbital character (l = 1). For a neutral element with all the lower-energetic shells and subshells filled completely, i.e. $1s^2 < 2s^2 < 2p^6 < 3s^2$, this element is silicon comprising altogether 12 electrons (and 12 protons).

1.4.3 Orbital occupancies *vs.* electron binding energies

Figure 1.15 illustrates how electrons fill the individual orbitals for the element calcium. The complete composition of this element of Z = 20 according to quantum number logic would $1s^2 < 2s^2 < 2p^6 < 3s^2 < 3p^6 < 3d^2$. However, this order is followed only for the first 18 electrons. The two outer electrons, i.e. electron numbers 19 and 20, prefer to occupy the $4s^2$ orbital instead of the 3d orbital! The explanation lies in the energy of the individual shells and subshells. It is a matter of fact that the $4s^2$ "comes first", i.e. shows a lower value of $E_{B(e)}$ relative to the binding energy of the

---

25 The Nobel Prize in Physics, 1945, was awarded to W PAULI "for the discovery of the Exclusion Principle, also called the Pauli Principle".

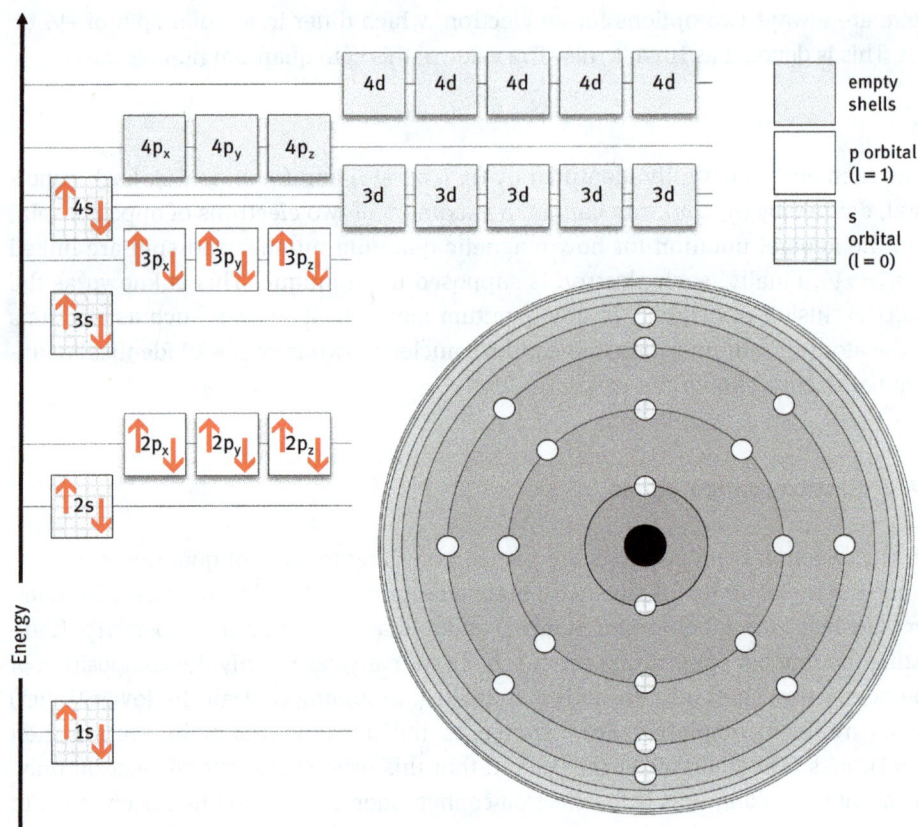

Fig. 1.15: Electron shells arranged according to the system of the four quantum numbers are filled by electrons according to the electron binding energy of each subshell. (left) Electron orbitals defined by quantum numbers and arranged with increasing electron binding energy. (right) The 20 electrons of the chemical element calcium filled into main shells (circles), subshells not indicated. The chemistry of calcium is defined by its outer electrons, which are the two 4s electrons instead of 3d electrons.

$3d^2$ subshell electrons.[26] This effect becomes more pronounced the closer the outer electron shells are in electron binding energy. Furthermore, because differences between individual subshells of higher main quantum number become smaller and smaller (see Fig. 1.7), the effect of overlapping subshell orbitals becomes more and more pronounced.

---

26 This order of electron subshell occupancies is reflected in the structure of the PTE, i.e. it perfectly correlates the experimentally observed systematics of chemical elements with electron structure parameters. Interestingly, the same will be discussed in the following chapters, where the overlap of nucleon shell structures will explain the stability of the nuclei of atoms.

The individual probabilities of electrons to exist in various shells, but in particular in the last, outer shell, are largely responsible for the chemical behavior of a given element and for the type of chemical bonding between two or more atoms in a chemical compound.

1.5 Mathematical explanations of the BOHR atom model

BOHR'S atom model postulates that electrons are located in various shells. This statement satisfies experimental evidence, such as line spectrum of hydrogen and positioning of chemical elements within the PTE. However, mathematical verification of the model's postulate was delivered only later. SCHRÖDINGER[27] developed the corresponding mathematics to calculate parameters such as the electron energy of the simplest model, the one electron of atomic hydrogen. The concept is to quantify electron parameters in terms of particles existing within a potential well.[28]

The key intention is to deduce fundamental parameters of the electron (and other particles), such as energy (potential and kinetic), velocity, and position. The mathematics itself in detail is beyond the scope of this book. The basic idea is to treat the electron as a wave, and to derive its wavefunction Ψ. Finally, the quantum mechanical parameters are correlated to the basic physics of the electron, such as the radius r of a certain electron shell, and its energy described as $E = f(r)$.

1.5.1 Potential well

A potential well considers a particle (here, an electron[29]) inside a well (or box), and defines a region defined by a (local) minimum of potential energy. In nature, a lake filling a valley surrounded by mountains represents a three-dimensional potential well.[30] For mathematical treatments, one-dimensional model systems are considered. The well localizes a "free" particle within the two walls of (infinite) high potential energy along an x-axis. Inside the box, the potential energy U is zero, while outside it is infinitely large.[31] The two walls are characterized by positions 0 and L

---

27 Nobel Prize in Physics, 1933, "for the discovery of new productive forms of atomic theory".
28 It is introduced here because it is also relevant to the quantification of nucleons within a potential well. It is thus applicable to particles of the atom's nucleus, for e.g. understanding the nucleon shell occupancies (see Chapter 2), and introducing the tunneling effect (see Chapter 9).
29 . . . could also be a gas molecule, an electron, or a nucleon.
30 The simplest model implies a single, negatively charged "free" electron existing in a box ("particle in a box") within the electrostatic attraction caused by the positively charged nucleus with a potential energy V > 0. This is a potential well, which does not, however, completely describe the whole truth of electron(s) within a real atom.
31 There are slightly modified considerations for a potential well.

(see Fig. 1.16). Within the well, i.e. between positions 0 and L, the particle moves with constant velocity and is reflected from each wall without losing energy. It can leave (or escape the well) only if it becomes energetically excited.

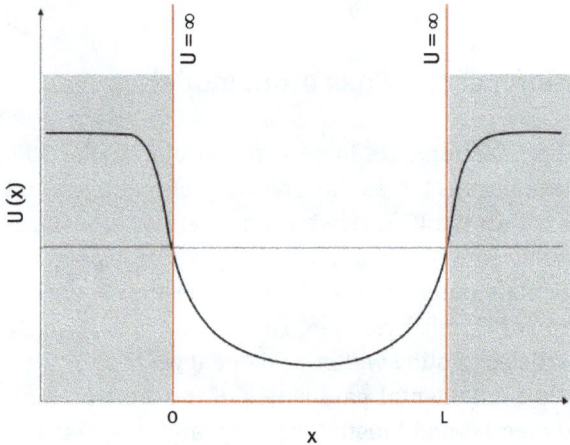

Fig. 1.16: Electron potential well. An electron inside a well (a box) exists within a specific region defined by a (local) minimum of potential energy. The well localizes the "free" electron within two walls of (infinitely) high potential energy along an x-axis of distance. Inside the well, the potential energy U is zero, while outside it is infinitely large. The electron moves with constant velocity between the two borders (walls), characterized by positions 0 and L.

In quantum physics, the electron is treated as a wavefunction with wavelength λ. The waves are reflected at the walls of the well. In order not to annihilate waves after reflection due to interference, particles may exist only if the wavelength fits within the well length L (or *vice versa*). The simple mathematical expression is L^{well} = n · λ/2, indicating that L should be an integer multiple of half of the wavelength. Only in this situation are the waves reflected in a way that reproduces the initial wave again (see Fig. 1.17).

Consequently, electrons are only allowed to exist in specific states (defined by n) where their energies are multiples of n. The "allowed" energies of the wave (i.e., the electron with mass m) are, according to eq. (1.6), proportional to the square of n. Each value of n corresponds to a wavenumber according to eq. (1.7):

$$E_n = \frac{n^2}{2L^2} \cdot \frac{\hbar^2}{m} = \frac{n^2}{2L^2} \cdot \frac{\hbar^2}{4m}, \quad \text{with} \quad n = 1, 2, 3, \ldots \tag{1.6}$$

$$v = \frac{\pi n}{L} \tag{1.7}$$

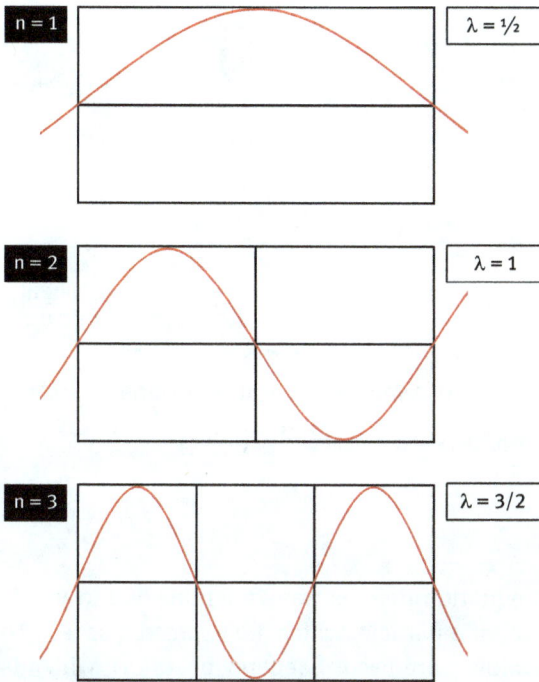

Fig. 1.17: Waves being reflected at the walls of a potential well. A wave may propagate "forever" only if half its wavelength λ is an integer multiple n of the well length L: $L^{well} = f(n \; \frac{1}{2}\lambda)$.

1.5.2 Harmonic oscillator

According to Nᴇᴡᴛᴏɴ's laws of classical mechanics, a classical harmonic oscillator describes a particle oscillating back and forth along a spring (see Fig. 1.18). There is a stable mass point of equilibrium with a corresponding potential, depending on a spring constant k and a restoring force F. The harmonic potential is $U \approx x^2$. Both the position and energy of the particle are exactly known.

$$U(x) = \frac{1}{2} k \cdot x^2 \tag{1.8}$$

$$F(x) = -\frac{\delta U(x)}{\delta x} = -k \cdot x \tag{1.9}$$

This is different in quantum mechanics. The position of an electron is not precisely determinable; the same is true for the absolute value of an electron's energy at a specific position. In contrast, a probability function is derived for the position of the electron (see Fig. 1.18).

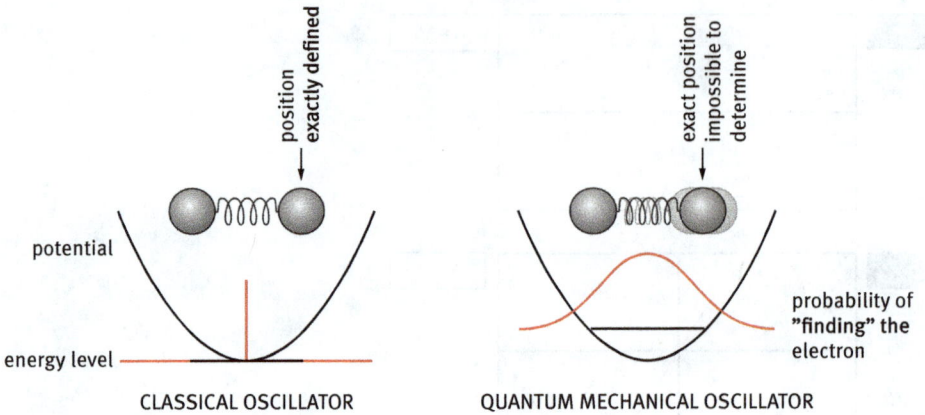

Fig. 1.18: The harmonic oscillator in classical physics and in quantum mechanics.

1.5.3 SCHRÖDINGER equation

The (time-dependent) SCHRÖDINGER equation defines wavefunctions Ψ and yields probabilities of finding (the residence) of a particle, which are expressed as Ψ^2. To solve these wavefunctions, two operators[32] are used: $\hat{E}$ (energy operator), which is relevant to the total energy of the wavefunction, and $\hat{H}$ (HAMILTONIAN operator), which represents the specifics of the object under consideration and the fields influencing the objects.

$$\hat{H}\Psi = \hat{E}\Psi \tag{1.10}$$

$$\hat{H} = -\frac{1}{2m}\frac{\hbar}{}\nabla^2 + U \tag{1.11}$$

$$\hat{E} = i\hbar\frac{\delta}{\delta t} \tag{1.12}$$

$$-\frac{1}{2m}\frac{\hbar}{}\nabla^2\Psi(x, y, z) + U(x, y, z)\Psi(x, y, z) = E\Psi(x, y, z) \tag{1.13a}$$

The HAMILTONIAN operator in its general form is expressed in eq. (1.11), and combines terms for both kinetic energy $\frac{1}{2}\hbar\nabla^2$ and potential energy U. For the kinetic energy, the reduced PLANCK constant $\hbar$ and the mass of the particle are considered. Further, the nabla operator ∇ is involved,[33] which represents the derivatives for all coordinates in space. Time is not considered here.

---

32 In mathematics, operators are procedures which apply to a whole function.
33 The nabla ∇ is a vector differential operator, similar to the LAPLACE operator (or Laplacian) Δ, used to calculate how the average value of a function f over spheres deviates from $f(r)$ as the radius of the sphere grows. $\nabla^2 = \Delta$.

The energy operator also contains $\hbar$. The imaginary unit i represents the square root of -1, which breaks wavefunctions into a real component and a virtual component. $\hat{E}$ includes a derivative for time only, and does not consider position (x, y, z). Solutions of the energy operator represent sine functions of characteristic frequencies.

Solutions to eq. (1.13a) are wavefunctions, Ψ, that satisfy the parameters of both operators. In the case of complex wavefunctions of type $\Psi(r, t)$, the harmonic potential correlates with the eigenfrequency ω of the harmonic oscillator. The corresponding HAMILTONIAN function H depends on the mass m and the impulse p and reflects the total energy of the system composed of kinetic and potential energy.

The time-independent (i.e., position-related) SCHRÖDINGER equation is given in eq. (1.13b). Here, the three coordinates (x, y, z) are represented by the radius r. In the specific case of the x-dimension, solutions to the eigenvalue equation for the stationary SCHRÖDINGER equation are the eigenvalues $\hat{E}_n$ as in eq. (1.14). In this case, Δ is the LAPLACE operator, where $\Delta = \delta/\delta x$ in a one-dimensional case.

In the case of a one-electron system, this equation finally yields a description of the electron's orbitals.

The SCHRÖDINGER equation yields the probability of finding (the residence) of a particle, which is expressed as Ψ^2. For the position of the electron within the harmonic oscillator, there is one probability only, which is the square of the wavefunction, i.e. $\Psi_n(x)^2$. Figure 1.19 reproduces wavefunctions of the electron in a harmonic oscillator for $n = 1–7$ and the corresponding probabilities of presence or "probability density".

1.5.4 Electron energies

The orbital energy E_n is found using $E_n = \hbar\omega$. The harmonic oscillator accepts only discrete energies where n is a whole number. The lowest possible state is $E_0 = \frac{1}{2}\hbar\omega$, where $E_0 > 0$. This is known as the zero-point energy, or the energy of a ground-state electron.

$$-\frac{1}{2}\frac{\hbar}{m}\Delta^2\Psi(r) + U(r)\Psi(r) = E\Psi(r) \tag{1.13b}$$

$$\hat{E}_n\Psi_n(x) = -\frac{1}{2}\hbar^2\Delta\Psi_n(x) + \frac{1}{2}m\omega^2 x\Psi_n(x) \tag{1.14}$$

$$E_n = \hbar\omega\left(n + \frac{1}{2}\right) \tag{1.15}$$

Unlike classical mechanics, which provides a particle's exact position and exact impulse p, in quantum mechanics the HEISENBERG uncertainty principle applies. This principle states that an electron's exact position x *and* exact impulse p cannot be simultaneously known: $\Delta x \Delta p \geq \hbar/2$.

$\Psi_7(x)$

$\Psi_6(x)$

$\Psi_5(x)$

$\Psi_4(x)$

$\Psi_3(x)$

$\Psi_2(x)$

$\Psi_1(x)$

$\Psi_0(x)$

(a)

$\Psi_n(x)^2$

(b)

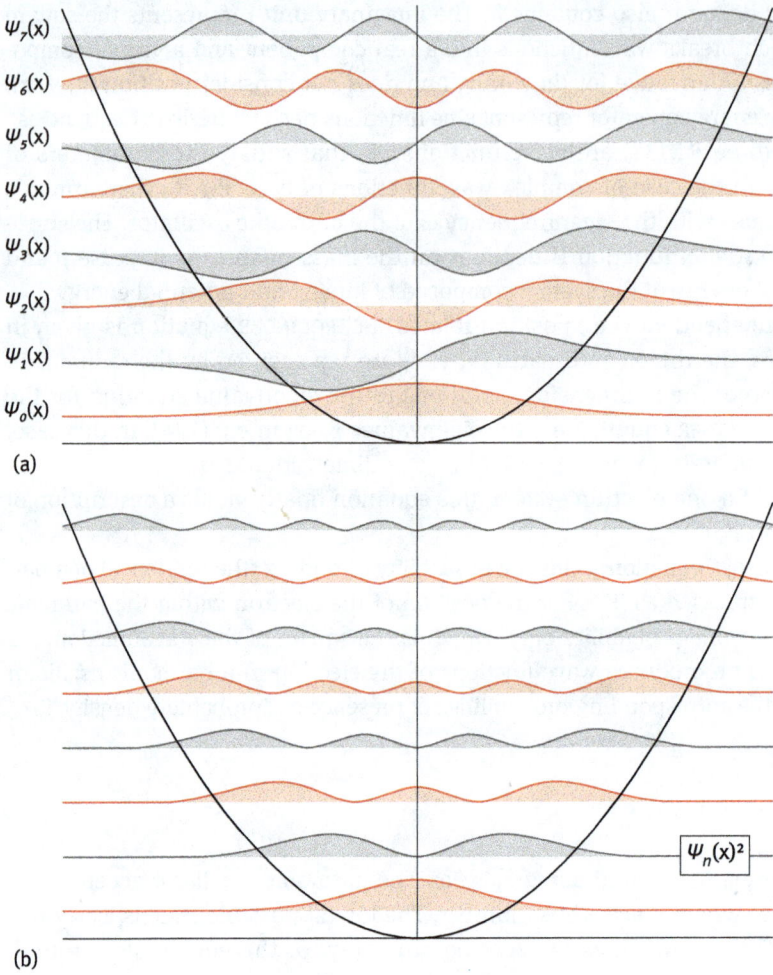

Fig. 1.19: Wavefunctions of a quantum harmonic oscillator (a) and probabilities of presence $\Psi n(x)^2$ (b).

When accounting for quantum mechanics, energy is expressed by eqs. (1.17) and (1.18):

$$\Delta x \Delta p \geq \frac{\hbar}{2} \tag{1.16}$$

$$E_n = \frac{1}{2}\frac{(\Delta p)^2}{m} + \frac{1}{2}m\omega^2\,(\Delta x) \tag{1.17}$$

$$E_n \geq \frac{1}{2}\frac{(\Delta p)^2}{m} + \frac{1}{8}\frac{m\hbar^2\omega^2}{(\Delta p)^2} \tag{1.18}$$

The harmonic oscillator models a particle in one dimension. In reality, the hydrogen electron surrounding its proton is three-dimensional. In this case, the SCHRÖDINGER equation uses a polar coordinate system instead of Cartesian coordinates. Any point in space is defined as the distance r (a vector) of a point from the center of the sphere, as well as two angles φ, θ between the vector and the directions of the Cartesian coordinates. These angles define a so-called solid angle Ω (see Fig. 1.20).

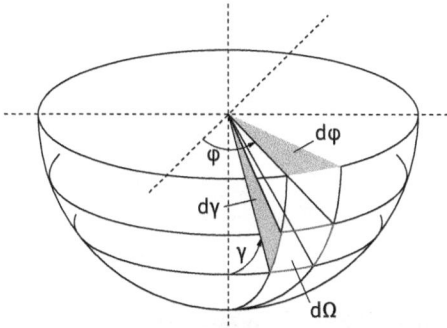

Fig. 1.20: The solid angle Ω describes a two-dimensional angle in three-dimensional space. Along with the distance vector r, it is part of a polar coordinate system.

These three polar variables (r, φ, θ) are treated separately by wavefunctions, which are divided into one part for radial distance $R_{n,l}$ and another one for angular positions $Y_{l,m}$. These are specific to one electron as characterized by its three quantum numbers n, l, and m.

$$\Psi_{nlm}(r,\varphi,\Theta) = R_{nl}(r) \cdot Y_{lm}(\varphi,\Theta) \tag{1.19}$$

Finally, a BOHR shell electron may be described with a set of "real" physical parameters, such as position, impulse, and energy. The radius r gives the radial distance between the electron and nucleus. The potential energy U of an electron thus depends on r. Because this radius depends on nuclear charge, another parameter needed is Z: the number of protons in the nucleus. The potential energy U(r) follows a ratio between Z and r, and is given in eq. (1.20) proportional to the electric charge density e_p.[34] Equation (1.21) then combines the kinetic energy $\frac{1}{2}\hbar(\nabla)^2$ and potential energy U in a simplified version following eq. (1.13b).

$$U(r) = -\frac{Ze_p^2}{n} \tag{1.20}$$

---

34 e_p = the electrical charge density. The charge of an electron is $-1.602 \cdot 10^{-19}$ C. The charge density is referred to as the volume of the electron, with units in C/m^3.

$$\left(E_{kin} - \frac{Ze_{n^2}}{n}\right)\Psi = E\Psi \qquad (1.21)$$

Expected eigenenergy values of the electron located within one of the main shells, denoted by E, are different between spherical orbitals ($l = 0$, i.e. s-electrons) and cases where $l > 0$. This equation finally corresponds to eq. (1.1) and provides the mathematical background of important electron parameters of the BOHR atom model.

$$E_{nl} = E_{n0} = \frac{n^2 h^2}{8mL^2} \qquad (1.22)$$

1.6 Outlook

1.6.1 Fine structures

The electron structure of the BOHR atom was deduced from the hydrogen line spectrum, and the SCHRÖDINGER equation calculates the parameters of the one-electron system. With an increasing number of electrons, however, the interactions between electrons and between electron and nucleus should be considered. Even for the one-electron system, there are a number of effects originating from a more sophisticated structuring of electrons inside the shells, with some of them being responsible for fundamental physico-chemical effects and analytical techniques. These are subsumed under "fine structures". Fine structures, more generally, are created by spin-orbit interactions and reflect the difference between nonrelativistic and relativistic treatments of the electron energies. There is a correction term for relativistic effects of the kinetic energy of the electron (to be considered in the case of objects moving with velocities close to the speed of light), and there is one for the potential energy of the electron (DARWIN term). This turns even the hydrogen line spectrum into a rather difficult, complex system.

Dimensions of fine structures (emission line splitting) follow a fine-structure constant α (where $\alpha = 7.297 \cdot 10^{-3}$) and correlate with the proton number Z through $Z^2\alpha^2$. In the case of hydrogen, the effects are very minor and could experimentally be resolved relatively late.[35] Fine structures of the α-line of the BALMER series (656.28 nm wavelength) amount to only 0.014 nm (in dimensions: 0.14 Å *vs.* 6562.8 Å).

---

35 WE LAMB: Nobel Prize in Physics, 1955, "for his discoveries concerning the fine structure of the hydrogen spectrum".

1.6.2 Spin-orbit coupling

In the context of this textbook, the spin-orbit coupling, which contributes to electron line fine structures, is of special interest.

Parameters like orbital angular momentum and spin angular momentum can be attributed to electrons. The two individual angular momenta may overlap within an electromagnetic field. This is called spin-orbit interaction or spin-orbit coupling. Orbital angular quantum l and spin angular quantum momentum s numbers create a total angular momentum quantum number j with $j = |l + s|$. The individual spin quantum number remains $\pm\frac{1}{2}$, but becomes arranged as illustrated by a rotating vector cone (see Fig. 1.21). The total angular momentum J is related to j through eq. (1.23).

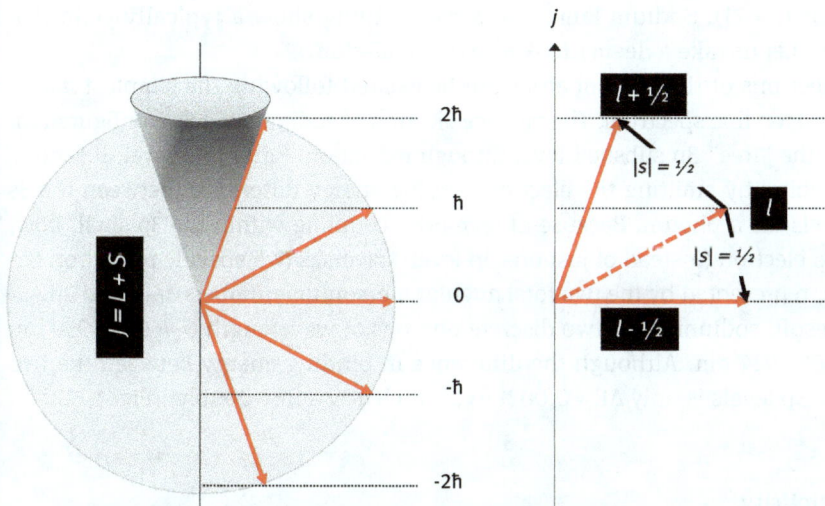

Fig. 1.21: Spin-orbit coupling. Total orbital angular (L) and total angular spin number (S) moments create a total angular momentum number J.

$$J = \hbar^2 j(j+1) \tag{1.23}$$

To demonstrate the calculation of j, consider the following example. For orbital quantum numbers $l = 1$ containing six p-electrons (assuming the three states of magnetic quantum numbers $m = 1, 0, -1$, (i.e. p_x, p_y, p_z) are each filled by two electrons of spin quantum number $+ 1/2$ and $- 1/2$, respectively), there are two possible values of j: $j = 1 + 1/2 = 3/2$ and $j = 1 - 1/2 = 1/2$. The electrons will reach two different levels, namely $j = 2 + 1/2 = 5/2$ and $j = 2 - 1/2 = 3/2$. There will always be two values of j, independent of the specific orbital quantum number considered, except when $l = 0$.

The number of electrons that fit into the two new suborbitals is $2j + 1$. For example, a fully filled p-orbit distributes its altogether 6 electrons in a ratio of 4:2, according to $2 \times 3/2 + 1 = 4$ and $2 \times \frac{1}{2} + 1 = 2$ electrons, respectively. For a completely occupied d-orbital, the balance is 6:4.

Interestingly, this induces a difference in energy between these newly arranged electron properties according to the total angular momenta obtained. Until this point, the 6 p-electrons in question were all considered to have identical binding energy (called "degeneracy"). Now, this degeneracy becomes obsolete and two new energetic levels occur with a characteristic value of ΔE in binding energy.

$$\Delta E_{ls} = U_{ls}(r) \frac{(2l+1)}{2V_{ls}(r)} \tag{1.24}$$

Again, this theoretical concept was proven by experimental data. Emission line splitting following $Z^2\alpha^2$ is, for instance, "obvious" in the case of emissions of sodium lamps ($Z = 21$). Sodium lamps for street lighting show a typically yellowish bright light. Let us take a deeper look into that emission.

The electrons of the sodium atom can be excited following the simplest model for the hydrogen line spectrum. The valence electron of sodium, in a $3s^1$ configuration, may reach the "free" 3p subshell level through excitation. From that level, it returns to the 3s orbital by emitting the electron binding energy difference (between the 3s and 3p levels) as a photon. Because of spin-orbit coupling within the 3p shell, however, the 3s electron (instead of just one 3p level) traverses two specific p-electron energy levels, represented by the two total angular momentum numbers $3p_{1/2}$ and $3p_{3/2}$.

As a result, sodium emits two discrete photons of wavelength $D_2 = 588.9950$ nm and $D_1 = 589.5924$ nm. Although the difference in binding energy between the two degenerate 3p levels is only $\Delta E = 0.0021$ eV, it is clearly a line doublet (Fig. 1.22).[36]

1.6.3 Multiplicity

Each individual s, p, d, and f suborbital may accept two electrons. The electron orbitals are filled by electrons according to the HUND rule. In the case where several degenerate levels (wavefunctions) are available, such as the three p_x, p_y, and p_z subshells, electrons are filled in a way to maximize overall spin values.

The different versions of occupancy of degenerate subshells are called multiplicity M. This quantum mechanical unit thus refers to the possible number of orientations of the angular spin:

$$M = (2S + 1) \tag{1.25}$$

---

[36] This experimental observation for the Na D-line and of similar effects in the case of other alkali (s^1 orbital) elements actually motivated SA GOUDSMIT and GE UHLENBECK in 1925 to postulate that electrons have an intrinsic spin angular momentum.

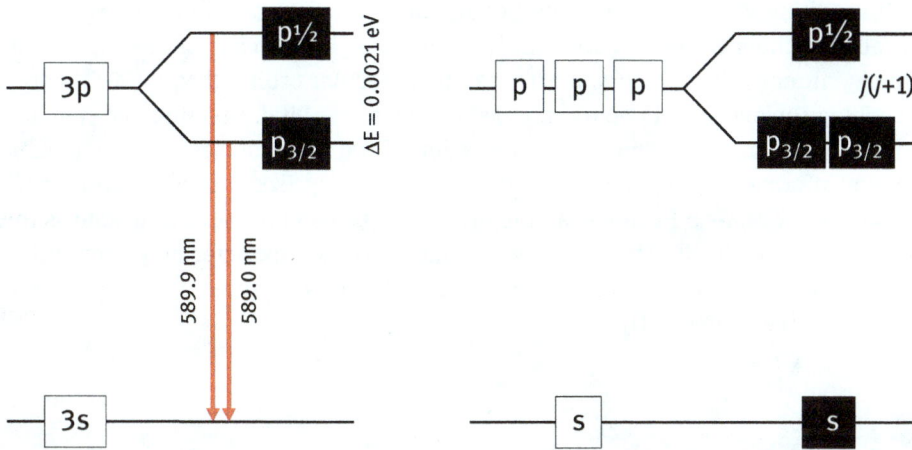

Fig. 1.22: Fine structure within the 3p level of sodium, caused by spin-orbit coupling, in terms of energy (a) and capacity (b).

Paired electrons, e.g. two electrons of spin $+\frac{1}{2}$ and $-\frac{1}{2}$ located within one filled orbital, compensate for their interaction. Consequently, only the number of unpaired electrons and their individual angular spins matter. According to eq. (1.25) and knowing S must be a multiple of $\frac{1}{2}$, multiplicities of 1, 2, 3 . . . (called singlet, doublet, and triplet states) are possible. Systems with no unpaired electrons always exist in a singlet state, those with one unpaired electron in a doublet state, and those with two exist in a triplet state. Higher multiplicities are also possible. Multiplicity notation is shown in Fig. 1.23.

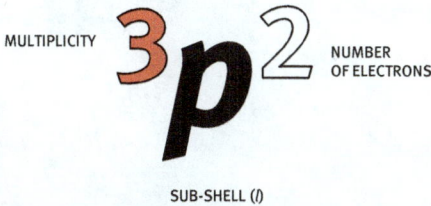

Fig. 1.23: Notation of multiplicity M. The multiplicity is indicated as left superscript for each subshell of quantum number l (in the case: $l = 1 = p$). The number of electrons involved is the right superscript.

Let's consider another example. The carbon electron shell $(1s^2 + 2s^2 + 2p^2)$ is filled pairwise for the 1s and the 2s shells, but the 2p shell is filled according to $2(p_x)^1 + 2(p_y)^1$ instead of $2(p_x)^2$. In the latter (hypothetical) case with all electrons paired, M would be $(2 \times 0) + 1 = 1$, indicating a singlet state. However, the real situation gives $M = (2 \times \frac{1}{2}) + 1 = 2$, indicating a triplet state.

Interestingly, multiplicity is applied not only to electrons within a single atom but also to chemical compounds and to nucleons. In the case of chemical compounds, atomic orbital theory is extended to molecular orbital theory (MO theory), describing the overlap of two (or more) single atomic orbitals creating one common molecular orbital. For example, the interaction of two s-orbital electrons from different atoms forms a sigma molecular orbital, which comprises a bonding and a non-bonding sub-MO. As a further example, hydrogen exists in its elemental state as the hydrogen molecule H_2. The two single electrons of $1s^1$ character may each either exist pairwise or unpaired in two molecule orbitals. This represents the well-known singlet or triplet states of H_2.

2 The atom's structure II: Nucleons and nucleus

Aim: While the first chapter recalled the principal composition of an atom, consisting of a nucleus and a shell with electrons of individual properties, this chapter introduces the principles and components of the atom's nucleus structure. The nucleus contains protons and neutrons, collectively called nucleons. Nuclei with the same number of protons but different numbers of neutrons create a rich world of isotopes of the same element. Every nucleus is described by a set of three numbers: the number of protons (identifying the chemical element), the number of neutrons (reflecting the specific isotope of that element), and the sum of protons and neutrons (forming the mass number). Analogous to the Periodic Table of Elements, the Chart of Nuclides arranges chemical properties of elements according to their electron shell parameters along with increasing mass of the element. The Chart of Nuclides correlates more than 3000 nuclei identified so far in an (x,y)-coordinate system, with x = neutron number, N and y = proton number, Z.

Mass and volume (or radius) are the physical properties of every set of nucleons. Mass divided by volume gives the density of a nucleus – the density of matter. Radius may refer to the (homogeneous) mass distribution within the nucleus' volume, but also to charge distribution (which is the distribution of protons within the nucleus).

The mass of individual nuclei is of ultimate interest since mass is proportional to energy. Exact knowledge of the precise mass of a nucleus provides access to the overall nucleon binding energy. The binding energy per nucleon determines, to a significant extent, the stability or instability of a given nucleus.

2.1 The neutron

With quantum mechanics, the one electron of the hydrogen atom was finally understood as energetically "surviving" in specific orbits of the shell. The electron is attracted by the one proton in the nucleus of the hydrogen atom. This particular atom has no other components. However, the atoms of all other chemical elements contain an additional basic constituent – the neutron.

In contrast to the electron and the proton, which were identified according to their charge and mass by the deviations they experience in magnetic fields, the neutron was identified much later, only in 1932. J CHADWICK[1] positioned an α-radium source close to a beryllium sample ($Z_{Be} = 4$), as RUTHERFORD had done with gold foil ($Z_{Au} = 79$). But unlike the RUTHERFORD experiment, the α-particles interacted with the

---

[1] Nobel Prize in Physics, 1935, "for the discovery of the neutron".

https://doi.org/10.1515/9783110742725-002

beryllium nucleus through a nuclear reaction, forming a (new) carbon nucleus and an unknown particle – the free neutron.

Both products were not easy to analyze: the newly formed atom, carbon, was uncharged, stable, did not emit any radiation, and the neutron released was also electrically neutral. The neutron was detected rather indirectly, namely in terms of a follow-up nuclear reaction between the neutron and a stable isotope, such as hydrogen or boron. Boron was an ideal element for this purpose because of its very high probability to catch a neutron (see Fig. 2.1).[2] The energy and mass of the neutron could be derived e.g. from the transfer of the impulse of the neutron to the proton.

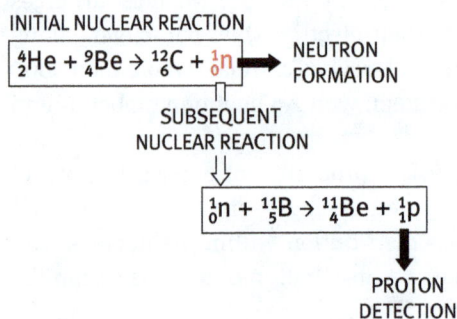

INITIAL NUCLEAR REACTION

$$\boxed{{}^{4}_{2}\text{He} + {}^{9}_{4}\text{Be} \rightarrow {}^{12}_{6}\text{C} + {}^{1}_{0}\text{n}} \longrightarrow \begin{array}{l}\text{NEUTRON}\\\text{FORMATION}\end{array}$$

SUBSEQUENT
NUCLEAR REACTION

$$\boxed{{}^{1}_{0}\text{n} + {}^{11}_{5}\text{B} \rightarrow {}^{11}_{4}\text{Be} + {}^{1}_{1}\text{p}}$$

PROTON
DETECTION

Fig. 2.1: Nuclear processes leading to the discovery of the neutron. The new radiation emitted from this process was historically called "beryllium radiation".

The complete picture was constructed retrospectively by assuming that (among other basic parameters) the number of nucleons within the two nuclear reactions should remain constant. They were assumed to be distributed differently between the reaction partners, in the same way that the number of electrons remains constant throughout a chemical reaction.

The neutron represents yet another subatomic particle, completing the set of three principal constituents of a chemical element: electron, proton, and neutron. The elementary charges are –1, +1, and 0 for the three particles, respectively. The two particles present in the nucleus, the proton and neutron, are collectively called nucleons. The two nucleons are of almost identical mass, i.e. $1.673 \cdot 10^{-27}$ kg (proton) *vs.* $1.675 \cdot 10^{-27}$ kg (neutron),[3] while the mass of the electron is significantly

---

2 See Chapter 13 for nuclear reaction mechanisms and cross-sections of neutron capture.

3 Interestingly, proton and neutron mass differ by a small fraction – the neutron is about 0.1% heavier than the proton. This seemingly negligible mass excess of the neutron, however, becomes the reason for a fundamental difference between the two particles. The "free" proton, i.e. one that is not bound in an atomic nucleus, is stable, while the analogue "free" neutron is not. Instead, it spontaneously transforms into a proton (and other products). For details, see Chapters 4 and 8.

smaller by a factor of ca. two thousand. The overall mass of an atom is thus predominantly localized in its nucleus.[4] On absolute scale, the mass of the three particles is extremely small. About 6×10^{23} protons or neutrons would weigh only one gram.

The classical properties of these particles, i.e. mass, charge, charge-to-mass ratios, but also relative mass in terms of atomic mass unit (see below) and mass expressed in terms of energy are summarized in Tab. 2.1. According to the equivalency of mass and energy, the energy is 938.272 and 939.566 MeV for the proton and the neutron, respectively.

Tab. 2.1: Summary of basic properties of the three basic atomic constituents: the electron, proton, and neutron. The elementary charges are − 1, + 1, and 0 for the three particles, but are given as values of the elementary (positive) charge unit C (coulomb). For the electron and the proton, the ratio between charge (q in C) and mass (m in g) was experimentally derived. The mass is given in units of kilograms and u (the atomic mass unit), as well as in energy in MeV via $E = mc^2$. (For energies relevant to nuclear physics, the unit MeV is preferred to the Joule J.[5])

Particle	q/m	Charge	q			m	
	C/g		C	kg	u		MeV
Electron	$-1.759 \cdot 10^8$	−1	$-1.602 \cdot 10^{-19}$	$9.109 \cdot 10^{-31}$	0.00055		0.511
Proton	$+9.58 \cdot 10^4$	+1	$+1.602 \cdot 10^{-19}$	$1.673 \cdot 10^{-27}$	1.007825		938.272
Neutron		0	0	$1.675 \cdot 10^{-27}$	1.008665		939.566

2.2 Nuclide notations and isotopes

2.2.1 Notations of nuclei

With the two constituents of an atomic nucleus, the proton and the neutron, it is now possible to describe an atom not only by its "atomic number" (which is the proton number, Z) and atomic mass (denoted in the PTE as g/mol) but also by its nuclear composition. The standard notation uses the element's symbol (E), adding the number of protons (Z) and neutrons (N) as subscripts (on the left- and right-

---

4 The mass number of natural gold is 197 and the proton number is 79. A simple estimate would say that there are 79 protons + (197 − 79 =) 118 neutrons in the nucleus, making a mass of $79 \times 1.673 \cdot 10^{-27}$ kg + $118 \times 1.675 \cdot 10^{-27}$ kg = $(132.167 + 197.650) \cdot 10^{-27}$ kg = $329.817 \cdot 10^{-27}$ kg for the nucleus versus 79 times the mass of an electron = $0.07 \cdot 10^{-27}$ kg. With an overall mass of $329.887 \cdot 10^{-27}$ kg, the nucleus of a gold atom represents about 99.98% of the entire atom's mass. (This simple mathematics is not completely true, as the mass of "bound" nucleons differs from the sum of individual, unbound nucleons; see later in this chapter.)
5 See Appendix for all the parameters listed in the book.

hand sides, respectively). Z is also denoted as "atomic number" or "charge number". The sum of proton and neutron numbers constitutes the overall mass number, A, of the nucleons (with $A = Z + N$), indicated as the left-hand superscript of the element's symbol. Figure 2.2 depicts this general notation.

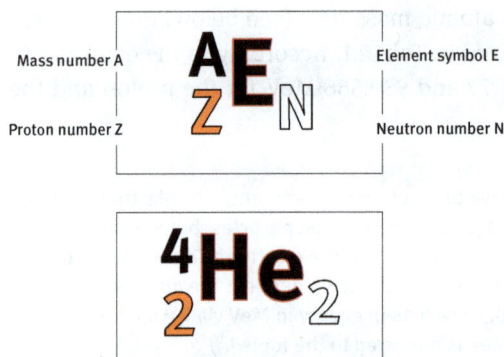

Mass number A $^{A}_{Z}E_{N}$ Element symbol E

Proton number Z   Neutron number N

$$^{4}_{2}He_{2}$$

Fig. 2.2: Notation showing the numbers of protons (Z), neutrons (N), and nuclear mass (A) of a chemical element (E). The nucleus of the helium-4 atom, i.e. the α-particle, is shown as an example.

In practice, it is sufficient to denote an isotope using only the mass number, because the chemical element and proton number are inherently linked. Thus, it is straightforward to calculate the corresponding neutron number N of the given isotope. For example, the α-particle, representing the nucleus of the helium atom and containing two neutrons, may simply be written as ^{4}He. It is composed of two protons plus two neutrons ($Z = 2$, $N = 2$, and $A = 2 + 2 = 4$). The α-particle thus represents the helium atom without its two electrons.[6]

Note that Z, N, and A do not represent absolute mass numbers. The real mass would result from multiplying the individual proton and neutron numbers with the absolute mass of each nucleon, as given in Tab. 2.1.

2.2.2 Isotopes

For the hydrogen atom discussed, the nucleus contained only one proton (and no neutrons). With the atom ionized, this nucleus is simply a "naked" proton (see Fig. 2.3). However, there exists another stable hydrogen nucleus, having one proton

---

6 It is created by ionizing the atom, i.e. by transferring enough energy to the helium atom to release both of its 1s electrons above the ionization potential. Similarly, the "naked" nucleus of atomic hydrogen is the proton itself (see Fig. 2.3).

The proton

$$^1_1P_0$$

The isotopes of hydrogen and their natural abundance:

HYDROGEN	DEUTERIUM	TRITIUM
99.9885%	0.0115%	10^{-15} %

$$^1_1H_0 \qquad ^2_1H_1 \qquad ^3_1H_2$$

The molecular weight of hydrogen of natural isotopic composition

$$1.00794\,^{}_1H$$

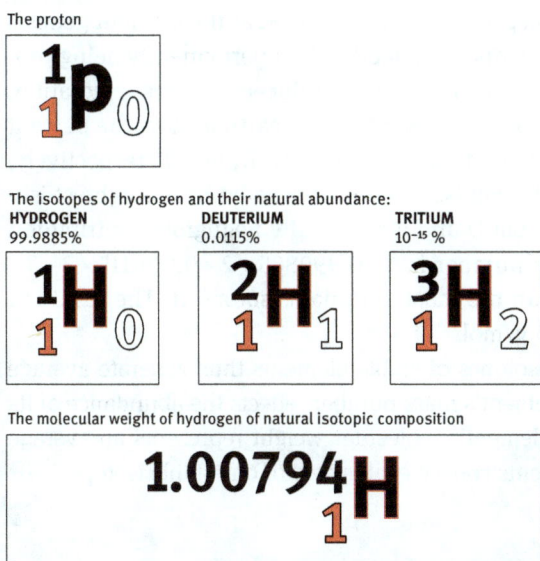

Fig. 2.3: Isotopes of hydrogen, expressed using the number of protons, neutrons, and mass for each of three isotopes. The hydrogen isotopes containing one or two neutrons are deuterium and tritium. The average molecular mass number of naturally occurring hydrogen is 1.00794 g/mol.

and one neutron. Helium also has two stable nuclei, having either two neutrons (the α-particle) or one neutron.

Thus, Z remains the same (Z = 1 for H, Z = 2 for He) since it characterizes the element, and (due to the electron configuration defined by Z) its chemical reactivity is also unchanged. However, these hydrogen and helium nuclei differ in their respective number of neutrons (by N = 0 and 1 for H, and N = 1 and 2 for He). Nuclei of constant Z but varying N are called isotopes, as they are "iso" (equal) in Z number. Consequently, isotopes differ by mass (as A = Z + N) according to A = 1 and 2 for H and A = 3 and 4 for He. In general, isotopes are written with the element symbol only (see Fig. 2.3).[7]

Because of the relative difference in mass (deuterium is roughly twice as heavy as hydrogen), the three hydrogen isotopes (and their chemical species) differ profoundly in some physico-chemical properties. Within the same element, these differences become smaller with increasing Z, due to the decreasing differences between the relative mass of the isotopes.[8]

---

7 In the case of hydrogen, unlike all the other chemical elements, the other stable (^{2}H) and the main radioactive (^{3}H) isotopes were given individual names and symbols: deuterium D (due to the mass number 2) and tritium T (due to mass number 3).
8 Still, there are detectable effects even for the heavy chemical element uranium. For example, the small differences in mass in uranium hexafluoride gas UF_6 are used to separate different isotopes

The two stable hydrogen isotopes, H and D, were formed at the element genesis in vastly different ratios. The third isotope, radioactive T, is permanently being produced in the earth's atmosphere by nuclear processes induced by cosmic radiation. The mixture of the three hydrogen isotopes present on our earth follows the percent distribution of 99.9885%, 0.0115%, and about 10^{-15}% for H, D, and T, respectively. The overall mixture of the three hydrogen isotopes is dominated by the lightest isotope (H), with small contributions from D and a practically negligible contribution from T. The resulting overall mass number is $1 \times 0.999885 + 2 \times 0.000115 \times 2 + 3 \times 10^{-17} = 1.000115$, i.e. a bit higher than the most abundant isotope H. The resulting average molecular weight is 1.00794 g/mol.

Mixtures of naturally occurring isotopes of stable elements thus generate average mass numbers. In other words: an element's mass number reflects the abundance of its constituent isotopes. Similarly, the element's molecular weight represents an average value. However, a few chemical elements consist exclusively of one stable isotope.[9]

2.3 Conventional parameters of the nucleus

2.3.1 Radius

After having described the three principle subatomic particles, the parameters of the atom can now be discussed in more detail. In general, there is a "spherical" model of the atom (see Fig. 2.4 for the main isotope of carbon, ^{12}C). The dimension of the (outer) electron shell generally defines the radius r of an atom.[10] Despite a few exceptions, there are systematic trends regarding radius size within the PTE. In the same group of the PTE, radius increases from top to bottom. Within the same period, the radius decreases from left to right. The latter is due to the increasing proton number Z within the same principal electron shell (i.e. the main quantum number n) and is called contraction. Hydrogen is thus understood to be the smallest atom, while the heaviest alkaline element (francium) is understood to be the largest. For hydrogen, r is around 32 pm; for a heavier element such as gold it is about 144 pm, and for cesium – the heaviest stable alkaline element and the lighter homologue of francium – it is

---

of uranium (e.g. the isotopes 235 and 238 needed for nuclear fission) out of the isotopic mixture of natural uranium by gas centrifuge cascades.

9 Beryllium (^{9}Be), fluorine (^{19}F), sodium (^{23}Na), aluminum (^{27}Al), phosphorus (^{31}P), scandium (^{45}Sc), manganese (^{55}Mn), cobalt (^{59}Co), arsenic (^{75}As), yttrium (^{89}Y), niobium (^{93}Nb), rhodium (^{103}Rh), iodine (^{127}I), cesium (^{133}Cs), praseodymium (^{141}Pr), terbium (^{159}Tb), holmium (^{165}Ho), thulium (^{169}Tm), gold (^{197}Au) and bismuth (^{209}Bi).

10 Note that the atom cannot be considered a geometric sphere with clearly defined radius, diameter, surface area, volume etc. because electron shell parameters are quantum mechanical. Chemistry thus defines the radius of a given element as half the distance between the nuclei of two neighboring atoms. Obviously, this depends on the type of chemical bond between these two atoms.

272 pm. Atom diameters $\varnothing$ thus range from ca. 60 to ca. 600 pm, i.e. approximately 6 to $60 \cdot 10^{-11}$ m. Atomic radius is often given in Ångströms (1 Å $= 10^{-10}$ m $= 100$ pm), i.e. approximately 3 to $30 \cdot 10^{-11}$ m or 30–300 Å.

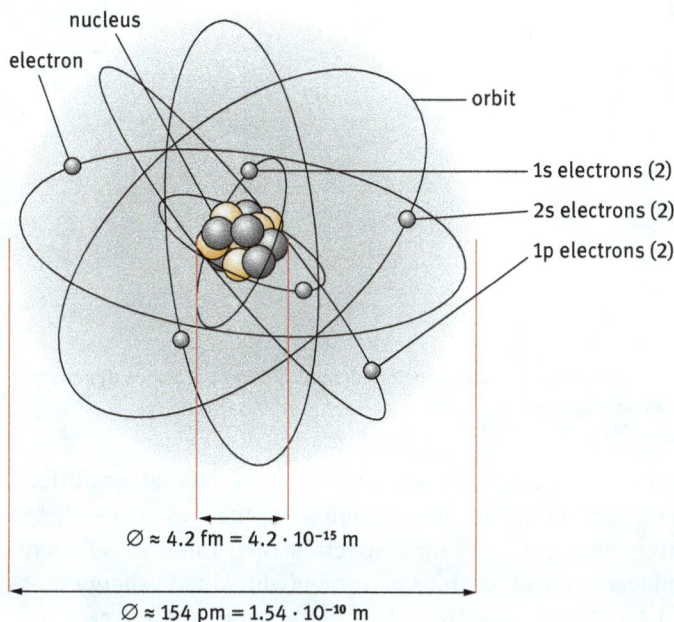

nucleus

electron

orbit

1s electrons (2)

2s electrons (2)

1p electrons (2)

$\varnothing \approx 4.2$ fm $= 4.2 \cdot 10^{-15}$ m

$\varnothing \approx 154$ pm $= 1.54 \cdot 10^{-10}$ m

Fig. 2.4: Simplified illustration of diameters of the carbon nucleus ^{12}C vs. carbon atom. The dimension of the nucleus is on an fm scale, which is a factor of about 1,000 smaller than the atom. The image dramatically overestimates the dimension of the nucleus. For carbon, the radius of the nucleus is $r \approx 2.1$ fm $= 2.1 \cdot 10^{-15}$ m and the diameter is $\varnothing \approx 4.2$ fm $= 4.2 \cdot 10^{-15}$ m. The nucleus of the carbon atom occupies around $5 \cdot 10^{-12}$% of the volume of the whole carbon atom, but collects more than 99.9% of the atom's total mass.

The radius of an atom's nucleus, in contrast, is on an fm scale, which is a factor of >1,000 smaller than the radius of the atom. For carbon, as exemplified in Fig. 2.4, the radius of the nucleus is $r \approx 2.1$ fm $= 2.1 \cdot 10^{-15}$ m, and the diameter $\varnothing \approx 4.2$ fm $= 4.2 \cdot 10^{-15}$ m. The nuclear diameter is smaller than the atomic diameter by about four orders of magnitude.

For the corresponding volumes via $V = 4/3 \, \pi r^3$, the difference becomes even more dramatic, as $V_{nucleus} / V_{atom} = (\varnothing_{nucleus} / \varnothing_{atom})^3$. The nucleus of the carbon atom occupies only $5 \cdot 10^{-12}$% of the volume of the whole carbon atom! This little piece of matter, on the other hand, accounts for more than 99.9% of the overall mass of the atom.

These extremely small dimensions explain why the inner structure of the atom was not resolvable by traditional spectroscopic methods. Figure 2.5 compares the dimensions of the atom with the wavelengths of visible light.

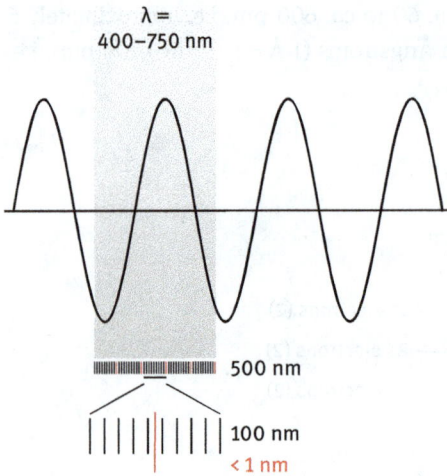

Fig. 2.5: Wavelength of visible light related to the dimension of an atom. Light provides dimensions much larger than the object of investigation.

However, similar to the dimension of the shell of the atom, it is rather difficult to exactly quantify the radius of a nucleus. Experimental approaches are needed to obtain data not accessible *via* e.g. visible light spectroscopy. Interestingly, sub-atomic particles are employed instead, such as neutrons (with kinetic energy ranging from $E_n = 14$ MeV – 1.4 GeV) and electrons for scattering experiments. These particles penetrate the atom and are scattered at the nucleus. The nuclei investigated are those of stable isotopes.

Therefore, the question is: what does a nucleus look like?
1. Is it a perfect sphere – or not?
2. Are atomic nuclei spherically shaped – or are they deformed?
3. Are the nucleons distributed homogeneously within the nucleus – or not?

Generally speaking, for spherical nuclei with homogeneously distributed nucleons, there should be a straightforward relationship between Z and A of the nucleus and its radius (and correspondingly also the volume and surface). Experimental data, however, provide different values for the size of the nucleus as illustrated in Fig. 2.6. The nuclear radius obtained through neutron scattering r^{mass} is a bit larger than that obtained by electron scattering for the same atom. Neutrons basically identify the distribution of nucleons. In contrast, electrons not only are sensitive to scattering matter in general, but can also discriminate between charged and uncharged matter (i.e. protons and neutrons). The nuclear radius r^{charge} obtained by electron scattering thus also reflect the distribution of the proton cluster within the nucleus.

The mass number A and volume V of a nucleus are correlated: the larger the number of nucleons, the larger the volume of the atomic nucleus. Consequently,

MASS RADIUS

neutron scattering

CHARGE RADIUS

electron scattering

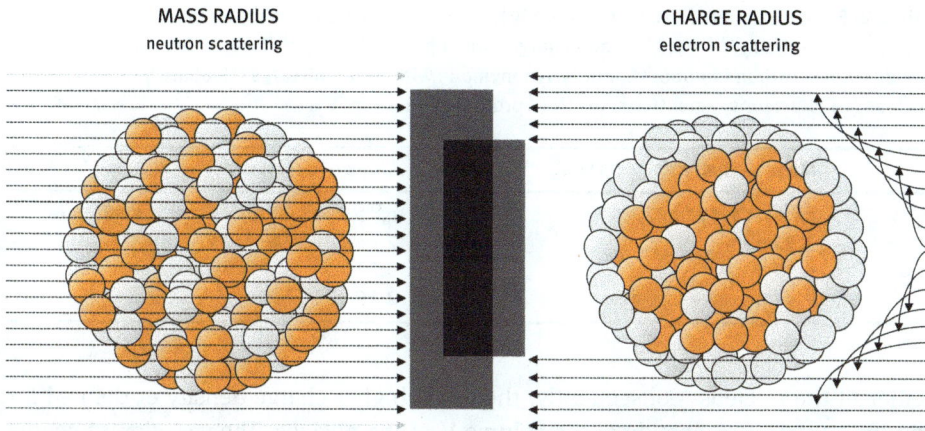

Fig. 2.6: Experimental determination of dimensions r^{mass} and r^{charge} of an atomic nucleus by neutron or electron scattering. Experimentally obtained sizes differ; neutron scattering indicates a larger size than that obtained by electron scattering for one and the same atom.

the radius also depends on the mass number A. This relation assumes the nucleus is at ground state energy and has a nondeformed, spherical shape, i.e. $V = 4/3\,\pi r^3$, and for its radius, $r \approx A^{\frac{1}{3}}$.

To account for the discrepancy between experimentally measured radii (r^{mass} and r^{charge}) and the theoretical radius (r), a proportionality factor is used: the so-called radius parameter r_0. The resulting quantitative correlations between radius and mass number are given by eqs. (2.2a) and (2.2b). When discussing the nuclear radius, two different values are used – a mass radius parameter r_0^{mass} and a charge radius parameter r_0^{charge}. The mass radius parameter is 1.3 fm $\leq r_0^{mass} \leq$ 1.4 fm, depending on the Z and N content of a given A. It represents the contribution of a single nucleon to the overall radius of a nucleus composed of a given number of nucleons. Likewise, the charge radius parameter reflects the contribution of each proton to the overall size of a nucleus. It is $r_0^{charge} = 1.23$ fm, which is a bit smaller than the mass radius parameter.

$$r \approx A^{\frac{1}{3}} \tag{2.1}$$

$$r^{mass} = f(A) = r_0^{mass} A^{\frac{1}{3}} \tag{2.2a}$$

$$r^{charge} = f(Z, A) = r_0^{charge} A^{\frac{1}{3}} \tag{2.2b}$$

Consequently, radii derived for identical A differ. Table 2.2 gives examples of the mass radius for three mass numbers: a small nucleus (A = 10), a larger nucleus (A = 100), and the carbon isotope ^{12}C.

One could say that, because the proton is lighter in mass than the neutron (see Tab. 2.1), the charge radius parameter should be smaller. In fact it is, but for a different reason; experimental evidence shows that protons are more tightly packed within the

Tab. 2.2: Radii of atoms of mass numbers 10, 12, and 100 calculated using an averaged mass radius parameters $r_o^{mass} = 1.35$ fm and charge radius parameters $r_o^{charge} = 1.23$ fm. Note that A = 12 stands for the carbon isotope ^{12}C, but is synonymous with other mixtures ("isobars") for A = Z + N, such as ^{12}B (5 protons, 7 neutrons) or ^{12}N (7 protons, 5 neutrons).

A	Mass radius: $r^{mass} = f(A) = r_o^{mass} A^{⅓}$	Charge radius: $r^{charge} = f(Z,A) = r_o^{charge} A^{⅓}$
	$r_o^{mass} = 1.35$ fm	$r_o^{charge} = 1.23$ fm
10	2.9 fm	2.64
100	6.3 fm	5.74
12	2.4 fm	2.19

center of the nucleus. Consequently, there is a higher charge density e_p defined for specific volume elements dv of the volume V of the nucleus. The variation of charge density in terms of $\rho(r)$ yields average radius $r_{1/2}$ values. The corresponding relationship utilizes ρ_0, the ultimate density of matter. Ideally, it would reflect the density of a single nucleon itself – but expanded into a larger dimension.

$$\rho(r) = \frac{\rho_0}{1 + \exp\left(\left(r - r_{1/2}\right)/a\right)} \tag{2.3}$$

Figure 2.7 illustrates the concept of a constant charge density changing along the radius of a nucleus. It remains constant (at $\rho/\rho_0 = 1$) starting from the center of the nucleus until an outer part of the spherical nucleus is reached. This outer region has a lower charge density (ranging from $\rho/\rho_0 = 0.9$ to 0.1, depending on the nucleus).

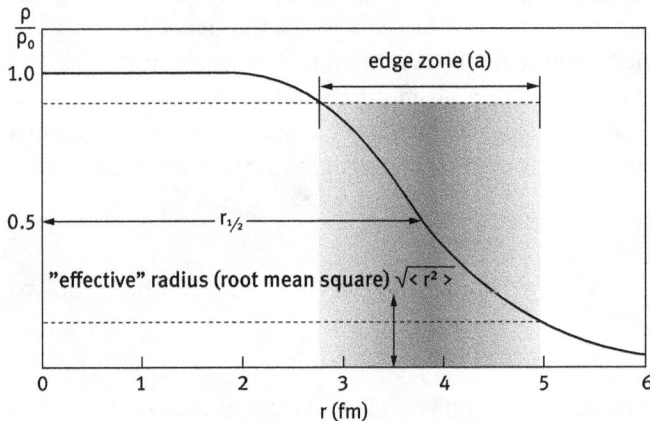

Fig. 2.7: Charge density distribution within an atom (FERMI distribution of charge density). The ratio of ρ/ρ_0 on the y-axis is expressed versus distance from the center of the nucleus. Resulting parameters include: a variable zone, a, of the edge; an average radius $r_{1/2}$ (ratio of $\rho/\rho_0 = 0.5$); and a radius r derived as an effective radius; r.m.s. = "root mean square" = $\sqrt{<r^2>}$.

The drop in charge density indicates that, although there is still a constant density of nucleons, the percentage of protons is decreasing within an edge zone of dimension a. The ratio of ρ/ρ_o equal to 0.5 defines the average radius $r_{1/2}$. Another noteworthy value is the effective radius, called r.m.s. = "root mean square" = $\sqrt{<r^2>}$. Figures 2.8 and 2.9 illustrate the charge density in terms of relative and absolute units.

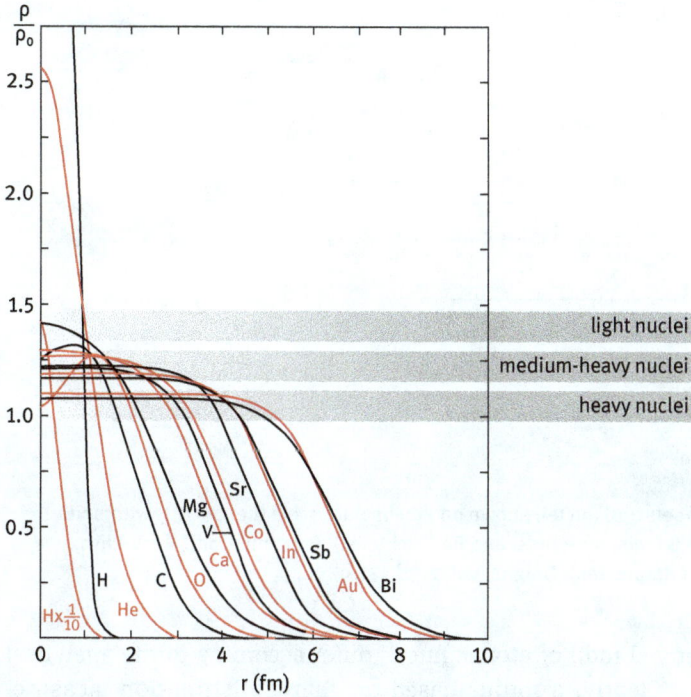

Fig. 2.8: Charge density of selected nuclei, shown on a relative scale (i.e. as the ratio of $\rho/\rho_o = f(r)$). With permission: R Hofstadter, Nuclear and nucleon scattering of high-energy electrons, Ann. Rev. Nucl. Sci. 7 (1957) 231.

The effective radius, $<r^2>$, depends on the charge density distribution along a radial or volume element according to:

$$\langle r \rangle = \int_0^\infty \rho(r) 4\pi\, r dr \tag{2.4a}$$

$$\langle r \rangle = \int_0^\infty \rho(r) 4\pi\, r dr \tag{2.4b}$$

There are some basic conclusions.

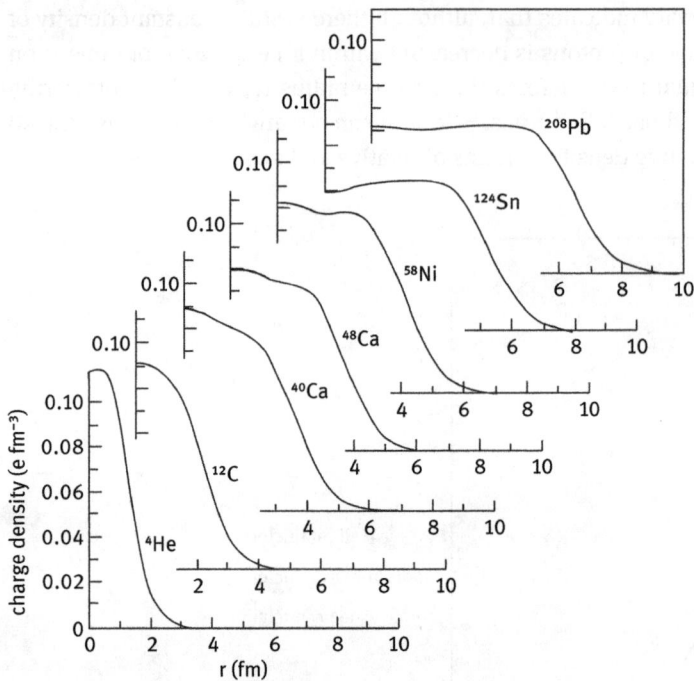

Fig. 2.9: Charge density of selected nuclei, shown on an absolute scale (i.e. $\rho = f(r)$) with units (e / fm³). With permission: B Frois, in: P Blasi and RA Ricci (eds.), Proc. Int. Conf. Nucl. Phys., Florence, 1983, Tipografia Compositoria Bolgna, vol. 2, p. 221.

1. Experimentally derived radii of atomic nuclei differ according to the analytical method employed. Clearly, a radius based on charge distribution measurements is smaller than one based on mass distribution.
2. Protons thus exist at a higher density in the core of the nucleus. For heavier atoms (with a significant excess of neutrons over protons in order to compensate for increasing coulomb repulsion between protons), neutrons in a way thus "dilute" the charge density.
3. The radius may be correlated to the mass number of the nucleus, and resulting values are different for the mass radius or charge radius parameter applied.
4. These correlations, however, are only true for ground state nuclei (i.e. not energetically excited).
5. The increase of radius with respect to Z, i.e. with the number of chemical elements, is not linear. The factor $A^{1/3}$ defines a moderate increase. Consequently, the radius of light and heavy nuclides ranges from about 1.23 fm for hydrogen ($A = 1$, $Z = 1$, and r_0^{charge} multiplied by $A^{1/3}$) and 6 to 7 fm for bismuth ($A = 208$, $Z = 83$), as shown in Fig. 2.8.

Precise determination of atomic radii remains a challenge, and not only because nuclei have no sharp edges/surfaces. Theoretical models are needed to describe the status of a nucleus in terms of its mass, which is obviously one key parameter that correlates with radius. However, measuring the mass of an individual nucleus with extreme precision remains an analytical problem. Moreover, theoretical models and recent measurements indicate that the shape of the nucleus – even in its ground state – is not necessarily a sphere. A nucleus may also be deformed in its ground state. Figure 2.10 illustrates three nuclei of natural elements (silver, gold, and neon) and their respective deformations: prolate for Ag, oblate for Au, and hexadecupol for Ne.

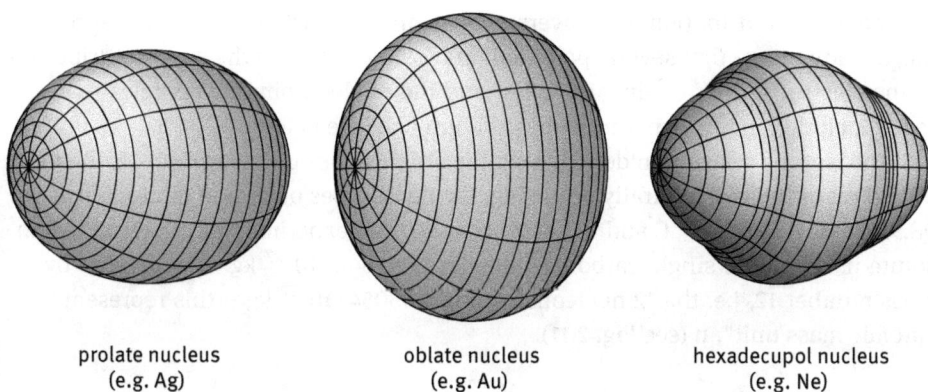

| prolate nucleus (e.g. Ag) | oblate nucleus (e.g. Au) | hexadecupol nucleus (e.g. Ne) |

Fig. 2.10: Shape of the ground states of select nuclei.

2.3.2 Mass of the atom and atomic mass unit

The two kinds of nucleons each have an individual absolute mass (see Tab. 2.1), which can be used to calculate the mass of the nucleus. The nucleus represents almost, but not all, of the mass of the atom. The mass of the shell is the mass of the electrons $Z \cdot m_e$. In addition, the binding energy $E_{B(e)}$ (n) of all the electrons must be considered. This will be different for each electron shell because of the different main quantum number, as discussed in Chapter 1. Overall, the mass of the shell components does not contribute significantly to the overall mass of the atom, but is definitely nonzero.[11]

---

11 The mass of a single electron with zero kinetic energy, called "rest mass", is only $m_e = 9.109382 \cdot 10^{-31}$ kg (Tab. 2.1). The overall binding energy of all the electrons (as estimated *via* the THOMAS–FERMI model) is found using eq. (1.2) $E_n = -R_y Z^2 \cdot 1/n^2$, or more precisely using $B_e(Z) = 14.4382 \, Z^{2.39} + 1.55468 \cdot 10^{-6} \, Z^{5.35}$ eV.

Tab. 2.3: Mass balances for an atom.

Mass of the atom	m^{atom}	$=$	$m^{nucleus}$	$+$			m^{shell}	(2.5)
	m^{atom}	$=$	$m^{nucleus}$	$+$	$Z \cdot m_e$	$+$	$E_{B(e)}(n)$	(2.6)
Mass of the nucleus	$m^{nucleus}$	$=$	m^{atom}	$-$	$Z \cdot m_e$	$-$	$E_{B(e)}(n)$	(2.7)

There is, however, a more practical approach to calculating atomic mass, using the "unified atomic mass unit" or "atomic mass unit" u. Instead of using the absolute numbers of mass and binding energy, it considers the mass of a stable isotope of a prominent atom as a reference value. This mass is very precisely known from experimental observations. The mass is divided by the number of nucleons, providing the mass contribution of an average nucleon to the whole atom. This approach has the advantage that several potential unknowns – such as the mass contribution of the electrons, their binding energies, and the nucleon binding energies – are already included in this experiment-based approach (see below).

The isotope selected in determining the atomic mass unit is carbon-12, the most abundant of the two naturally occurring, stable isotopes of carbon (abundances are 98.93% and 1.07% for ^{12}C and ^{13}C, respectively). The experimentally determined absolute mass of one single carbon-12 atom is $19.92648 \cdot 10^{-27}$ kg. It is divided by its mass number 12, i.e. the 12 nucleons, giving $1.66054 \cdot 10^{-27}$ kg – this represents the "atomic mass unit", u (see Fig. 2.11).

$$u = \frac{M^{nuclide}\left(^{12}_{6}C\right)}{12} \tag{2.8}$$

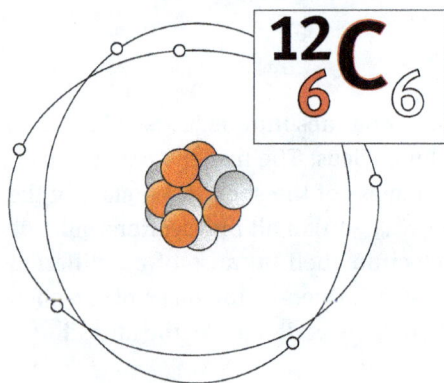

Fig. 2.11: Composition of the carbon-12 isotope, which serves as the reference atom in calculating the atomic mass unit, u. The isotope is "symmetrically" composed of the same number of the three subatomic particles: 6 protons (orange), 6 neutrons (white), and 6 electrons. The "real" weight of the atom is well known from experiments.

With this parameter in hand, the absolute mass of every other isotope is easily estimated[12] by multiplying the mass number A of the given isotope by the atomic mass unit. In addition, the atomic mass unit may serve not only as a proportionality factor to calculate the mass of atoms; it also serves as a measure of subatomic particle mass. For the electron, proton, and nucleon, mass can be expressed as fractions of u.[13] The proton and neutron mass are thus 1.008665 u and 1.007825 u, i.e. slightly larger than 1. These values reflect the contributions of electron mass, electron binding energy, and nucleon binding energy to the overall mass of the atom. For the electron, the value is $m_e = 5.4858 \cdot 10^{-4}$ u (see also Tab. 2.1).

2.3.3 Density

Table 2.4 shows the density of four typical metals, determined for atoms packed in a metallic state. Their densities range from 2.70 to 11.34 kg/m³. If the radius, volume (see eq. (2.2a)), and mass of a given nucleus ($m \approx u \cdot A$) are known, it is possible to calculate the nuclear density as V/m (see eq. (2.9)).

Tab. 2.4: The density of nuclei *vs.* the density of atoms. The chemical element aluminum has a densities of or 2699 kg/m³, respectively (metallic state of the natural elements at 20 °C temperature). For the nucleus ^{30}Al or aluminum-30, the density is estimated using the mass number A, mass (in kg) and size (radius and volume) of the nucleus. The resulting density of the aluminum nucleus is $\approx 2.8 \cdot 10^{+17}$ kg/m³, compared to $2.70 \cdot 10^3$ kg/m³ for the aluminum atom.

Parameters		
Al atom	Density	$2699 \cdot 10^3$ kg/m³
^{30}Al nucleus	Mass ($\approx u \cdot A$)	≈ 30 u $\approx 5 \cdot 10^{-26}$ kg
	Radius ($= r_0^{mass} A^{1/3}$)	$\approx 3.5 \cdot 10^{-15}$ m
	Volume ($= 4/3 \pi r^3$)	$\approx 1.8 \cdot 10^{-43}$ m³
	Density	$\approx 2.8 \cdot 10^{+17}$ kg/m³

---

12 However, since every isotope varies in its nucleon interactions, and the electron binding energy varies (albeit weakly) with Z, calculated and experimental mass may differ slightly.

13 The atomic mass unit, u, thus becomes a proportionality factor for very small mass values. This is similar to the AVOGADRO constant used in chemistry to define an ensemble of a large number of atoms or molecules: 1 mol = $6.022141 \cdot 10^{23}$ atoms or molecules. Although unusual, one may express any large number in molar fractions. Other large-scale units, such as those measuring distance, are the *light-year* and the *parsec*. A light-year is the distance that light travels in vacuum in one year and is 10^{16} m. In astronomy, the parsec is used, which is equal to approximately 3.26 light-years.

$$\rho^{\text{nucleus}} = \frac{V}{m} \approx \frac{\frac{4}{3}\pi \left(r_0^{\text{mass}} \cdot A^{\frac{1}{3}}\right)^3}{u \cdot A} = \frac{\frac{4}{3}\pi \left(r_0^{\text{mass}}\right)^3}{u} \tag{2.9}$$

For aluminum, for example, the calculated nuclear density is $\rho_{\text{nucleus}} \approx 2.88 \cdot 10^{+17}$ kg/m^3 compared to 2.70 kg/m^3 for the aluminum atom. That is a difference of 17 orders of magnitude! It again speaks to the fact that an atom's mass is almost exclusively located in the nucleus, while the nuclear volume is an almost negligible fraction of the whole atom. The atom in general thus appears to be an almost empty space, which subatomic particles and photons can easily penetrate.

In fact, the density of each nucleus is more or less the same, because the nucleus density ρ_{nucleus} ultimately depends on the relationship between radius and mass number. With a mean distance of about 1.8 fm between nucleons, there are about 0.13 nucleons per fm^3 volume in the nucleus. The most dense nuclei have around 0.17 nucleons per fm^3. Furthermore, the density of nucleons packed within an atomic nucleus is the highest density of matter achievable on earth.[14]

The dimensions of an atom are extremely small (see Chapter 1); the dimensions of the atom's nucleus are even smaller. Since the scale of a 10^{-43} m^3 volume or a 10^{+17} kg/m^3 density is hard to picture, let us therefore try to translate those numbers into a more familiar scale.

A cube of aluminum containing 1 mole of aluminum atoms has a weight of 26.98 g (obtained from the PTE). In the form of a cube, and with the density of aluminum of 2.70 g/cm^3, the cube's volume and edge length are 9.99 cm^3 and 2.15 cm, respectively. Suppose that this same cube consisted of aluminum nuclei instead of atoms; its weight would be an incredible $2.76 \cdot 10^{12}$ kg or $2.76 \cdot 10^{9}$ t! The weight of the complete Cheops pyramid is of the same magnitude – it weighs about $6 \cdot 10^{6}$ t.

2.4 Mass *vs.* energy of the nucleus: The mass defect

Why can we not simply add the absolute mass of each neutron and proton[15] to determine the absolute mass of a nucleus? The answer is: the (absolute) mass of an unbound nucleon differs from the (real) mass of a <u>bound</u> nucleon. Fantastically, $m^{\text{"bound" nucleon}} \neq m^{\text{"free" nucleon}}$. In the case of nuclei, the whole is not the sum of its parts.

---

14 The ultimate density of nucleons packed tightly inside the nucleus at "infinitely diluted state" according to the drop model of the atom nucleus (see Chapter 3) is $\approx 2.96 \cdot 10^{+17}$ kg/m^3.
15 Masses of the proton and the neutron are from Tab. 2.1 and valid in the case of mass at rest energy (m_0).

Let us for example consider the α-particle again, i.e. the nucleus of the helium isotope ^{4}He. It consists of two protons and two neutrons (see Fig. 2.2). The experimental value for the mass of the He <u>atom</u> is 4.00260325415 u.

2.4.1 Mass defect

The corresponding mass of the He <u>nucleus</u>, 4.00150 u, is obtained by subtracting the mass and binding energy contribution of two electrons (as per eq. (2.7)) from the experimental mass above. The calculated (i.e. expected) mass of the ^{4}He nucleus is $2m_p$ (u) + $2m_n$ (u) = 4.03188 (u). For the four nucleons bound together within the nucleus, the total mass of 4.00150 u is thus smaller than the sum of the four unbound nucleon masses, i.e. 4.03188 u (see Fig. 2.12). The problem of $m^{\text{“bound” nucleons}} \neq m^{\text{“free” nucleons}}$ in fact turns into $m^{\text{“bound” nucleons}} < m^{\text{“free” nucleons}}$. The nucleus is lighter than its individual components! The difference in mass is always negative when expressed as $m^{\text{nucleus}} - m^{\text{sum of individual, nonbound nucleons}}$. A fraction of the initial mass, i.e. of the unbound nucleons, is missing in the end. This is called "mass defect":

$$\Delta m^{\text{defect}} = m^{\text{nucleus}} - m^{\text{sum of individual, non − bound nucleons}} = m^{\text{nucleus}} - \left(Zm_p + Nm_n\right) \quad (2.10)$$

For the He nucleus, the mass defect value is Δm = 4.00150 u − 4.03188 u = −0.030377 u. This relative difference is not dramatic, being only <1% in the ^{4}He example. However, the impact of mass defect on the understanding of matter and energy sources is one of the most dramatic effects of our material world.[16]

Using the experimental mass of the nucleus, along with the precise rest mass of the proton and neutron, values of mass defect are tabulated[17] for all the stable nuclei.

---

[16] To understand this effect takes one to the philosophic question of " . . . Dass ich erkenne, was die Welt // Im Innersten zusammenhält, . . . " JOHANN WOLFGANG GOETHE, *Faust: The First Part of the Tragedy*, 1808: " . . . So that I may perceive whatever holds // The world together in its inmost folds."

[17] If not otherwise indicated, data used throughout the first edition of this textbook were from M Wang et al.: The AME2012 atomic mass evaluation (II). Tables, graphs and references, Chinese Physics C 36, (2012) 1603–2014. The database is regularly updated, with new editions published in 2016 and 2020. This textbook refers to the 2020 database by M Wang et al., Chinese Phys. C 45 (2021) 030003 (https://iopscience.iop.org/article/10.1088/1674-1137/abddaf/pdf) For nuclides relevant to the present textbook, data are listed in Table 14.4 with full precision.

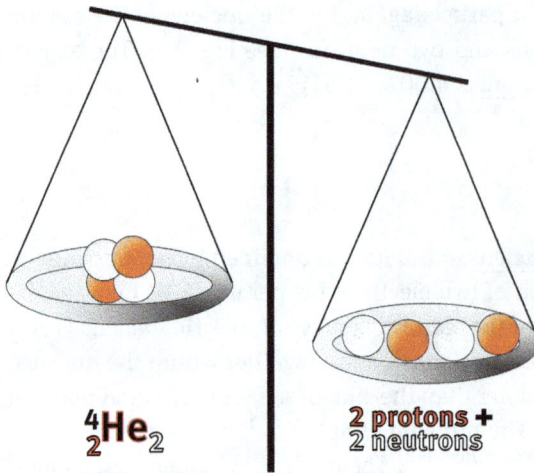

$${}^{4}_{2}\text{He}_{2} \qquad \text{2 protons +}\\ \text{2 neutrons}$$

Fig. 2.12: Nucleons bound together within the nucleus form a nucleus of an atom. This nucleus is lighter in weight (and consequently, lower in energy) than the sum of its identical, unbound components. When converting a number of individual, "free" nucleons into an atomic nucleus, the nucleus weighs less than the sum of its nonbound components. There is a "mass defect"!

Tab. 2.5: Mass balance and Δm^{defect} for the ^{4}He nucleus.

Mass of nucleus = f(Z,N)		m^{nucleus}	<	$Z \cdot m_p$	+	$N \cdot m_n$		
			=	$Z \cdot m_p$	+	$N \cdot m_n$	−	Δm
Example								
Mass (units of u)	experimental	4.002603						
	calculated	4.032980	=	2.015650	+	2.017330		
	difference							0.030378

2.4.2 Mass excess

Another important expression is the so-called mass excess Δm^{excess} (or simply Λ). It is the difference between the "real", experimentally determined mass of the nuclide m^{nucleus} (i.e. its "atomic mass"), and the "relative" mass number A of the same nuclide:

$$\Delta = m^{\text{nucleus}}(\text{in u}) - A \cdot u \tag{2.11}$$

Δ is typically expressed in units of the atomic mass unit, u. The mass excess value, obtained by eq. (2.11), yields energy when multiplied by c^2. For ^{12}C (carbon-12), *by definition* the value of Δ in eq. (2.11) must equal 0, according to eq. (2.8). While the mass defect has an obvious physical meaning (as the equivalent binding energy within an atomic nucleus), the mass excess is to be regarded as a useful auxiliary computational quantity.

Some nuclides have a "positive" mass excess relative to carbon-12 (all stable isotopes with A < 16), while others have a "negative" mass excess (see Tab. 2.6). The value of Δ ranges from about +16 MeV to – 90 MeV. Figure 2.13 illustrates the trend in mass excess relative to proton number, Z.

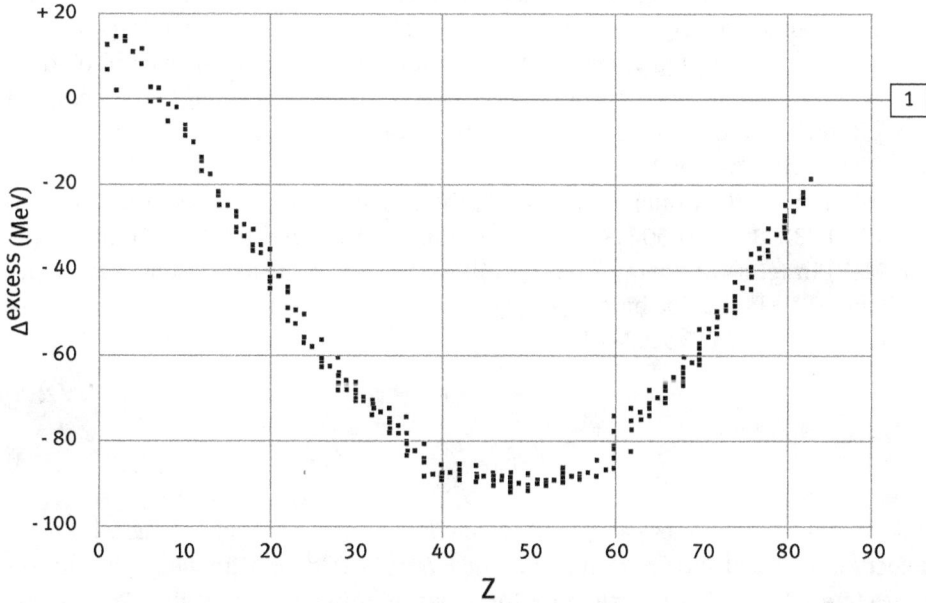

Fig. 2.13: Correlation between mass excess and the proton number Z of stable nuclei. [1] carbon-12 reference line.

Nevertheless, the key to understanding nucleon binding lies in the precise correlation between absolute atomic mass and mean nucleon binding energy. Both mass defect Δm and mass excess Δ are intermediates in this direction.

2.4.3 Binding energy

Where does that mass, "the mass defect Δm", go? Of course, mass cannot disappear – it is converted to energy. Obviously, there is a reason for nucleons to bind together: in doing so, they reach a lower final energy state compared to the initial (unbound) state. This amount of energy, ΔE, is equivalent to the mass defect Δm^{defect}, which appears to be "lost":

$$\Delta E = \Delta m^{\text{defect}} c^2 \qquad (2.12)$$

As for mass, the energy equivalent ΔE is negative:

$$\Delta E = E^{\text{nucleons bound in nucleus}} - E^{\text{individual nucleons (at rest energy)}} \qquad (2.13)$$

Once the nucleons are within a very small distance of each other (on the order of fm, i.e. the dimension of the atom nucleus), they are attracted to each other by a "strong force" – the strongest force known in our universe. The energy saved ($-\Delta E$) when nucleons bind together, compared to their unbound state, is called "overall binding energy". The equivalents of Δm^{defect} and ΔE thus reflect the overall binding energy, E_B, of the nucleus. Accordingly, $E_B \approx A$. E_B corresponds not only to mass in general, but also to the number of protons and neutrons, and is given by $E_B = f(Z, N)^{A=\text{constant}}$ (see Chapter 3).

Returning to the example in Tab. 2.5, the overall binding energy of the ^{4}He nucleus is 0.03038 u $= 0.05045 \cdot 10^{-27}$ kg in terms of mass and $4.53 \cdot 10^{-12}$ J (or 28.295 660 MeV) in terms of energy.[18] The resulting average or mean binding energy per nucleon is $\bar{E}_B = E_B / A$. The binding energy per nucleon within the ^{4}He nucleus (where $A = 4$) is thus $\bar{E}_B = 7.073\ 916$ MeV.

$$E_B \overset{\wedge}{=} \Delta E \qquad (2.14)$$

$$E_B = \frac{E_B}{A} \qquad (2.15)$$

How are binding energies calculated? Nucleon binding energy correlates with mass defect *via* eq. (2.12) and (2.14), (i.e., $E_B = \Delta E = \Delta m^{\text{defect}} c^2$), and with mean binding energy *via* eq. (2.15). Let us calculate the mean binding energy of the 12 nucleons within a carbon-12 nucleus. The absolute experimental mass of the <u>atom</u> ^{12}C is 12.0000 u $= 12 \times 1.66054 \cdot 10^{-27}$ kg $= 19.92648 \cdot 10^{-27}$ kg. Subtracting the atomic mass and binding energy of 6 electrons, the mass of the ^{12}C nucleus becomes 11.99676 u. On the other hand, the sum of 6 protons and 6 neutrons (in u) is 12.0957 u. Thus, the mass defect becomes $\Delta m^{\text{defect}} = 11.996761$ u $- 12.0957$ u $= -0.098940$ u. In terms of energy (1 u $= 931.494$ MeV), it becomes $\Delta E = 92161.728$ MeV. This is the nucleon binding energy E_B of the whole nucleus. Finally, the mean nucleon binding energy $\bar{E}_B$ is E_B divided by 12:

$$\bar{E}_B(^{12}_{6}\text{C}) = \frac{E_B}{12} = \frac{(6m_p + 6m_n - m(^{12}_{6}\text{C}))\,931.494\ \text{MeV}}{12} = 7.680\ \text{MeV} \qquad (2.16)$$

Compared to the mean nucleon binding energy of the ^{4}He nucleus, $\bar{E}_B(^4\text{He}) = 7.074$ MeV, the 12 nucleons of the ^{12}C nucleus are bound more strongly. Table 2.6

---

18 u $=1.66054 \cdot 10^{-27}$ kg $= 931.494$ MeV, 1 eV $= 1.602177 \cdot 10^{-19}$ J.

lists for a light nucleus (^{4}He), for ^{12}C, for a medium mass number nucleus (^{56}Fe), and a very heavy nucleus (^{238}U) the values of experimental atomic mass, mean binding energy, and mass excess.

Tab. 2.6: Experimental atomic mass, mass excess and mean binding energy of ^{4}He, ^{12}C, ^{56}Fe, and ^{238}U.

Nucleus	(Atomic) Mass m	Mass excess Δ	Binding energy $\bar{E}_B$
	Experiment (MeV)	(u)	Mean per nucleon (MeV)
^{1}n	1.008665	+ 8.071317	0
^{1}p	1.007825	+ 7.288970	0
^{4}He	4.002603	+ 2.4291587	7.073916
^{12}C	12.000000	0	7.680145
^{56}Fe	55.934935	− 60.60716	8.790356
^{238}U	238.0507869	+ 47.3077	7.570126

Overall binding energy ranges from about 28 MeV for ^{4}He to about 1800 MeV for ^{238}U. Maximum values for mean nucleon binding energy are $\bar{E}_B = 8.790$ MeV for ^{56}Fe, 8.792 MeV for ^{58}Fe, and 8.794 MeV for ^{62}Ni. Interestingly, the $\bar{E}_B$ values are very similar compared to the strongly varying mass numbers and atomic weights, at least for most nuclei of A > 10. In this range of mass numbers, average values for $\bar{E}_B$ are 8.18 ± 0.62 MeV for the ca. 300 stable and more than 3000 unstable nuclei. The appendix contains a table (14.4) listing relevant "atomic mass data" for the nuclides discussed throughout this textbook. It should be noted that nuclear data compilations prefer to report mass excess, Δ, instead of mass defect, Δm. These terms correlate according to eq. (2.17):

$$\Delta m = N \cdot m_n + Z \cdot m_p - A \cdot u - \Delta/c^2 \qquad (2.17)$$

2.5 Outlook

With the numbers discussed above, one may summarize the rather classical parameters of an atom's nucleus, i.e. charge, size (radius and volume), mass, and density. An individual atom's size and weight are extremely small, and thus they appear to be invisible and undetectable by the human senses. Nuclei are even smaller than atoms, by several orders of magnitude. Nuclear mass and volume can be converted into an impressive, unimaginable density. The contributions made by philosophers more than 2000 years ago and by scientists about 150 years ago regarding existent but invisible atomic and subatomic particles earn great respect!

However, in the case of atoms, size is not nearly as important as mass and energy, particularly of the atomic nucleus. How mass (*via* nucleons) is arranged within an atomic nucleus not only affects the characteristic energy of that nucleus, but also its stability or instability. Chapter 3 continues with the binding energy of an atomic nucleus and how this experimental parameter challenges theoretical models of nuclear composition.

Finally, in addition to the "conventional" properties of atomic nuclei such as mass and radius, there are far more individual parameters. These were identified within the last century when quantum physics entered the field. The relevant identities are spin, isospin, electric moment, magnetic moment, and parity (see Tab. 2.7), which will be introduced in the following chapters. They are relevant to the fundamental processes of transforming an unstable nucleus into a stable one, and for the process of energetic de-excitation of nuclear states into less energetic (i.e. more stable) states.

Tab. 2.7: Conventional and quantum physical parameters of a nucleus of an atom.

Conventional parameters	Quantum mechanical/ electromagnetic parameters
Size, radius, volume	Spin
Mass	Electric moment
Density	Magnetic moment
Charge	Parity
Energy	Isospin

3 Nucleons: Binding energy and shell structure

Aim: Although it is energetically expensive for two nucleons to approach each other, they are ultimately able to create a bound state with a "win" in energy relative to the unbound nucleons. Above all, it is the enormous energy density that makes the nucleus of an atom an extraordinary object of research. The total and mean nucleon binding energies (E_B and $\bar{E}_B$) are key parameters in nuclear science. The amount of energy a nucleus wins depends on the number of nucleons involved, A.

To understand the systematics of stable (and unstable) nuclides, nuclear science structures mass numbers according to the number of protons and neutrons. The Chart of Nuclides correlates neutron number N with proton number Z in a Cartesian coordinate system, revealing how parameters of stability (and processes of nuclei transformation) relate to both mass number and varying Z + N combinations.

The localization of the stable nuclides and their experimental mass values along a set of parameters (A, Z, N) results in a correlation of type $\bar{E}_B = f(A, Z, N)$. The classical theory used to explain this correlation utilizes the "liquid drop model" (LDM) of the atomic nucleus, postulating that all protons are identical and all neutrons are identical, and all nucleons are distributed homogeneously within the nucleus – like H_2O molecules within a drop of liquid water. The semi-empiric mathematics quantifying this experimental dependency is the so-called WEIZSÄCKER equation.

Within the existing sets of (A, Z, N), there is a surprising overexpression of a few numbers: namely 2, 8, 20, 28, 50, 82 (both for Z and N), and 126 for N only. Nuclei expressing these numbers of protons and/or neutrons are generally more stable than predicted by the LDM. Consequently, another theory accompanies the liquid drop model theory: the "nuclear shell model" (NSM). Similar to the orbital theory of electrons, both protons and neutrons are assumed to exist at characteristic shell levels with specific quantum numbers.

3.1 Stable nuclides and the Chart of Nuclides

Atomic hydrogen is defined by the one proton in its nucleus and the one electron in its shell. Isotopes of hydrogen originate from an increasing number of neutrons (see Fig. 2.2). Nucleon mixtures for hydrogen are thus composed of proton-to-neutron ratios of 1:0, 1:1 (deuterium), 1:2 (tritium), etc. However, only two of the mixtures result in a stable isotope, namely at proton-to-neutron ratios of 1:0 and 1:1, i.e. the ^{1}H and ^{2}H isotopes. The same is true of helium; the helium isotopes exist as combinations of its two protons with 1 neutron (^{3}He), 2 neutrons (^{4}He), or more neutrons (^{5}He, ^{6}He, ^{7}He, etc.). Only two of these mixtures guarantee a stable nucleus, namely ^{3}He and ^{4}He. Likewise, lithium has only two stable isotopes: ^{6}Li and ^{7}Li. For the next chemical element, there is only one stable isotope: ^{9}Be.

https://doi.org/10.1515/9783110742725-003

A Cartesian coordinate system, where the x-axis represents the number of neutrons N and the y-axis shows the number of protons Z, creates the Chart of Nuclides.[1] It starts with the lightest nuclide at $Z = 1$ and $N = 0$, i.e. with the hydrogen isotope ^{1}H. (From an abstract point of view, one could also start from $Z = 0$ and $N = 1$, representing an unbound neutron instead of a nuclide.) Figure 3.1 shows this correlation for the stable isotopes of the first four chemical elements. It demonstrates that for these elements, only a few possible combinations of protons and neutrons result in stable nuclides. Isotopes like ^{3}H, ^{4}H, ^{2}He, ^{5}He, ^{6}He, ^{3}Li, ^{4}Li, ^{5}Li, ^{8}Li, ^{9}Li, etc. do exist – but are not stable! These will appear later in this textbook, starting in Chapter 6.

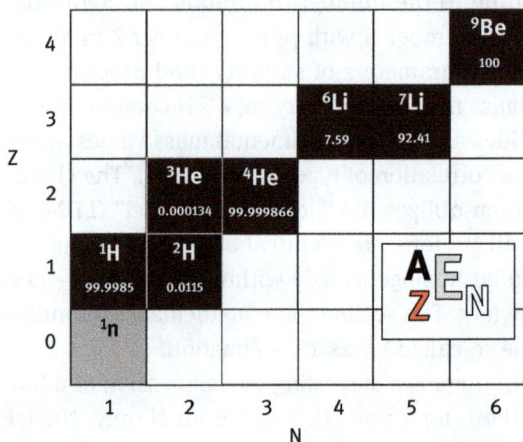

Fig. 3.1: The Chart of Nuclides plots the proton number (Z) versus the neutron number (N) for stable nuclides on a Cartesian coordinate system (x, y). Stable nuclides are typically labeled in black, alongside their mass number A. This figure is a simplified excerpt from the Chart of Nuclides, showing the stable isotopes of the four lightest elements: hydrogen, helium, lithium, and beryllium. The lower number in each field represents the natural abundance (%) of that isotope within a given element. This means hydrogen has a natural abundance (^{natural}H) of 99.9985% (^{1}H) and 0.0115% (^{2}H), totaling 100%. Note that the (1, 0) combination representing a single, unbound neutron is in gray: it is not stable.

The PTE and the Chart of Nuclides thus represent two different perspectives on chemical and nuclear sciences (see Fig. 3.2). The PTE predicts the *chemical* behavior of an element and reveals the system of electron distribution within the inner structure of the atomic shell. It is a periodic system, despite the fact that chemical elements in the PTE (generally) appear in order of increasing mass. The main information is on the chemistry of an element E following the number of electrons, which in turn are representative of Z.

---

[1] In the following, the word "nuclide(s)" is preferred when referring to the whole unstable atom, rather than the nucleus alone. Isotopes are nuclides of one and the same chemical element.

In the Chart of Nuclides, the "chemical reactivity" of a single isotope or element is not important at all. Instead, its purpose is to show how the Z and N numbers correlate with the corresponding theoretical and experimental mass of the atomic nucleus. It is not a periodic system, as it starts at $(N, Z) = (0,0)$ and continues along the x- and y-axis. For nuclear and radiochemistry, it also reveals fundamental information on the types of nuclear processes an isotope can undergo, in particular when turning unstable nuclides into stable ones.

Since the overall number of electrons of a neutral chemical element is identical to the number of protons in its nucleus, the PTE and the Chart of Nuclides are linked virtually. Another property common to both systems is atomic mass.[2]

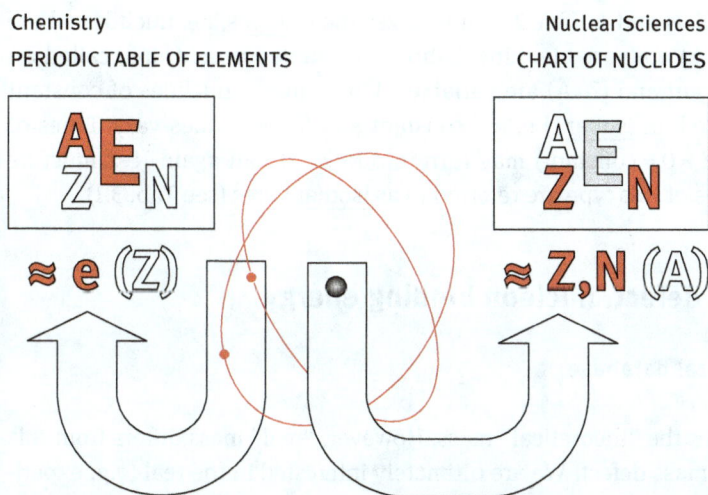

Fig. 3.2: The main parameters used by the Periodic Table of Elements and the Chart of Nuclides.

The Chart of Nuclides is a coordinate system with the number of neutrons and protons on the x- and y-axes, respectively. The y-axis ranges from $Z = 1$ (representing the hydrogen isotopes) to $Z = 118$, encompassing all known chemical elements, stable or unstable (at least so far). The x-axis scales from $N = 0$ (the hydrogen isotope ^1H) to $N > 170$ for the artificially produced super-heavy, transactinide elements. This results in a matrix of about $N \times Z = 170 \times 120 = 20,400$ possible combinations. Yet, little more than 1% (281)[3] of

---

2 Chemists typically handle stable isotopes and may not need the Chart of Nuclides. Likewise, nuclear scientists interested in the properties of an atom's nucleus or in nuclear transformations may not look at the chemistry of the nuclides involved. Nevertheless, many scientific research topics require both the chemical and nuclear parameters of unstable nuclides.
3 The total number of stable nuclides varies between textbooks. This is due in part to the somewhat philosophical question: what is stability? If "absolute" stability is meant, then with increasing Z, the "last" stable nuclides are those of lead $(Z = 82)$ and bismuth $(Z = 83)$: ^{206}Pb, ^{207}Pb, ^{208}Pb and

all the theoretical combinations are "realized" on earth as stable nuclides. Within the Chart of Nuclides, the black squares denote stable nuclides.

Abstractly, the squares can represent the surface of a box which on all six faces may offer characteristic information. At the surface, which is depicted in the Chart of Nuclides, there are data on the isotope notation, the abundance of a stable isotope within a given element, and the cross-section σ of the most relevant nuclear reaction[4] process.

Figure 3.3 illustrates the Chart of Nuclides summarizing the stable isotopes (black squares). Four reference lines are marked. Line one represents nuclides with Z and N symmetry (N = Z).

Three more lines indicate situations where one parameter is constant ("iso"-) among the values of Z, N and A. Line 2 characterizes the isotopes, i.e. nuclides where Z = constant and (N, A) are variable. Line 3 shows the same for neutrons, called an isotone, i.e. N = constant and (Z, A) are variables. Line 4 marks nuclides of constant mass number A, called an isobar, i.e. A = constant and (Z, N) values vary. Lines of type 1 (i.e. when N–Z = 0 = constant) may turn into N–Z > 0, but again N–Z must be aconstant value. Lines of this type are referred to as isodiapheres (see Tab. 3.1).

3.2 Mass, mass defect, nucleon binding energy

3.2.1 The experimental database

The value of A denotes the "theoretical" mass. However, "real" mass differs from this value because of the mass defect. We are ultimately interested in the real (e.g. experimental) mass of individual nuclides because mass is tantamount to energy. Only knowledge of the precise mass of a nucleus can open access to the overall nucleon binding energy. Finally, knowledge of the overall and mean nucleon binding energies is the key to correlating experimental values with theoretical models, ultimately to address the question: how and why do nucleons exist inside an atomic nucleus?

---

[209]Bi. Beyond bismuth, chemical elements only have (more or less) unstable isotopes. In terms of "relative" stability, nuclides may be considered whose half-life ($t\frac{1}{2}$) is greater than the age of our planet: about $4.5 \cdot 10^9$ years (see Chapter 5). Translated into a human scale, these isotopes decay by a negligible fraction (<0.00000178%) over the average life-span of 80 years. Over the observation period of a human, such a nuclide is "quasi-stable". This is true even referenced against the age of mankind, about $2 \cdot 10^6$ years. Over this period, an isotope of $t\frac{1}{2} > 4.5 \cdot 10^9$ years will have changed by 0.044% – representing *de facto* stability. However, even those nuclides having half-lives of several thousand years – the thorium isotope ^{232}Th ($t\frac{1}{2} = 1.405 \cdot 10^{10}$ years) and the three uranium isotopes (^{234}U, ^{235}U and ^{238}U with $t\frac{1}{2} = 2.455 \cdot 10^5$, $7.038 \cdot 10^8$ and $4.468 \cdot 10^9$ years – are not absolutely stable. Nevertheless, in the Chart of Nuclides they are denoted by (partially) black fields.

4 See Chapter 13 for nuclear reactions.

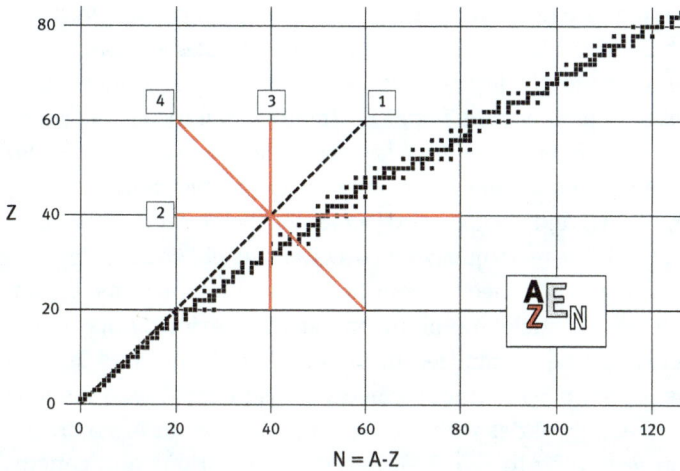

Fig. 3.3: The Chart of Nuclides indicating the stable isotopes (black squares). Overall, the nuclides do not follow a straight line of type Z = N (line 1). Line 1 is followed only for mass numbers <40, i.e. proton and neutron numbers <20. Above Z = 20, the distribution is in favor of neutron excess (N > Z). Three more lines are indicated, each marking the line of a certain constant ("iso"-) parameter among the variations of Z, N, and A. Line 2 characterizes the isotopes, i.e. nuclides of Z = constant, with variable N and A values. Line 3 shows an isotone, where N = constant and (Z, A) are variables. Line 4 marks nuclides of constant mass number A, called an isobar, i.e. A = constant and Z and N varying. Isodiapher lines (of N−Z > 0 = constant) are not shown.

Tab. 3.1: Straight lines displayed in the Chart of Nuclides.

Notation	Characteristics	Examples
Isotopes	Z = constant	e.g. Z = 1: hydrogen isotopes: ^{1}H (stable), ^{2}H (stable), ^{3}H, ^{4}H, ^{5}H, ^{6}H, ^{7}H (all unstable)
Isotones	N = constant	e.g. N = 20: ^{36}S, ^{37}Cl, ^{38}Ar, ^{39}K, ^{40}Ca
Isobars	A = constant	e.g. A = 12: $_4$Be$_8$, $_5$B$_7$, $_6$C$_6$, $_7$N$_5$, $_8$O$_4$
Isodiaphers	N − Z = constant, (N = Z)	e.g. $_1$H$_1$, $_2$He$_2$, $_3$Li$_3$, $_5$B$_5$, $_6$C$_6$, $_7$N$_7$, $_8$O$_8$, . . .
	(N > Z)	$_{92}$U$_{142}$, $_{90}$Th$_{140}$, $_{88}$Rg$_{138}$, $_{86}$Rn$_{136}$, $_{84}$Po$_{134}$, $_{82}$Pb$_{132}$

But first, what is meant by "precise mass of a nucleus"? It requires that the real mass of an atom's nucleus should be experimentally determined as precisely as possible.[5] This

---

5 There are several methods to do so, with laser spectroscopy being the most recent and accurate. The accuracy of mass determination today is comparable to weighing the mass of a passenger airplane, such as an Airbus 380–800, and detecting whether 1 out of >600 passengers carries

is practically achievable in the case of stable nuclides.[6] So, what information is included in the experimental database of stable nuclides? This database includes sets of nucleon numbers in terms of three parameters (Z, N, and A), as well as experimentally determined atomic masses. The number of protons runs from Z = 1 (representing the chemical element hydrogen) to Z = 83 (bismuth) and includes a few very long-lived isotopes at Z = 90 and 92 (thorium and uranium). The number of neutrons covers a larger range, starting from N = 0 (corresponding to ^{1}H) and approaches N = 146 (for the heaviest "relatively" stable isotope of uranium, ^{238}U). The mass number A covers a range of $1 \le A \le 238$.

This is the database to develop a theory about the nucleon interactions responsible for mass defect and overall nucleon binding energy. As previously discussed, the <u>overall</u> binding energy can be normalized to the number of nucleons in each nucleus, resulting in the <u>mean</u> nucleon binding energy.[7] It is interesting to see how the mean nucleon binding energy varies with increasing mass number A. Figure 3.4 arranges this correlation according to $\bar{E}_B = f(A)$. Several observations and conclusions can be drawn from this correlation; these are listed in Tab. 3.2.

With the experimental database in hand, basic theories about the atomic nucleus relying on the number of nucleons and the experimental mass of a nucleus can be addressed (see Tab. 3.3).

There are two principal models to explain the observations and the additional facts in Tab. 3.2 and the questions in Tab. 3.3: the "liquid drop model" (LDM) and the "nuclear shell model" (NSM).[8] The main features of these two models are summarized in Tab. 3.4.

The LDM postulates that all protons are identical and all neutrons are identical: one proton cannot be distinguished from another, the same holding true for neutrons. It further posits that the nucleons are distributed homogeneously throughout the nucleus (similar to the early RUTHERFORD plum pudding model of the entire atom, which expected to find protons and electrons distributed throughout the atom). All these identical nucleons distributed homogeneously within the nucleus should behave like H_2O molecules within a drop of liquid water. The nucleons represent an ensemble of identical particles equidistant from each other. This distance is fixed and cannot be modified by e.g. trying to deform the "liquid drop". In fact, it is impossible to compress the liquid drop-like atomic nucleus. It may only change its shape, while the volume remains constant – like a drop of liquid water.

---

a single one-cent coin or not. Suppose the airplane weighs 560 tons and the coin 1 g; the precision is of $\Delta m/m \le 10^{-9}$.

6 Stable isotopes are much easier to analyze than unstable, radioactive ones.

7 Table 2.6 lists these numbers for four selected nuclei. Standard publications in the nuclear sciences list those and other parameters for almost all the nuclides; see Table 2.7.

8 Both concepts appeared in the 1930s. These concepts have been verified, yet still today there is an ongoing process to further qualify these theories.

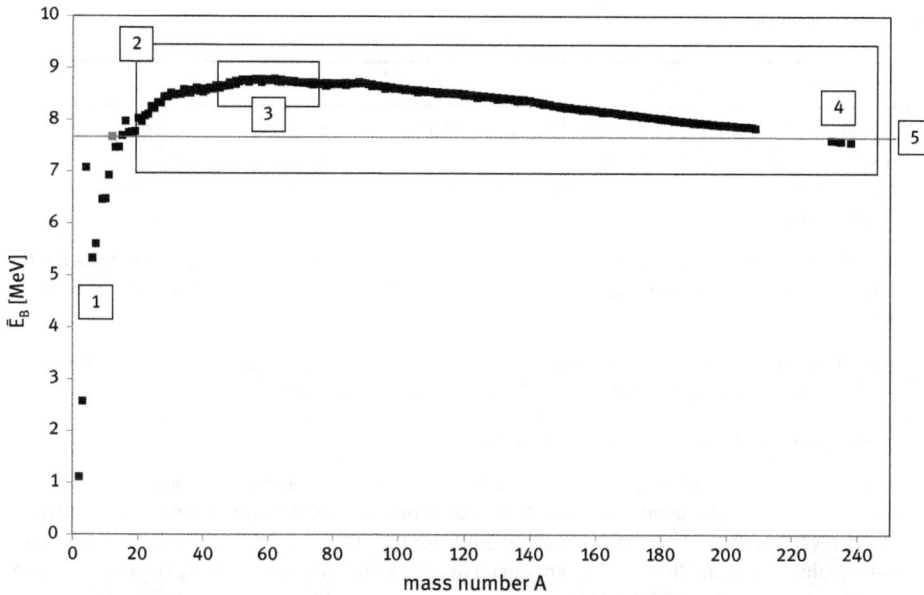

Fig. 3.4: Mean nucleon binding energy $\bar{E}_B$ as a function of mass number A, shown for the 281 stable nuclei. $\bar{E}_B$ values range from about 1–9 MeV per nucleon. (1) Light nuclides show low values of $\bar{E}_B$. (2) At A > 50 until about 90, the $\bar{E}_B$ values become relatively similar (about 8.18 ±0.62 MeV). (3) The maximum of $\bar{E}_B$ appears around A = 50 (where $\bar{E}_B$ = 8.7 to 8.8 MeV), and subsequently the values decrease. (4) Very heavy nuclides of thorium and uranium (A = 234, 235 and 238, 7.570 MeV for ^{238}U). (5) Value of $\bar{E}_B$ for 12C ($\bar{E}_B$ = 7.680 MeV).

Tab. 3.2: Experimental observations from Fig. 3.4, useful for understanding the distribution of stable nuclei in the Chart of Nuclei and for theoretically explaining correlations of type $\bar{E}_B$ = f(A) of stable nuclei. Some additional statistical facts needed to understand the WEIZSÄCKER equation are also provided.

Observations from Fig. 3.4:
1. There is a general trend in mean nucleon binding energy versus mass number A. $\bar{E}_B$ first increases from light nuclei to medium mass nuclei, stays relatively constant between ca. 50 < A < 90, reaching maximum values for isotopes of iron, cobalt, and nickel and finally decreases slightly with further increasing mass numbers·
2. Values of $\bar{E}_B$ are relatively similar – at least when mass number A > 15. Except the two stable nuclei ^{2}H and ^{3}He ($\bar{E}_B$ = 1.112 MeV and 2.573 MeV, respectively), mean nucleon binding energy ranges from about 7 to 8.8 MeV. This is a relatively narrow range for atomic nuclei of mass numbers spanning about two orders of magnitude (^{4}He to ^{238}U).

Tab. 3.2 (continued)

Observations from Fig. 3.4:

3. Relative to the "reference" nucleus ^{12}C ($\bar{E}_B = 7.680$ MeV), there are nuclei with lower mean nucleon binding energy in the mass range of 2–14, but also nuclei with higher binding energy per nucleon for mass number A ≥ 235, while most of the nuclei have higher mean nucleon binding energy.

4. The stable nuclei do not distribute within the Chart of Nuclei along a straight line of type Z = N. This is true only for mass numbers <40. At A > 40, the distribution is in favor of neutron excess (N > Z).

5. In the low mass region, there are several "peaks" along the virtual "line". Obviously, some mass numbers create higher values for $\bar{E}_B$ than "usual" or "expected".

"Statistical" facts not directly seen in Fig. 3.4

6. When proton and neutron numbers of a nucleus are counted separately, either even or odd numbers for Z and N appear. Mass numbers A thus reflect combinations of type (Z, N) of (even, even), (even, odd) or (odd, even) and (odd, odd). For the stable nuclei, those combinations are not equally abundant. The (even, even) nuclei dominate and the ratios among the four possible combinations are about 30:10:10:1. Thus, there are about 30 times more stable nuclei representing (even, even) numbers compared to (odd, odd) versions.

7. The ranges of proton and neutron numbers for stable nuclei are 0 < Z < 93 and 0 < N < 147, respectively. One may expect that the complete range is found more or less balanced within the combinations reflecting the range of mass numbers, i.e. 1 < A < 239. However, statistics for the abundance of numbers of Z and N for stable nuclei reveal an unusually high abundance of the following six numbers. These numbers are 8, 20, 28, 50, and 82 for Z; and 8, 20, 28, 50, 82, and 126 for N. These numbers are called "magic".

8. In contrast to the high abundance of these numbers (8, 20, 28, 50, 82, and 126), numbers corresponding to a magic number ±1 (e.g. N −1 or N +1) appear surprisingly seldom.

Tab. 3.3: Questions concerning the atomic nucleus and the mean binding energy of its nucleons.

What model best explains the difference between theoretical and experimental mass – the mass defect?
What theory explains how much energy is needed to make a nucleus out of individual nucleons – and how much energy is released when atomic structures are transformed into free nucleons?
What theory can describe and ultimately predict the combinations of type (N, Z) reflected in the structure of a nucleus?
Some nuclei are stable, while others are not – why?
Some stable nuclei are more abundant than others – why?
What theory can explain and predict the transformation processes from an unstable nucleus to a stable one?

Tab. 3.4: A comparison of the principal theories on the atomic nucleus.

Liquid drop model	Nuclear shell model
Nucleons are equivalent: Protons and neutrons differ in charge, but all protons are equal, and all neutrons are equal	Nucleons differ not only by charge: Not a single nucleon is identical to another one present in a nucleus
Nucleons are distributed homogeneously	Nucleons occupy discrete, quantum mechanical orbits
Electric charges are distributed uniformly	Electric charges may be distributed nonuniformly
Key parameters: $\bar{E}_B$ as f(A, Z, N) Nucleon separation energy	Key parameters: $\bar{E}_B$ as f(quantum mechanics) Excitation energy Overall spin of the nucleus

Since the protons are distributed homogeneously, charge distribution within the nucleus is assumed to be uniform.

Within the LDM, one would expect that the specific Z and N numbers do not matter – it is all about the overall nucleon number, A. Cataloguing the stable nuclei, however, reveals a surprising overexpression (in terms of their statistical abundance) of certain, limited numbers for (interestingly) both Z and N. Namely, these are 2, 8, 20, 28, 50, 82 and (for N only) 126. Because these protons and/or neutrons numbers are more commonly expressed in stable nuclei than other values of Z and N, they seem to be a source of higher stability of a nucleus. In the absence of a theory to explain this overexpression, the numbers were called "magic".

This oddity is addressed by the nuclear shell model theory. The nuclear shell model does not contradict the LDM, but rather accompanies it. It correlates experimental evidence on the increased number of stable nuclei having "magic" nucleon numbers with the internal structure of nucleons within the nucleus.

Similar to the shell model used for electrons, protons and neutrons are understood to exist at characteristic shell levels with individual quantum number and a resulting numbers of nucleons per shell. It applies the concept of quantum numbers introduced for electrons to nucleons. Similar to how each electron of an atom is different (i.e. each is unique in its set of quantum numbers), now each nucleon is treated "individually" as well. As an example, in the stable isotope gold-198, all 79 protons and 119 neutrons are different – each of the 198 nucleons is unique.

3.3 The liquid drop model

The mathematics quantifying mass defect or nucleon binding energy in a correlation of type $E_B = f(A, Z, N)$ is called the WEIZSÄCKER equation.[9] It describes the overall nucleon binding energy as a derivative of (a) the mass number A and (b) the individual numbers of protons Z and neutrons N. The approach is rather classical, i.e. not involving quantum mechanical interpretations. It is an attempt to semi-empirically fit the 281 data points of Fig. 3.4.

$$E_B = \alpha^{LDM} A - \beta^{LDM} A^{2/3} - \gamma^{LDM} \frac{Z^2}{A^{1/3}} - \zeta^{LDM} \frac{(N-Z)^2}{A} \pm \frac{\delta^{LDM}}{A^{3/4}} \tag{3.1}$$

VOLUME · SURFACE · COULOMB · ASYMMETRY · PAIRING

The basic equation considers five different terms, although LDM theory is still being developed and rethought today. Equation (3.1) reproduces a rather standard version, while an expanded version is given, for example, in the Handbook of Nuclear Chemistry.[10]

The equation may be divided into several parts, also listed in Tab. 3.5. Each term of this equation has a physical rationale, describing the various ways the two different types of nucleons contribute to the binding energy. Some terms depend exclusively on mass number A, while others reflect the individual contributions of either protons or neutrons. Finally, each term has a coefficient, which overall resembles a polynomial equation. The absolute values of the five coefficients – α^{LDM}, β^{LDM}, γ^{LDM}, ζ^{LDM}, and δ^{LDM} – are fitted to the trend in "experimental" values of mean nucleon binding energy in Fig. 3.4.

---

9 . . . although a number of other great scientists have contributed to its conception and modifications, such as BETHE, GAMOV, BOHR AND FERMI. Equation (3.1) is also called the BETHE–WEIZSÄCKER equation.
10 A Vértes, S Nagy, Z Klencsár, RG Lovas, F Rösch (eds.), Handbook of Nuclear Chemistry, second edition, Springer 2011.

Tab. 3.5: Constituents of the WEIZSÄCKER formula used to semi-empirically describe correlations of type $E_B = f(A, Z, N)$ for stable nuclei.

Term	Variables of A, Z, N	Coefficient	
		Symbol	Sign
Volume	A	α^{LDM}	+
Surface	$A^{2/3}$	β^{LDM}	−
Coulomb	$Z^2/A^{1/3}$	γ^{LDM}	−
Asymmetry	$(N-Z)^2/A$	ζ^{LDM}	−
Pairing	$1/A^{3/4}$	δ^{LDM}	+, − or 0

3.3.1 Volume

Among all the terms listed for eq. (3.1), the "volume term" is the principal one, and the one showing *per se* a positive sign. It is positive because energy is gained when the nucleons are "bound";[11] bounded nucleons have lower potential energy than unbounded ones. Figure 3.5 depicts an illustration of protons and neutrons within a nucleus. The more interactions per individual nucleon, the larger the gain in energy for that particular nucleon.

This increase in mean nucleon binding energy depends exclusively on A according to a direct proportionality. The term does not depend on Z nor on N.

Consequently, larger nuclei have higher overall binding energies. The volume of a sphere is $4/3\pi r^3$ and volume $\approx$ number of nucleons A.[12]

The volume term expresses E_B as proportional to A, regardless of the composition of A in terms of Z and N. Because A is a unitless number (for stable nuclei ranging from mass number 2 to about 200),[13] the coefficient α^{LDM} is also needed to introduce a unit: MeV. According to recent modifications of the WEIZSÄCKER equation, its value is $\alpha^{LDM} = 15.15$ MeV. When only the volume term of the LDM is considered, the expression would be $E_B = \alpha A$. The hydrogen isotope ^2H and uranium isotope ^{238}U thus should have overall nucleon binding energies of $E_B = 30.30$ MeV and $E_B = 3605.70$ MeV, respectively.When expressing these values in terms of mean nucleon binding energy, each value of E_B is divided by A; consequently, for both nuclei, the result is

---

11 It reflects the strong interaction of nucleons, mediated by the strong force (see Chapter 7). This strongest force of our universe attracts nucleons approaching each other. The range of this strong nuclear force is extremely small, similar to the dimension of nucleons (i.e. at fm scale).

12 Relationships between size and radius of an atomic nucleus have been discussed in Chapter 2.

13 Note that the range is not $A \geq 1$, but $A > 1$. The hydrogen isotope ^1H cannot be considered as there is only one nucleon; there are no additional nucleons with which the proton can interact.

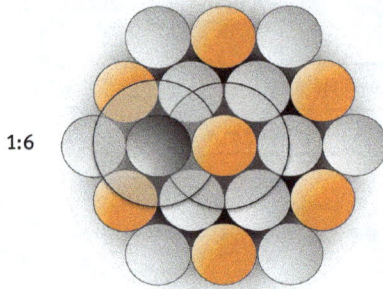

1:6

Fig. 3.5: Volume term: protons (orange) and neutrons (white) within a nucleus attracted by the strong force. In this simplified two-dimensional drawing of a circle instead of a sphere, each nucleon within the volume of a sphere is surrounded by six other nucleons. In reality, i.e. in three dimensions, there are more.

$\bar{E}_B$ = 15.15 MeV. This would be true for all stable nuclei. Figure 3.10 plots the result of the LDM-based mean nucleon binding energy as $\bar{E}_B = f(A)$. The horizontal line at 15.15 MeV is the resulting hypothetical expression for the volume term only.

3.3.2 Surface

The second constituent of the WEIZSÄCKER equation is the "surface term".

Its definition follows the concept of the volume term; however, only nucleons *inside* an atomic nucleus are maximally surrounded by other nucleons, in the sense of saturation. In contrast, protons or neutrons located at the surface of the sphere are lacking some potential partners. Consequently, those nucleons are less bound (see Fig. 3.6). Thus, not all nucleons (A) are equal – those at the surface of the atomic nucleus are different from the others. In terms of the overall nucleon binding energy, the energy contributed by nucleons at the surface is less than those in the center of the sphere. The fraction of nucleons constituting the surface of the sphere thus reduces the overall nucleon binding energy of a given nucleus, reflected in the volume term.

1:4

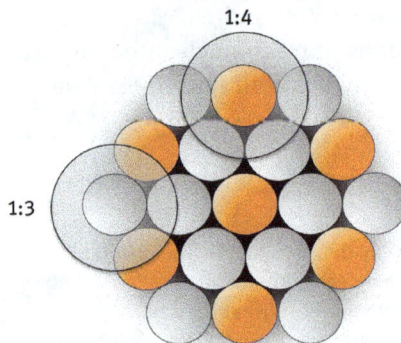

1:3

Fig. 3.6: Surface term: nucleons at the surface of the nucleus are less bound because they are surrounded by fewer other nucleons than those located inside the nucleus volume. In this simplified two-dimensional drawing, each nucleon located at the periphery is surrounded by three or four other nucleons – compared to six for the center, as shown in Fig. 3.5.

The surface term calculates the fraction of these nucleons, and varies by the ratio between volume and surface of a sphere. The surface area of a sphere is $S = 4\pi r^2$ and is proportional to $A^{2/3}$. The surface term thus reads as $-\beta^{LDM} A^{2/3}$. Note that the sign is negative, so this term subtracts a certain value from the volume term.

The mean nucleon binding energy corrected this way is given in Fig. 3.10. Clearly, this type of correction is most dramatic for light nuclei.[14] Consequently, the surface term correction dominates for nuclei of about $A < 10$. For increasing A, the ratio between volume and surface (i.e. of nucleons inside the volume and nucleons at the surface) becomes smaller. Thus, the fraction of nucleons that are more weakly bound (at the surface) increases with A. The value of (volume term − surface term) for $A = 50$ becomes 11.823 MeV, and approaching heavy nuclei it further decreases. For ^{238}U, the mathematical outcome for $\bar{E}_B$ is 12.578 MeV, instead of 15.15 MeV for the volume term alone.

3.3.3 Coulomb interaction

Both the volume and surface terms of the liquid drop model take into account the total number of nucleons, A. However, the additional impact of the positively charged nucleons must be considered. This is captured in the next two terms.

The first is the "coulomb term". It realizes that protons (unlike neutrons) undergo coulombic interactions. Protons "classically" should try to achieve a distance between each other to minimize electrostatic repulsion. Protons are thus distributed within the nucleus in a way that minimizes repulsion (see Fig. 3.7).

Overall, coulombic interactions tend to lower the overall nucleon binding energy. While the volume is represented mathematically as proportional to $A^{1/3}$, coulombic interaction depends on Z^2. In more detail, the coulombic energy is $E_{COUL} = 3/5 Z^2 e^2 / r$, where e represents the elementary charge unit and r is the radius of the nucleus ($r = r_o A^{1/3}$). E_{COUL} thus is proportional to Z^2/r and with r proportional to $A^{1/3}$ it becomes a proportionality to $Z^2/A^{1/3}$. The resulting coulomb term is $-\gamma^{LDM} Z^2/A^{1/3}$. The message is that the more protons there are within a given volume, the more dramatically nucleon binding energy is affected.[15] However, the absolute amount

---

14 The larger the volume of a sphere, the smaller the relative fraction of nucleons at the surface becomes relative to the overall number of nucleons. For example, a light nucleus such as ^2H contains all its nucleons at the "surface".

15 Example: The reference nucleus ^{12}C is composed of 6 protons and 6 neutrons. The constituents of the coulomb term are $Z^2 = 6 \times 6 = 36$ and $A^{1/3} = 12^{1/3} = 2.28943$. In this case, $Z^2/A^{1/3}$ equals 15.724. Its isobar with 7 protons (where $A = 12$ and $N = 5$) is ^{12}N. Here, since $Z^2 = 7 \times 7 = 49$, the term $Z^2/A^{1/3}$ becomes 21.403. The difference in $Z^2/A^{1/3}$ is 36%. This term decreases the ^{12}N nucleon binding energy (until now exclusively determined by volume and surface term). As a result, ^{12}C is stable, while ^{12}N is not. In general for a given mass number, the larger the proton excess, the lower the remaining nucleon binding energy.

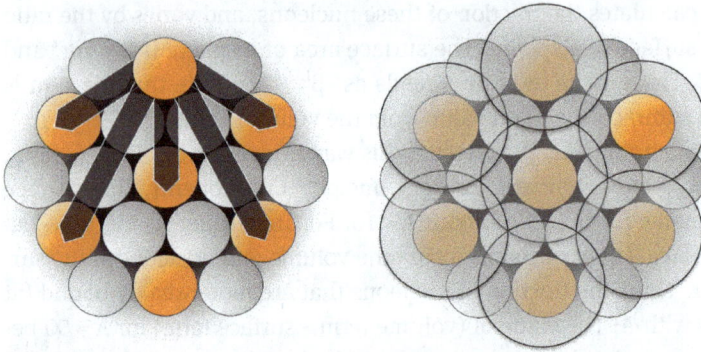

Fig. 3.7: Coulomb term: protons (orange) are distributed in a way that minimizes coulombic repulsion. In this simplified sketch, there are no protons in close contact with each other (left). Because there is no "empty space" within a typical nucleus,[16] direct contact between protons is avoided by "buffering" with neutrons (right).

subtracted from the volume and surface term is relatively low, as the coefficient γ^{LDM} is only 0.665 MeV.

Figure 3.10 illustrates the impact this third term has on the trend in mean nucleon binding energy. Its effect is more pronounced for large mass numbers. For $A = 50$ and $A = 238$, for example, values calculated so far *via* volume and surface term of 10.823 and 12.578 MeV are reduced to 9.076 and 8.762 MeV, respectively.

3.3.4 Symmetry and asymmetry

The database for light stable nuclei indicates that whenever one goes from one chemical element to the next heavier one (i.e. from Z to Z +1), the number of neutrons follows in the same manner. There are several prominent examples among the light nuclei, such as ^{2}H (1 proton + 1 neutron), ^{4}He (2 protons + 2 neutrons), ^{6}Li (3 protons + 3 neutrons), . . ., ^{12}C (6 protons + 6 neutrons), . . ., ^{40}Ca (20 protons + 20 neutrons), etc. These nuclei are "symmetric" in the sense that Z = N. However, experimental database reveals that this is true for a very limited number of stable nuclei only. When A > 40, this symmetry is lost. All stable nuclei at Z > 20 and N > 20 tend to deviate from this symmetry (line 1 in Fig. 3.3). The relationship between Z and N becomes asymmetric (Z ≠ N, or more precisely: N > Z). Asymmetry also means that whenever protons and neutrons are taken artificially as pairs, one or more neutrons remain solo (see Fig. 3.8). This "excess" of neutrons over protons increases with increasing mass number A. For heavy stable nuclei (such as ^{238}U), the excess

---

16 There are exceptions to the rule for several light nuclei, called halo nuclei.

reaches an impressive neutron-to-proton ratio of 146:92 = 1.587:1, as opposed to 1:1 for the line 1 in Fig. 3.3.[17]

It is interesting to compare the meanings of the coulomb term and the asymmetry term. The coulomb term limits the increase of the number of protons per given mass number (i.e. volume) of a nucleus by Z^2. The asymmetry term "allows" a nucleus to go beyond that proton number finally, but requires an adequate number of additional neutrons.

Obviously, with an increasing number of protons per given atomic nucleus, neutrons appear to have a significant value in providing stability. The meaning in a classical physical sense is that the electrically uncharged neutrons may compensate (chemically speaking "buffer") for the increasing coulombic repulsion of protons in a given atomic nucleus. Mathematically, the WEIZSÄCKER equation considers asymmetry as $(N - Z)^2/A$.[18] The corresponding coefficient is $\gamma^{LDM} = 21.57$ MeV. The asymmetry term further corrects $\bar{E}_B$ down to lower values. As the excess of neutrons increases exponentially with increasing mass number A (Fig. 3.3), the impact of the asymmetry term is more pronounced at higher mass number A (see Fig. 3.8).

Fig. 3.8: Asymmetry term: protons and neutrons may virtually be considered as pairs inside a nucleus if the symmetry of Z = N holds true. Whenever all virtual pairs of proton + neutron are marked, one or more neutrons remain. In this simplified two-dimensional drawing, five neutrons remain in "excess".

---

17 For ^{238}U, pairing the 92 protons with neutrons would altogether give 184 nucleons. However, the mass number is 238, i.e. the remaining 54 nucleons (238−184 = 54) are "excess" neutrons.

18 For symmetric nuclei, N − Z becomes zero and the entire term vanishes. Increasing the excess of neutrons over protons creates numbers such that N − Z is >0. For example, the stable fluorine nucleus ^{19}F has 10 neutrons minus 9 protons, making N − Z = 1 and $(N - Z)^2 = 1$. Likewise, ^{238}U has 146 neutrons minus 92 protons, meaning and N − Z = 54 and $(N - Z)^2 = 2916$. The neutron excess alone results in the ^{238}U nucleus being more than three thousand-fold stronger than ^{19}F. However, in the end, that square of neutron excess is less impactful, since the asymmetry term relates the neutron excess to the overall number of nucleons. For ^{19}F, the expression $(N - Z)^2/A$ is 1/19 = 0.05263168 MeV; for ^{238}U it is 13.65126 MeV. This is still a significant difference of more than two orders of magnitude.

3.3.5 Pairing

The final term of the original WEIZSÄCKER equation is the pairing term. In this case, protons and neutrons are counted separately, resulting in an odd or even proton number, and an odd or even neutron number. In terms of $A = Z + N$, the possible combinations of (Z, N) nuclei are (even, even), (odd, odd), (even, odd), and (odd, even). Any mass number A thus reflects one of these four combinations. The question, nevertheless, is: why should one deal with this kind of statistic? The answer refers to the experimental database (see point 6 in Tab. 3.2). Strikingly, there is a very strange statistic regarding the stable nuclei. The (even, even) nuclei dominate and the ratios among the four possible combinations (even, even), (even, odd), (odd, even), and (odd, odd) are about 30:10:10:1. There are about 30 times more stable (even, even) nuclei compared to (odd, odd) versions.[19] Table 3.6 lists the statistics.

Tab. 3.6: Statistics of (Z, N) combinations of all the stable nuclei in terms of even or odd proton or neutron numbers and sign of δLDM, the coefficient of the pairing term in the WEIZSÄCKER equation.

Combination of (Z, N)	Number of stable nuclei	Sign of δ^{LDM}
(even, even)	166	+
(even, odd)	55	0
(odd, even)	52	0
(odd, odd)	8 (see[20])	−

Mathematically, the pairing term depends on the mass number according to $1/A^{3/4}$. Pairing further "corrects" the calculated mean nucleon binding energy and is efficient over the whole range of mass number A (see Fig. 3.9).

Two protons (or similarly, two neutrons) may be considered to form a pair inside the nucleus.[21] From the dominating fraction of (even, even) nuclei in the universe, it is concluded that (even, even) combinations result in increased stability, i.e. higher nucleon binding energy. The coefficient δ^{LDM} of the pairing term thus gets a positive sign.

---

19 This is mirrored even in the presence of the chemical elements in the universe.

20 $^{6}_{3}Li_{3}$, $^{10}_{5}B_{5}$, $^{14}_{7}N_{7}$, $^{50}_{23}V_{27}$, $^{176}_{71}Lu_{105}$, $^{180}_{73}Ta_{107}$.

21 For the WEIZSÄCKER equation, the pairing term is rather semi-empirical in character. At this stage, its physical meaning remains uncertain; here, it simply reflects the status of the database. A rational explanation will arise only later, in the context of quantum mechanics; see the nuclear shell model attributing each nucleon with the quantum numbers known for electrons. "Paired" nucleons minimize overall spin according to wave function characteristics, and thereby improve binding energy.

In contrast, (odd, odd) combinations of (Z, N) are very rare, and "seem" to create a problem for forming stable nuclei. In this case, nucleon binding energy is interpreted as "negatively" affected. Unpaired nucleons (like the single "unpaired" proton in Fig. 3.9) contribute to nucleon interactions in a different way and tend to decrease nucleon binding energy. The coefficient thus has a negative sign.

If one sort of nucleon, i.e. either the protons or the neutrons, is of even number while the other is of odd number, the "positive" and "negative" impacts cancel out and the coefficient becomes $\delta^{LDM} = 0$. The pairing term for those ca. 100 stable nuclei then simply disappears.

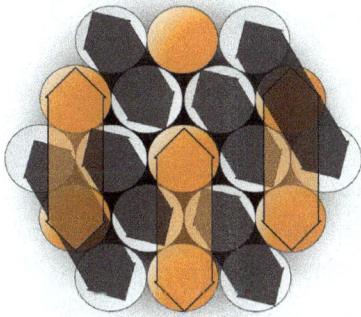

Fig. 3.9: Pairing term: two protons and two neutrons may be considered as forming pairs inside a nucleus. Existing virtually as pairs, this may optimize overall spin quantum numbers (as introduced for electrons in the Hund rule). Unpaired nucleons, like the "unpaired" proton shown here, thus contribute to a weaker nucleon binding energy.

3.3.6 The coefficients

The aim of the WEIZSÄCKER equation was to reproduce the experimental data of type $\bar{E}_B = f(A)$ shown in Fig. 3.4. Based on the LDM, the equation makes use of five (or more) terms reflecting some intercorrelations found in the stable nuclei. The equation itself is a function of mass number and mathematically is a polynomial. The terms of the WEIZSÄCKER equation are various multiples of individual powers of A. The specific values of the individual coefficients of each term (α^{LDM}, β^{LDM}, γ^{LDM}, ζ^{LDM}, and δ^{LDM}) are finally derived from a mathematical fit of the polynomial with the $(A, \bar{E}_B)$ dataset. Each of the coefficients carries the unit of energy (MeV) and a corresponding sign. Table 3.7 summarizes the values of these coefficients.

Figure 3.10 shows the effect of including subsequent terms of the WEIZSÄCKER equation, weighted by the coefficients. The impact of the pairing term is illustrated for the "minus" effect, and had the "plus" effect been modeled, the final curve would be above the yellow line. Likewise, for most of the unstable nuclides with the "zero" effect, the curve would overlap with the yellow line.

Finally, Fig. 3.11 compares the mathematical model of mean nucleon binding energy versus mass number A with the experimental data, indicated by individual values of $\bar{E}_B$ for the stable nuclei. Clearly, the LDM is able to reproduce the increase

Tab. 3.7: Coefficients of the WEIZSÄCKER equation given in the Handbook of Nuclear Chemistry.[22]

Coefficient	α^{LDM}	β^{LDM}	γ^{LDM}	ζ^{LDM}	d^{LDM}
Value (MeV)	15.15	15.94	0.665	21.57	22.4

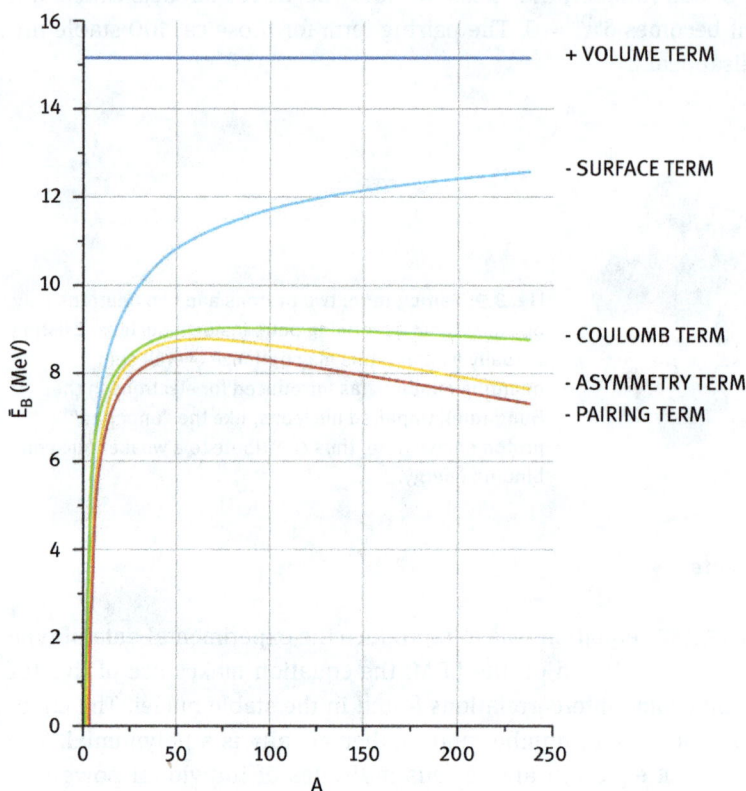

Fig. 3.10: Output of the WEIZSÄCKER equation when successively using the five basic terms (shown in simplified and smoothed version).

in $\bar{E}_B$ with increasing mass number A, a trend seen until ca. A = 50. It also captures the relatively constant values of $\bar{E}_B$ for mass number ranging from A about 50 to about 100. Further, it nicely illustrates the impact of the last two terms, asymmetry

---

22 More terms: The classical WEIZSÄCKER equation included the five terms discussed. As nuclear science research is an ongoing process, and the experimental database is improving (modern atomic mass measurements provide more precise values of mass excess and nucleon binding energy), there are several modifications of the original version. This concerns both the value of the coefficients and additional terms, such as "promiscuity".

and pairing, in order to reproduce the decrease in mean nucleon binding energy at mass numbers of A > 100.

Table 3.8 compares the experimental values of $\bar{E}_B$ with those calculated by eq. (3.1) and gives the deviations $\Delta\bar{E}_B$ between calculated and experimental values, both in MeV and in percentage. For most isotopes of A > 20, there is excellent agreement between experiment and theory. This is a mark of success for any theory and indicates that rather classical physical relationships are working well.

However, some nuclei with specific mass numbers are not well fitted at all. This is true in particular for the low mass number region (see Fig. 3.11 and the values e.g. ^{4}He in Tab. 3.8). Obviously, there is a limitation in the LDM's ability to model the mean nucleon binding energy.

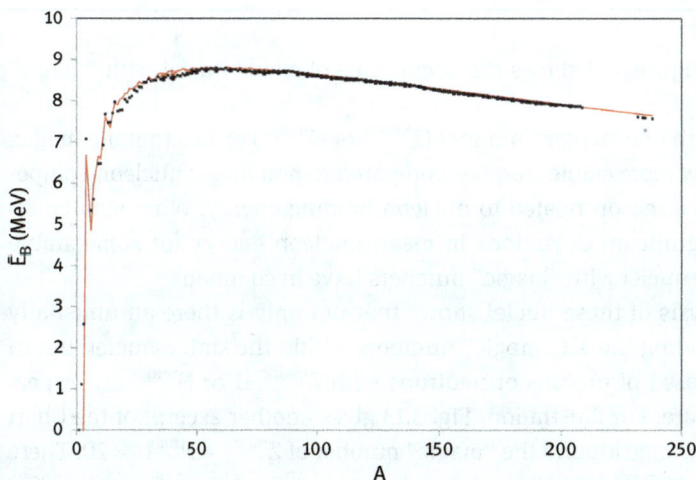

Fig. 3.11: Mathematical model of the WEIZSÄCKER equation versus mass number A (red line), shown alongside "real" mean nucleon binding energies $\bar{E}_B$ (black squares). Experimental $\bar{E}_B$ values are shown for stable nuclei and the three uranium isotopes 234, 235 and 238 and ^{232}Th.

3.4 The nuclear shell model

Accordingly, another theory is needed to understand the relatively strong deviation of light isotopes (such as ^{4}He) from the WEIZSÄCKER equation, and the dominance of (even, even) proton and neutron numbers among the stable nuclei. The WEIZSÄCKER equation tried to account for this fact within the pairing term, but the term had no physical meaning. In addition, the unusually high abundance of nuclei having one of the six "magic" numbers for N (8, 20, 28, 50, 82 and 126) and Z (8, 20, 28, 50 and 82)

Tab. 3.8: Experimental values of $\bar{E}_B$ compared to those calculated via the liquid drop model and the WEIZSÄCKER equation of type (3.1). Deviations $\Delta\bar{E}_B$ between calculated and experimental values are expressed both in MeV and in percent. In most cases, the liquid drop theory is predictive of mean nucleon binding energy, i.e. fits well with experimental data.

Nucleus	Mean nucleon binding energy (MeV)			
	$\bar{E}_B$ experimental	$\bar{E}_B$ calculated by eq. (3.1)	$\Delta\bar{E}_B$ calculated – experimental	
^{4}He	7.074	6.669	−0.405	6.1%
^{12}C	7.680	7.605	−0.075	1.0%
^{16}O	7.976	7.994	−0.0	0.4%
^{56}Fe	8.790	8.795	+0.005	0.05%
^{238}U	7.570	7.653	+0.083	1.1%

went unexplained.[23] Figure 3.12 shows the occurrence of stable nuclei with "magic" numbers.

Because nuclei with one "magic" number (Z^{magic} or N^{magic}) or two (double-magic, Z^{magic} and N^{magic}) show more stable isotopes compared to non magic nucleon compositions, there must be a reason related to nucleon binding energy. What may be responsible for such significant deviations in mean nucleon energy for some stable nuclei? What do these nuclei with "magic" numbers have in common?

A statistical analysis of these nuclei shows that not only is there an unusually high probability of finding the six "magic" numbers within the stable nuclei, but in parallel, nuclei composed of protons or neutrons with $Z^{magic} \pm 1$ or $N^{magic} \pm 1$, respectively, are unusually rare. For illustration, Fig. 3.13 gives another excerpt of the Chart of Nuclei. It covers the range around the "magic" number of $Z^{magic} = N^{magic} = 20$. There are six stable nuclei for $Z^{magic} = 20$ (the calcium isotopes, ^{40}Ca, ^{42}Ca, ^{43}Ca, ^{44}Ca, ^{46}Ca, ^{48}Ca), but only three for $Z = 19$ (^{39}K, ^{40}K, ^{41}K), and only one for $Z = 21$ (^{45}Sc). There are five stable nuclei for $N^{magic} = 20$ (^{36}S, ^{37}Cl, ^{38}Ar, ^{39}K, ^{40}Ca), but none for $N = 19$ and only one for $N = 21$ (^{40}K).

The LDM is not able to explain this "magic" number effect. Explanations arise from a completely different point of view – the nuclear shell model (NSM). The two most conclusive postulates are:

1. The nucleons are not distributed homogeneously, but in specific "shells".
2. In contrast to the LDM, the NSM supposes that all the protons and all the neutrons are different from each other, i.e. they have individual characteristics that make each nucleon in an atomic nucleus unique. This is the PAULI principle (introduced for electrons) now applied to nucleons.

---

23 There is no "magic" number 126 for protons – such a chemical element does not exist.

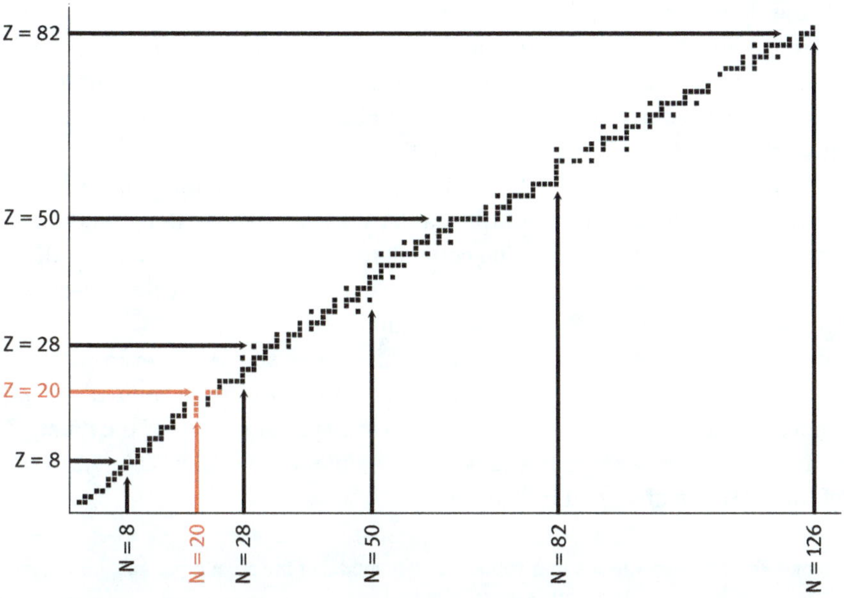

Fig. 3.12: Expression of stable nuclei with "magic" proton and neutron numbers indicated by arrows. The "magic" numbers are 8, 20, 28, 50, 82 for Z and N and, in addition, 126 for N. Stable nuclei of Z = 126 are not (yet) known. The N = 20 and Z = 20 section is enlarged in Fig. 3.13.

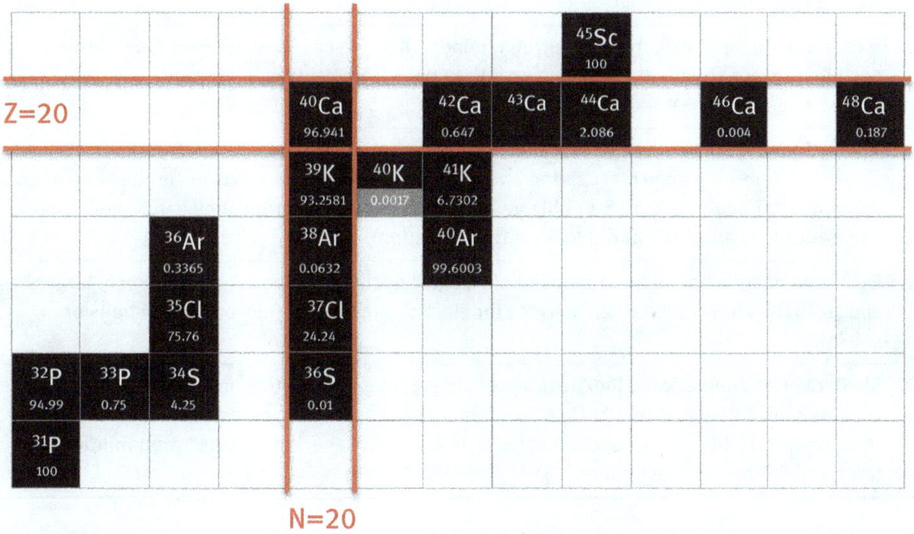

Fig. 3.13: Excerpt from the Chart of Nuclei covering stable nuclei around the "magic" number 20. There are six stable nuclei for Zmagic = 20, but only three for Z = 19 and one for Z = 21. There are five stable nuclei for Nmagic = 20, but none for N = 19 and only one for N = 21. (^{40}K is not "really" stable; its half-life is $1.28 \cdot 10^9$ years; comparable with ^{238}U).

Recall the development of the atomic model in general, and in particular the change from the RUTHERFORD atom to the BOHR atom. It was quantum mechanics that finally addressed this "individualism" of every electron by introducing shells and shell occupancies derived from quantum numbers. The same theory must now be applied to the protons and neutrons inside the nucleus.

For example, the experimental database says that nuclei having 2 or 8 protons and/or neutrons appear to be particularly stable. Why not believe that 2 and 8 nucleons each "fill" a "nucleon shell"? The equivalent would be electrons filling the K shell with maximum 2 electrons (first period of PTE) and the L shell with 8 electrons (second period of the PTE). This would be very interesting – if true!

Beyond a solely statistical distribution of numbers of stable nuclei in terms of an excess along Z^{magic} and N^{magic}, or a deficit in terms of $Z^{magic} \pm 1$ and $N^{magic} \pm 1$, there are several experimental facts that point to the special impact of the "magic" numbers. In a sense, they extend the experimental database listed in Tab. 3.2 in the context of the LDM and are listed in Tab. 3.9.

Tab. 3.9: "Experimental" observations that support a shell model of the atomic nucleus. Continued from Tab. 3.2.

Observation
9.   Especially in the low mass region, there are several "peaks" along the fit of $E_B = f(A)$. Obviously, some mass numbers create higher values of $\bar{E}_B$ than predicted by the LDM (see Fig. 3.11).
10.   Nuclei with "magic" proton and neutron numbers (and the chemical element behind these nuclei) are more abundant in the universe than nuclei/elements with proton or neutron numbers, e.g. $Z^{magic} \pm 1$ and $N^{magic} \pm 1$.
11.   The nucleon separation energy of "magic" nuclei is extremely large (see Fig. 3.14). This is similar to the known excitation and ionization energies for electrons in the shell of an atom (it is "difficult" to ionize a noble gas because its outer electron shell is full, and thus energetically favored compared to an e-1 composition).
12.   Excitation energies of nuclei composed of "magic" nucleon number(s) are extremely large, similar to the known excitation energies for electrons located in fully occupied shells or subshells.
13.   Similarly, it is energetically inexpensive to add e.g. one neutron to a nucleus of $N^{magic} - 1$ composition. This refers to a nuclear reaction process and is expressed as increased nuclear reaction probability ("cross-section value"). In contrast, "neutron capture" probabilities are low for N^{magic} nuclei (see Chapter 13).

3.4.1 Nucleon separation energy

Argument no.11 of Tab. 3.9 concerns the separation energy of nucleons. This parallels the discussion on ionization energies of chemical elements (see Chapter 1). What is separation energy? While there is (in general) a gain in binding energy when single nucleons enter the bound state of a nucleus, this is not an endless process. Ultimately, the mixture of A, Z, and N becomes too unbalanced for one more neutron or proton to be added. In a sense, the terms of the WEIZSÄCKER equation (excluding the volume term) reduce the nucleon binding energy below some critical level. Thus, this "last" nucleon is relatively easy to reject from the ensemble of existing nucleons. Separation energy describes the situation where the balance of energy is in favor of releasing a nucleon instead of keeping it. This is comparable to the low first ionization energy of the alkali elements, and the high ionization energies of the noble gases (see Fig. 1.11).

The partial separation energy E_s of nucleons is defined by eqs. (3.2) and (3.3).

Separation energy $\quad$ Initial nucleus * K1 $\quad$ Final nucleus K2

$$E_S(p) = (m_{K1}(A, Z) \quad - \quad (m_{K2}(A-1, Z-1) \quad + \quad m_p)c^2 \tag{3.2}$$

$$E_S(n) = (m_{K1}(A, N) \quad - \quad (m_{K2}(A-1, N-1) \quad + \quad m_n)c^2 \tag{3.3}$$

Values of E_s are high for a nucleus with a well-balanced mixture of protons and neutrons, and in particular for constellations of "magic" nucleon numbers. In contrast, the separation energy is low in nuclei of one proton or neutron above a "magic" number. Figure 3.14 illustrates this pattern for nuclei around $N^{magic} = 8$ and $Z^{magic} = 8$.

3.4.2 Neutron capture energy

Argument no.13 of Tab. 3.9 may be considered as the reverse of nucleon separation, namely the capture of one more nucleon. This is particularly relevant to the capture of neutrons. Figure 3.15 illustrates this case for two of the stable strontium isotopes, ^{87}Sr and ^{88}Sr. The two neutron capture reactions are $^{87}Sr + {}^1n \rightarrow {}^{88}Sr$ and $^{88}Sr + {}^1n \rightarrow {}^{89}Sr$. The nucleus ^{88}Sr contains 50 neutrons, representing a "magic" number. When receiving another neutron, the "magic" number 50 turns into 51. Correspondingly, the "reaction constant" of this nuclear reaction is low (0.005 b).[24] In contrast, ^{87}Sr may turn its "non magic" neutron number of 49 into "magic" 50 when capturing one additional neutron. This reaction is much more probable than the other, and the reaction constant is 16 b, i.e. larger by a factor of 3200.

---

24 See Chapter 13 for nuclear reaction constants expressed as cross-section values (σ), given in units of barn (b). The cross-section σ is expressed in units of 10^{-28} m^2, which is 1 barn.

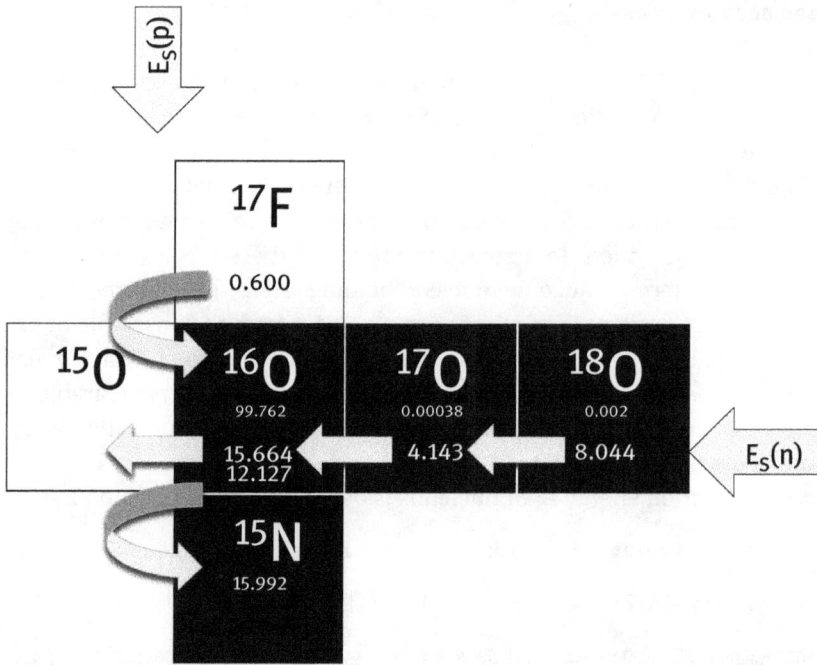

Fig. 3.14: Separation energies of nuclei around $N^{magic} = Z^{magic} = 8$. Neutron separation energies from $^{18}O \rightarrow ^{17}O \rightarrow ^{16}O \rightarrow ^{15}O$ are 8.044 MeV, 4.143 MeV, and 15.664 MeV, respectively. Proton separation energies from $^{17}N \rightarrow ^{16}O \rightarrow ^{15}N$ are 0.600 MeV and 12.127 MeV. An unusually high amount of energy is needed to separate a proton or a neutron from the double-magic nucleus ^{16}O: 15.664 MeV to remove one neutron, and 12.127 MeV to separate one proton. In contrast, (N^{magic} +1) and (Z^{magic} +1) nuclei such as ^{17}O and ^{17}F are eager to separate the one nucleon which is in excess of a "magic" nucleon number. For example, the nucleus ^{17}F will separate one proton given only 0.6 MeV.

Fig. 3.15: The probability of neutron capture via a nuclear reaction process for two strontium isotopes. The stable isotope ^{87}Sr is of (Z, N) composition (38, 49), and the isotope ^{88}Sr is of (38, 50). N = 50 is a "magic" number. Both stable isotopes may accept a further neutron, which creates a (Z, N +1) composition: ^{87}Sr +1n = ^{88}Sr, and ^{88}Sr +1n = ^{89}Sr. The cross-sections are $\sigma = 16 \cdot 10^{-28}$ m^2 and $\sigma = 0.0058 \cdot 10^{-28}$ m^2, respectively. Obviously, ^{87}Sr is much more likely to capture a neutron than ^{88}Sr. This is because ^{87}Sr would *reach* a "magic" neutron number, while ^{88}Sr would *lose* its "magic" number.

3.4.3 The PAULI principle applied to nucleons

Obviously, the observations in Tab. 3.9 lead one to assign protons and neutrons to specific shells within the nucleus according to a set of quantum numbers – similar to electrons, populating specific orbits in the shell of an atom according to characteristic quantum numbers. However, there should be some specific aspects for electrons and nucleons, as, of course, the type and the energy of interactions are different for electrons and nucleons. For nucleons, some features are listed in Tab. 3.10.

Tab. 3.10: The concept of quantum numbers applied to nucleons.

Quantum number (symbol)	Meaning
Main (n)	The main quantum number n defines the shell that the protons or neutrons exist in, depending on the energy of the shell.
Orbital (l)	Nucleon shells are structured into subshells of orbital quantum number l, and there are $0 \le l \le (n-1)$ subshells. Identical to electron orbitals, two nucleons can fit in each s-shell. Likewise, there are 6, 10, 14, 18, 22, and 26 positions available for the $l =$ p, d, f, g, h, and i shells, respectively.
	However:
	(i) there is a significantly greater overlap of low n subshells compared to electron orbitals;
	(ii) when naming nucleon orbitals, notation may follow the convention used for electrons, or it may follow a different nomenclature system. In this new system, the first occupied subshell l (according to increasing energy) gets the number 1. This results in 1p and 1d notations – for an example, see Fig. 3.19(a) *vs.* (b).
Magnetic (m)	The magnetic quantum number m reflects the projection of the orbital angular momentum of the nucleons within a magnetic field. Possible values are $m = -l, \ldots, l$.
Spin (s)	The spin quantum number of nucleons is either $+\frac{1}{2}$ or $-\frac{1}{2}$.[25] Every energetic and angular momentum state of an electron characterized by quantum numbers n, l, and m thus does not accept more than two either protons or neutrons. Both get the same set of (n, l, m), but differ in angular spin of $+\frac{1}{2}$ or $-\frac{1}{2}$. While the HUND rule arranges electrons in subshells by first filling electrons of the same spin into available orbitals (maximizing the total spin), nucleon subshells are filled by initially pairing nucleons with opposite spin (minimizing the total spin of the nucleus).

---

25 This was already seen for electrons in the p subshell, for example. Although of different orbital quantum numbers (p_x, p_y, p_z), their energies are identical.

Tab. 3.10 (continued)

Quantum number (symbol)	Meaning
Total angular moments (j)	Interaction of orbital and angular momentum within the magnetic field of the nucleus results in a <u>total</u> angular quantum number j. This is comparable to spin-orbit coupling. The possible values are $j = l \pm \frac{1}{2}$. The total angular momentum defines differences in energy. Levels initially defined by (nl) quantum numbers form levels of ΔE_{ls}.
	While this is (in principle) the same effect introduced for electrons, the consequences are much more pronounced for nucleons. Fine structures become more important.

With this set of quantum numbers, each nucleon is fully identified by charge (+1 or 0) and energy, according to parameters described by quantum numbers $(n, l, m, +\frac{1}{2})$ and $(n, m, l, -\frac{1}{2})$, or $(j, +\frac{1}{2})$ and $(j, -\frac{1}{2})$, as well as by $s = \pm\frac{1}{2}$. Thus, PAULI'S exclusion principle applies: in any quantum mechanical system (electrons in the atom's shells, nucleons in the atom's nucleus), two particles with identical quantum mechanical parameters should not exist.

3.4.4 Potential well considerations

Protons and neutrons occupy separate shells. This is because the two different nucleons have different potential well considerations – see Fig. 3.16, which illustrates the "particle in a box" model for nucleons, also called "infinite potential well" or "infinite square well". The well identifies, similar to the concept introduced for electrons (Fig. 1.16), the distance over which a body exists at a local energy minimum. The dimension of the well is very narrow, reflecting the size of the atom's nucleus. In contrast to a particle, which may "travel" unrestricted within a wide potential well, nucleons may occupy certain positive energy levels only within a very limited dimension. In classical physics, a particle is analyzed mathematically in terms of its motion (or oscillation) when acted on by an external force. The parameters are sinusoidal with constant amplitude and constant frequency. In quantum mechanics, the particle is described as a wave function. The particle in a box model can be solved analytically, depending on the model and its parameters, such as the well width and the nucleon energies.

This situation considers nucleons already bound within the nucleus. A different approach is needed to consider the moment a "free" nucleon enters the nucleus. Here, the nucleon must approach a certain distance to the well, where it becomes attracted by the strong force. This process is different for neutrons and protons due to coulombic repulsion; an incoming proton is repelled by the positive charge of other protons in the nucleus (see Fig. 3.16). Finally, the potential energy of the nucleons decreases when they enter the nucleus (the well). In this "bound" case, the previous discussions on mass defect and nucleon binding energy apply.

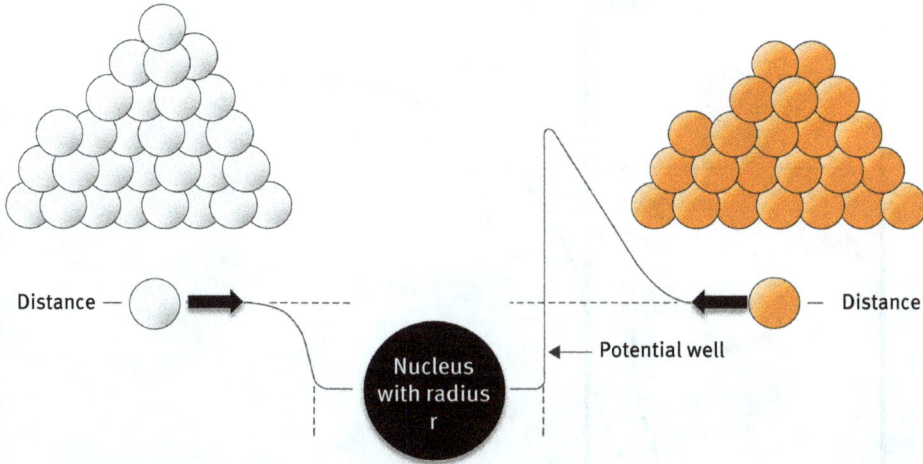

Fig. 3.16: The particle-in-a-box model for a "free" nucleon entering the nucleus. The task is to fill neutrons (white) and protons (orange) into shells and subshells of a nucleus in such a way that shell occupancy mirrors "magic" nucleon numbers.

The potential well concept is rather abstract, particularly for wells with an infinite square potential. To more accurately reproduce the potential energy of bound nucleons, modifications must be made. One alternative to the "infinite square" potential is the "harmonic oscillator" potential. In addition, there is also a "realistic" potential. Figure 3.17 schematically compares the three approaches.

The concept of the "infinite square" potential reproduces the observations indicated in Tab. 3.9, where subshells are defined by "standard" quantum numbers, leading every shell and subshell to have a specific (quantized) energy. Building from this framework, the "harmonic oscillator" potential also allows specific energy values to degenerate. Degeneration describes the situation when particles such as nucleons (although of different quantum states) have the same energy. Discrete energies of the "infinite square potential" become equal. Quantum mechanics handles this effect mathematically by the Hamiltonian for the particular system of more than one linearly independent eigenstate with the same eigenvalue. As a result, the number of available orbitals is reduced, and nucleons should exist at lower, discrete energy levels with higher probability. Some initial quantum states subsume into one energy level, which is filled by nucleons with equal probability. It no longer matters which subshell the nucleon was in originally.

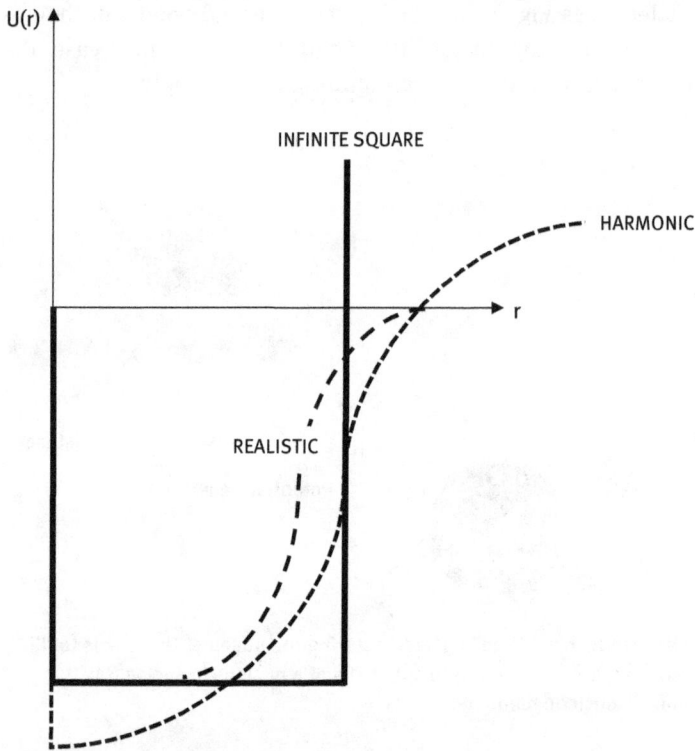

Fig. 3.17: Variations on the potential well: infinite square potential, harmonic oscillator potential, and "realistic" potential.

3.4.5 Correlating shell occupancy with "magic" numbers

One may now start to puzzle the nucleons together. Let us see what happens when the concept of quantum numbers is applied to nucleon shells, and a certain number of protons and neutrons are put into the potential well. The aim is to see which combinations of Z and N result in fully occupied shells and subshells, mirroring the "magic" number isotopes (see Fig. 3.18 for the strategy).

Like electrons, nucleons are "filled" into the lowest-energy shells first. However, note that the arrangement of shells according to energy level is different, following: 1s ⇨ 2p ⇨ 3d ⇨ 2s etc. (as shown in column (a) of Fig. 3.19). The subshell nomenclature used in nuclear sciences differs from quantum number descriptions used for electrons. Every time a new (sub)shell appears, it is indicated by consecutive numbers: 1s ⇨ 1p ⇨ 1d ⇨ 2s etc. (column (b)). The energy difference between shell numbers decreases with increasing shell number.

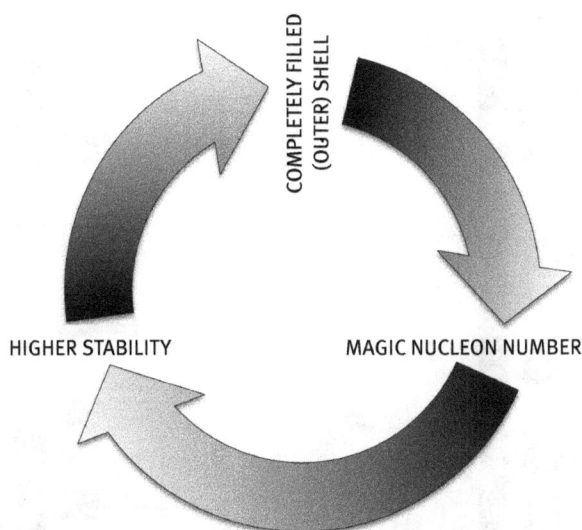

Fig. 3.18: Strategy of the correlation of "magic" nucleon numbers with a nucleon shell model. A completely filled (outer) nucleon shell should correspond to "magic" nucleon numbers (and vice versa). A completely filled (outer) nucleon shell then represents higher stability of the nucleus.

This approach is exemplified for proton levels in Fig. 3.19.[26] Two model potentials are tested – the infinite square and the harmonic oscillator – resulting in different shell occupancies. The infinite square model considers the energy of all the nucleons, as derived from their quantum numbers.

In contrast, the harmonic oscillator model allows for different sets of quantum numbers to be identical in energy. It shifts the orbital energy levels known from the infinite square approach. Second, nucleons within these "degenerate" levels have identical energy. Finally, the number of energetically equivalent nucleons within a common degenerate shell level is often larger than the shell occupancies of the other model.

What matters here is the cumulative number of proton orbitals with a closed shell. Similar to the concept that electron main shells are composed of individual subshells, nucleon shell closures should be defined for shell levels separated by a "sufficient" energy gap. For both models, these numbers are derived and given in the black boxes of Fig. 3.19.

---

26 The model refers to spherically shaped nuclei which are also at ground state (i.e. not energetically excited).

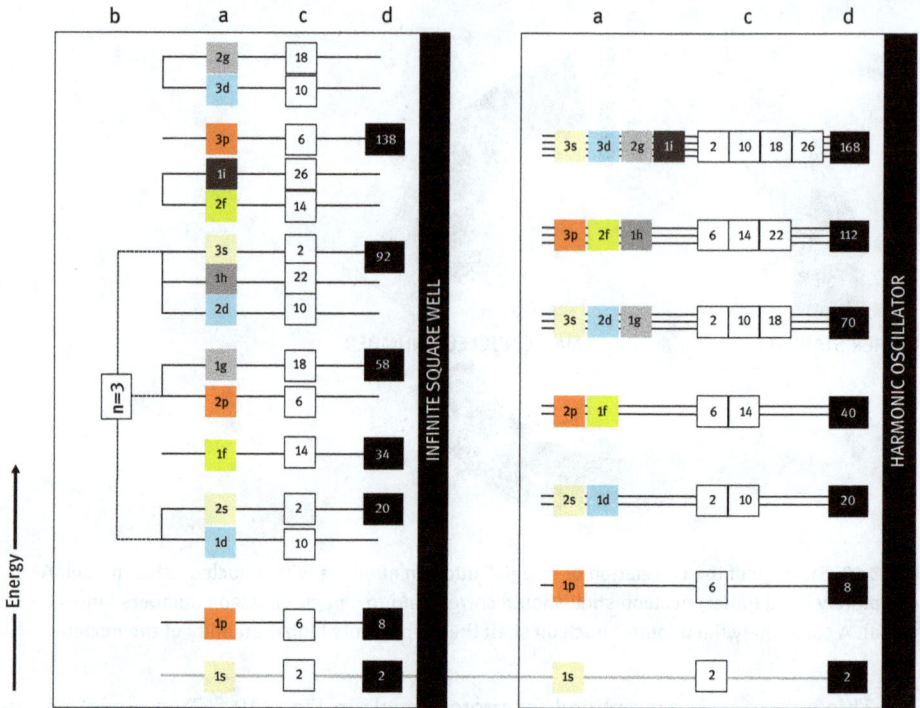

Fig. 3.19: Proton shells arranged according to the infinite square potential (left) and harmonic oscillator potential (right) wells. The subshell arrangement used in nuclear sciences differs from the ones used for electrons (a). For the sake of comparison, (b) indicates different energetic levels for the n = 3 main shell. The theoretical maximum subshell occupancies are shown in white boxes (c). Black boxes (d) show the resultant number of nucleons when the shells or subshells are filled cumulatively. Obviously, both models are able to reproduce the "magic" numbers 2, 8, and 20 in column (d). However, the shell closure numbers which follow (34, 58, 92, and 138 for the infinite square well potential, and 40, 70, 112, and 168 for the harmonic oscillator model) are "non magic".

Obviously, both models are able to reproduce the "magic" numbers 2, 8, and 20, as shown in column (e). However, the following numbers for completely filled shells (shell closures) are 34, 58, 92, and 138 for the infinite square well potential, and 40, 70, 112, and 168 for the harmonic oscillator. Both concepts thus fail to reproduce the "magic" numbers >20.

Consequently, another approach is needed. Fortunately, the physics already exists: however, one must think beyond (sub)shell structures and (non)harmonic oscillation to the effect of fine structure, which relates to spin-orbit coupling (s-o-c).[27] The interactions of orbital and angular momentum within the magnetic field of the

---

[27] which was experimentally already known and described theoretically in the case of electron shell structures and shell fine structures: the coupling of orbital and angular moments, see Fig. 1.21.

nucleus results in a total angular momentum J. The total orbital quantum numbers are $j = l \pm \frac{1}{2}$. The total angular momentum defines differences in energy relative to the initial, i.e. not spin-orbit coupled energy levels. Levels initially defined by (nl) quantum numbers form new energetic levels of ΔE_{ls}. For example, the initial p-orbital ($l = 1$) with altogether six protons (according to the three options of $m = -1$, 0, and $+1$, each filled with two protons of $s = +\frac{1}{2}$ and $-\frac{1}{2}$) forms two new levels of $j = 1 + \frac{1}{2} = 3/2$ and $j = 1 - \frac{1}{2} = \frac{1}{2}$. These are filled with four and two protons, respectively (shell occupancies after s-o-c are according to $2j + 1$.) Likewise, the d-orbital splits into $j = 5/2$ (6 protons) and 3/2 (4 protons).

Spin-orbit coupling results in almost twice as many individual shells of characteristic energy. All l-quantum numbers except $l = 0$ (the s-orbital) split into two levels, i.e. double the number of shells. Moreover, their energy levels are rearranged. Initially, the 1d-level was found above the 2s-shell, and protons had to be filled accordingly: first two protons in the 2s-shell, followed by 10 protons in the d-shell. After spin-orbit coupling, the 1d-derived $j = 5/2$ level is filled first (with 6 protons), followed by the 2s-shell (2 protons), and finally the $j = 3/2$ shell (4 protons). This principle has the smallest effect on low energy level shell configurations. Differences in energy for the first shells are relatively large, at least larger than the effect of splitting one initial level into two new ones. Thus, the effects of spin-orbit coupling are operative only at higher shells.

In fact, spin-orbit coupling creates an extreme pattern of fine structure at higher energy shells – exactly where they are needed to "correct" shell occupancy numbers to better match the "magic" numbers. Consequently, the three cumulative effects of new and energy shell open a great chance to "puzzle" again. Indeed, Fig. 3.20 finally gives the optimum result: this model reproduces all the "magic" numbers.

When comparing proton and neutron levels, there are two facts to note. First, the energy levels of identical proton and neutron shells are slightly different. Second, there are more shells available for neutrons to occupy – mainly because there are more neutrons than protons in heavy nuclei. This corresponds to the increasing excess of neutrons over protons at mass numbers of $A > 40$, and is a constituent part of the WEIZSÄCKER equation.

3.4.6 Energies of nucleon separation and excitation: from ground state to excited state nuclei

Figure 3.14 illustrated the nucleon separation energy, E_S, of stable nuclei around $N^{\text{magic}} = Z^{\text{magic}} = 8$. The individual values of E_S ranged from 0.6 MeV to 15 MeV. This is a representative range for the process of nucleon separation, which is equivalent to atomic ionization for electrons (the latter are of keV energy only).

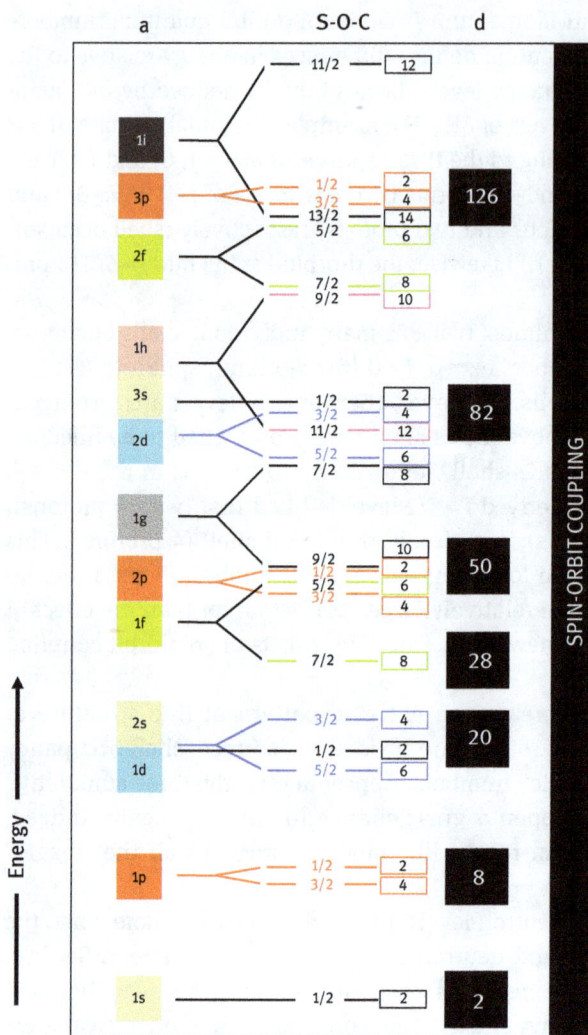

Fig. 3.20: Proton shells arranged according to spin-orbit coupling (s-o-c) relative to initial shell occupancies (b). The white boxes show the maximum number of protons that can occupy an individual shell/subshell after spin-orbit coupling. The black boxes (e) again show the resulting number of nucleons when shells/subshells are filled cumulatively – these now represent the magic numbers!

Similar to the processes of excitation known for electrons, nucleons may not escape from the nucleus, but occupy higher energy levels as a consequence of excitation.[28] This important effect is typical for nuclei where Z or N > 20, as spin-orbit coupling creates an increasingly dense overlap of upper subshell levels.

Nucleon excitation energies are thus lower than nucleon separation energy, but again cover a wide range. Unlike electrons, excited nucleon levels may even persist for relatively long periods. This results in differences between nuclei in the ground state and excited states.

3.5 Outlook

3.5.1 PTE *vs.* Chart of Nuclides

The LDM treats protons and neutrons as identical particles, differing only by charge. Based on this model, the semi-empirical WEIZSÄCKER equation used rather classical ideas to describe nucleon interactions. Nevertheless, the calculated mean nucleon binding energies fit very well with experimental data – except for a few nuclei composed of "magic" proton and neutron numbers. The NSM, in contrast, considers all nucleons to be unique according to a set of quantum numbers. Once again, quantum mechanics (in particular spin-orbit coupling) is key to modeling the "correct" arrangement of shells in terms of energy and occupancy, reproducing all the "magic" numbers.

Suppose each field of a stable nuclide within the Chart of Nuclides represents a black box, with only one face indicated. Here we get the parameters discussed so far. However, both LDM and NSM provide much more information on each stable nuclide, such as proton or neutron separation energies. Furthermore, for unstable nuclides, much more information will become relevant – filling every surface of the virtual box.

These two nuclear models explain most properties of our material world at the atomic scale. For example, it explains why we live in a world with an oxygen atmosphere. Why? There are several stable isotopes of oxygen (^{16}O, ^{17}O, and ^{18}O – see Fig. 3.21). The most abundant is ^{16}O, representing 99.762% of the mixture of the three stable isotopes. The mean molecular weight of oxygen is 15.992 g/mol, as indicated in the Periodic Table of Elements. This is because ^{16}O is a double-magic nucleus (8 protons, 8 neutrons). Its abundance on earth is much higher than expected for non magic (Z, N) nuclei.

---

28 The fate of excited nucleons will become particularly important in Chapter 12, when discussing post-effects of transformation (decay) processes of radioactive nuclides.

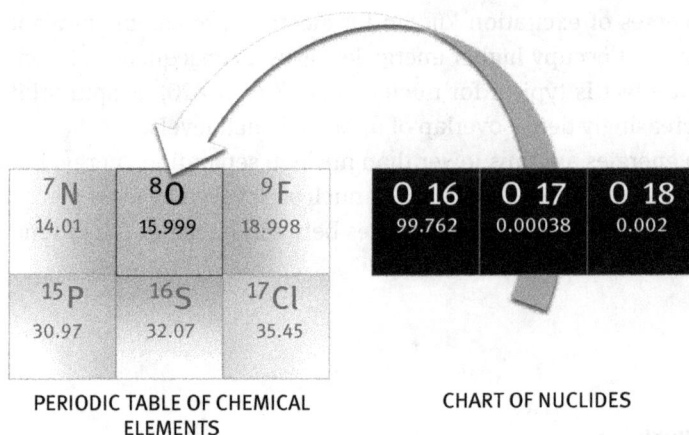

^{7}N	^{8}O	^{9}F
14.01	15.999	18.998

^{15}P	^{16}S	^{17}Cl
30.97	32.07	35.45

O 16	O 17	O 18
99.762	0.00038	0.002

PERIODIC TABLE OF CHEMICAL ELEMENTS   CHART OF NUCLIDES

Fig. 3.21: The Periodic Table of Elements and the Chart of Nuclides provide complementary information on an element and its isotopes. As shown above, the molecular mass of oxygen indicated in the PTE corresponds to the stable oxygen isotopes and their mixture.

3.5.2 From spherical to deformed nuclei

The concepts discussed in this chapter are simplified, assuming that nuclei are spherical and energetically non excited (ground state). Nucleons themselves are considered to be solitary units, with no inner structure inducing specific behavior. They are also assumed to have quasi-inert behavior and are not dependent on the individual properties of neighboring nucleons. The mathematics describing nucleons in this way is non relativistic.

Tab. 3.11: Extensions of the nuclear shell model.

Additional effects considered	Model
Special binding energy for outer-shell (valence) nucleons, particularly unpaired nucleons. This assumes that they are different from the nucleons of inner shells, which form an "inert" core	"Independent" particle model
Interactions between (several) outer shell nucleons	Residual interaction
Interactions between all nucleons, including those in lower shells and in pairs	"Collective" model
Experimental evidence on deformation of nuclei and the effects on shell energy	NSM of deformed nuclei ("NILSSON" model)

However, in reality, each nucleon occupying a characteristic nuclear shell "feels" the ensemble of other nucleons. Electromagnetic forces and the deformation of nuclei must also be considered. For example, Fig. 3.22 explains the effects of deformation of a spherical nucleus on the energy of the subshell nucleon orbitals.

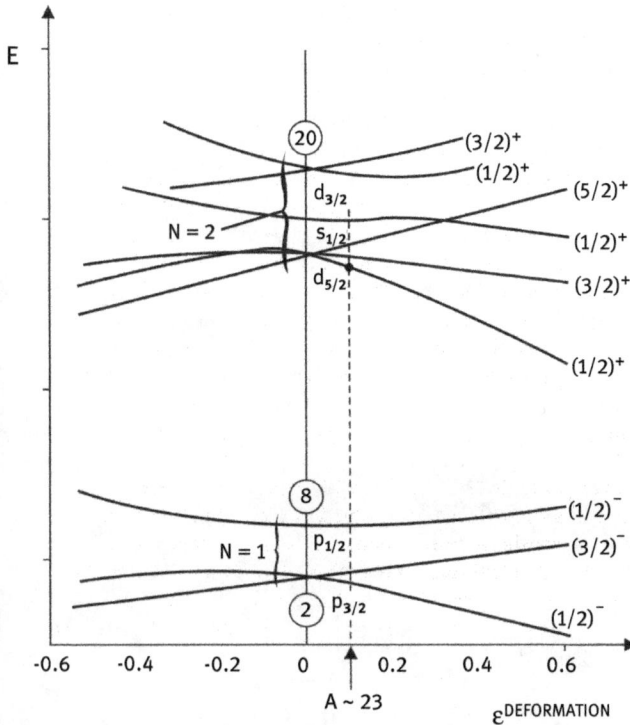

Fig. 3.22: A Nilsson diagram, showing the dependence of single nucleon energy (E) on the degree of deformation ($\varepsilon^{DEFORMATION}$) for nuclei of Z = N < 20. A spherical nucleus is represented by $\varepsilon = 0$, and deviations toward negative or positive values indicate deformation along one axis. It results in oblate or prolate shapes of the nucleus (see Fig. 2.9). The energy levels of all subshells change with ε, but in different ways. Depending on ε, this may result in new shell configurations and ground states for deformed nuclei. The circled numbers are magic nucleon numbers 2, 8, and 20. With permission: SG Nilsson, Kgl. Danske Videnskab. Selskab. Mat.-Fys. Medd 29 (1955) 16.

Thus, theories on atomic nuclei are under continuous revision. Consequently, there are many ongoing discussions on alternatives to the LDM and NSM. Important extensions are listed in Tab. 3.11. In general, for a deeper understanding of the properties and behavior of a nucleus, the composites of the nucleons (elementary particles) must be considered, as well as the forces that allow subatomic particles to interact. These interactions are needed to describe processes of transformations between different states of an atomic nucleus and the atom in general. Primary and secondary nuclear transfomations, as well as post-processes, will be considered later in the corresponding chapters. Finally, combining features of enhanced nuclear models, such as the "independent" particle model and the "collective" model, leads to a "unified" model.

4 From stable to unstable nuclides

Aim: Until now, more than 3200 nuclide species have been identified, and less than 10% of them belong to the class of stable nuclides. Therefore, approximately 3000 known nuclides are unstable. They show compositions of proton and neutron numbers that do not provide mean nucleon binding energies sufficient to keep the ensemble of nucleons stable forever. Instead, unstable nuclides try to convert into stable ones by either changing the ratio between protons and neutrons or the absolute number of nucleons.

This transformation is always driven by the gain in $\Delta \bar{E}_B$ between an initial (K1) and a formed (K2) nucleus. Processes of transformation of an unstable nucleus into a stable (or at least a "more stable") one are always exothermic. Transformation processes start with primary processes (changing nucleon compositions). In many cases, primary processes populate excited nuclear states of K2, which either directly or with some "delay" further stabilize via secondary processes. Both primary and secondary transformations may induce follow-up (post) processes.

Actually, the primary, secondary and post-transformation processes are the source of what is called "radioactivity" and unstable nuclei are the "radioactive" ones.

This chapter aims to introduce a kind of systematics of/between the three main primary transformations. The following chapters then will focus on the individual processes in detail.

4.1 Unstable nuclei and the Chart of Nuclides

4.1.1 Limited options for stability, many possibilities for instability

The strong interaction responsible for nucleon binding energy is the driving force behind nuclide stability. This nucleon binding energy correlates with numbers of A, Z, N of a nucleus (theoretically well described by the LDM) and the internal structure of its nucleons (covered by the NSM) – based on the data available for stable nuclides. Figure 3.11 for example shows that for each mass number A there is a corresponding value of $\bar{E}_B$ representing stable nuclei. Nuclei with these values of $\bar{E}_B(A)$ show a certain mixture of protons and neutrons – guaranteeing stability. So what if the mixture deviates from that level? The WEIZSÄCKER equation will yield values of $\bar{E}_B$ less than the corresponding maximum value $\bar{E}_B(A)$. Consequently, such nuclei will not be stable.

Let us see how changes in nucleon composition develop and finally leave stable configurations. Figure 4.1 gives an excerpt of the Chart of Nuclides for the segment

https://doi.org/10.1515/9783110742725-004

of $N = 3 - 17$, $Z = 6 - 11$ with the stable nuclides marked in black. Neighboring fields indicate unstable (N, Z) configurations.

This is illustrated for isotopes of oxygen in more detail. There are three stable isotopes as discussed in Chapter 3, all characterized by $Z = 8$, but different neutron numbers: 8, 9, and 10, respectively, $^{16}O_8$, $^{17}O_9$, and $^{18}O_{10}$. There may be isotopes of oxygen of $N < 8$, and others of $N > 10$. They fill the squares indicated with different intensities of gray. The closer the isotopes are to the stable nuclides, the darker the gray. The further the isotopes are from the stable core, the lighter the gray. Oxygen isotopes very far from the stable ones have hardly any color. This decreasing intensity of the color follows the decreasing mean nucleon binding energy and conceptually corresponds to the decreasing stability of the nuclei. Oxygen isotopes very far from the stable ones are not only less stable: they simply cannot exist (see below).

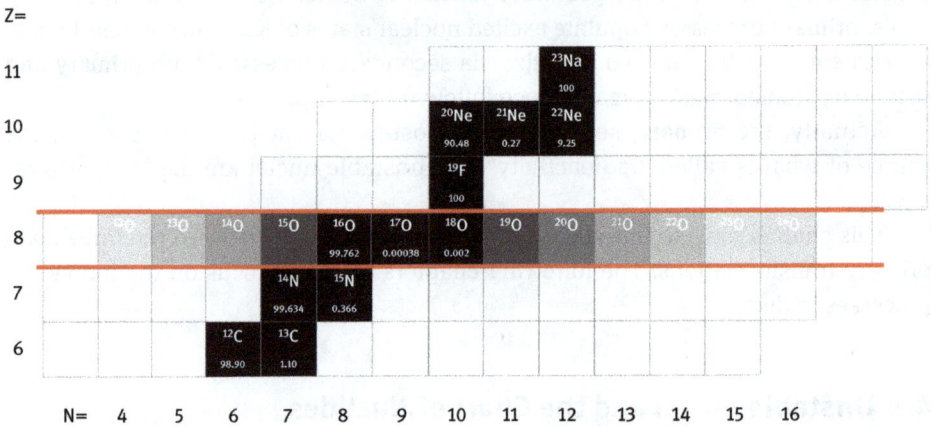

Fig. 4.1: Segment of the Chart of Nuclides. Stable nuclides are marked in black. White fields indicate "vacancies" available for isotopes of corresponding (N, Z) configuration. Lower numbers are the relative abundances of the stable isotopes. The filling of these vacancies is illustrated for isotopes of oxygen. Theoretically, there may be isotopes of oxygen of either $0 < N < 8$ or $N > 10$. They are filled with different intensities of gray. The closer the isotopes are to the stable nuclides, i.e. ^{15}O close to stable ^{16}O, and ^{19}O close to stable ^{18}O, the darker the gray. The further the isotopes are from the stable core, the lighter the gray. This decreasing color intensity follows the decreasing mean nucleon binding energy and schematically corresponds to the decreasing stability of the nuclides.

4.1.2 Nucleon drop lines

Figure 4.1 indicates that there may be isotopes of oxygen of very different neutron number. Theoretically, this could be of either less than 8 ($8 > N > 0$), which is "neutron deficit", or more than 10, which is "neutron excess". However, not all the theoretical (N, $Z = 8$) combinations are feasible. Qualitatively, this is obvious. It would be an extreme situation to imagine an oxygen nucleus composed of its 8 protons, but without any neutron! Or just one or two neutrons! One would already "feel" that this could never create a nucleus that could exist even for a short period of time. A semi-quantitative explanation comes from the WEIZSÄCKER equation, in particular from the coulomb term.

To quantify the feasibility, the concept is best illustrated for the process of adding more and more neutrons to the 8 protons of the nucleus of oxygen. Yes, each individual neutron added increases the overall nucleon binding energy partially $(\delta \bar{E}_B)$, defined by the volume term of the WEIZSÄCKER equation: $K1_{N-1} \rightarrow K2_N$.

However, the $(N - Z)^2$ part of the WEIZSÄCKER equation increases with every further neutron – and thus reduces the value of the overall nucleon binding energy of the nucleus. Such nuclei may prefer to eliminate an "excess" neutron instead. This is the neutron separation energy $E_S(n)$ needed to eliminate one neutron out of the nucleus: $K1_N \rightarrow K2_{N-1}$. The energy is calculated by eq. (4.1) using absolute masses of the two nuclei and the neutron.

$$\delta \bar{E}_B(n) = (^{A-1}m^{K1} + {^A}m^{K2})c^2 \tag{4.1}$$

$$E_S(n) = -\delta \bar{E}_B \tag{4.2}$$

Conceptually, there will be a situation in which the partial nucleon binding energy gained by capturing another neutron through the volume term $(\delta \bar{E}_B)$ is the same or even less than neutron separation energy $E_S(n)$, eq. (4.2).

This correlation allows one to discriminate between isotopes that theoretically may exist and those that are simply not allowed to exist. For oxygen isotopes, Fig. 4.1 graphically indicates the general and simplified[1] tendency of nucleon binding energy among the isotopes of neutron excess and deficit. The same can be done for neutron and proton separation energies $E_S(n)$ and $E_S(p)$.[2] For the isotopes of oxygen and carbon they are illustrated in Fig. 4.2.

There is a general conclusion: the more "in excess" one type of nucleon is to the other, the easier it is to separate. And vice versa: the stronger the deficit of one sort of nucleon compared to the other, the more energy is needed to separate it.

For oxygen, for example, the regions in which isotopes simply cannot exist at all are at mass numbers $A(_8O) \leq 12$ and $A(_8O) \geq 25$. The crosses at the left and right in Fig. 4.3 thus indicate the impossibility of the existence of these isotopes.

---

1 There is an alternating effect of nucleon binding energies for odd and even isotopes, discussed in Chapter 3. Oxygen isotopes with an even neutron number are superior in mean nucleon binding energy compared to isotopes with an odd neutron number.
2 These values can be calculated according to eq. (4.2), and are tabulated in nuclear science literature.

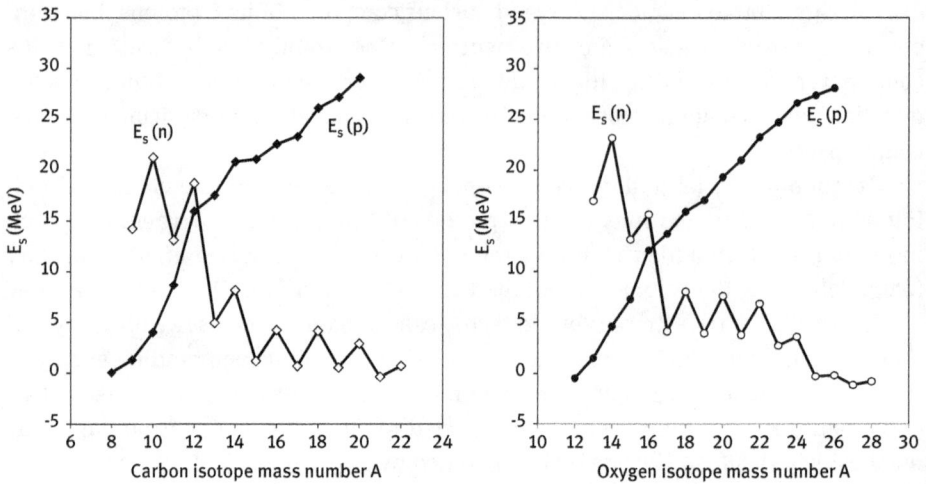

Fig. 4.2: Neutron and proton separation energies $E_S(n)$ and $E_S(p)$ for the isotopes of carbon and oxygen.

Consequently, the number of unstable isotopes allowed to exist is less than the theoretically possible ones.[3] Figure 4.3 shows the resulting isotopes of oxygen and carbon. For oxygen, it indicates that among the light (neutron-poor) isotopes, four (N, Z) mixtures remain, i.e. at $4 \leq N < 8$. For the heavier (neutron-rich) isotopes of oxygen, the last one representing $E_S(n) < -\delta \bar{E}_B$ is ^{28}O, i.e. $N = 22$. Altogether, there are (only) 4 "light" and 10 "heavy" unstable isotopes of oxygen. For carbon, there are 4 and 9, respectively. (The reason for generally more unstable isotopes in the case of neutron excess compared to deficiency can also be explained by the LDM and NSM.)

The same principle applies to protons, i.e. the processes of $K1_{N-1} \rightarrow K2_N$ vs. $K1_N \rightarrow K2_{N-1}$ along an isotone line. The concept, nevertheless, is better illustrated by following an isotone line on the Chart of Nuclides. Another option is to discuss the isobar line of the Chart of Nuclides. Figure 4.4 shows the same segment of the Chart of Nuclides as in Fig. 4.3. This time, the isobar nuclides of mass number $A = 16$ are highlighted. It demonstrates that only 6 nuclides may exist along an isobar line. Close to the only one stable nucleus ^{16}O ($Z = 8$) at $A = 16$ ($\bar{E}_B = 7.976$ MeV), there are two at larger Z ($Z = 9 = {}^{16}F$ and $Z = 10 = {}^{16}Ne$), and three with lower proton number ($Z = 7 = {}^{16}N$, $Z = 6 = {}^{16}C$, $Z = 5 = {}^{16}B$). Further combinations of Z:N, such as ^{16}Na and ^{16}Be, are energetically impossible.

When the crosses in Fig. 4.3 are connected, two lines will appear: the one at neutron excess is called the "neutron drip line", the other one the "proton drip

---

3 Furthermore, the number of unstable isotopes experimentally identified is even smaller. Research towards new, i.e. not yet detected nuclei is ongoing.

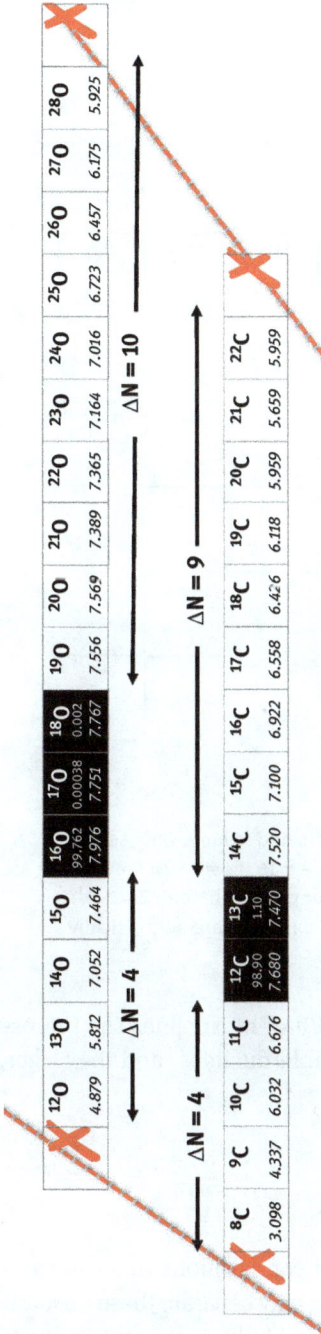

Fig. 4.3: Unstable isotopes of oxygen and carbon theoretically allowed to exist, according to the balance of protons:neutrons, partial nucleon binding energy and neutron separation energy − $\delta\bar{E}_B$ vs. $E_S(n)$. $\bar{E}_B$ values are indicated in MeV. The crosses at the left and right indicate that these isotopes cannot exist at all. When connecting the crosses at the neutron-poor branch and the neutron-rich branch, two lines will appear: The one at neutron excess is called the "neutron drip line", the other one the "proton drip line".

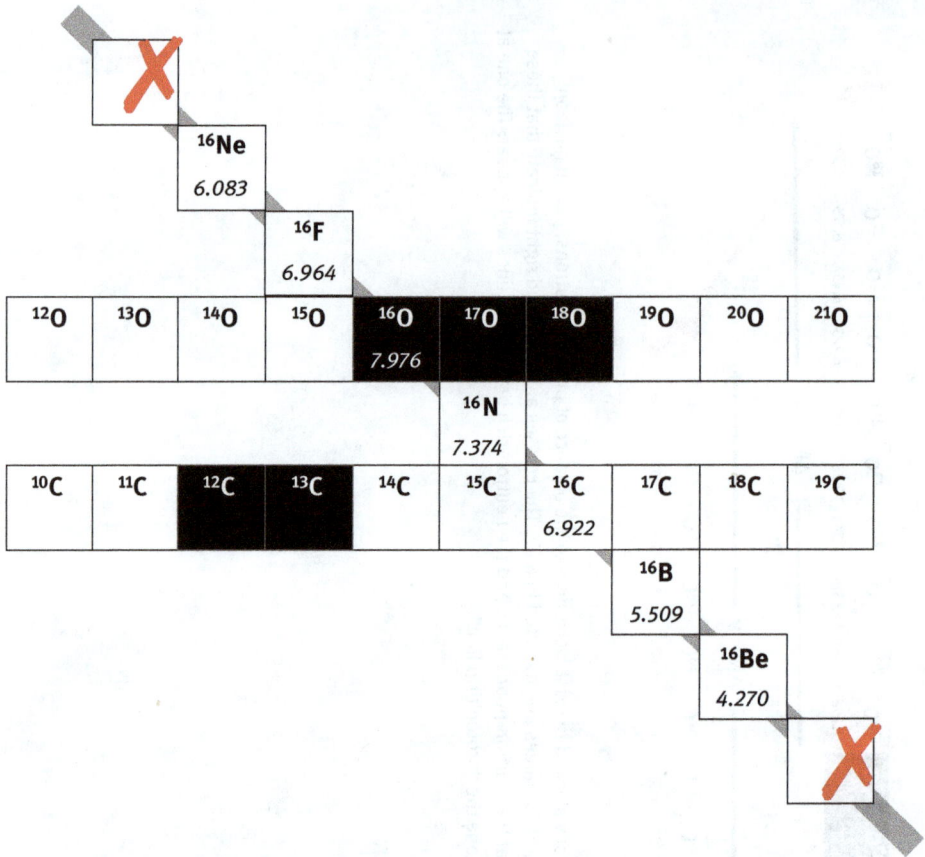

Fig. 4.4: Isobar nuclides of mass number A = 16. $\bar{E}_B$ values are indicated in MeV. Only seven nuclides may exist. Close to the only stable nucleus ^{16}O (Z = 8) at A = 16 ($\bar{E}_B$ = 7.976 MeV), there are two at larger Z (Z = 9 = ^{16}F, and Z = 10 = ^{16}Ne), and three with lower proton number: Z = 7 = ^{16}N, Z = 6 = ^{16}C, Z = 5 = ^{16}B. Further combinations of Z:N, such as ^{16}Na and ^{16}Be, are energetically impossible (red crosses).

line". Figure 4.5 shows the two resulting areas of the Chart of Nuclide. The intense gray area indicates the experimentally identified unstable nuclides, and the larger, light gray area the theoretically possible ones.

4.2 Unstable nuclides on earth

When discussing the huge number of unstable nucleon compositions that can theoretically exist (relative to the number of stable ones), one may separate these unstable nuclides into two classes. One class has a more "natural" character and includes

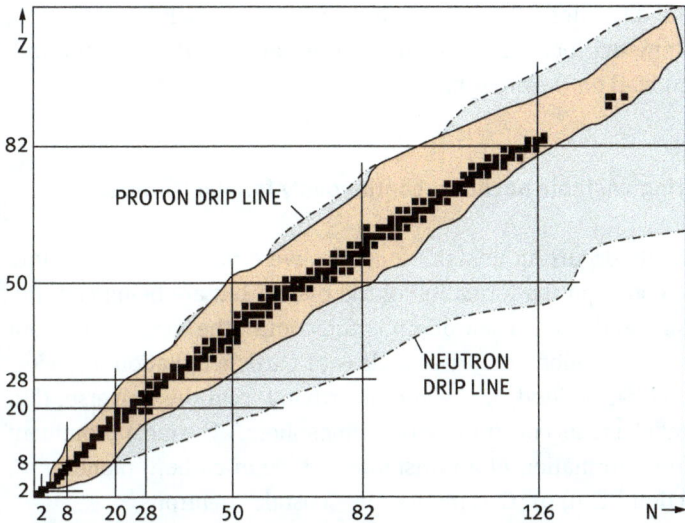

Fig. 4.5: The Chart of Nuclides indicating nucleon drip lines of $E_S \approx 0$. The dotted line limiting the lower area at neutron excess is the "neutron drip line", the upper one the "proton drip line". The (red) area along the stable nuclides (solid lines) indicates the area of experimentally identified unstable nuclides.

unstable nuclides that exist on our earth without human help, while the other class covers nuclides that are created "artificially" by man-made nuclear reactions.

4.2.1 Naturally occurring unstable nuclides: Ancient ones

The first class of unstable nuclides includes a relative small number that are of practical relevance in many fields (scientifically, analytically, industrially, medically). Although of natural origin, there are three very different subtypes according to the process of origin and the "age" of the unstable nuclides.

The first sort includes a subgroup of nuclides of ancient age – the origins of these unstable nuclides goes back to the formation of our planet and beyond. The nuclides thus are of half-life of at least (or close to) the age of the earth ($4.6 \cdot 10^9$ years). This group includes light nuclides such as ^{40}K ($t\frac{1}{2} = 1.28 \cdot 10^9$ a), but mainly very heavy nuclides such as isotopes of thorium and uranium, namely ^{232}Th ($t\frac{1}{2} = 1.405 \cdot 10^{10}$ a), ^{235}U ($t\frac{1}{2} = 7.038 \cdot 10^{08}$ a), ^{238}U ($t\frac{1}{2} = 4.47 \cdot 10^9$ a). These nuclides transform into others in order to stabilize (see Chapter 5 for "Naturally occurring transformation chains"). They permanently and naturally generate a number of unstable nuclides until they finally reach stable nucleon configuration at isotopes of lead or bismuth. This is another subgroup comprising the successors of the three mentioned nuclides ^{232}Th, ^{235}U and ^{238}U (sometimes also called "daughters" of the three "parent" nuclides). The daughter subgroup contains about 40 nuclides belonging to all the chemical elements located between uranium and thorium on the one hand, and bismuth and lead on the other hand.

4.2.2 Naturally occurring unstable nuclides: continuously formed isotopes

The other sort of naturally occurring unstable nuclides were not incorporated into the earth's crust in the course of the formation of the planet, but are being continuously produced by cosmic radiation. Their origin is thus not in the earth's crust, but in its atmosphere. Cosmic radiation comprises a class of particle radiation and electromagnetic radiation, either emitted by the sun or arriving from the universe, that reacts with the chemical elements constituting our atmosphere.[4] The most prominent example is the permanent formation of an unstable isotope of carbon, namely ^{14}C, within the nuclear reaction $^{14}N(n, p)^{14}C$ as induced by "cosmic" neutrons.[5]

What may be concluded from this sort of unstable nuclides is: unstable (i.e., radioactive) nuclides, transforming in the context of "radioactivity", are a natural part of our ecological system. Furthermore, radionuclides such as ^{40}K and ^{14}C co-exist among stable isotopes of potassium (^{39}K and ^{41}K) and carbon (^{12}C and ^{13}C), respectively, and are incorporated in our own biological matrix! A characteristic (although mass-wise low) amount of our biological matter (for example for carbon: amino acids, proteins, glucose etc.) is "radioactive". So – this is what we are in part ourselves: radioactive by nature.

4.2.3 Artificially-produced unstable nuclides

None of the chemical elements beyond lead offer stable isotopes, so all of them are radioactive. A few of them are generated in a natural way as successors of the families of the long-lived nuclides of ^{232}Th, ^{235}U and ^{238}U. All the isotopes of the chemical elements beyond uranium have been man-made.

The same is true for all the radioactive isotopes of proton number 82 (Pb) > A > 90 (Th) not involved in the natural transformation chains; and almost all of the unstable isotopes of the chemical elements ranging from hydrogen to lead.

---

4 The processes are quite complex and are in part covered by the Chapter 13 on "Nuclear reactions", but mainly in Chapter 2 on "Radiation dosimetry" and Chapter 3 "Nuclear dating" of Volume II.
5 The character of nuclear reactions is described in Chapter 13, and some relevant features of ^{14}C are covered in Volume II in the context of "nuclear dating", Chapter 3.

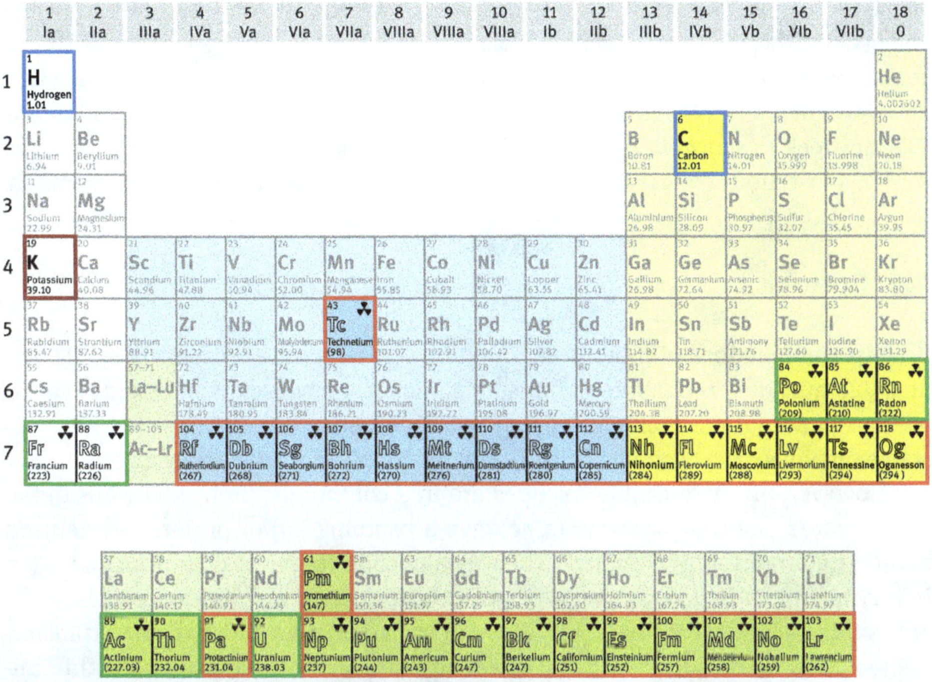

Fig. 4.6: Radioelements and selected chemical elements co-existing with unstable isotopes.
Red: Radioelements are those elements existing on earth *only* in the form of unstable isotopes when produced by artificial, man-made nuclear reactions, namely Tc, Pm, Pa and all the transuranium elements. **Green:** Elements such as U, Th, Ac, Ra, Fr, Rn, At and Po exist on earth exclusively in the form of more or less unstable isotopes. These are either permanently generated from long-lived nuclides of uranium and thorium and thus are of endogenous origin, or may also be produced artificially. **Blue:** Carbon has stable isotopes (^{12}C and ^{13}C), but co-exists with an unstable isotope (^{14}C), which is, firstly, permanently being generated by cosmic radiation in the earth's atmosphere and, secondly, incorporated into biological material. Hydrogen with its unstable isotope tritium (^{3}H) is another example. **Brown:** Potassium belongs to the "normal" chemical elements and was part of the earth's formation. However, one isotope (^{40}K) is in fact unstable (although of extremely long half-life), is radioactive and thereby emits radiation – but nevertheless constitutes an endogenous part of biological matter.

This may be no of great concern for "conventional" chemistry, as almost all the chemical elements ranging from hydrogen to lead offer stable isotopes – sufficient to investigate chemical properties and to cater to practical applications. However, there are a few exceptions. Some of the chemical elements listed in the PTE *do not* have stable nucleon configurations and do not exist at realistic amounts in nature.[6] Chemical

---

6 This explains – at least in part – why the original Periodic Tables of Mendelevev and Meyer (established on the basis of impressive systematic experimental evidence) revealed some "empty spaces" or "blank spots".

elements such as technetium (positioned along the series of Mn / Tc / Re), francium and astatine (heaviest homologues of the alkali metal and halogen groups) thus simply could not appear in the initial table. Today, it is obvious that these chemical elements do exist – but not in the form of "stable" nuclides. Taking this into consideration, the PTE may look as shown in Fig. 4.6. Such a representation reveals the whole system according to electron configuration, but indicates instability due to nucleon configuration.

4.3 Outlook

Various combinations of protons and neutrons create a little universe of stable and unstable nuclides. The number of unstable nuclides, however, is limited due to the balance of partial nucleon binding and separation energies. So far, about 3000 unstable nuclides have been identified compared to less than 300 stable ones.

However, all the unstable nuclides are in a continuous process of transformation towards one final destiny: to achieve a combination of proton and neutron numbers guaranteeing "stability". The way a nuclide "organizes" its transformation is very complex. Usually, these types of transformation are also called "decay",[7] but nothing actually "decays": it is all about transformations! Actually, the transformation itself is the source of what is called "radioactivity" and the unstable nuclides involved are the "radioactive" ones. All these unstable nuclides thus are also referred to as "radionuclides". The effects that accompany this transformation, fundamentally particle or electromagnetic emission, are called "radiation".

As in nature in general, physical and chemical objects tend to minimize their overall energy. This is also true for nucleon binding energy – and thus for nucleon configurations of a nucleus of an atom. According to the WEIZSÄCKER equation, mean nucleon binding energy changes whenever compositions of (A, Z, N) are modified. This transformation is driven by the gain in $\bar{E}_B$ between an initial (*K1) and a formed (K2) nucleus: $\bar{E}_B(*K1) = (Z, N) \rightarrow \bar{E}_B(K2) = (´Z, ´N)$. It proceeds spontaneously and is irreversible! The transformation processes are exothermic, as the newly formed nuclide K2 has a smaller overall nucleon binding energy than the initial one. A more stable nuclide will never spontaneously convert into a less stable descendant.

The various versions to realize these processes require separate and detailed description in the following chapters. The different classes of transformation of one unstable nuclide into another, more stable one and the kind of emissions that accompany the nuclear processes, nevertheless, all obey the same mathematical concepts. These are introduced in the next chapter.

---

7 This textbook prefers the wording "transformation" to decay, although both expressions are equivalent. The basic feature of unstable nuclei is to proceed into more stable states.

5 From stable to unstable nuclides: Mathematics

Aim: Radioactive transformation processes of an unstable nuclide *K are exothermic and proceed spontaneously. Although there are different types of transformation process, their velocities can be described by identical mathematics. The transformation velocity follows an exponential function. Key parameters are the transformation (or "decay") constant λ in units of reciprocal time (s^{-1}) and, synonymously, the half-life $t^{1/2}$ (in units of seconds, minutes, hours or years). Both λ and $t^{1/2}$ depend basically on the absolute amount of $\Delta \tilde{E}_B$ gained in terms of $\tilde{E}_B(K2) - \tilde{E}_B(K1)$, which may be modified by specific parameters, and cover many orders of magnitude.

The transformation constant (or half-life) links the number $\acute{N}$ of transforming nuclides to the corresponding radioactivity $\acute{A}$. The mathematics discriminates three classes of transformation.

The simplest case consists in a single transformation step of an unstable nuclide *K1 to a stable nuclide: $^*K1^{UNSTABLE} \rightarrow K2^{STABLE}$. In many cases, the nuclide K2 is not (yet) stable and undergoes a final transformation step: $^*K1^{UNSTABLE} \rightarrow {}^*K2^{UNSTABLE} \rightarrow K3^{STABLE}$. Thus, an initial radioactive nuclide "generates" a second radioactive one. There are three possible ratios of the two transformation constants involved: $\lambda_1 > \lambda_2$, $\lambda_1 \approx \lambda_2$, and $\lambda_1 < \lambda_2$. Consequently, the interplay of the two transformation rates must be taken into account by individual mathematics. Finally, most of the unstable nuclides undergo multiple transformation steps, called a "transformation chain", unless a stable nucleon configuration is reached in a new nuclide. In this case, more than two individual velocities must be considered.

5.1 Transformation parameters

Knowing the principal reason of nuclide transformation, namely to optimize nucleon binding energy, the next question is: How fast does this transformation occur? Despite the different types of transformation process, their velocities can be described by similar mathematics. There are two main features. First, nuclear transformation (decay) processes follow an exponential function. Second, mathematics complies with statistical considerations and applies to a large number of nuclides.

5.1.1 Exponential laws

Let us consider a number $\acute{N}$ of one sort of an unstable nuclide *K1 which transforms into a new nuclide K2: $^*K1^{UNSTABLE} \rightarrow K2$. This number will decrease because a new nuclide is formed. A fraction of nuclides ($d\acute{N}$) will transform in a fraction of time (dt), the sign of which is negative: $- d\acute{N}/dt$. The value of $d\acute{N}/dt$ represents energetic

https://doi.org/10.1515/9783110742725-005

parameters of the unstable nuclide, mainly the amount by which the mean nucleon binding energy $\Delta \bar{E}_B$ can be improved when transforming into a more stable nuclide. For each nuclide, there is a probability for how fast the process is – the kinetics of the transformation are specific to each sort of unstable nuclide.

Now it is possible to consider the fraction of unstable nuclides that have transformed over a period of time, relative to the initial number of nuclides (at $t = 0$): $- d\acute{N}/\acute{N} = f(t)$. If the number of nuclides transformed is proportional to the period of time, $d\acute{N}/\acute{N} \approx dt$. Next, a proportionality factor λ is introduced: $d\acute{N}/\acute{N} = -\lambda dt$. In order to guarantee constant units on both sides of the equation, the unit of λ should by $1/t$, a reciprocal of time. This is typically expressed in terms of inverse seconds, i.e. s^{-1}.

When integrating eq. (5.1), a quantitative correlation between the number of unstable nuclides ($\acute{N}$) still existing after any period of t and $\acute{N}_o$, the number existing initially (at $t = 0$), can be calculated. Actually, the parameter "time t" in eq. (5.3) should read as $(t - t_o)$ or Δt, with $\acute{N}_o$ identifying the number of nuclides at time t_o. However, t_o may be taken as $t = 0$. In analogy to monomolecular reaction kinetics in chemistry, this is a mononuclear kinetics of transformation of unstable nuclides: nuclear transformations follow a first-order exponential rate law, see Figure 5.1.

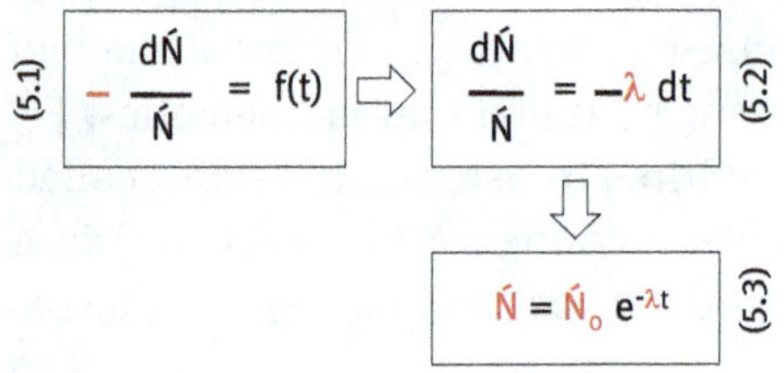

$$(5.1) \quad -\frac{d\acute{N}}{\acute{N}} = f(t) \quad \Rightarrow \quad \frac{d\acute{N}}{\acute{N}} = -\lambda \, dt \quad (5.2)$$

$$\acute{N} = \acute{N}_o \, e^{-\lambda t} \quad (5.3)$$

Fig. 5.1: Basic consideration to derive the mononuclear transformation kinetics of unstable nuclides: A first-order exponential rate law. $\acute{N}$ is the number of unstable (radioactive) nuclides existing at any time t, relative to the number of unstable nuclides which had existed initially at time $t = 0$. The proportionality factor λ is the transformation (or "decay") constant.

Because the process of transformation is referred to as "radioactivity", eq. (5.4) quantifies radioactivity as: $-d\acute{N}/dt \approx \acute{A}$, where radioactivity or "activity" is symbolized by $\acute{A}$.[1] This reveals another proportionality: the higher the count of radioactive nuclides, the higher the radioactivity: $\acute{N} \approx \acute{A}$. Analogously, no unstable nuclides = no radioactivity. The factor of proportionality is again the transformation constant λ, as shown in eq. (5.5). The corresponding unit of radioactivity is thus $1/time$, i.e.

---

1 It may be confusing to see that the symbols N and A are getting ambiguous: N in the earlier chapters symbolized "neutron number", now also "number of nuclei". A was introduced as "mass number", and now it is used for "activity" or "radioactivity" too. This is the reason why $\acute{N}$ and $\acute{A}$ are used instead in this textbook for radioactivity and number of transforming nuclides.

s^{-1}. In parallel with the decreasing number of unstable nuclides over time, radioactivity also decreases exponentially with $e^{-\lambda t}$, as per eq. (5.6), see Figure 5.2.

$$-\frac{d\acute{N}}{dt} \triangleq \acute{A} \tag{5.4}$$

$$\acute{A} = \lambda\,\acute{N} \tag{5.5}$$

$$\lambda = \frac{\acute{A}}{\acute{N}} \tag{5.6}$$

$$\boxed{-\frac{d\acute{N}}{dt} \triangleq \acute{A}} \Rightarrow \boxed{\acute{A} = \lambda\acute{N}} \Rightarrow \boxed{\lambda = \acute{A}/\acute{N}}$$

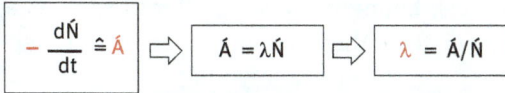

Fig. 5.2: Radioactivity $\acute{A}$ is derived from the kinetics of transformation of unstable nuclides into more stable ones. $\acute{N}$ and $\acute{A}$ are directly proportional. The proportionality constant is λ. The corresponding unit of radioactivity is 1/time $= s^{-1}$.

5.1.2 Transformation (or "decay") constant

The decay (or transformation) constant λ has units of reciprocal time (s^{-1}).[2] The transformation constant λ lends its unit to that of activity, shown in eq. (5.5). The IUPAC system adopts this unit in the context of radioactive transformation, defining it as 1 Bq (Becquerel) $= 1\,s^{-1}$. The value of 1 Bq signifies exactly one event of spontaneous radioactive nuclear transformation within 1 second. Because in many areas of research or daily life the scaling of radioactivity is much higher, it is common to use units such as MBq or GBq (mega-Becquerel or giga-Becquerel, respectively) – see also Fig. 5.8.

5.1.3 Half-life

There are two other options to quantify the velocity of (radioactive) transformation. One is "mean lifetime" τ as sometimes used in nuclear physics.[3] This other one is

---

2 An identical unit is known from another area of physics to quantify "frequency" with the unit Hz (Hertz), i.e. the number of oscillations (or other cycles) per second. For example, 1000 oscillations per seconds = 1000 Hz = 1000 s^{-1}. This describes a constant process, i.e. 1000 oscillations per seconds = 1000 Hz = 1000 s^{-1}. In contrast, kinetics of radioactivity follows a first-order exponential law, so the meaning is different.

3 Nuclear physics may use an expression well-established in general mathematics, called "exponential time constant", "mean lifetime" or simply "lifetime". It quantifies the period of time t,

"half-life", which is much more common in nuclear and radiochemistry. The half-life identifies the period of time t during which exactly half of the initial unstable nuclides have transformed, as shown in eq. (5.7).[4]

$$\frac{\acute{N}}{\acute{N}_0} = \frac{1}{2} \qquad (5.7)$$

It is the most common parameter to quantify the kinetics of radioactive transformation ("decay"). It describes radioactive transformation kinetics in a more "user-friendly" unit, namely in fractions of time (instead of reciprocal time as for λ). The units of half-life $t\frac{1}{2}$ vary (seconds, minutes, hours or years). Similar to λ, $t\frac{1}{2}$ is a value specific to each type of unstable nuclides – there are more than 3000 different half-lives!

Figure 5.3 illustrates how the quantity $t\frac{1}{2}$ is derived from the equations already introduced. It starts with $\acute{N} = \acute{N}_0/2$ (which is equivalent to $\acute{A} = \acute{A}_0/2$) and transforms further *via* eq. (5.3). With $\acute{N}/\acute{N}_0 = \frac{1}{2} = 0.5$, time t converts into half-life $t\frac{1}{2}$. Now the transformation constant λ and half-life $t\frac{1}{2}$ do not depend on $\acute{N}$ anymore, but are linked *via* the natural (EULER'S) constant e (e = 2.71828183) (eq. (5.8)), or its version of the natural logarithm of 2 (ln2 = 0.693147181) (eq. (5.9)). Consequently, the two parameters λ and $t\frac{1}{2}$ are inversely proportional according to ln2 by $t\frac{1}{2} = \ln 2 / \lambda$ or $\lambda = \ln 2 / t\frac{1}{2}$. The larger the value of the transformation constant, the smaller the half-life.

The continuing process of radioactive nuclide transformation (of one sort) shows a continuous (exponential) decrease in the number of unstable nuclides existing at longer times. When expressed in terms of periods of half-life, a general trend is obtained (see Fig. 5.4). Consider the initial number $\acute{N}_0$ or corresponding radioactivity $\acute{A}_0$ of an unstable nuclide species in existence at an initial time-point of t = 0; the fraction of nuclides (or activity) still present at subsequent numbers of

---

within which an initial number of nuclides $\acute{N}_0$ (or any other species) had transformed down to a fraction of the reciprocal value of e (1/e = 0.367879411), i.e. $\acute{N} = 1/e \; \acute{N}_0$. Similar to the use of "transformation constant", the corresponding exponential equation is $\acute{N}(t) = \acute{N}_0 \; e^{-t/\tau}$. The correlations are $\tau = 1/\lambda = t\frac{1}{2} / \ln(2)$, and $t\frac{1}{2} = \tau \ln(2) = 0.693147181 \; \tau$. Consequently, all three parameters are interchangeable. In the case of the carbon-14 isotope, for example, the transformation constant is $\lambda = 5.775 \cdot 10^{-4} \; s^{-1}$, the half-life is $t\frac{1}{2} = 5730$ years, and the mean lifetime is $\tau = 8267$ years. However, the use of "lifetime" is not applied to radio- and nuclear chemistry, and may be limited to elementary particles. For example, for free neutrons the mean lifetime is usually given as $\tau = 881.5 \pm 1.5$ s (= 14.69 min), which is $t\frac{1}{2} = 611 \pm 1.0$ s (= 10.18 min).

4 What, for example, is described by a value of $t\frac{1}{2} = 1$ h? Suppose there are $\acute{N}_0$ unstable nuclides at any time-point t_0 (which we measure according to their radioactivity $\acute{A}_0$). Exactly one hour later, the number of unstable nuclides still in existance is exactly half: $\acute{N}_{(t+1 \, h)} = \acute{N}_0/2 = 0.5 \; \acute{N}_0$. The same is true of the radioactivity "belonging" to these nuclides: $\acute{A}_{(t+1 \, h)} = \acute{A}_0/2 = 0.5 \; \acute{A}_0$.

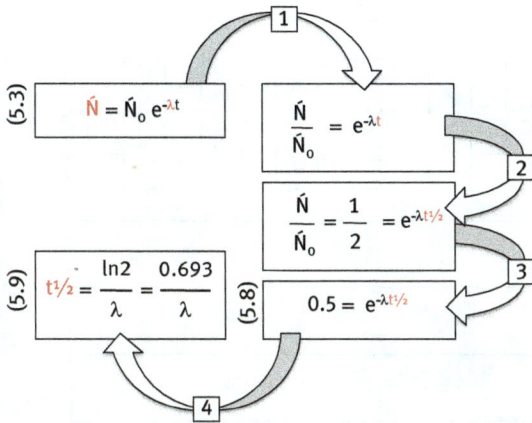

Fig. 5.3: Conversion of transformation constant λ into half-life $t^{1/2}$. At $t = t_{1/2}$ there is $\acute{N} = \acute{N}_o/2$. $\acute{N} = \acute{N}_o e^{-\lambda t}$ transforms (2) into a modified version of eq. (5.3). This proceeds via (2) $\acute{N}/\acute{N}_o = \frac{1}{2} = 0.5$, and substitutes time t by half-life $t^{1/2}$. Via (3) and (4) the transformation constant λ and half-life $t^{1/2}$ correlate.

half-lives steadily decrease by a factor of 2. After one half-life, exactly 50% of $\acute{N}_o$ (or $\acute{A}_o$) remains. After another half-life, this is 25%. The values of $\acute{N}_o$ (or $\acute{A}_o$) are halved after each additional half-life. Following 10 half-lives, the nuclide population is less than 0.1% (0.09765625%) of its initial value. Mathematically, the fraction or percentage of the number of initial nuclides remaining after n half-lives is $1/2^n$ or $100/2^n$, respectively.

Exponential correlations can be converted from a linear scale to a logarithmic scale, yielding instead of eq. (5.3) the linear relationships of type $\log(\acute{N}) = \log(\acute{N}_o) - \lambda t$, as per eq. (5.10). This is true for the number of unstable nuclides, as well as their absolute radioactivity $\acute{A}$ and the (relative) count rate $\acute{C}$ of measured radioactive events (see below).

$$\log(\acute{N}) = \log(\acute{N}_0) - \lambda t \qquad (5.10a)$$

$$\log(\acute{A}) = \log(\acute{A}_0) - \lambda t \qquad (5.10b)$$

Figure 5.5 identifies two elementary features. The linear function crosses the y-axis at $\log(\acute{N}_o)$, indicative of the number of unstable nuclides $\acute{N}_o$ at time-point 0 (or may be extrapolated to any earlier time-point.) The line's slope is negative and represents the value of λ. Thus, experimentally obtained trends like this quantify both the half-life of a single unstable nuclide and its absolute numbers.

Among the unstable nuclides detected so far, there is an impressive range of half-lives (or transformation constants). There are unstable nuclides of half-lives $>10^9$

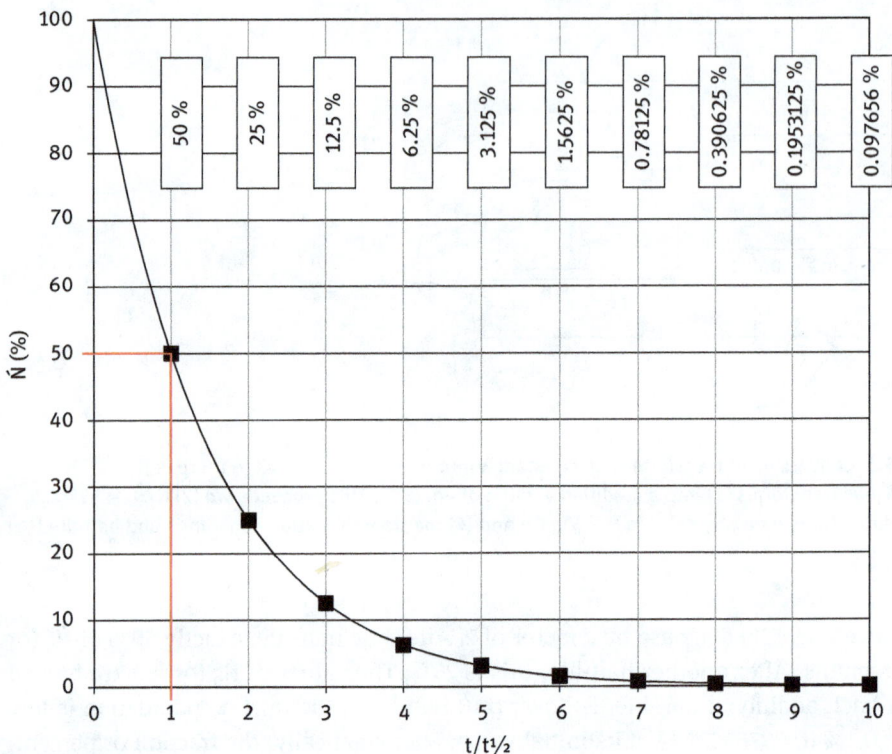

Fig. 5.4: General trend in transformation of unstable radioactive nuclides *vs.* half-lives: Exponential law of the kinetics of radioactive transformation. The x-axis representing time is on a linear scale and is divided by half-life. The y-axis is on a linear scale and represents Ń. If the initial number Ń$_o$ or the corresponding radioactivity Á$_o$ of a sort of unstable nuclide in existence at an initial time-point of t = 0 is considered, the fraction of nuclides (or activity) still present after subsequent half-lives steadily decrease by a factor of 2. After one half-life, exactly 50% of Ń$_o$ (or Á$_o$) remains. After another half-life, this is 25% etc.

years, which preceeds the existence of our planet. Extremely short half-lives approach the dimensions of fractions of a second. Table 5.1 collects some selected unstable nuclides, which are relevant for specific reasons,[5] together with their half-lives and transformation constants.

---

[5] Individual radionuclides will be discussed in detail in the various chapters of this Volume and in Volume II of this textbook.

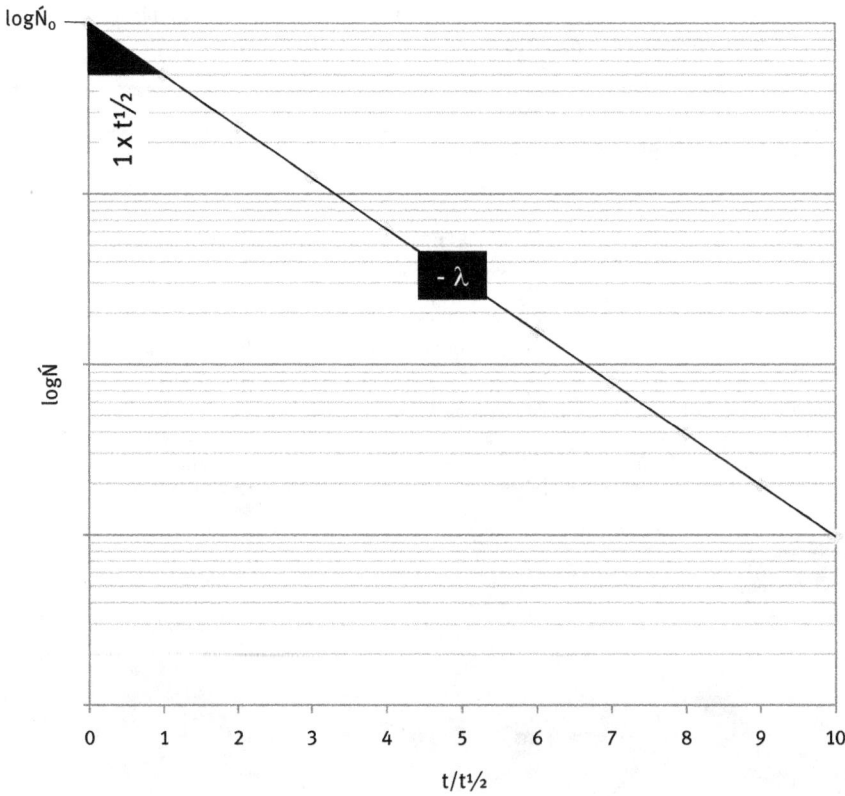

Fig. 5.5: Linearized version of the first-order rate law of radioactive transformation kinetics. The x-axis representing time is (still) on a linear scale (as in Fig. 5.4) and gives multiples of half-lives. The y-axis is plotted on a logarithmic scale as $\log(\acute{N})$ instead of $\acute{N}$. According to $\log(\acute{N}) = \log(\acute{N}_o) - \lambda t$, there is a linear line with a negative slope, and its absolute value is λ. The intercept of the linear line at y-axis gives $\log(\acute{N}_o)$. Instead of $\log(\acute{N})$, the y-axis could represent $\log(\acute{A})$ or $\log(\acute{C})$.

5.2 Correlations between radioactivity, number of unstable nuclides and masses

Radioactivity and number of unstable nuclides correlate *via* eq. (5.5): more unstable nuclides of one sort represent more radioactivity. The transformation constant or the half-life quantifies this proportionality. Any of the three parameters can be calculated from the other two, as shown in Fig. 5.6.

Tab. 5.1: Selected unstable nuclides representing a broad range of half-lives and typical areas of use in nuclear science research and applications.

Nuclide	t½	λ (s⁻¹)	Application/relevance
^{213}Po	4.2 μs	$1.65 \cdot 10^{+5}$	Member of a natural transformation chain, emitting high-energy α-particles (8.376 MeV)
^{253}Rf	1.8 s	$3.85 \cdot 10^{-1}$	Super heavy element research
^{11}C	20 min	$5.78 \cdot 10^{-4}$	Medicine: Positron emission tomography (PET)
^{99m}Tc	6.0 h	$3.21 \cdot 10^{-5}$	Medicine: Single photon emission tomography (SPECT)
^{131}I	8.02 d	$1.00 \cdot 10^{-6}$	Medicine: Scintigraphy and tumor treatment
^{3}H	12.323 a	$1.78 \cdot 10^{-9}$	Chemistry, *in vitro* assays
^{14}C	5730 a	$3.85 \cdot 10^{-12}$	Permanently formed in the earth's atmosphere/Chemistry, *in vitro* assays, nuclear dating
^{40}K	$1.28 \cdot 10^{9}$ a	$1.72 \cdot 10^{-17}$	Endogenous
^{238}U	$4.47 \cdot 10^{9}$ a	$4.92 \cdot 10^{-18}$	Parent of a natural transformation chain/Nuclear energy
^{232}Th	$1.405 \cdot 10^{10}$ a	$1.56 \cdot 10^{-18}$	Parent of a natural transformation chain/Nuclear energy

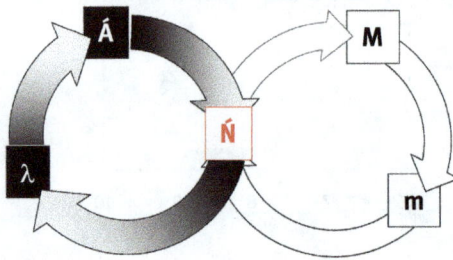

Fig. 5.6: Correlations between radioactivity, number of unstable nuclides (and via the molar mass M and mass m of the nuclides), and transformation constant.

5.2.1 From Ń to Á

Table 5.2 illustrates this situation for a fixed number of 10^6 unstable nuclides, but various half-lives ranging from 1 s to 1 min, 1 h, 1 day and 1 year. Whereas in the case of a short half-life (e.g. 1 min) the radioactivity is high ($1.155 \cdot 10^4$ Bq), a long half-life (e.g. 1 h) gives a radioactivity of almost 2 orders of magnitude less ($1.925 \cdot 10^2$ Bq).

5.2.2 From Á to Ń

Vice versa, the same level of radioactivity may be created by a very different number of radioactive nuclides. Table 5.3 compares the corresponding values for two different radionuclides of carbon, namely the isotopes carbon-11 and carbon-14. Both differ significantly in terms of λ and t½. The values of λ are $5.78 \cdot 10^{-4}$ s⁻¹ and $3.85 \cdot 10^{-12}$ s⁻¹,

Tab. 5.2: Radioactivity levels of a constant number Ń = 106 of unstable nuclides depend on the half-life (or transformation constant).

t½	λ (s⁻¹)	Á (Bq)
1 second	$6.931 \cdot 10^{-1}$	$6.93 \cdot 10^{+5}$
1 minute	$1.155 \cdot 10^{-2}$	$1.155 \cdot 10^{+4}$
1 hour	$1.925 \cdot 10^{-4}$	$1.925 \cdot 10^{+2}$
1 day	$8.023 \cdot 10^{-6}$	8.023
1 year	$2.204 \cdot 10^{-8}$	$2.204 \cdot 10^{-2}$

respectively. As mentioned above, it is more convenient to orientate on the half-lives, which are 20 min and 5730 years, respectively.[6] The same number of unstable isotopes thus transform with quite different rates. Already after 20 min, half of the ^{11}C isotopes are no longer present (having transformed into ^{11}B with a more stable proton:neutron configuration of 5:6 instead of 6:5 for ^{11}C). In the case of ^{14}C, after 20 min 99.9999995397% of the unstable isotope is still unchanged. Thus, nucleon mixtures of 6 protons + 8 neutrons (^{14}C) only slowly turn into a more stable configuration of 7 protons + 7 neutrons (^{14}N.) It takes more than 57 centuries to decrease the number of ^{14}C nuclei by 50%.

Tab. 5.3: Half-lives t½ and transformation constants λ of two different unstable, radioactive isotopes of the chemical element carbon, namely carbon-11 and carbon-14. Both differ significantly in terms of λ and t½.

isotope	t½	λ (s⁻¹)	Ń	mol	g
			Á = 1·10⁶ Bq = 1 MBq	6.022·10²³ Ń = 1 mol = x g	
^{11}C	20 min	$5.78 \cdot 10^{-4}$	$1.73 \cdot 10^{9}$	$2.9 \cdot 10^{-15}$	$3.2 \cdot 10^{-14}$
^{14}C	5730 a	$3.85 \cdot 10^{-12}$	$2.59 \cdot 10^{17}$	$4.3 \cdot 10^{-7}$	$6.0 \cdot 10^{-6}$

Following Á = λŃ, the number of radioactive nuclides for the same Á depends on the value of the transformation constant λ. For larger λ values (or smaller t½ values, according to t½ ≈ 1/λ), fewer transforming nuclides are necessary to create the same radioactivity in terms of transformations per second. Figure 5.7 illustrates the

---

6 One may say that ^{11}C is a "short-lived" isotope, and ^{14}C is a "long-lived" isotope. However, as half-lives of the many unstable nuclei range from a fraction of a millisecond to billions of years – everything is relative.

correlation of the half-life of a radioactive isotope and the corresponding mass in the case of a constant radioactivity of 10^6 Bq (1 MBq). The half-lives listed in Tab. 5.2 are shown separately and also the two longer half-lives of 100 and 10^9 years. The x-axis gives values of t½ ranging from 10^0 to 10^{18} s, i.e. over 18 orders of magnitude! The y-axis gives the corresponding number Ń of radioactive nuclides according to half-life. It ranges from 10^6 to 10^{21} nuclides. Again, showing 15 orders of magnitude requires a logarithmic scale. The red line is the correlation of t½ vs. Ń for the level of Á = 1 MBq.

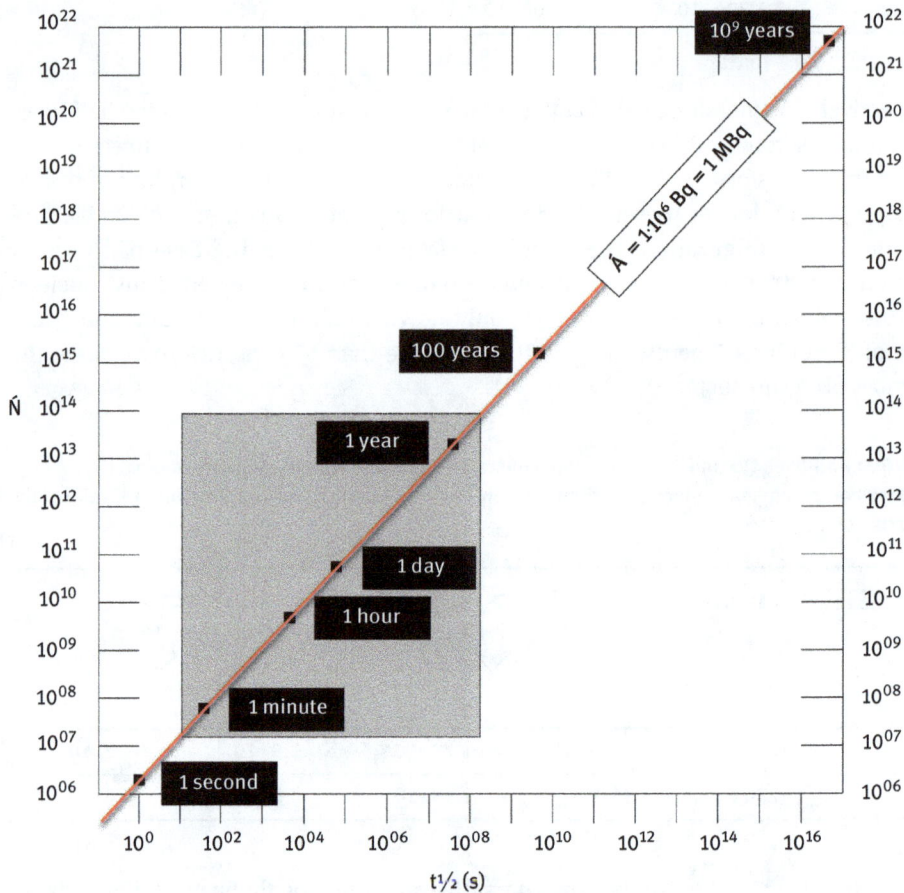

Fig. 5.7: Correlation between half-life t½ and number Ń of unstable nuclides for a constant radioactivity. Half-lives of 1 second, 1 min, 1 h, 1 day, 1 year, 100 years and 10^9 years are shown separately. Both the x-axis and y-axis are logarithmic. The line correlates t½ with N for the level of Á = 1 MBq. (Other levels of radioactivities would create lines parallel to the one shown.) The square indicates the range of half-lives typically handled in research institutions, i.e. from minutes to years. Consequently, typical numbers of radioactive nuclides (for the Á = 1 MBq level selected) range from 10^7 to 10^{16}, i.e. many orders of magnitude less than, e.g., the 1 mole level of $6.022 \cdot 10^{23}$.

5.2.3 From Á and Ń to mass

Among the many special and unique features of radiochemistry compared to conventional chemistry is the correlation between an unstable isotope's radioactivity level and its minuscule mass. The number of radioactive isotopes Ń is related to the molar mass of a chemical element: $6.022 \cdot 10^{23}$ atoms (nuclides) represent 1 mole, as per Fig. 5.6. Subsequently, the mass of a given number of radioactive isotopes can be calculated. 1 mole of ^{12}C involves $6.022 \cdot 10^{23}$ nuclides, and the mass of this one mole is 12.000 g.

Let us apply this correlation to calculate the real mass of a typical level of radioactivity of ^{11}C and ^{14}C, again for example Á = 1 MBq. The mass of this 1 MBq sample varies significantly for different isotopes and molar masses: 1 MBq of ^{11}C contains only $1.7 \cdot 10^9$ nuclides. This is an incredibly small fraction of a mole, namely only $2.9 \cdot 10^{-15}$ moles. Mass-wise, it corresponds to the ultra-low amount of $3.2 \cdot 10^{-14}$ g only, which is $3.2 \cdot 10^{-11}$ mg or $3.2 \cdot 10^{-8}$ μg or $3.2 \cdot 10^{-5}$ ng or 320 pg. For the much longer-lived ^{14}C, a pure 1 MBq sample still only weights 6 μg,[7] yet this mass is already larger by about eight orders of magnitude compared to ^{11}C.

5.2.4 Statistics: A short note

Velocities of radioactive transformation have been introduced as following exponential laws. One consideration is then a very large ensemble of nuclides, another the mono-atomic scale ("one-atom-at-a-time").

5.2.4.1 Large number of unstable nuclides

These may be treated by "probabilities". The mathematical approach is based on the laws of statistics, including very large numbers of participating bodies. The mathematics is not discussed here. It ultimately expresses precise "probabilities" of how a large number of unstable nuclides behave, instead of predicting the fate of one specific nuclide.

---

7 Remember that "normal" preparative chemistry is used to handle chemical compounds at molar scale, which are typically analyzed gravimetrically. This appears to be very difficult if not impossible in the case of radioactive substances. In many cases, radiochemists are thus handling radioactive compounds of tiny, almost negligible mass, which cannot be seen by the human eye and are difficult to analyze by conventional chemical methods!.

Figure 5.8 shows an initial upper line of 16 orange circles, each representing a number (e.g. 10^6 nuclides per circle)[8] of unstable nuclides *K1. The kinetics could be illustrated similar to Fig. 5.4, but adapted to the value of $t\frac{1}{2} = 1.0$ h over four hours.

Fig. 5.8: Example of the statistics of transformation of unstable nuclides. The upper line shows 16 orange circles, each representing a huge number (e.g. 10^6 nuclides per circle) of unstable nuclides of one sort *K1 with a half-life of $t\frac{1}{2} = 1.0$ h. Within one half-life, e.g. after 1 h, half of these nuclides have transformed into a stable nuclide K2 (black circle). After 4 h, there is only one orange circle remaining. It represents the 10^6 *K1 nuclides still in existance. (Actually, the effect is simplified because nuclear transformations will proceed among all the 10^6 nuclides of each circle at the same time).

5.2.4.2 A single unstable nuclide

However, what about the situation of a very limited number of unstable nuclei? Let us take Fig. 5.8 not as an ensemble of many nuclei (10^6 per circle) but just one unstable nucleus per circle. The challenge is to quantify the half-life of just 16 unstable nuclei. And what if there are not many unstable nuclides, but only one? This is illustrated in Fig. 5.9 (left).

For every single unstable nuclide (from a thermodynamic point of view) the transformation is irreversible, and from a time point of view it is stochastic. The transformation of a single nuclide thus occurs randomly. It may be possible to determine this event experimentally, but the time-point at which a specific nuclide transformed is not related to a reliable half-life value. In another experiment, another specific nuclide of the same isotope may convert at another time-point.

---

8 This corresponds to the common situation in radiochemistry, which is the handling of radioactivities representing large numbers of nuclides. Figure 5.7 indicates a range of half-lives typically handled in research institutions, i.e. at half-life levels ranging from minutes to years. The typical numbers of radioactive nuclei (for the $Á = 1$ MBq level selected) ranged from $Ń = 10^7$ to $Ń = 10^{16}$. For Fig. 5.7, 16 times the circles of 10^6 nuclei per circle just made $Ń = 1.6 \cdot 10^7$.

Consequently, it is impossible to predict when a single given atom will trans-form.[9] There is one strange observation and one follow-up theoretical concept.[10] The observation is that within four half-lives, 93.75% of the initial numbers of *K1 did transform – but 6.25% did not! This fraction still remains with the nucleon composition of *K1. So, what about the half-life of this 6.25% of the sample?

Remember that Fig. 5.8 was designed to show the situation for a very large number of unstable nuclides, i.e. 16 times 10^6. In fact, eq. (5.1) reflects the probabil-ity of transformation, and in this sense it provides a true number – statistically. And yes, there are some nuclides that "survive" for longer, pretending a half-life longer than 1.0 h. However, there are also nuclides that transform much faster.

Ultimately, the value of the half-life is an average, a statistically derived and experimentally proven value. Any experiment with a very large number of unstable nuclides involved will reveal the same value of t½ again, as per Fig. 5.9 (right). Any detector registering a count rate (Ć ~ Á and thus Ć ~ Ń) will register a decrease in count rate over time according to Fig. 5.4 with a half-life of t½ = 1.0 h.

---

9 This is the concept behind the virtual experiment of SCHRÖDINGER's cat. It was originally meant to showcase the ideological differences between quantum physics and conventional physics. It is cen-tered on the fate of a single unstable (radio)nuclide, which is packed together with a cat and a poi-son vial in a box. The box should remain closed for a certain period of time (which may be understood as e.g. one half-life of the particular radionuclide). The radionuclide's transformation would induce a signal, releasing the poison – and the cat would die. When the box is closed, the fate of the cat is unknown. However, when the box is opened after a certain time, you can observe whether the cat is dead or alive. So: there is nothing to predict.

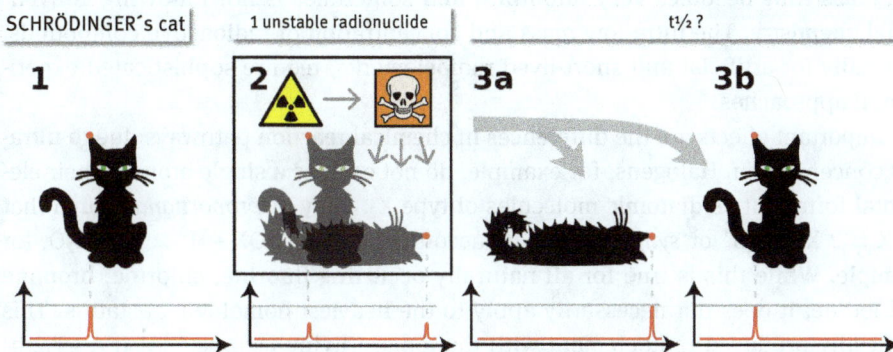

SCHRÖDINGER's cat | 1 unstable radionuclide | t½ ?

1 **2** **3a** **3b**

Courtesy: http://commons.wikimedia.org/wiki/File:Catexperiment.svg?uselang = de

10 Or let us take an ultimate example: two unstable nuclei. When a measurement is performed at time-point t, if the two nuclei are both still in the form of *K1, one would say that they are stable. In contrast, suppose one nucleus had transformed; its half-life would then equal the time at which the measurement was taken.

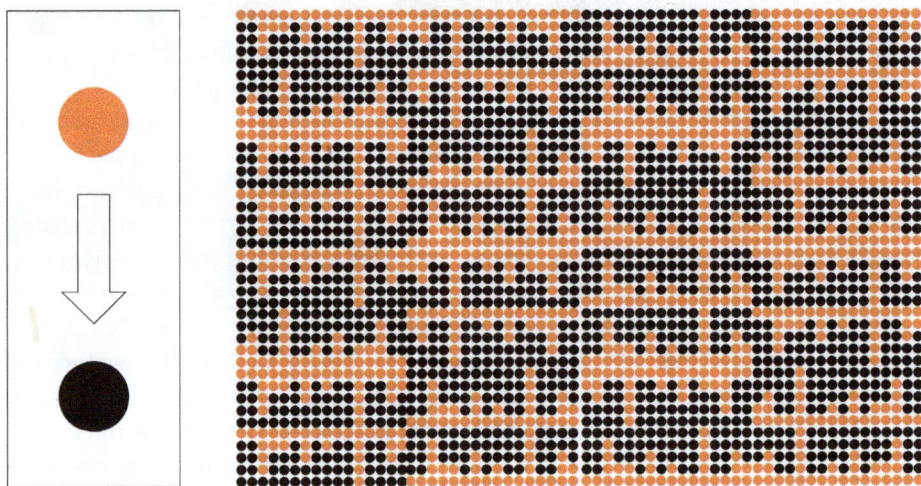

Fig. 5.9: Different approaches to half-lives. (left) One unstable (orange) nuclide transforms into one stable (black) nuclide. Since the transformation of unstable nuclides occurs randomly, it is hard to predict at what time it will occur. (right) The same is true for each unstable nuclide among a larger sample. However, the signal from many transformation events, which can be measured at any time-point, provides experimental values of the half-life (or transformation constant). This in turn describes the probability of these many nuclides to transform over any unit of time.

5.2.5 From extremely low masses of unstable nuclides to their chemistry

Conventional chemical reactions and their equilibrium constants deal with a large number of reagent atoms and/or molecules. The relationship between radioactivity and mass thus becomes very important and sometimes is not following conventional chemistry. The ultra-low mass and concentration of radioactive compounds, especially for artificial and short-lived radioelements, require sophisticated experimental approaches.

Important effects are the differences in chemical reaction pathways due to ultra-low concentration. Halogens, for example, do not exist as a single atom in their elemental form, but as diatomic molecules of type X_2. They *disproportionate* such that $2\,HX \rightleftarrows 2\,X^- + 2\,H^+$ or *synproportionate* according to $X^- + HOX + H^+ \rightleftarrows X_2 + H_2O$, for example. While this is true for all naturally occurring fluorine, chlorine, bromine and iodine, it does not necessarily apply to the heaviest homologue, astatine. This artificially produced radioelement with the longest-living isotope ^{211}At ($t\frac{1}{2} = 7.2$ h) exists exclusively at ultra-low concentration. An activity of 1 MBq for ^{211}At corresponds to $3.77 \cdot 10^{+8}$ atoms only ($<5 \cdot 10^{-17}$ mol). At such a low concentration, it is not able to form a $^{211}At_2$ molecule because the probability of two single ^{211}At atoms meeting by chance is almost zero; it cannot disproportionate nor synproportionate for the same reason.

Other features concern reaction kinetics, which are – unlike analogous conventional chemical reactions – typically pseudo first-order in nature.[11] These and many more special aspects are covered by the science of radiochemistry.

5.3 Units of radioactivity

5.3.1 Absolute radioactivity

The unit of radioactivity is 1/second. The SI unit of radioactivity (since 1975) is the Becquerel: 1 Bq = 1 s^{-1}. Although the use of SI units is recommended, there are some cases in natural sciences where "older" units are used in parallel. Since the discovery of radioactivity by HENRY BECQUEREL in 1896, several other units have been suggested. The most popular alternative (still commonly used today) is the "Curie", (Ci), named after the pioneering contributions of MARIE SKŁODOWSKA CURIE to radiochemistry and nuclear sciences.[12] Historically (1930), it was derived from the radioactivity represented by exactly 1 gram of the isotope ^{226}Ra of the chemical element radium (one of the two new elements discovered by M. CURIE). According to Fig. 5.2 the radioactivity of 1 g of ^{226}Ra ($t\frac{1}{2}$ = 1600 a) corresponds to $3.7 \cdot 10^{10}$ s^{-1}. Thus, 1 Ci = $3.7 \cdot 10^{10}$ s^{-1} (or 1 Ci = $37 \cdot 10^{9}$ Bq = 37 GBq). This unit is more convenient for quantifying large levels of radioactivity.

The SI unit of Bq expresses activities in rather large numbers. For example, 1 Ci = $37 \cdot 10^{9}$ Bq, or *vice versa*: 1 Bq = $2.7027 \cdot 10^{-10}$ Ci, 1 MBq = 27.027 µCi. Typical levels of radioactivity in radiochemical laboratories may thus be given in either multiples of kBq, MBq or GBq or in multiples of µCi or mCi. This relationship is illustrated in Fig. 5.10.

---

11 The velocity (i.e., the reaction rate R of a chemical reaction) typically correlates the concentrations of e.g. two reactants A and B with a reaction rate constant k. The units of the latter depend on the reaction order: zero, first, second. For a chemical process of e.g. aA + bB → cC with given concentrations of the species A, B and C and their stoichiometric coefficients a, b and c, respectively, the rate equation is R = k $[A]^a$ $[B]^b$. Over the course of the reaction, the concentration of A decreases, following –d[A]/dt = k$[A]^a[B]^b$. In the case where a = b = 1, it is –d[A]/dt = k[A][B]. Here, both concentrations must be considered, making the kinetics second-order. For a (pseudo) first-order reaction, the concentration of radionuclide reactant (A) is extremely low compared to non radioactive species (B). Due to its excess, the concentration of B does not change remarkably. The new rate equation is –d[A]/dt = k′[A] with R = k′[A] and k′ reflecting a parameter of k′ = k[B].

12 Both units thus honor the groundbreaking contributions of the pioneering scientists who were awarded the Nobel Prize in 1903. The Nobel Prize in Physics, 1903, was divided: one half awarded to BECQUEREL "in recognition of the extraordinary services he has rendered by his discovery of spontaneous radioactivity", the other half jointly to PIERRE CURIE and MARIE SKŁODOWSKA CURIE "in recognition of the extraordinary services they have rendered by their joint researches on the radiation phenomena discovered by Professor HENRI BECQUEREL".

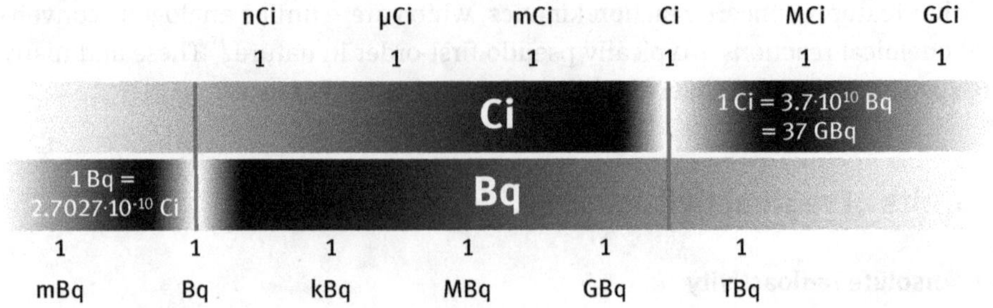

nCi	μCi	mCi	Ci	MCi	GCi
1	1	1	1	1	1

Ci

$1\,Ci = 3.7 \cdot 10^{10}\,Bq$
$= 37\,GBq$

$1\,Bq =$
$2.7027 \cdot 10^{-10}\,Ci$

Bq

1	1	1	1	1	1
mBq	Bq	kBq	MBq	GBq	TBq

Fig. 5.10: Units of radioactivity. The SI unit is the Becquerel: $1\,Bq = 1\,s^{-1}$. In parallel, the unit Curie is used: $1\,Ci = 3.7 \cdot 10^{10}\,s^{-1}$. Both units correlate by $1\,Bq = 2.7027 \cdot 10^{-10}\,Ci$. It is practical to treat it as $1\,mCi = 37\,MBq$, or *vice versa*: $1\,MBq = 27.027\,\mu Ci$. The Curie may be better suited to describe large levels of radioactivity; the Becquerel is more convenient for smaller levels.

5.3.2 Absolute *vs.* measured radioactivity

The radioactivity Á introduced so far is the "absolute" radioactivity. In fact, quantification of radioactivity usually requires the experimental measurement of radioactivity. There are many options to do so.[13] Typically, a transformation rate is measured by counting the specific radiation emitted per transformation. Laboratory units are count rates, Ć, registered for example as "counts per second" (cps). One count per second, however, is not equivalent to the absolute transformation rate of $Á = 1\,s^{-1}$.

Suppose there is a "source" of radioactivity (which is a sample of a radioactive species) and a radiation detector, as shown in Fig. 5.11. Suppose the radioactive sample is of point-like geometry (which in itself is experimentally difficult.) The count rate will obviously be affected by the experimental design, including (among other things):

- the distance between the source and the detector,
- the efficacy of the detector (its material, dimension etc., or: does one event induce one count?),
- the shielding of the detector (for example, the air between source and detector will already make a difference compared to vacuum).

The challenge is to correct for all potential losses in counting the initial events.[14] In the case of radiation measurements, several coefficients are considered, addressing losses of count rate caused by the aforementioned effects. All types of radioactive

---

13 "Radiation measurement" is the specific topic of Chapter 1 of Volume II.
14 This is comparable to analytical techniques applied in conventional chemistry that, for example, rely on the detection of radiation, such as UV detectors. To turn qualitative data into quantitative data, individual calibrations are needed.

radiation to be measured are interacting with matter – though to a strongly varying degree. Even ordinary air is a type of matter. Interaction results in weakening of the radiation intensity. The larger the distance, the more relevant the interaction with the matter between source and detector – and the lower the count rate. In general, the decrease of radiation intensity caused by interaction follows an exponential law again. In most cases, an increase in distance by a factor of 2 results in an attenuation of count rate by a potency of 2.

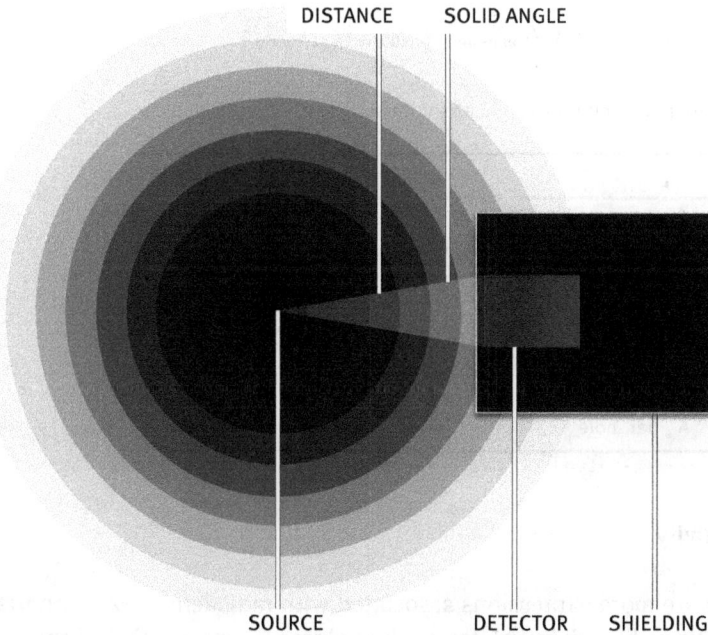

Fig. 5.11: Absolute radioactivity *vs.* experimental count rate. The count rate will be affected by experimental design, including (among other things) the distance between source and detector, the detector efficacy (its material, dimension etc.) and the detector shielding. Decreasing intensity of the black color indicates decreasing intensity of radiation emitted from the source with increasing distance. Note that the effect shown here in a two-dimensional situation is in reality of 4π geometry. This also applies to the solid angle, which relates to the dimension of the detector surface. The larger the surface, the larger the solid angle, the larger the fraction of radiation arriving at the detector.

Experimental count rates thus significantly depend on the set-up of a measurement; absolute activities do not. The same radioactivity source will result in different count rates for different detectors. Count rates are registered as counts per second (cps), but other expressions may be used as well (counts per minute) or others. Finally, the experimentally obtained count rate $\acute{C}$ needs to be converted to absolute activity, as shown in Tab. 5.4. This involves coefficients ε which account for:

–   the effect of the branching of various types of emissions per one event of nuclide transformation ($\varepsilon_{emission}$);
–   the efficacy with which the detector registers each radiation ($\varepsilon_{detector}$), and;
–   the effect of distance between the source and detector, as well as their dimensions ($\varepsilon_{geometry}$).

For the latter coefficient, for example, the closer the source and detector are, the higher the $\varepsilon_{geometry}$. Each coefficient ranges from 0 to 1, but is typically far below 1.

$$Á = Ć/\varepsilon_{emission}/\varepsilon_{detector}/\varepsilon_{geometry} \tag{5.11}$$

Tab. 5.4: Absolute activity *vs.* count rate, and absolute activity *vs.* specific activity.

Parameter			Units
Count rate	$Ć = Ć_o\,e^{-\lambda t}$ dependent on the experimental design (detector geometry, distance, . . .)		cps, cpm
Relationship	$Ć \approx Á$	$Á = Ć/\varepsilon_{emission} / \varepsilon_{detector}/\varepsilon_{geometry}$	$0 < \varepsilon \leq 1$
Specific activity	$Á_s$ per volume, $Á_s$ per mass, $Á_s$ per mole		Bq/mL Bq/g Bq/mol

5.3.3 Specific activity

In addition, there are more expressions associated with radioactivity. An important term is the absolute radioactivity related to the volume (or mass, etc.,) of the sample. This gives the "specific" activity $Á_s$. For example, the same activity $Á$ may be available in a small volume of 1 mL or diluted in a volume of 1 l. In the latter case, $Á_s$ is lower by a factor of 1000.

   More important is the specific activity related to mass. This is highlighted in the case where radioactivity is solely related to one radionuclide species. For example, 1 MBq of ^{14}C corresponds (according to $Á = (\ln 2 / t_{1/2})\,Ñ$ and with $t_{1/2} = 5730$ years) to a mass of $m = 6 \cdot 10^{-6}$ g, shown in Tab. 5.3. The specific activity $Á_s = Á/m$ of carbon-14 thus becomes 1.65 GBq/g. This is also called the "theoretical" or "maximum specific" activity. It is easy to calculate. However, theoretical specific activities are calculated exclusively for artificial and mono-isotopic radioelements, which do not exist on earth as stable isotopes (such as astatine and francium), as well as for all the transuranium elements.

At laboratory scale, it is rather impossible to exclude stable isotopes from the radioactive source.[15] In most cases, the real specific activities are less. Why is this? The precise equation to calculate specific activities needs the masses m_i of all relevant isotopes of the element, including the radioactive one, as shown in eq. (5.12). Ń is the number of unstable nuclides, where $Ń^{(STABLE1)} + Ń^{(STABLE2)} + \ldots$ stands for the number of stable isotopes (all of one and the same chemical element).

$$Á_s = \frac{Á}{\sum m_i} \qquad (5.12)$$

Consider the following example. The activity of 1 MBq of artificially produced ^{11}C corresponds to $3.2 \cdot 10^{-14}$ g, as shown in Tab. 5.3. Suppose this fraction is in contact with air, plastic tubes, glass vessels, chemical compounds etc., which usually are composed of carbon to a significant part, then ^{11}C and $^{12}C+^{13}C$ will mix. Figure 5.12 takes a (low[16]) mass of 1 μg of natural carbon, which co-exists with ^{11}C. What about the specific activity now? The overall mass of carbon atoms would increase significantly from $3.2 \cdot 10^{-14}$ g to $3.2 \cdot 10^{-14} + 1 \cdot 10^{-6}$ g $= 1.000000032 \cdot 10^{-6}$ g. The specific activity would decrease by seven orders of magnitude.

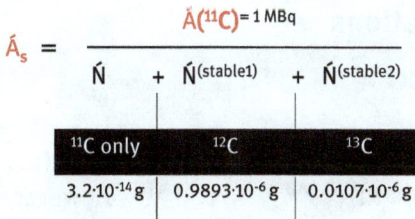

Fig. 5.12: "Real" specific activity of an ^{11}C sample containing altogether 1 μg of stable carbon. The corresponding masses of ^{12}C and ^{13}C (with the 1.0 μg attributed to the two stable carbon isotopes according to their percentages of 98.93:0.107) yield a real specific activity of $1 \cdot 10^{+6}$ Bq / $1.000000032 \cdot 10^{-6}$ g $\approx 1 \cdot 10^{+12}$ Bq/g. The theoretical specific activity of ^{11}C according to eq. (5.5) is $^{THEORY}Á_s = 3.15 \cdot 10^{19}$ Bq/g (or $8.4 \cdot 10^8$ Ci/g).

Table 5.5 summarizes special categories used in terms of specific activity related to mass. The categories are "carrier free", "no carrier added", and "carrier added". The wording "carrier" refers to any amount of stable isotope(s) of a chemical element of a

---

15 What happens is a kind of "dilution" of the radioactive isotope by more-or-less significant amounts of stable isotopes. This may dramatically lower the value of $Á_s$. These effects are important in the context of radiometric dating *via* ^{14}C or by molecular imaging utilizing ^{11}C (see Chapters 5 and 13 of Volume II for details).

16 Most conventional chemists perform chemical reactions with carbon compounds at typical mass scales of e.g. gram, and would not at all care about the 1 μg which may exist uncontrolled in solution.

specific mass, which is much higher than the mass of the unstable isotope ($\hat{N} \ll \hat{N}^{(STABLE1)} + \hat{N}^{(STABLE2)} + \dots$), in particular for relatively short-lived unstable nuclides.

Tab. 5.5: Absolute activity *vs.* specific activity.

Category	Typically applies to		Examples
Carrier free	Radioelements	Stable isotopes of that chemical element that do not exist	Astatine, francium, heavier lanthanides, transuranium elements
No carrier added (nca)	Artificially produced radionuclides	Stable isotopes of that chemical element that exist and whose presence in the fraction of the radioactive isotope is unavoidable	^{11}C as produced for radiopharmacy and molecular imaging is accompanied by omnipresent ^{12}C
Carrier added	Any radionuclide	Defined addition of weighable amounts of stable isotopes	To influence or direct chemical reactions of the radionuclide

5.4 Classes of radioactive transformations

5.4.1 Types of successive transformations

Until now the transformation of an unstable nuclide (*K1) into a stable nuclide (K2) was discussed within one transformation step: $^{*}K1^{UNSTABLE} \rightarrow K2^{STABLE}$. However, the process of stabilizing an unstable nucleon configuration may proceed in several steps, and the first transformation may result in a "more stable", but not "fully stable", nuclide. Thus, this first transformation yields a second unstable nuclide *K2, which undertakes a further transformation step to obtain stable nucleon composition: $^{*}K1^{UNSTABLE} \rightarrow {}^{*}K2^{UNSTABLE} \rightarrow K3^{STABLE}$. This constellation is particularly relevant since the first radioactive nuclide "generates" a second radioactive one. This intermediate is both being formed and transformed simultaneously. The constellation is called "equilibrium" and in cases where the intermediate nuclide *K2 is of practical interest, it is referred to as "radionuclide generator".

Most of the unstable nuclides, however, are so far from having stable nucleon compositions that several transformation steps are needed until a final, stable configuration is reached. This is called a "transformation chain" (or "decay chain"). Some of the intermediate nuclides may undergo different transformations in parallel. In this case, an initial chain branches into two or more parallel chains.

The three types mentioned are illustrated in Fig. 5.13.

SINGLE TRANSFORMATION

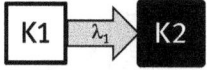

EQUILIBRIUM

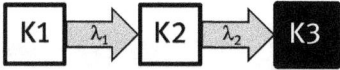

CHAIN

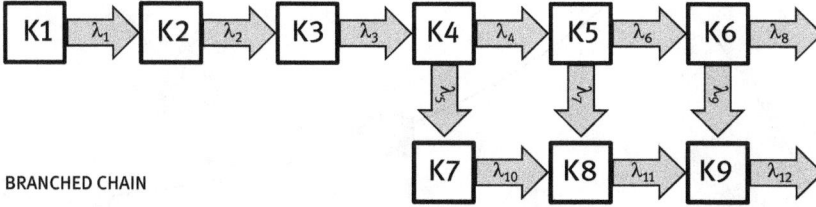

BRANCHED CHAIN

Fig. 5.13: Classes of radioactive transformation, which require separate mathematical treatment. White and black boxes denote unstable and stable nuclides, respectively. The simplest case is to transform one unstable nuclide into a stable one: *K1 → K2. There is one transformation constant only: λ_1. If the initial process yields one unstable intermediate *K2, a second transformation step is required to obtain stable nucleon composition: *K1 → *K2 → K3. This involves two different transformation constants, λ_1 and λ_2. For a "chain", there may be a number of successive transformation steps with their individual transformation constants. Some of the intermediate nuclides may undergo different transformations in parallel, for example at stage *K4. Here, two different transformation types proceed simultaneously, and two different unstable nuclides *K5 and *K7 are formed with specific λ_4 and λ_{10} constants, respectively. Both continue to transform. In this case, an initial chain which started from *K1 continues along two (or more) parallel "branched chains".

5.4.2 Single transformation

"Single radioactive transformation" was introduced by the exponential law of type (5.1) or (5.3) for the number of nuclides or their radioactivity, respectively. Figure 5.14 depicts the corresponding change in the number of stable "product" nuclides K2 as well. With every nuclide of *K1 that transforms, a stable nuclide of K2 appears. The overall number of nuclides, i.e. *K1 + K2, remains constant. (Again, there is no real "decay", but a "transformation"). While the number of *K1 changes by $- d{*}Ń_1/dt$ and $Ń = Ń_o\,e^{-\lambda t}$, it is opposite for K2:

$$-\frac{d{*}Ń_1}{dt} = +\frac{dŃ_2}{dt} \tag{5.13}$$

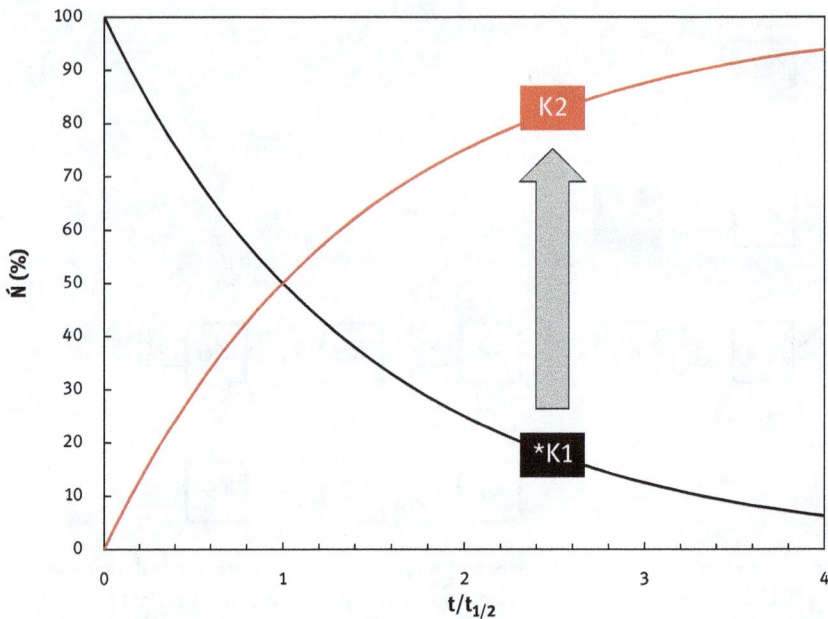

Fig. 5.14: Kinetics of radioactive transformation. ($Ń = Ń_o\, e^{-\lambda_1 t}$). An unstable nuclide *K1 "turns" into a stable nuclide K2. After one half-life of the nuclide *K1, 50% of unstable *K1 had transformed into stable K2.

5.4.3 Transformation equilibrium

A process of stabilizing an unstable nucleon configuration may not directly yield a stable nuclide. Instead, the newly formed nuclide may be more stable than the initial one, but not yet stable enough. A first transformation step thus "generates" a second unstable nuclide *K2, which undergoes a next and final transformation step to obtain stable nucleon composition: *K1 → *K2 → K3. The intermediate *K2 is both being formed and is transforming simultaneously. The line shown in Fig. 5.14 thus would not be correct for the number of nuclei of *K2 existing at a given time-point.

So, what is the radioactivity of *K2 at a specific time-point? It depends on two different transformation constants, λ_1 and λ_2. While λ_1 is responsible for generating *K2 and results in an increasing number and radioactivity of *K2, λ_2 determines the decrease of these values due to the transformation velocity of *K2 into stable K3. It is the ratio between λ_1 and λ_2 that will determine the radioactivity of *K2 at any time-point, as shown in Tab. 5.6. The mathematics classifies three situations according to three possible ratios of the two transformation constants involved: $\lambda_1 \ll \lambda_2$, $\lambda_1 < \lambda_2$, and $\lambda_1 > \lambda_2$. Figure 5.15 illustrates the three versions according to the ratio of the two transformation constants (or half-lives).

5.4.3.1 Equilibrium

The wording "equilibrium" refers to the situation that the two radioactivities of *K1 and *K2 reach a certain ratio that remains constant over time: the radioactivity of *K2 is in "equilibrium" with the radioactivity of *K1. "Equilibrium" will occur if the half-life of *K1 is larger than that of *K2. More precisely, two subtypes may be discussed: a secular ($\lambda_1 \ll \lambda_2$) and a transient ($\lambda_1 < \lambda_2$) equilibrium. (The same is true when considering half-lives, and the cases will be $t\frac{1}{2}(1) \gg t\frac{1}{2}(2)$ or $t\frac{1}{2}(1) > t\frac{1}{2}(2)$.) Mathematically, there are relations of type $\lambda_1 = x\,\lambda_2$ with $x > 1$. In order to decide between the two categories $\lambda_1 \ll \lambda_2$ or $\lambda_1 < \lambda_2$, one may fix the ratio according to the factor x as either $x > 100$ or $1 < x < 100$, as shown in Table 5.6.[17]

5.4.3.2 Non equilibrium

Supposing that the half-life of *K1 is shorter than that of *K2, there will be no equilibrium at all. The radioactivity of *K2 will not change proportionally to that of *K1.

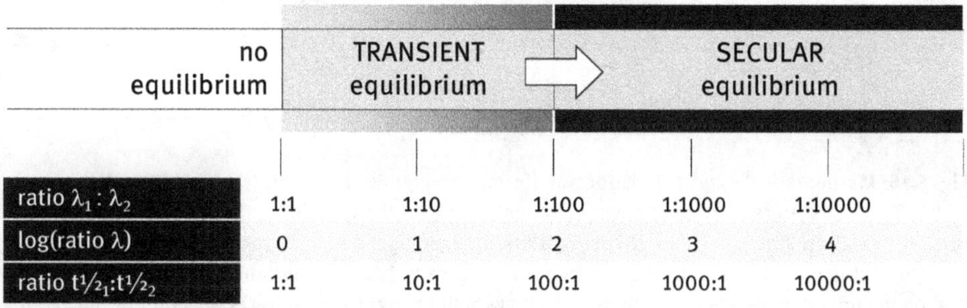

	no equilibrium	TRANSIENT equilibrium		SECULAR equilibrium	
ratio $\lambda_1 : \lambda_2$	1:1	1:10	1:100	1:1000	1:10000
log(ratio λ)	0	1	2	3	4
ratio $t\frac{1}{2}_1 : t\frac{1}{2}_2$	1:1	10:1	100:1	1000:1	10000:1

Fig. 5.15: Categories of transformation of type *K1 (*via* λ_1) → *K2 (*via* λ_2) → K3, related to the ratio between transformation constants. The two equilibria are categorized according to λ_1 being smaller than λ_2 for ratios of $\lambda_1:\lambda_2$ ranging from 1:1 to 1:100 (transient) and $\lambda_1:\lambda_2$ being larger than 1:100 (secular).

Tab. 5.6: Versions of radioactive equilibrium of type *K1 → *K2 (*via* λ_1) and *K2 → K3 (*via* λ_2).

Ratio of $\lambda_1:\lambda_2$	Ratio of $t\frac{1}{2}(*K1):t\frac{1}{2}(*K2)$	Type of equilibrium
≪	≫	>100	Secular equilibrium
<	<	1 – 100	Transient equilibrium
>	<	<1	No equilibrium

---

17 The case of $\lambda_1 < \lambda_2$ would mean that the value of $t\frac{1}{2}(1)$ is "a bit" greater than the one for $t\frac{1}{2}(2)$. This appears to be somehow arbitrary, as illustrated in Fig. 5.14, because there is actually no definitive borderline between the ratios of the two constants.

5.4.3.3 General mathematics

Figure 5.16 illustrates the development of mathematical expressions for *K1 (via λ_1) $\rightarrow$ *K2 (via λ_2) $\rightarrow$ K3. The first step reflects the known exponential law for the transformation of an unstable nuclide *K1. The number of nuclides decreases ($-d*\hat{N}_1/dt$) according to $\lambda_1\hat{N}_{1(o)}$. Nuclide by nuclide, *K1 transforms and generates *K2 via λ_1. Analogously to eq. (5.13), the decrease in numbers of $\hat{N}_1$ is the increase in numbers of $\hat{N}_2$, i.e. $-d*\hat{N}_1/dt = + d*\hat{N}_2/dt$.

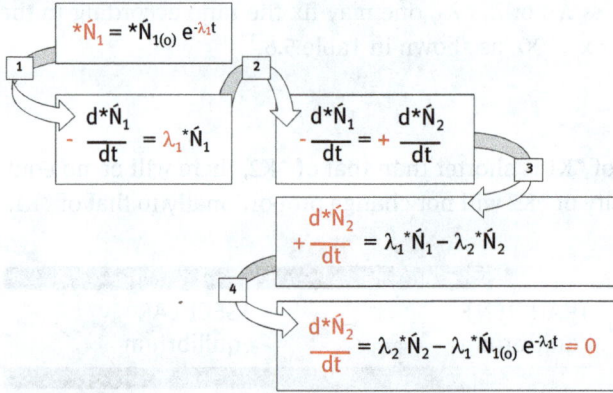

Fig. 5.16: Mathematical expressions for transformation equilibria of type *K1 (via λ_1) $\rightarrow$ *K2 (via λ_2) $\rightarrow$ K3. The first step reflects the known exponential law for the transformation of an unstable nuclide *K1. Step 2 thus says $-d*\hat{N}_1/dt = + d*\hat{N}_2/dt$. Each nuclide of *K2, once formed, transforms into the stable nuclide K3. Step 3 describes the kinetics of *K2 as simultaneously being formed and decomposed. In phase 4, the number of $\hat{N}_2$ (which changes over time) is in "equilibrium" with the two processes of formation and transformation. This depends on the individual transformation constants λ_1 and λ_2 and more specifically, on the ratio between λ_1 and λ_2, as discussed below.

Each nuclide of *K2, once formed, transforms into a stable nuclide of K3 via λ_2. The overall kinetics then follow eq. (5.14) in a differential form of $d*\hat{N}_2/dt = + \lambda_1*\hat{N}_1 - \lambda_2*\hat{N}_2$, i.e. as formation and transformation (decay). Equation (5.15) expresses the same concept, now with a specific number of unstable nuclides *K1 known at a given time-point t = 0. With the initial number of *$\hat{N}_{1(o)}$ known at any time-point t = 0, the change in *$\hat{N}_2$ over time (which is radioactivity) is obtained.

$$\frac{d*\hat{N}_2}{dt} = +\lambda_1*\hat{N}_1 - \lambda_2*\hat{N}_2 \tag{5.14}$$

$$\frac{d*\hat{N}_2}{dt} = +\lambda_1*\hat{N}_{1(0)}e^{-\lambda_1 t} - \lambda_2*\hat{N}_2 \tag{5.15}$$

Now, what is "equilibrium"? Suppose velocities of formation and transformation of *K2 are identical, as per eq. (5.16). An equilibrium thus defines a situation where the number of intermediate nuclides *$\hat{N}_2$ (i.e. the radioactivity of the daughter) changes in a well-defined ratio between the transformation constants (or half-lives) of the two radioactive nuclides *K1 and *K2. Altogether, this yields eq. (5.17):

$$+ \lambda_1 {}^*\hat{N}_1 = - \lambda_2 {}^*\hat{N}_2 \tag{5.16}$$

$$0 = \frac{d{}^*\hat{N}_2}{dt} + \lambda_2 {}^*\hat{N}_2 - \lambda_1 {}^*\hat{N}_{1(0)} e^{-\lambda_1 t} \tag{5.17}$$

Since eq. (5.17) is a differential one, integration results in eq. (5.18). The main message is: with the initial number of ${}^*\hat{N}_1$ known at time-point $t = 0$, the number of ${}^*\hat{N}_2$ is mathematically derived according to the time-point t considered and the values of the two transformation constants λ_1 and λ_2. While eq. (5.18) refers to a situation in which there were no nuclides of *K2 present at $t = 0$, eq. (5.19) considers this case, that a certain number ${}^*\hat{N}_2$ of nuclides of *K2 was present, as per Fig. 5.17. The total number of ${}^*\hat{N}_2$ existing is thus composed of those that existed earlier (i.e. at $t = 0$), and the new ones generated by the transformation of *K1.

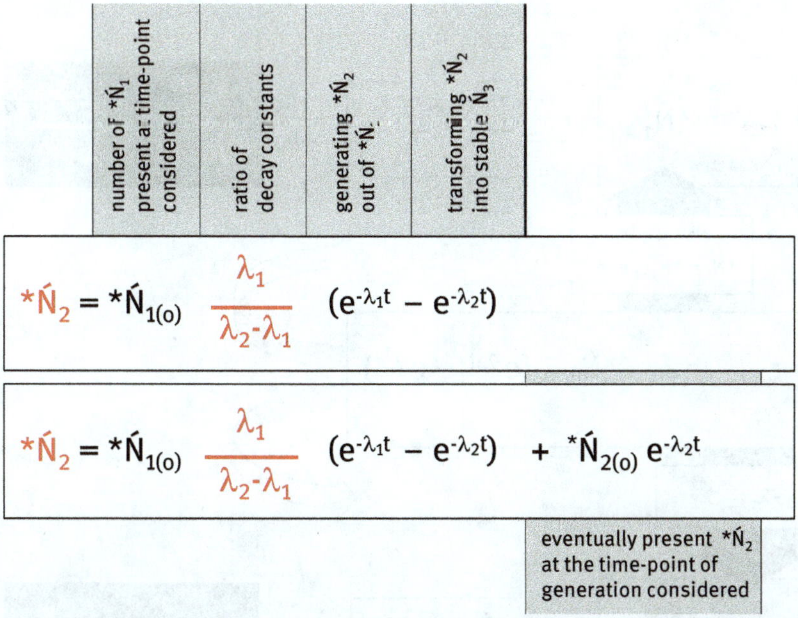

number of ${}^*\hat{N}_1$ present at time-point considered | ratio of decay constants | generating ${}^*\hat{N}_2$ out of ${}^*\hat{N}_1$ | transforming ${}^*\hat{N}_2$ into stable $\hat{N}_3$

$${}^*\hat{N}_2 = {}^*\hat{N}_{1(0)} \quad \frac{\lambda_1}{\lambda_2 - \lambda_1} \quad (e^{-\lambda_1 t} - e^{-\lambda_2 t})$$

$${}^*\hat{N}_2 = {}^*\hat{N}_{1(0)} \quad \frac{\lambda_1}{\lambda_2 - \lambda_1} \quad (e^{-\lambda_1 t} - e^{-\lambda_2 t}) \quad + \quad {}^*\hat{N}_{2(0)} e^{-\lambda_2 t}$$

eventually present ${}^*\hat{N}_2$ at the time-point of generation considered

Fig. 5.17: Integral version of eq. (5.17). With the initial number of ${}^*\hat{N}_1$ known at time-point $t = 0$, and with the number of ${}^*\hat{N}_2$ at this moment being zero, the number of ${}^*\hat{N}_2$ is mathematically derived according to the time-point t considered and the values of the two transformation constants λ_1 and λ_2. There are four terms which influence the number and the kinetics of ${}^*\hat{N}_2$. First, of course, it is the number of ${}^*\hat{N}_1$ existing at time-point $t = 0$. Next, is the ratio of the two transformation constants, namely $\lambda_1/(\lambda_2 - \lambda_1)$ and their individual contributions according to $(e^{-\lambda_1 t} - e^{-\lambda_2 t})$. Finally, eq. (5.19) considers the case where the total number of ${}^*\hat{N}_2$ existing are composed of new ones generated by the transformation of ${}^*\hat{N}_1$ plus the ones that existed earlier (i.e. at $t = 0$) and which transform separately via ${}^*\hat{N}_{2(0)} e^{-\lambda_2 t}$.

$$*\hat{N}_2 = \frac{\lambda_1}{\lambda_2 - \lambda_1} *\hat{N}_{1(0)} \left(e^{-\lambda_1 t} - e^{-\lambda_2 t}\right) \tag{5.18}$$

$$*\hat{N}_2 = \frac{\lambda_1}{\lambda_2 - \lambda_1} *\hat{N}_{1(0)} \left(e^{-\lambda_1 t} - e^{-\lambda_2 t}\right) + *\hat{N}_{2(0)} e^{-\lambda_2 t} \tag{5.19}$$

There are several tasks related to the mathematical description of this type of transformation and in particular to the activity of *K2 available at various moments. Suppose the initial values for $\hat{A}_1$ and $*\hat{N}_1$ are known:

1. At any time: What is the radioactivity of *K2 given the initial activity of *K1?
2. At equilibrium: How does the radioactivity *K2 relate to the radioactivity of *K1?
3. At what time is the maximum activity of *K2 available in the system?

The exact calculation is based on eq. (5.18). However, in the case of either $\lambda_1 < \lambda_2$ or $\lambda_1 \ll \lambda_2$, two simplifications will modify eq. (5.18). The concept is illustrated in Fig. 5.18.

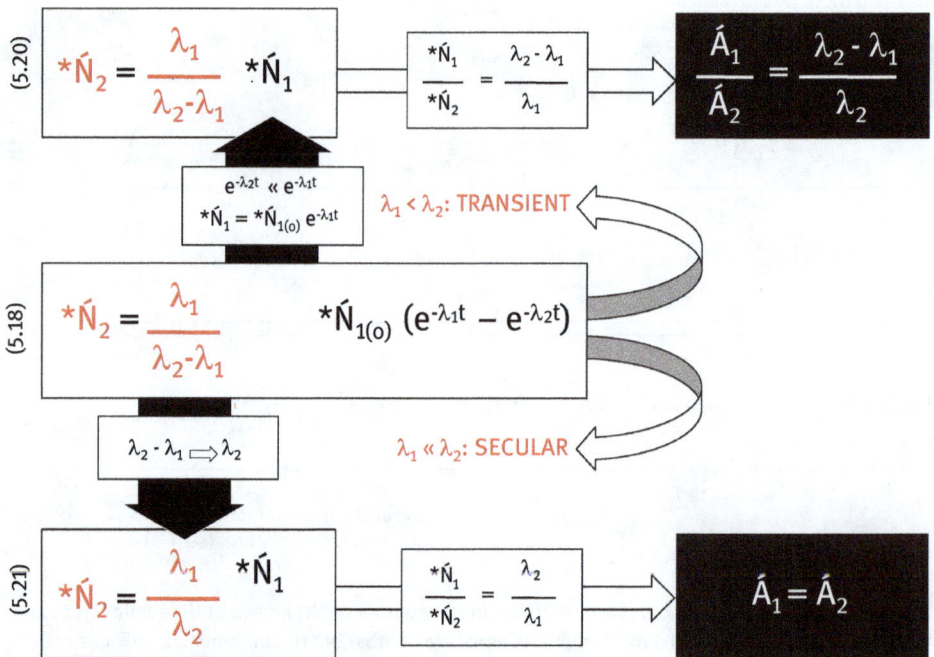

Fig. 5.18: Two simplifications of the general equation of transformation equilibria of type *K1 (via λ_1) → *K2 (via λ_2) → K3. When $\lambda_1 < \lambda_2$, it yields eq. (5.20) due to the simplifications of $(e^{-\lambda_2 t} \ll e^{-\lambda_1 t}) \approx e^{-\lambda_1 t}$ and $*\hat{N}_{1(0)} e^{-\lambda_1 t} = *\hat{N}_1$. When $\lambda_1 \ll \lambda_2$, this continues to simplify due to $\lambda_1/(\lambda_2 - \lambda_1) \approx \lambda_1/\lambda_2$ and yields eq. (5.21).

5.4.3.4 Transient equilibrium

At $\lambda_1 < \lambda_2$, the difference in exponential form becomes significant: $e^{-\lambda_2 t} \ll e^{-\lambda_1 t}$ and $e^{-\lambda_2 t}$ may be neglected. For the transient system, eq. (5.18) turns into eq. (5.20). For the two nuclides *K1 and *K2 at equilibrium, the relationship between the two unstable nuclides and their activities yields eqs. (5.20a) and (5.20b), respectively. The maximum radioactivity of *K2 is observed at a time-point t^{max}, eq. (5.20c).

$$*\acute{N}_2 = *\acute{N}_1 \frac{\lambda_1}{\lambda_2 - \lambda_1} \tag{5.20a}$$

$$*\acute{A}_2 = *\acute{A}_1 \frac{\lambda_2}{\lambda_2 - \lambda_1} \tag{5.20b}$$

$$t_{*K2max} = \frac{1}{\lambda_2 - \lambda_1} \ln \frac{\lambda_2}{\lambda_1} \tag{5.20c}$$

Figure 5.19 illustrates the most relevant features of a transient transformation equilibrium for a situation in which an initial number $*\acute{N}_{1(o)}$ of unstable nuclides *K1 is present without any nuclides of *K2. Let the ratio between the transformation constants be 10, i.e. $\lambda_2 = 10 \cdot \lambda_1$, indicating the half-life of *K1 is longer than *K2 by a factor of 10. The difference in the exponential form becomes significant: $e^{-\lambda_2 t} \ll e^{-\lambda_1 t}$ and $e^{-\lambda_2 t}$ may be neglected. $*\acute{N}_{1(o)} e^{-\lambda_1 t}$ remains and is written as $*\acute{N}_1$.

From $t = 0$ on, *K1 starts to transform according to λ_1 and generates *K2. Within the 10 periods of the half-life of *K2, which is exactly 1 half-life of *K1, the value of $\acute{A}_1$ drops to 50%.

$\acute{A}_2$ is defined by both continuous formation *K1 → *K2 and subsequent transformation *K2 → K3. Each *K2 generated adds a given activity $\acute{A}_2$ on top of $\acute{A}_1$. Both values of activity combine to a cumulative $\acute{A}_1 + \acute{A}_2$, which is larger than either $\acute{A}_1$ or $\acute{A}_2$ alone. After several periods of half-life, the cumulative activity parallels $\acute{A}_1$ – finally meaning what? *K2 is "in equilibrium" with *K1.

Equation (5.20c) allows for the time-points of two maxima to be determined; one for the cumulative activity, and the other (more importantly) for the maximum activity $\acute{A}_2^{max}$. This parameter can be very important in the practical use of *K2, since it may define a time-point when *K2 is radiochemically separated from the mixture of *K1 + *K2. Once isolated, it will transform individually according to its own half-life.

5.4.3.5 Secular equilibrium

At $\lambda_1 \ll \lambda_2$, the difference in $e^{-\lambda_2 t} \ll e^{-\lambda_1 t}$ is even more significant and $e^{-\lambda_2 t}$ may be neglected again. Again, $*\acute{N}_1(o) e^{-\lambda_1 t}$ is written as $*\acute{N}_1$. In addition, at $\lambda_1 \ll \lambda_2$ the

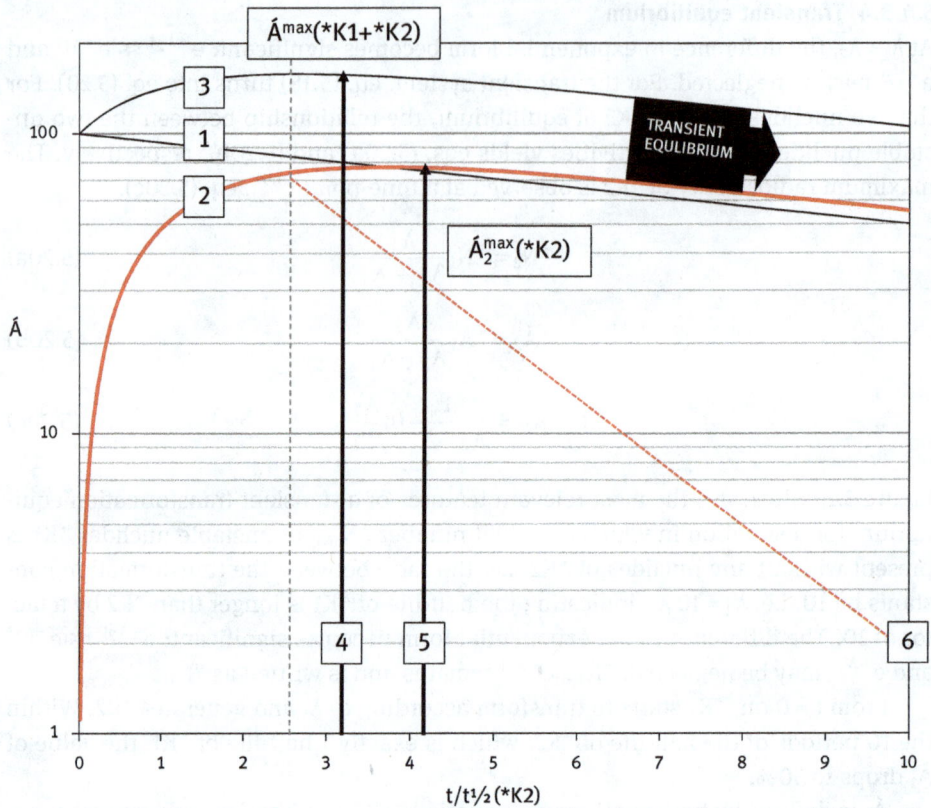

Fig. 5.19: Kinetics of transient equilibrium showing activities of *K1 and *K2 depending on the period of half-lives of *K2. The ratio between the transformation constants is fixed to $x = 10$, i.e. $\lambda_2 = 10 \cdot \lambda_1$. (1) Transformation of *K1 according to λ_1 expressed at $Á_1$, (2) increasing activity $Á_2$ starting at $Á_2 = 0$ at $t = 0$. This corresponds to eq. (5.12) and would be different if a given fraction of *K2 ws already present at $t = 0$, as expressed in eq. (5.19) by $+ \; {}^*\acute{N}_2(o)e^{-\lambda_2 t}$. (3) Both values of activity combine as $Á_1 + Á_2$, and this cumulative activity is larger than either $Á_1$ or $Á_2$ alone. (4) The time-point of maximum value for cumulative activity; in this case, 2.637 $t/t_{1/2}$. (5) Maximum activity of $Á_2$ achievable in the equilibrium according to $t_{*K2max} = [1 \, / \, (\lambda_2 - \lambda_1)] \, \ln(\lambda_2/\lambda_1)$, eq. (5.20c); 3.697 $t/t_{1/2}$ in this case. (6) Once *K2 is radiochemically separated from the mixture of *K1 + *K2, it will transform according to its own half-life.

term $\lambda_1 \, / \, (\lambda_2 - \lambda_1)$ is approaching λ_1/λ_2 because λ_1 is very small compared to λ_2.[18] For the two nuclides *K1 and *K2 at equilibrium, the relationships simplifies according to eqs. (5.21a) and (5.21b):

---

18 For example, a ratio of $\lambda_1{:}\lambda_2$ of 1:1000 would be either $1 \, / \, (1000 - 1)$ in the original version of eq. (5.20) or $1 \, / \, 1000$ in the version of eq. (5.21). The two values of 0.001001001 and 0.001000000 differ by about 0.1% only. This is not the case for e.g. a ratio of $\lambda_1{:}\lambda_2$ of 1:10. This is the reason to classify transient and secular systems by ratios of $\lambda_1{:}\lambda_2$ of 1 to 100 and >100, respectively.

$$*\hat{N}_2 = *\hat{N}_1 \frac{\lambda_1}{\lambda_2} \qquad (5.21a)$$

$$\acute{A}_2 = \acute{A}_1 \qquad (5.21b)$$

Figure 5.20 illustrates the kinetics of secular transformation equilibrium. Let the ratio between the transformation constants be x = 1000, i.e. $\lambda_2 = 1000\lambda_1$, indicating the half-life of *K1 is longer than *K2 by a factor of 1000. Again, an initial number $*\hat{N}_1(o)$ of unstable nuclides *K1 is present without any nuclides of *K2, and starts to transform according to λ_1 generating *K2. This time, the activity $\acute{A}_1$ does not change significantly over the time period considered. After ten periods of $t\frac{1}{2}(*K2)$, just 0.1% of one half-life of $t\frac{1}{2}(*K1)$ have passed, and the fraction of $*\hat{N}_1$ still present (according to $1/2^n$) is 99.309%.

Each *K2 generated adds a given activity $\acute{A}_2$ on top of $\acute{A}_1$. After several half-life periods, the cumulative activity again parallels $\acute{A}_1$. At equilibrium both activities are the same: $\acute{A}_2 = \acute{A}_1$. Both values of activity combine to a cumulative $\acute{A}_1 + \acute{A}_2$, which is exactly double the individual values of $\acute{A}_1$ or $\acute{A}_2$. Again, *K2 is "in equilibrium" with *K1, but this time the individual activity $\acute{A}_2$ also remains (practically) constant – at least for the 10 half-lives of *K2 shown in Fig. 5.20. Unlike the transient system, no true maxima are reached, neither for the cumulative activity, nor for the maximum activity $\acute{A}_2$ (the latter is a saturation effect). Starting from t = 0, the fraction of *K2 generated will exponentially increase. After one half-life, this is 50%, after two half-lives it is 75%, after three half-lives 87.5%. The activity $\acute{A}_2$ generated approaches a maximum (asymptotic) level, and then will remain at this value.

For practical applications, *K2 may radiochemically be separated from the mixture of *K1 + *K2 at a time-point when a sufficient value of $\acute{A}_2$ is generated. This could be a reasonable time-point to separate *K2. Once isolated, it will transform individually according to its own half-life.

5.4.3.6 No equilibrium

Let's assume the ratio between the transformation constants is $\lambda_2 = 0.1 \cdot \lambda_1$, i.e. the half-life of *K1 is *shorter* than *K2 by a factor of 10. Figure 5.21 illustrates the transformation kinetics in this case. Again, *K1 starts to transform according to λ_1 and generates *K2. Both values of activity combine to a cumulative $\acute{A}_1 + \acute{A}_2$, which is larger than either $\acute{A}_1$ or $\acute{A}_2$ alone. However, the cumulative activity never parallels $\acute{A}_1$ – this means there is "no equilibrium".

5.4.3.7 Radionuclide generators

The idea of an equilibrium between two radioactive nuclides – caused by a longer-lived "parent" and shorter-lived "daughter" – may be quite convenient for some practical applications. It gives access to radioactive nuclide *K2 by sequentially separating it radiochemically from its source of *K1. This is usually referred to as a "radionuclide

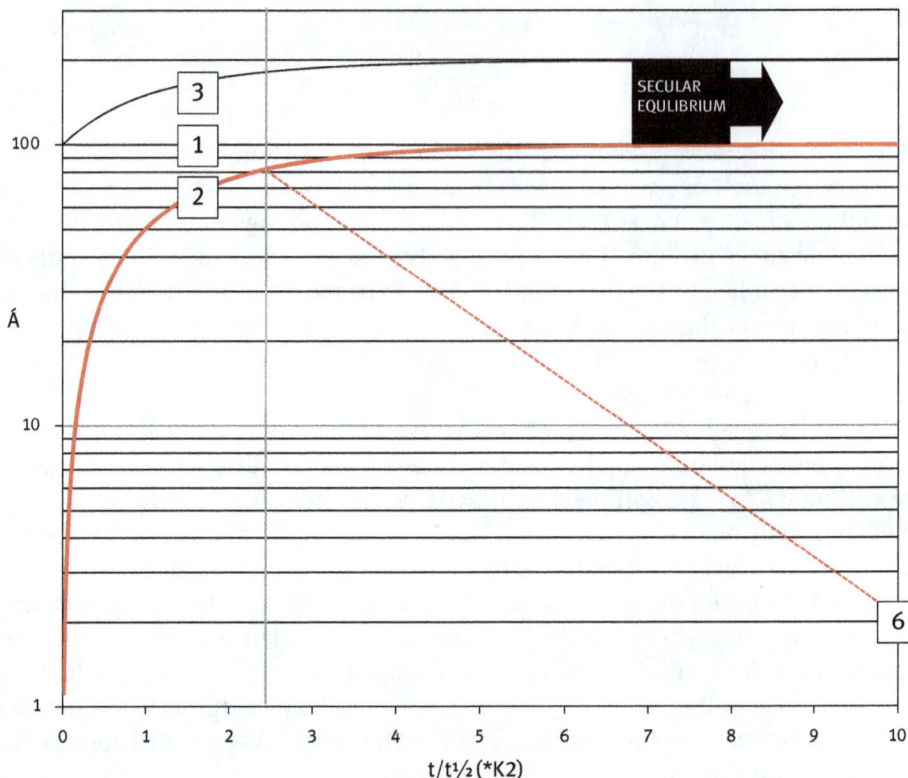

Fig. 5.20: Kinetics of secular equilibrium showing activities of *K1 and *K2 depending on the period of half-lives of *K2. The ratio between the transformation constants is fixed to x = 1000, i.e. $\lambda_2 = 1000 \cdot \lambda_1$. (1) Transformation of *K1 according to λ_1 expressed at $\acute{A}_1$, staying practically constant for the 10 periods of half-life of *K2 considered. (2) Activity $\acute{A}_2$ generated starting at $\acute{A}_2 = 0$ at t = 0. This would be different supposing that at t = 0 a given fraction of *K2 would already have been present in the system, as expressed in eq. (5.19) by $+ *\acute{N}_2(o)e^{-\lambda_2 t}$. At equilibrium, $\acute{A}_2$ approaches $\acute{A}_1$. (3) Cumulative activities at equilibrium are $\acute{A} = \acute{A}_1 + \acute{A}_2$, and this cumulative activity is larger by a factor 2 than either $\acute{A}_1$ or $\acute{A}_2$. (6) Once *K2 is radiochemically separated from the mixture of *K1 + *K2, it will transform individually according to its own half-life. Starting from t = 0, the fraction of *K2 generated relative to its maximum value will exponentially increase.

generator".[19] The longer-lived "precursor" may be installed at any laboratory and provide the shorter-lived offspring "on demand". Thus, it is radiochemical expertise that

---

19 However, the number of generator systems available is rather limited. This is mainly due to the fact that nuclear transitions of type $*K1 - \lambda_1 \rightarrow *K2 - \lambda_2 \rightarrow K3$ with $\lambda_1 < \lambda_2$ are rather "atypical". As differences in nuclear mass, mass excess and mean nuclear binding energy become smaller for transitions towards the finally stable nucleus, the half-lives of the successive nuclei increase and the transformation constant decreases. Thus, the "typical" situation for $*K1 - \lambda_1 \rightarrow *K2 - \lambda_2 \rightarrow K3$ is $\lambda_1 < \lambda_2$ and $t\frac{1}{2}(*K1) > t\frac{1}{2}(*K2)$. See the detailed discussions of β- and α-transformation processes.

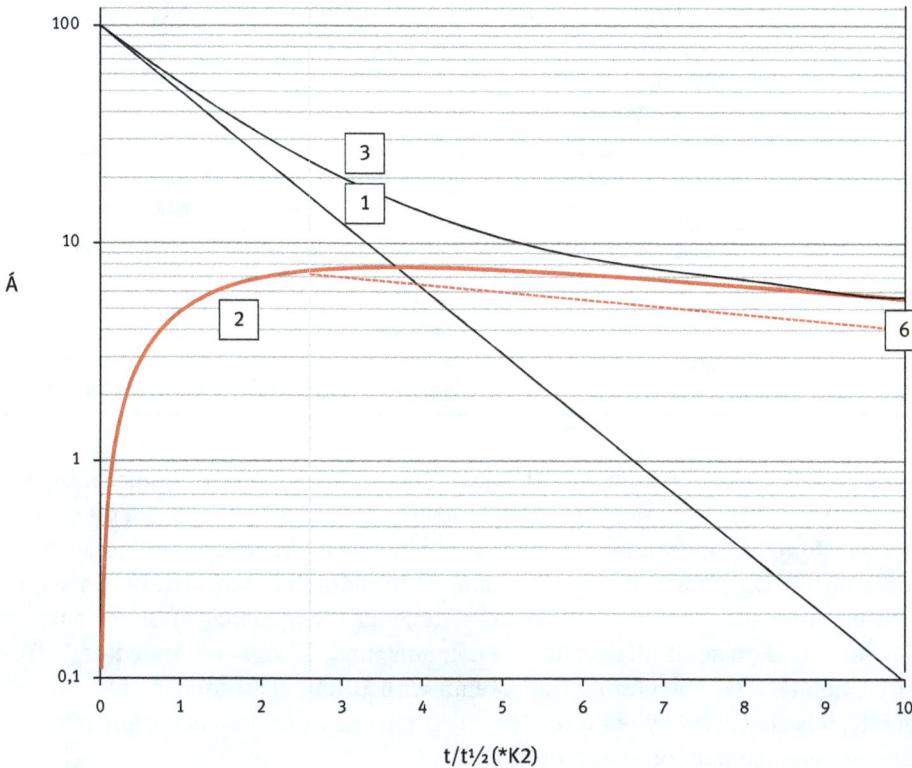

Á

t/t½ (*K2)

Fig. 5.21: No equilibrium is reached for transformation of type *K1 (*via* λ_1) → *K2 (*via* λ_2) → K3. Let the ratio between the transformation constants be x = 0.1, i.e. $\lambda_2 = 0.1 \cdot \lambda_1$, iindicating the half-life of *K1 is shorter than *K2 by a factor of 10. (1) Activity $Á_1$ changes significantly over the time period considered. After ten periods of t½(*K2), 100 half-life periods of *K1 have passed. The fraction of *$Ń_1$ still present is $1/2^n = 1/2^{100} = 7.9 \cdot 10^{-31}$. This is negligible. (2) $Á_2$ is defined by both continuous formation *K1 → *K2 and subsequent transformation *K2 → K3. (3) Cumulative activity $Á_1 + Á_2$ has no equilibrium. (6) Activity $Á_1$, once *K2 has been radiochemically isolated from the mixture of *K1 + *K2.

turns a formal transformation system of type *K1 (*via* λ_1) → *K2 (*via* λ_2) → K3 into a "generator" of practical relevance. The two different unstable nuclides *K1 and *K2 most often represent isotopes of different chemical elements. When designing a system to exploit the significant differences in chemical behavior between the parent and daughter nuclides (elements), the goals are threefold:

1. Achieve high separation efficacy when isolating the chemical fraction of *K2 from the chemical fraction of *K1. This is called "separation" yield.
2. Achieve low turnover of parts of *K1 still present in the separated fraction of *K2. This is called "breakthrough".
3. Guarantee long shelf life of the generator system, allowing many subsequent separation cycles of *K2.

Table 5.7 lists some of the radionuclide generator systems of practical relevance.

Tab. 5.7: Some of the radionuclide generator systems of practical relevance, arranged according to increasing ratios of $t\frac{1}{2}(^*K1)$ / $t\frac{1}{2}(^*K2)$.

Parent nuclide ($^*K1$)	$t\frac{1}{2}(^*K1)$	Daughter nuclide ($^*K2$)	$t\frac{1}{2}(^*K2)$	Ratio of $t\frac{1}{2}(^*K1)/t\frac{1}{2}(^*K2)$	Type of equilibrium
^{140}Ba	12.75 d	^{140}La	40.272 h	7.6	transient
^{99}Mo	66.0 h	^{99m}Tc	6.0 h	11.0	transient
^{62}Zn	9.13 h	^{62}Cu	9.74 min	56.2	transient
^{68}Ge	271 d	^{68}Ga	67.7 min	$5.8 \cdot 10^3$	secular
^{82}Sr	25.34 d	^{82}Rb	1.27 min	$2.9 \cdot 10^4$	secular
^{90}Sr	28.64 a	^{90}Y	64.1 h	$2.4 \cdot 10^5$	secular

The most prominent example of a radionuclide generator applied to nuclear medicine diagnostics is the ^{99}Mo/^{99m}Tc system. Many millions of patients are diagnosed for organ function and tumor detection annually using Single Photon Emission Tomography (SPECT). Each preparation of a ^{99m}Tc radiopharmaceutical is based on the separation of ^{99m}Tc from a ^{99}Mo-based generator. Commercial systems provide ^{99}Mo adsorbed on solid-phase alumina columns, and ^{99m}Tc is separated as ^{99m}Tc-pertechnetate TcO_4^- by eluting the column with 10 mL of isotonic saline. Subsequently, it is converted by redox and coordination chemistry into medically relevant compounds – the radiopharmaceuticals.[20]

5.4.4 Transformation (decay) chains

The last means of transforming unstable, radioactive nuclides into more stable ones considers nuclides with mean nucleon binding energies far from being stable. Here, transformation proceeds stepwise, including formation of many intermediate unstable nuclides. The most prominent examples of transformation chains are the so-called "naturally occurring decay chains". Theoretically, there are four chains, which originate from four individual and long-lived unstable heavy nuclides. Table 5.8 summarizes the four "parent" nuclides, all belonging to the actinide elements: one thorium isotope, one neptunium isotope, and two uranium isotopes: ^{232}Th, ^{237}Np, ^{238}U and ^{235}U. They are located near the very end of the Chart of Nuclides and form an "island" of semi-stable nuclides, as per Fig. 5.22.

The final destinations of the four chains are four corresponding individual stable nuclides, namely ^{208}Pb, ^{209}Bi, ^{206}Pb and ^{207}Pb, respectively. These final nuclides are

---

20 See Chapter 7 on Radiochemical separations and Chapter 13 on Molecular imaging, Vol. II.

characterized by their high mean nucleon binding energy, which are additionally stabilized due to magic nucleon numbers: $Z = 82$ for the lead isotopes, $N = 126$ for the bismuth isotope. Every chain covers many unstable nuclides located between $A = 232{-}238$ and $A = 206 - 208$.

Sometimes each chain is called a "family". (Remember the case of the two-step transformations creating radionuclide generators: the first unstable nuclide *K1 again is a "parent", the subsequent unstable nuclide *K2 is a "daughter".) Similarly, ^{232}Th, ^{237}Np, ^{238}U, and ^{235}U induce daughters and granddaughters etc., thereby forming a genetically consistent family. The "genes" in this case lie in the mass number A, and the four chains are unique in forming follow-up nuclides characterized by mass number of $A = 4n + i$, with i being 0, 1, 2, or 3 and thus individual for each "family". For example, "daughters" of ^{238}U belong to the (i = 2) family. ^{238}U itself is $A = 4 \times 59 + 2 = 238$. Corresponding daughter nuclides must have mass numbers of $A = 4 \times 58 + 2 = 234$. Next-generation transformation creates nuclides of mass

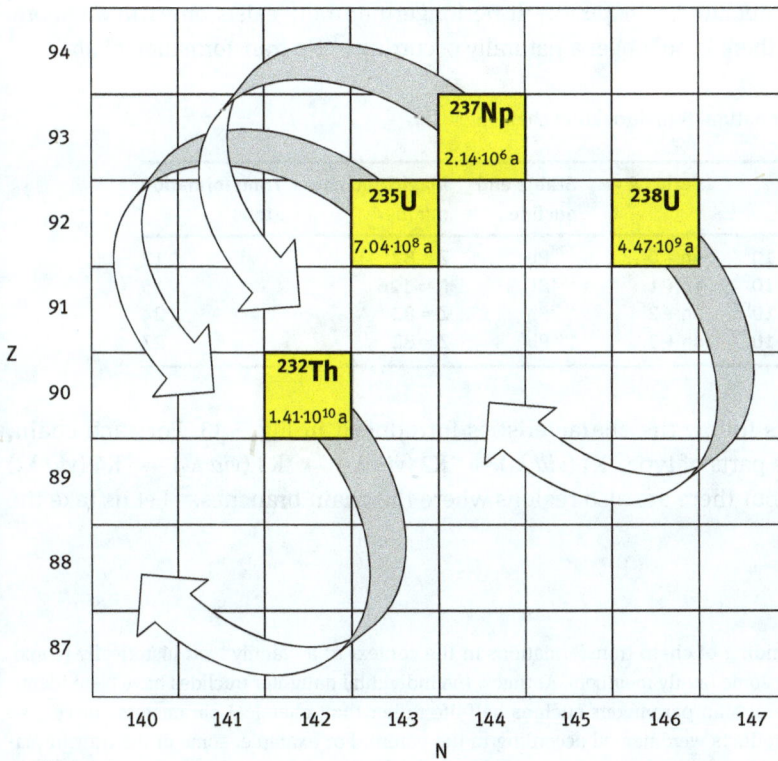

Fig. 5.22: Position of the four parent nuclides of the naturally occurring transformation chains in the Chart of Nuclides. Each of the four nuclides starts with an α-transformation and continues with α- and/or β-transformations.

number $A = 4 \times 57 + 2 = 230$ and so on. The final, stable nuclide is of mass number $A = 206$, which is $4 \times 51 + 2$, representing ^{206}Pb.[21]

What does a "naturally occurring" chain mean? It may mean that the transformations are "natural", going on since the planet earth was formed, ca. $4.55 \cdot 10^9$ years ago. Well, this is true – more or less. Table 5.9 compares the half-lives of the four parent nuclides with the age of the earth. The fraction of unstable parent nuclides still present after a number n of half-lives is $1/2^n$. If related to the age of the earth, a fraction of n' may be defined as the ratio of $n' = t^{earth}/t\frac{1}{2}(*K1)$ and the fraction still in existance is $1/2^{n'}$. In the case of ^{238}U, for example, its half-life is almost the age of the earth and $n' = t^{Earth}/t\frac{1}{2} = 4.55 \cdot 10^9 \, a / 4.47 \cdot 10^9 \, a = 1.0179$. Since $1/2^{1.0179}$ is 0.492835, about 49% of ^{238}U nuclides are still present on earth today compared to earth's genesis ($= 100\%$). For ^{232}Th, it is even more: $1/2^{0 \cdot 3,238,434} = 0.7989386$, i.e. 79.9%. These two nuclides form daughters and granddaughters – naturally, during the past $4.55 \cdot 10^9$ years and still today. The situation is qualitatively the same for ^{235}U, but quantitatively different: with $n' = 6.4649$ for ^{235}U, only 1.132% is still present. However, with $n' > 2000$ for ^{237}Np, this nuclide has expired – it no longer naturally exists on earth anymore. Consequently, there is no longer a naturally occurring ^{237}Np transformation chain.

Tab. 5.8: The four natural transformation chains..

	t½ (a)	Series A =	Stable end nuclide	Magic nucleon number	Transformation steps
^{232}Th	$1.405 \cdot 10^{10}$	$4n + 0$	^{208}Pb	$Z = 82$	12
^{237}Np	$2.144 \cdot 10^{06}$	$4n + 1$	^{209}Bi	$N = 126$	15
^{238}U	$4.468 \cdot 10^{09}$	$4n + 2$	^{206}Pb	$Z = 82$	24
^{235}U	$7.038 \cdot 10^{08}$	$4n + 3$	^{207}Pb	$Z = 82$	27

All four chains follow the characteristics introduced in Fig. 5.13. For each chain, there are some parts of type $*K1$ (via λ_1) $\rightarrow$ $*K2$ (via λ_2) $\rightarrow$ $*K3$ (via λ_3) $\rightarrow$ $*K4$ (via λ_4) $\rightarrow$ and so on, but there are also regions where the chain branches.[22] Let us take the

---

[21] The understanding of chain transformations in the context of a "family" led historically to specific notations for some family members. At times, the individual daughter nuclides have been identified in terms of emission parameters such as half-life rather than chemical parameters, successive transformation products were named according to the parent. For example, some of the transformation products of ^{226}Ra were considered "post-radium-nuclides" by e.g. RaA ($= ^{218}Po$), RaB ($= ^{214}Pb$), RaC ($= ^{214}Bi$), and, because ^{214}Bi branches, even RaC′ ($= ^{214}Po$), and RaC″ ($= ^{210}Tl$), etc.

[22] The mathematics involved in generating several unstable nuclides in a row, particularly when branching chains are considered, are more complex than the equilibrium between two unstable nuclides. This mathematics is not discussed here. Instead, the typical performance of a natural transformation chain is described.

^{238}U chain, for example. Its complete series is illustrated in Fig. 5.23, containing the excerpt of the Chart of Nuclides ranging from Z = 80 (U) to 93 (Np) and N = 124 to 146.

There are two directions of transformations within the chain. Namely, there is a long step from upper right to lower left crossing one field within the Chart of Nuclides, and a short one from lower right to upper left. These transformations belong to the emission types of either β^- or α; see Chapters 8 and 9.

The linear part of the chain, the isodiaphere line, ranging from ^{238}U to ^{218}Po and the beginning of the first branching of the chain at ^{218}Po, which simultaneously transforms into ^{214}Pb and ^{218}At, is highlighted. Four of the following nuclides branch too, namely ^{218}At, ^{214}Bi, ^{210}Pb, and ^{210}Bi. ^{226}Ra is also highlighted.[23]

Tab. 5.9: Fractions of the four natural transformation chain parent nuclides still existing today.

	t½ (a)	Ratio n˙t^{earth}/t½	Fraction remaining primordial
^{232}Th	$1.405 \cdot 10^{10}$	≈0.3	≈80%
^{237}Np	$2.144 \cdot 10^{6}$	≈0.25	0
^{238}U	$4.468 \cdot 10^{9}$	≈1	≈47%
^{235}U	$7.038 \cdot 10^{9}$	≈6.5	≈1.1%

Finally, Fig. 5.24 illustrates the pathways of all the four chains. Among the members of the four families, there is no overlap because of the different "genetic code" of A = (4n + i). Because of the many steps involved when transforming the unstable, long-lived parent nuclide into the final stable one, the four families populate a significant number of intermediate nuclides, showing a broad range of half-lives. Table 5.10 summarizes all the nuclides involved in the four chains.

Although different from the radionuclide generators of type *K1 (*via* λ_1) → *K2 (*via* λ_2) → K3 described above, radionuclide transformation chains provide family members of scientific and practical interest that may be isolated radiochemically. Many of these nuclides became relevant in fundamental nuclear science, significantly contributing to the progress in nuclear and radiochemistry, but also in modern applications in medicine and technology. Some are listed in Tab. 5.11. One example was already mentioned: the ^{226}Ra isolated by M. CURIE from the ^{238}U family, which existed in uranium ores at equilibrium. Another important example,

---

23 Because this member of the ^{238}U natural transformation chain was the first to be separated from ^{238}U (according to the equilibrium within the whole chain) and identified as a new chemical element. It later became the traditional unit of radioactivity (see above: 1 g of ^{226}Ra = 1 Ci).

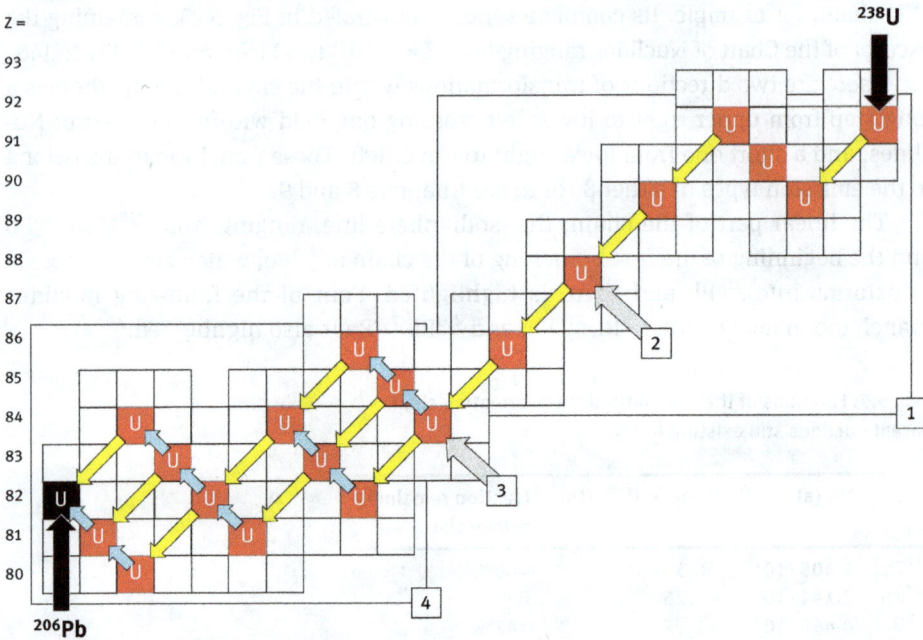

Fig. 5.23: The naturally occurring ^{238}U transformation chain. The excerpt from the Chart of Nuclides ranging from Z = 80 (U) to 93 (Np) and N = 124 to 146 identifies fields for the members of the A = 4n + 2 family. (They all are indicated by "U" to clarify they all are uranium´s daughter's and granddaughters.) The final black square is for the stable terminal nuclide, ^{206}Pb. Several aspects are highlighted: (1) the first part of the chain ranging from ^{238}U to ^{218}Po including α linear line starting from ^{234}U. (3) The beginning of the first branching of the chain at ^{218}Po, which simultaneously transforms into ^{214}Pb continuing the α-transformation, and ^{218}At, according to β$^-$ transformation. (4) Four of the following nuclides branch too, namely ^{218}At, ^{214}Bi, ^{210}Pb and ^{210}Bi. The final stable nuclide ^{206}Pb is thus being populated by ^{210}Po and by ^{206}Tl. (2) ^{226}Ra.

originated from the same chain, was the use of the . . . $\rightarrow$ ^{210}Pb — $t^{1/2}$ = 22·3 a $\rightarrow$ ^{210}Bi — $t^{1/2}$ = 5·013 d $\rightarrow$ . . . [24]

5.5 Outlook

The various mathematical concepts for following transformation processes of unstable nuclides are all based on exponential laws and can be treated accordingly – supposing the number of nuclides involved is large enough to obey the laws of

---

[24] It was isolated and applied by G. HEVESEY for fundamental physico-chemical and pioneering medical applications (Nobel Prize in Chemistry, 1943, " . . . for his work on the use of isotopes as tracers in the study of chemical processes").

Z =

Z\N	124	125	126	127	128	129	130	131	132	133	134	135	136	137	138	139	140	141	142	143	144	145	146
93																					Np		
92																		Np	U	Ac			U
91																	Ac		Np	U			
90												Ac	Th	Np	U	Ac	Th				U		
89											Np			Ac	Th								
88										Ac	Th	Np	U			Th							
87									Np		Ac												
86							U	Ac	Th			U											
85					Ac	Th	Np	U	Ac														
84		U	Ac	Th	Np	U	Ac	Th			U												
83		Np	U	Ac	Th	Np	U	Ac															
82	U	Ac	Th	Np	U	Ac	Th			U													
81		U	Ac	Th	Np	U																	
80			U																				

N = 124 125 126 127 128 129 130 131 132 133 134 135 136 137 138 139 140 141 142 143 144 145 146

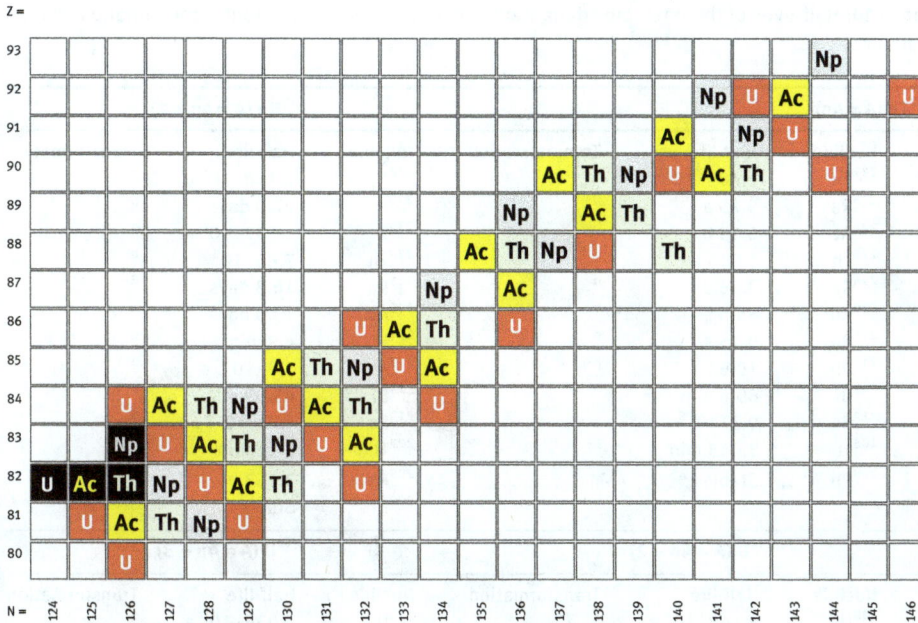

Fig. 5.24: Individual pathways of the four chains, starting from the "parents" Ac, U, Th, Np. The "family tree" of each parent is shown superimposed on the chart of nuclides. There is no overlap among the members of the families because of the different genetic code of A = 4n + i.

statistics. Key parameters are the half-life $t_{1/2}$ or its inverse pendant, the transformation constant λ.

An important message in the context of radio- and nuclear chemistry is the correlation between the number Ń of radioactive nuclei and their radioactivity Á. It reveals that in many cases, a relatively high level of radioactivity is expressed by a rather low number of nuclei. Mathematics is relevant for various practical applications including the design and the handling of radionuclide generator systems.

However, the way a nuclide "organizes" its transformation is itself very complex. The different classes of transformation of one unstable nuclide into a more stable one and the kind of emissions that accompany the nuclear processes are introduced in the next chapters.

Tab. 5.10: Half-lives of the unstable "daughter" nuclides involved in the four transformation chain "families".

^{232}Th (A = 4n)			^{237}Np (A = 4n + 1)		
Nuclide	Half-life	Transformation	Nuclide	Half-life	Transformation
^{232}Th	$1.405 \cdot 10^{10}$ a	α	^{237}Np	$2.144 \cdot 10^{6}$ a	α
^{228}Ra	5.75 a	β–	^{233}Pa	27.0 days	β–
^{228}Ac	6.13 h	β–	^{233}U	$1.592 \cdot 10^{5}$ a	α
^{228}Th	1.913 a	α	^{229}Th	$7.88 \cdot 10^{3}$ a	α
^{224}Ra	3.66 d	α	^{225}Ra	14.8 days	β–
^{220}Rn	55.6 s	α	^{225}Ac	10.0 days	α
^{216}Po	0.15 s	α	^{221}Fr	4.9 min	α
^{212}Pb	10.64 h	β–	^{217}At	$3.2 \cdot 10^{-3}$ s	α
^{212}Bi	60.6 min	α, β–	^{213}Bi	45.59 min	α, β–
^{212}Po	$0.3 \cdot 10^{-6}$ s	α	^{213}Po	4.2×10^{-6} s	α
^{208}Tl	3.053 min	β–	^{209}Tl	2.16 min	β–
^{208}Pb	Stable		^{209}Pb	3.253 h	β–
			^{209}Bi	Stable	

^{238}U (A = 4n + 2)			^{235}U (A = 4n + 3)		
Nuclide	Half-life	Transformation	Nuclide	Half-life	Transformation
^{238}U	$4.468 \cdot 10^{9}$ a	α	^{235}U	$7.038 \cdot 10^{8}$ a	α
^{234}Th	24.1 days	β–	^{231}Th	25.5 h	β–
^{234m}Pa	1.17 min	β–	^{231}Pa	$3.276 \cdot 10^{4}$ a	α1), β–
^{234}Pa	6.7 h	β–	^{227}Ac	21.773 a	α
^{234}U	$2.455 \cdot 10^{5}$ a	α	^{227}Th	18.72 days	α1), β–
^{230}Th	$7.54 \cdot 10^{4}$ a	α	^{223}Fr	21.8 min	α
^{226}Ra	1.600 a	α	^{223}Ra	11.43 days	α, β
^{222}Rn	3.825 days	α	^{219}At	0.9 min	α
^{218}Po	3.05 min	α, β	^{219}Rn	3.96 s	β–
^{214}Pb	26.8 min	β–	^{215}Bi	7.7 min	α, β
^{218}At	$\gg$ 2 s	α, β	^{215}Po	$1.78 \cdot 10^{-3}$ s	β–
^{218}Rn	$35 \cdot 10^{-3}$ s	α	^{211}Pb	36.1 min	α
^{214}Bi	19.9 min	α, β–	^{215}At	$0.1 \cdot 10^{-3}$ s	α, β
^{214}Po	0.164×10^{-3} s	α	^{211}Bi	2.17 min	α
^{210}Tl	1.3 min	β–	^{211m}Po	25.2 s	α
^{210}Pb	22.3 a	α, β–	^{211}Po	0.516 s	β–
^{206}Hg	8.15 min	β–	^{207}Tl	4.77 min	β–
^{210}Bi	5.013 days	α, β–	^{207}Pb	Stable	β–
^{206}Tl	4.2 min	β–			
^{210}Po	138.38 days	u			
^{206}Pb	Stable				

Tab. 5.11: Examples of relevant members of the natural transformation chains, radiochemically isolated from the chain equilibrium.

Parent isotope	Nuclide or generator pair	$t\frac{1}{2}$	Ratio of $t\frac{1}{2}(^*K1) / t\frac{1}{2}(^*K2)$	Type of equilibrium
^{238}U	^{226}Ra	1600 a	–	–
^{232}Th	^{228}Th	1.93 a	$1.9 \cdot 10^2$	secular
	^{224}Ra	3.66 d		
^{238}U	^{210}Pb	22.3 a	$1.6 \cdot 10^3$	secular
	^{210}Bi	5.013 d		
^{237}Np	^{225}Ac	10.0 d	$3.2 \cdot 10^2$	secular
	^{213}Bi	45.59 min		

6 Processes of transformations: Overview

Aim: Unstable nuclei convert into stable ones by minimizing the absolute mass of a nuclide, which is typically expressed in terms of nucleon binding energy. The common feature of all primary transformations is their exothermic character, and the corresponding energy is introduced in this chapter in a general approach.

There are three principal approaches as derived from the LDM: (a) "simply" changing the ratio between protons and neutrons, thereby retaining the same mass number; (b) reducing the mass number by emitting single nucleons or clusters of nucleons; or (c) dividing the whole nucleus into (mainly) two pieces. The corresponding pathways of transformation are for type (a) the class of β-processes, for type (b) the α-emission processes and for type (c) spontaneous fission.

The dominant route of transformation, occurring in most unstable nuclides, is the β-process. Its essence is: turning a neutron into a proton, or *vice versa*. This class of transformation involves emission or capture of an electron. Therefore, the mass number remains the same, and the nuclides involved are isobars. It includes three individual subtypes: the β⁻-process, β⁺-process, and electron capture (EC or ε).

Among the heavy nuclides, α-emission and spontaneous fission become additional options.

Primary transformations create a new nucleon composition of the nucleus formed, which does not necessarily represent the ground state of that nucleus. Instead, excited nuclear levels are populated and these undergo de-excitation through secondary transitions.

6.1 Transformation processes overview

6.1.1 Transformation = radioactivity

Unstable nuclei convert into stable ones by minimizing their absolute mass, which is typically expressed in terms of nucleon binding energy. The absolute mass of the transformation product(s) should be less than the absolute mass of the initial unstable nuclide. The difference in mass (Δm) is expressed as an absolute mass value, but typically it is converted into values of energy (ΔE)[1] according to $\Delta E = \Delta m c^2$. Transformations proceed exothermically and, similar to conventional chemical reactions, this yields energy: there is less energy "located" in the newly formed nuclide compared to the initial, less stable one.

---

[1] The amounts of energy of these transformation reactions are much higher than energy balances of conventional chemical reactions.

https://doi.org/10.1515/9783110742725-006

$$m_{K1}^{nuclide} > m_{K2}^{nuclide} \tag{6.1}$$

Parallel to the changes within the nuclei, one or two small components are ejected during the transformation. The corresponding subtypes of primary transformation processes thus also differ in terms of the radiation[2] emitted within the course of the transformation. The primary step of transformation of an unstable nuclide into a stable (or at least a more stable) one is accompanied by the emission of another component. It is denoted here as "x" and/or "y".

The following is valid in all cases:
- spontaneous transformations are exothermic,
- the sum of the mass of the transformation products is less than the mass of the initial nuclide,
- a characteristic amount of energy is obtained as ΔE.

Figure 6.1 illustrates the general concept. Equation (6.1) summarizes the three aspects: change in nucleon configuration from *K1 to K2, release of components x (and/or y), and gain in energy ΔE.

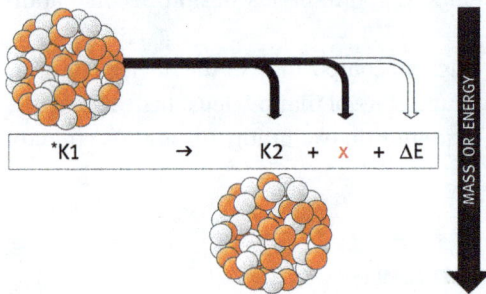

Fig. 6.1: Simplified scheme of primary transformation of an unstable nuclide *K1 into a stable nuclide K2.
Mass numbers may remain the same (A2 = A1, the β-processes) or become lower
(A2 < A1, in the case of α-emission and spontaneous fission). Transformations are accompanied by the emission of one or two components x and y. This "x" is typically a particle, such as a ^{4}He nucleus (the α-particle) or β-particle (β-processes).

However, eq. (6.2a) is the general representation of the whole process. For the individual primary transformation of an unstable nuclide *K1 into a stable (or at least a more stable) nuclide K2, eqs. (6.2b–g) apply. Accordingly, transformation processes

---

2 Actually, the transformation itself is the source of what is called "radioactivity" and the unstable nuclei are the "radioactive" ones. Usually, this type of transformation is called "decay". The way "radioactivity" is detected usually refers to the components x and y that are responsible for most of the effects attributed to the phenomenon of radioactivity.

are in most cases accompanied by the *emission* of (small) components x (electrons e, n, p, α, . . .) and in some case by a further (even smaller) component y (an electron neutrino, photons, X-rays). In one version, a shell electron is needed to initiate the transformation, as shown in eq. (6.2f). Equation (6.2g) indicates another pathway called spontaneous fission, where an initial, large nuclide *K1 splits into two smaller nuclei.

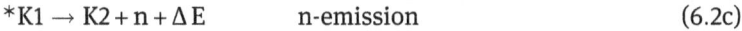

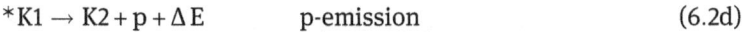

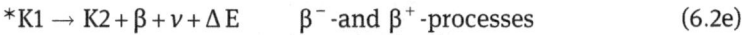

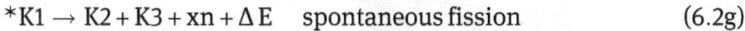

$$^*K1 \rightarrow K2 + x + y + \Delta_E \tag{6.2a}$$

$$^*K1 \rightarrow K2 + \alpha + \Delta E \qquad \text{α-emission} \tag{6.2b}$$

$$^*K1 \rightarrow K2 + n + \Delta E \qquad \text{n-emission} \tag{6.2c}$$

$$^*K1 \rightarrow K2 + p + \Delta E \qquad \text{p-emission} \tag{6.2d}$$

$$^*K1 \rightarrow K2 + \beta + \nu + \Delta E \qquad \text{β}^-\text{-and β}^+\text{-processes} \tag{6.2e}$$

$$^*K1 + e \rightarrow K2 + \nu + \Delta E \qquad \text{electron capture} \tag{6.2f}$$

$$^*K1 \rightarrow K2 + K3 + xn + \Delta E \qquad \text{spontaneous fission} \tag{6.2g}$$

6.1.2 Primary transformations *vs.* secondary transitions and post-effects

The "primary" transformation results in the formation of a new nucleus as defined by mass number A, proton number Z and neutron number N. The transformation consists of pathways changing the proton number Z from $_{Z1}{}^*$K1 to $_{Z2}$K2. The difference of Z2 – Z1 = ΔZ is either >0 or <0, i.e. ≠0. These are called "primary" processes. There are several options for the impact of ΔZ ≠ 0 on the change in mass numbers A of the two nuclides involved: ^{A2}K2 – ^{A1*}K1 = ΔA can be ≤0, but never >0. In the ΔA > 0 case, the main changes in mass number are either –4 or between ca. –100 and ca. –140. The corresponding subtypes of primary transformations are β-processes, α-emission and spontaneous fission, respectively.

Typically, the proton and/or neutron shell occupancies of the newly formed nucleus K2 are not necessarily identical to those of the ground state nucleus. Immediately after the initial primary transformation, one or more nucleons find themselves in higher shells than in the ground state of the product nucleus. This newly formed nucleus may exist – for shorter or longer periods of time – in an "excited" state, but must de-excite to levels of lower energy according to the shell model of the nucleus. In many cases, this is a multi-step process similar to the relaxation of electrons within the shell of an atom (see Fig. 1.6). Likewise, de-excitation of nucleons releases the difference in energy between the nucleon shells involved; in most cases it is emitted as electromagnetic radiation.[3] Thus, secondary processes proceed

---

[3] These complex processes are discussed as "secondary" transformations in Chapter 9.

within one and the same nucleus K2, i.e. at both ΔZ and $\Delta A = 0$ and may thus be classified as "transitions" rather than transformations.

In addition, some of the primary and secondary transformation mechanisms, in particular the electron capture process, leave a hole within the electron shell of the nuclide involved. Once the new *nucleus* K2 has been formed, the vacancy in its *electron shell* must be filled. Finally, the particle emissions x (released in primary and secondary transformations) interact with atoms surrounding the newly formed nuclide K2. The latter two independent effects are referred to as "post-effects". Figure 6.2 illustrates the three process classes.

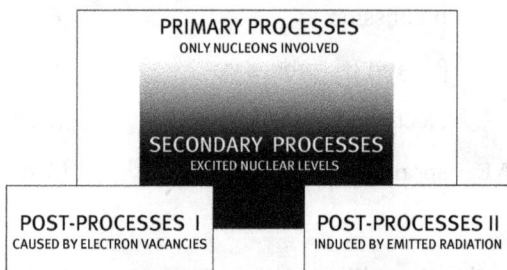

PRIMARY PROCESSES
ONLY NUCLEONS INVOLVED

SECONDARY PROCESSES
EXCITED NUCLEAR LEVELS

POST-PROCESSES I
CAUSED BY ELECTRON VACANCIES

POST-PROCESSES II
INDUCED BY EMITTED RADIATION

Fig. 6.2: Sketch of the interplay of primary transformations, secondary transitions and post-effects. The transformation characteristics of a certain unstable nuclide, in particular the different types of its "radiative" emissions, can only be explained as the sum of the corresponding processes.

6.1.2.1 Primary transformations

The basic concept of an unstable nucleus is to optimize its nucleon composition. The three main options are covered by eqs. (6.2c) and (6.2e, f) and (6.2g) and balances are listed in Fig. 6.3. The three pathways thus effect very different changes for mass number, proton number and neutron number. While A remains constant for all the beta processes (eqs. (6.2e, f)), i.e. only proton and neutron numbers change by ± 1, the α-transformation affects A, Z and N. The nucleon composition of K2 relative to K1 is $\Delta A = -4$, $\Delta Z = -2$, $\Delta N = -2$. This effect is even more "dramatic" for spontaneous fission according to eq. (6.2g).

6.1.2.2 Secondary transitions

In some cases, the re-arrangement of nucleons directly yields the ground state of the new nucleus K2. In many other cases, excited nuclear states of K2 are generated in parallel or exclusively. Excited electron states require de-excitation mechanisms. The principal approach for the nucleus is similar: de-excitation to lower-lying or ground state levels occurs through the emission of electromagnetic radiation, i.e. photons. However, two other options exist: internal conversion and pair formation.

PROCESSES	ΔZ	ΔA	type
PRIMARY	≠0	0	β
		-4	α
		≈ -100 + ≈ -140	sf
SECONDARY	=0	=0	γ, IC…

Fig. 6.3: Categorizing transformation routes of unstable nuclides *K1 → K2 into primary and secondary processes according to the balances of ΔZ and ΔA for the most dominant processes. The corresponding subtypes differ in terms of the radiation co-emitted within the course of the transformation and represent β-processes, α-emission or spontaneous fission, respectively. For secondary processes, the most relevant case is emission of electromagnetic radiation (photons).

6.1.2.3 Post-processes

Electron capture means capturing an s-shell electron from the electron shell of K2 to allow the primary transformation of eq. (6.2f). After the new nucleus K2 has been formed, the electron vacancy remains and must be filled by follow-up processes ("post-processes I"). Independently, the components x released as particle emission in eqs. (6.2c–e) interact with atoms surrounding the newly formed nuclide K2. The effects induced by these interactions are discussed as "post-processes II".

6.1.3 Nuclear *vs.* chemical transformations

Transformation processes going on within a nucleus of an atom are not dependent on the chemistry of the chemical element behind that nucleus. Typically, the electron shells are far apart in distance and electron energies are less than nucleon binding energies by many orders of magnitude. Consequently, nuclear transformation processes do not depend on the experimental parameters typically relevant to chemical processes, such as temperature, pressure, concentration, etc.[4]

However, effects on the chemistry of the element may arise later, as a consequence of a nuclear transformation. Suppose the transformation of an unstable nuclide to a more stable one involves a change in proton number (which is typical for primary transformations). The newly formed nuclide immediately stimulates the arrangements of the corresponding proton and electron numbers: the new atom is of

---

4 There is one exception, valid in a few cases of electron capture processes; this is shown in eq. (6.1f) and Chapter 10.

different chemistry. Thus, chemistry is usually not important for nuclear processes, but nuclear processes affect chemistry.

6.2 Primary transformation pathways

6.2.1 Mechanism of primary transformation processes

The corresponding subtypes of primary processes differ in terms of the way nucleon compositions change within the nucleus of the atom. Consequently, unstable nuclides convert into stable ones by either changing the absolute number of nucleons (changing A) or by modifying the ratio between protons and neutron (changing Z:N for constant A), as per eqs. (6.2e, f). In the latter cases, an "excess" neutron "just" converts into a proton (supposing the nucleus has an excess of neutrons over protons) or *vice versa*. In other cases, the nucleus releases a number of nucleons, typically as a small cluster of 2 neutrons and 2 protons – the α-particle – thereby lowering mass number A (see eq. (6.2b)). For a limited number of very heavy nuclides, there is another option to lower mass, namely a spontaneous split of the large nucleus into (mainly two) fractions as per eq. (6.2g). This is called spontaneous fission (sf).

Interestingly, many unstable nuclides realize two or even more options; they stabilize in parallel *via* α- *and* β-emission, or α-emission *and* sf, etc. This is because for a given *K1, there is more than one option for fulfilling the ultimate prerequisite of $\Delta E > 0$. Table 6.1 illustrates the various transformation processes of unstable nuclides.

For a very limited number of extremely neutron-rich or proton-rich unstable nuclides, located close to the neutron drop line or proton drop line, a single "excess" nucleon is emitted (eqs. (6.2c, d)).

However, for all naturally occurring unstable nuclides and most artificially produced unstable nuclides, β-emission, α-emission and sf are the most relevant primary transformation pathways. This textbook will concentrate on the β-processes, α-emission[5] and spontaneous fission.

---

5 The reason the β-processes are named "β" is historical and refers to the three sorts of emissions that originate from naturally occurring unstable nuclides. These radiations were traditionally analyzed in terms of charge and deviation in mostly magnetic fields, which work according to the ratio of q/m_o. This approach not only allowed to draw direct conclusions on the charge of the emission, but also to derive its mass according to the deviation of the radiation. This is summarized in the well-known picture of the three emissions showing radiation emitted from unstable nuclides and analyzed in terms of charge and deviation in electromagnetic fields. Following the Greek alphabet, the first radiation characterized was the α-emission. The next generation of emissions of electrons analyzed this way was named β. They were completely different to the α-emissions: charged negatively (–1), i.e. half of absolute charge (α-particles are of + 2 charge), but of about 4000 times less mass. Consequently, deviation was in the opposite direction, but significantly more pronounced. (The final emission categorized this way was the γ-emission: no charge, no mass and thus not affected by the electromagnetic field at all.)

Tab. 6.1: The primary transformation processes of unstable nuclides.

Type	Balance in mass number ΔA	Principal emission x	Symbol	*K1 mass area
β-processes	0	electron	β⁻	Light, medium, heavy and very heavy nuclides
		positron	β⁺	
			EC, ε	
Single nucleon emission	−1	proton	p	Nuclides at the edge of proton-rich isotopes
		neutron	n	Nuclides at the edge of neutron-rich isotopes
Cluster emission (x = 4, 14, . . .)	−4	α-particle	α	Large nuclides
	−14	carbon cluster		Very rare
Spontaneous fission	Various, peaking around −100 and −140	various	sf	Very large nuclides

6.2.1.1 ΔA = 0

In most cases for a given nucleus, a proton just converts into a neutron or *vice versa*. This type of transformation, the β-process, occurs throughout the Chart of Nuclides, but is excusive for the light and medium heavy nuclides. In fact, β-transformation processes are dominant for most unstable nuclides. More than 2000 nuclides are known to "utilize" this class of transformation. For nuclides of mass number $2 \leq A \leq 144$, it is the only possible class of transformation; the other two main transformation types of $\Delta A \neq 0$, namely α-emission and spontaneous fission, do not occur. For the ca. 3000 unstable nuclides, the proportion of nuclides preferring β-, α- or sf processes is roughly 100:10:1, as seen in Fig. 6.4.

6.2.1.2 ΔA ≠ 0

In the remaining cases, a nucleus releases one nucleon or (more likely) a small cluster of 2 neutrons and 2 protons, the α-particle. This is a popular option for many heavy unstable nuclides. For a number of very heavy nuclei, a third option appears, namely to split the large nucleus into fractions. This only takes place above mass number 232.

Figure 6.5 identifies the areas within the Chart of Nuclides, where the three main primary transformation pathways dominate.

α-TRANSFORMATIONS ——————┐   ┌——————————— SPONTANEOUS FISSION

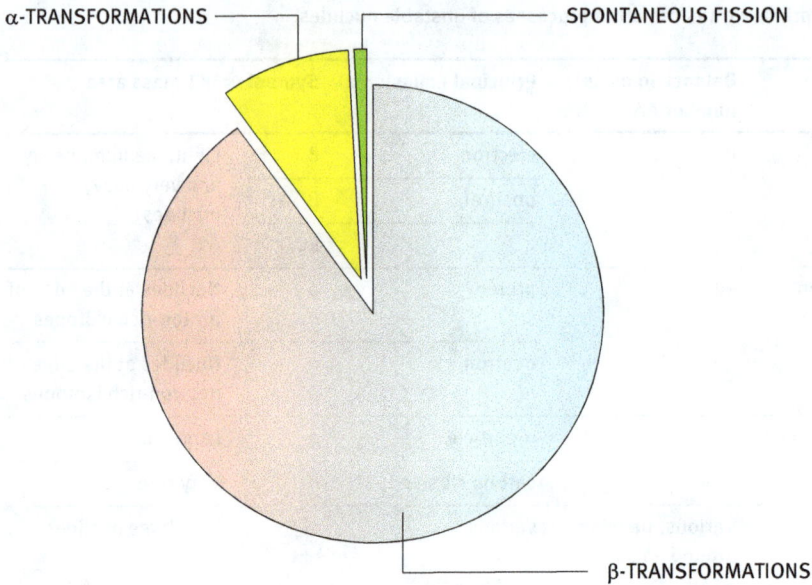

β-TRANSFORMATIONS

Fig. 6.4: Frequency of the three primary transformation processes among naturally occurring and most artificially produced unstable nuclides. The colors follow the color coding of the Karlsruhe Chart of Nuclides. While yellow stands for α-emission and green for spontaneous fission, the β-transformations are colored in either red or blue.

6.3 Energetics

The transformation of unstable nuclides proceeds exothermically, resulting in a state with less mass and less energy. In all cases, the absolute mass of the transformation product(s) must be less than the mass of the initial nuclide.

The energetics of nuclide transformation involve the whole nuclide (i.e., the nucleus), the electrons, and the binding energy of the electrons. The balance of the exothermic process is in any case thus defined following eq. (6.1): mass of the new *nuclide* K2 < the mass of the initial *nuclide* *K1. To turn this inequality (<) into an equation (=), two components are needed, as shown in eqs. (6.2). The first is easy: it is ΔE, basically representing the fraction Δm of mass by which the nuclide K2 is less than the nuclide K1. In addition – and extremely important to the phenomenon of radioactivity – there are one or two smaller components involved as well.

The value of Δm correlates with the energy of the transformation process. Similar to exothermic chemical reactions, ΔE has a positive sign as energy is gained. In nuclear sciences it is usually denoted by ΔE = Q in units of MeV. This fraction of energy is not released as heat like one may suppose, as in many exothermic chemical reactions. Instead, it reflects the difference in mass between the two nuclides

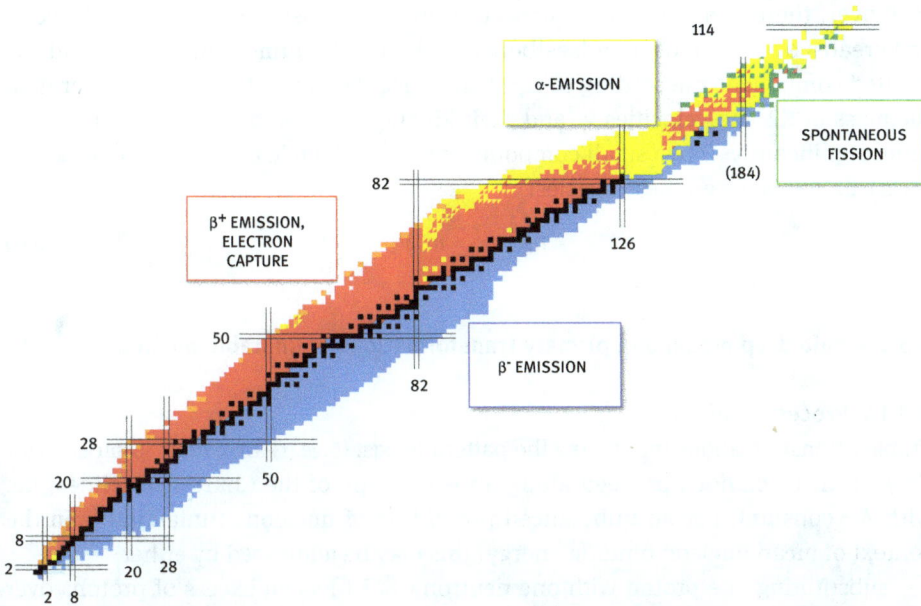

Fig. 6.5: Distribution of radionuclides in the Chart of Nuclides undergoing the β-processes, α-emission and spontaneous fission. According to the Karlsruhe Chart of Nuclides, β-processes are indicated by either blue (β⁻) or red (β⁺ or ε), α-emission by yellow, and spontaneous fission by green.

and correlates with nucleon binding energy. The Q-value may also be calculated using the mass defect of the components involved rather than the absolute masses.

However, something else is truly "emitted". This other constituent is essential for the impact of radioactive transformations in science and technology: the "radiation" released by and accompanying the transformation processes. These emissions are responsible for the effects generally associated with "radioactivity". At this stage, it is called "x" and/or "y", and subsumes the various kinds of "radiation" to be discussed later in detail.[6]

$$m_{*K1}^{nuclide} = m_{*K2}^{nuclide} + x\,(+y) + \Delta m \tag{6.3}$$

---

6 Later, it will be specified that "radiation" consists of elementary particles (such as an electron), subatomic particles (such as a positron and neutrino), or composites (such as an α-particle or neutron(s)). Even later, in the context of secondary transformation processes, "x" will mainly denote electromagnetic radiation such as photons and X-rays. For the latter, "X-ray" exactly reproduces the historical context: emission of a kind of radiation not known that time (i.e. when ROENTGEN performed his "X"-ray studies.)

Moreover, there is an "impulse" aspect of nuclear transformation. The release of "x" creates a balance of impulses between the newly formed nuclide K2 and the emitted component x according to eq. (6.4). Since the mass of x is much lower than the mass of K2, the velocities v (and their kinetic energies) of the two components differ significantly. If two small components x and y are released, the kinetic energy is distributed between these species.

$$m_{K2}\, v_{K2} = m_x\, v_x \qquad\qquad (6.4)$$

6.3.1 Liquid drop model and primary transformations of unstable nuclides

6.3.1.1 Processes of $\Delta A = 0$
Primary transformations that follow the pattern of eqs. (6.2e, f) are called "β-processes". These transformations proceed along an isobar line of the Chart of Nuclides, i.e. with A = constant. For an unbalanced proportion of nucleons (unfavorable in the context of mean nucleon binding energy) this may be addressed by either:
- substituting one proton with one neutron (if *K1 has an excess of protons over neutrons) or
- substituting one neutron with one proton (if *K1 has an excess of neutrons over protons).

The new coulomb term would be $(Z \pm 1)^2/A^{\frac{1}{3}}$. The asymmetry term would count for both nucleons as $[(N-1)-(Z+1)]^2/A$ or $[(N+1)-(Z-1)]^2/A$ (simplified to $[(N-Z)-2]^2/A$ and $[(N-Z)+2]^2/A$), respectively. Figure 6.6 illustrates the impact of these two terms of the WEIZSÄCKER equation on the "improvement" of mean nucleon binding energy along lines of constant mass number. The message is that *internal* changes of proton to neutron (or *vice versa*) significantly affect the energetic – although the overall number of nucleons remains the same. Each of these conversions must yield a new nuclide of (slightly) less absolute mass.

6.3.1.2 Processes of $\Delta A \neq 0$
Primary transformations following this pattern are the α-processes (eq. (6.2b)) and spontaneous fission (eq. (6.2g)). For a given mass number A and nuclei with increasing proton number, the coulomb repulsion becomes critical. Two terms of the WEIZSÄCKER equation explicitly quantify the impact of the proton number directly, namely the coulomb term $Z^2/A^{\frac{1}{3}}$ and the asymmetry term $(N-Z)^2/A$. In the case of an α-emission, the numerator Z^2 turns into $(Z-2)^2$ and $(N-Z)^2$ turns into $((N-2)-(Z-2))^2$, despite changes in mass number from A to A – 4 for the denominator. This tendency is reflected in Fig. 6.7, showing $\bar{E}_B$ values depending on mass number A. For heavier nuclei, each move towards lower A with $\Delta A = -4$ slightly improves mean nucleon binding energy.

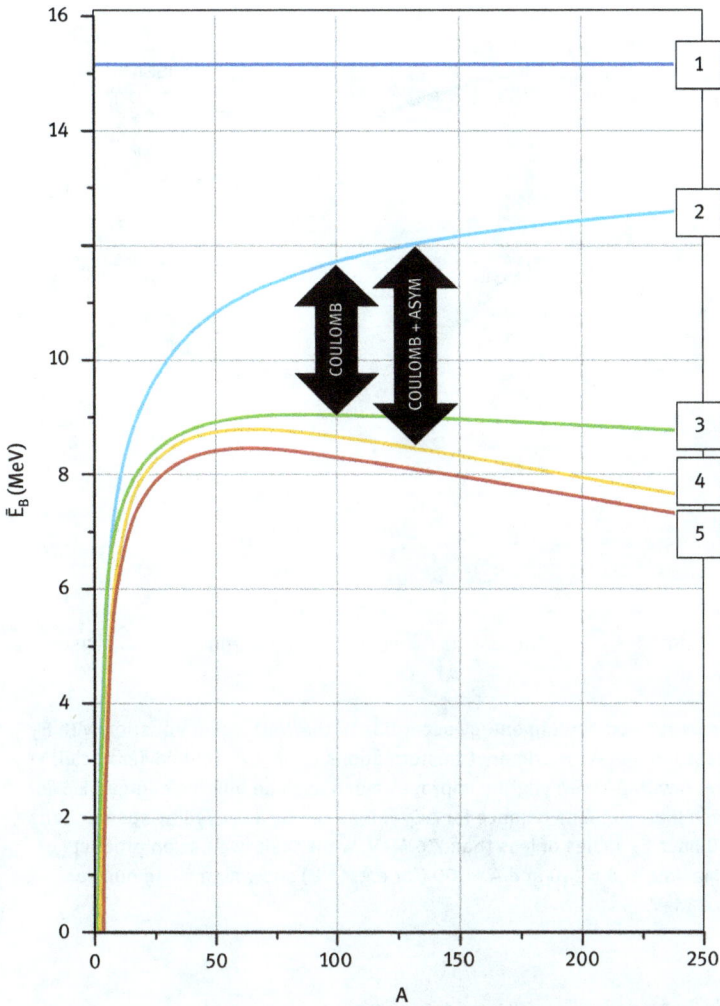

Fig. 6.6: The principal effects of the various terms of the Weizsäcker equation on $\bar{E}_B$ values versus mass number A. These terms reflect the LDM of the nucleus of the atom: volume (1), surface (2), coulomb (3), asymmetry (4) and pairing (5). Suppose that one sort of nucleon converts into the other while mass number A remains the same; the coulomb term alone and the coulomb + asymmetry terms together strongly modify the $\bar{E}_B$ values.

An even more pronounced balance for $\Delta\bar{E}_B$ values is achieved in the case of fissioning a large nucleus. Nuclei of A > 230 have $\bar{E}_B$ values less than 7.6 MeV, while their fission products of mass numbers around $\Delta A \approx -140$ and $\Delta A \approx -100$ achieve values of $\bar{E}_B > 8$ MeV.

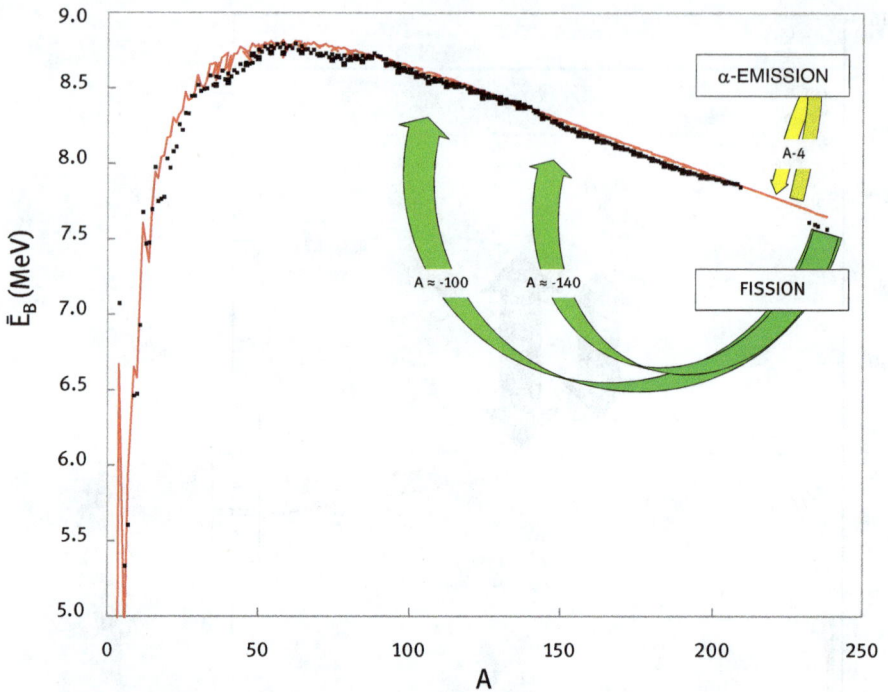

Fig. 6.7: The trend in mean nucleon binding energy according to the Weizsäcker equation with $\bar{E}_B$ values depending on mass number A for primary transformations of $\Delta A \neq 0$. For heavier nuclei, each move towards lower A with $\Delta A = -4$ slightly improves mean nucleon binding energy $\Delta\bar{E}_B$. For very heavy nuclei, an even more extreme balance for $\Delta\bar{E}_B$ values can be achieved by spontaneous fission. Nuclei of A > 230 have $\bar{E}_B$ values of less than 7.6 MeV, while their two fission products of mass numbers around maxima at A ≈ 140 and A ≈ 100 (for e.g. ^{238}U) show high mean nucleon binding energies of $\bar{E}_B$ > 8 MeV.

6.4 β-Transformation processes ($\Delta A = 0$)

6.4.1 Stable *vs*. unstable isotopes

Let us consider a line of isotopes, as in Fig. 6.8. Suppose that the black square involves a stable nucleon composition and isotopes to the right and left show increasing or decreasing numbers of neutrons (at constant proton number). This will (sooner or later) create proton to neutron ratios reducing overall and mean nucleon binding energy as quantified by the WEIZSÄCKER equation, as per Figs. 4.1 and 4.3.

What happens next? Let us select the neutron-rich unstable isotope to the right of to the stable nuclide. What should this A neutron-rich nuclide do? It may try to lower the number of neutrons to get rid of the excess. Elimination of a neutron thus seems to be a good idea. However, what does "excess" of neutrons mean? It is tantamount to a "deficit" of protons. So, converting an excess neutron into a deficient proton would solve the problem in a more elegant way. The same applies to the neutron deficient isotopes to the left of the stable one, which are proton-rich. They could convert a proton into a neutron. The point of the β-process is to convert the sort of nucleons that are in excess into the other sort that are scarce. In doing so, the mass number will remain constant.

Conversion of a neutron to a proton follows the process $_z^*K1 \rightarrow _{z+1}K2$. This is accompanied by emission of a *negatively charged electron* and is called the β⁻-process. Conversion of a proton to a neutron results in the opposite case, $_z^*K1 \rightarrow _{z-1}K2$.

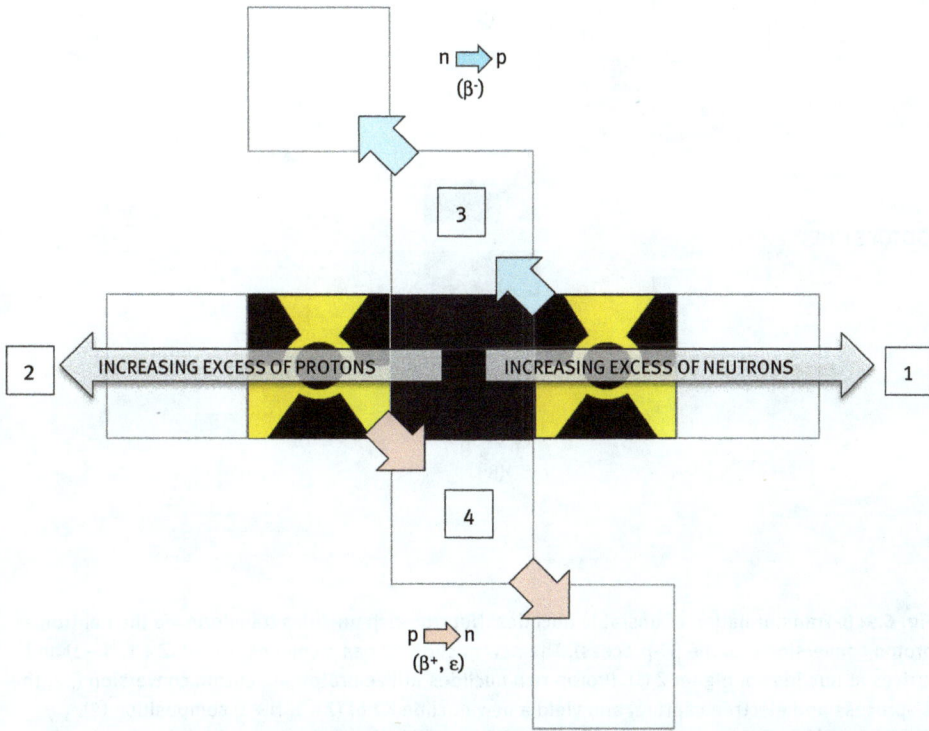

Fig. 6.8: Schematic view on the origin and pathway of β-processes matching the Chart of Nuclides. The middle horizontal line shows the isotopes of a given element. The black field in the middle indicates stable nucleon composition, isotopes to the right (1) and left (2) show increasing or decreasing number of neutrons (at constant proton number). These proton-to-neutron ratios tend to reduce overall and mean nucleon binding energies as quantified by the Weizsäcker equation, and (sooner or later) these nuclides are unstable. However, "excess" of neutrons is synonymous with "deficit" of protons. Converting an excess neutron into a deficient proton (3) would thus address the problem simultaneously. The same applies to neutron-deficient isotopes, which are proton-rich, by converting a proton into a neutron (4). The mass number will always remain constant. The successive steps of transformation thus proceed along the lines of isobars. Conversion of a neutron into a proton at A = constant is the β⁻-process. Radioisotopes preferring this pathway are color-coded in blue. Conversion of a proton into a neutron has two versions, both color-coded in red: the β⁺-process and "electron capture".

While there is only one approach for the $_ZK1 \rightarrow _{Z+1}K2$ conversion, there are two options for the $_ZK1 \rightarrow _{Z-1}K2$ process. The one accompanied by *emission* of a positively charged electron, the β^+ particle, is called the β^+-process. Alternatively (or in parallel), neutron-deficient nuclides may transform by *capture* of an electron from the K electron shell. This type of β-process is aptly named "electron capture".

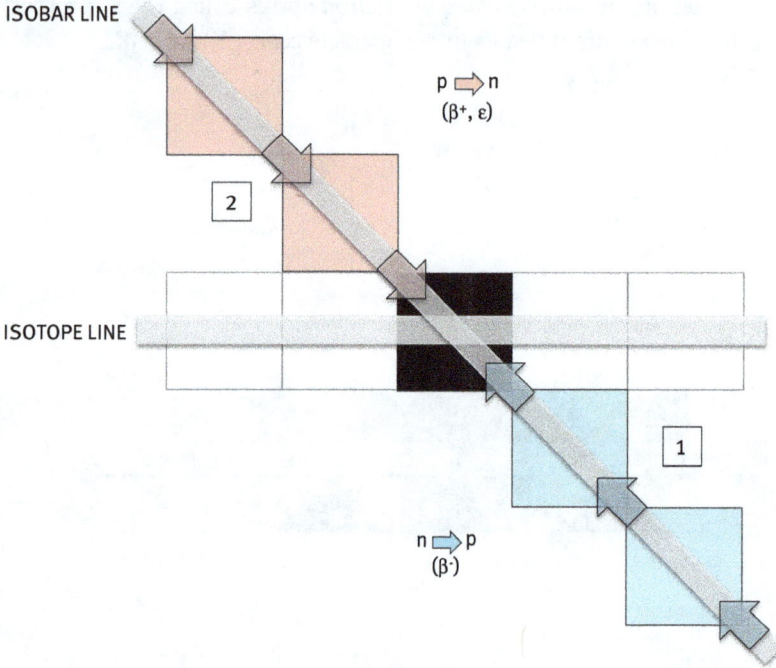

Fig. 6.9: β-Transformation of unstable nuclides. Neutron-rich nuclides transform *via* the neutron $\rightarrow$ proton conversion (i.e. the β^--process). The new nuclide K2 has a composition of $(Z+1, N-1)$ and arrives at nuclides of higher Z (1). Proton-rich nuclides utilize proton $\rightarrow$ neutron conversion (i.e. the β^+-process and electron capture) and yield a new nuclide K2 of $(Z-1, N+1)$ composition (2). Transformation may continue along the line of the same isobar stepwise, unless the Z to N ratio represents a stable nuclide. It will have adequate mean nucleon binding energy according to the liquid drop and shell models.

6.4.2 From isotopes to isobars

All β-transformations of unstable nuclides proceed with A = constant. Neutron-rich nuclides transform *via* the neutron $\rightarrow$ proton conversion. The new nuclide K2 has a composition of $(Z+1, N-1)$, representing a heavier chemical element.[7] Proton-rich nuclides utilize proton $\rightarrow$ neutron conversion and yield a new nuclide K2 of $(Z-1, N+1)$ composition. Here, K2 represents a chemical element of lower Z. Transformation may continue stepwise as *K1 $\rightarrow$ *K2 $\rightarrow$ *K3 etc., until the Z-to-N ratio reached represents a stable nuclide (see Fig. 6.9).

This is illustrated in more detail in Fig. 6.10. An isobar line is indicated at $A = 18$ with ^{18}O being the stable nuclide. The β^+- and electron capture processes approach ^{18}O from the proton-rich nuclides ^{18}Ne and ^{18}F, while the β^--processes accumulate at ^{18}O via the ^{18}B $\rightarrow$ ^{18}C $\rightarrow$ ^{18}N cascade.

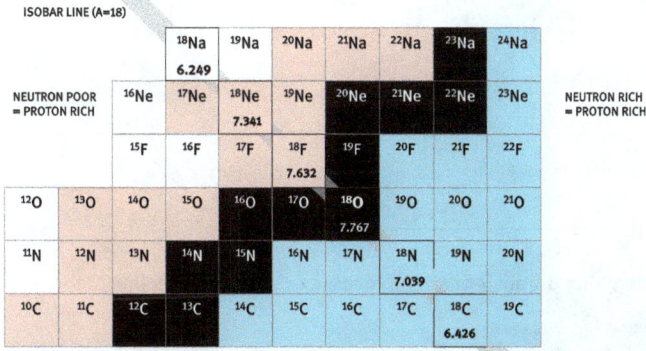

ISOBAR LINE (A=18)

	N=4	N=5	N=6	N=7	N=8	N=9	N=10	N=11	N=12	N=13
Na				^{18}Na (6.249)	^{19}Na	^{20}Na	^{21}Na	^{22}Na	^{23}Na	^{24}Na
Ne (NEUTRON POOR = PROTON RICH)			^{16}Ne	^{17}Ne	^{18}Ne (7.341)	^{19}Ne	^{20}Ne	^{21}Ne	^{22}Ne	^{23}Ne (NEUTRON RICH = PROTON RICH)
F			^{15}F	^{16}F	^{17}F	^{18}F (7.632)	^{19}F	^{20}F	^{21}F	^{22}F
O	^{12}O	^{13}O	^{14}O	^{15}O	^{16}O	^{17}O	^{18}O (7.767)	^{19}O	^{20}O	^{21}O
N	^{11}N	^{12}N	^{13}N	^{14}N	^{15}N	^{16}N	^{17}N	^{18}N (7.039)	^{19}N	^{20}N
C	^{10}C	^{11}C	^{12}C	^{13}C	^{14}C	^{15}C	^{16}C	^{17}C	^{18}C (6.426)	^{19}C

Fig. 6.10: β-transformation of unstable nuclides along the $A = 18$ isobar line with $\bar{E}_B$ values in MeV. The β^+- and electron capture processes approach ^{18}O from the proton-rich nuclides ^{18}Ne and ^{18}F, while the β^--processes accumulate at ^{18}O via ^{18}B $\rightarrow$ ^{18}C $\rightarrow$ ^{18}N $\rightarrow$.

There are many isobar lines across the Chart of Nuclides, ranging from short ones (e.g. $A = 3$ with the two nuclides ^{3}H and ^{3}He) to very long ones (e.g. $A = 100$ includes 15 nuclides). An isobar line then represents directions of β^--transformation and β^+- and/or EC processes (see Fig. 6.11).

For β^--transformation, β^+- and/or EC processes, there is $\Delta Z = +1$ and $\Delta Z = -1$, respectively. The trend in $\Delta\bar{E}_B = f(Z)$ at $A = $ constant is shown in Fig. 6.12. The maximum mean nucleon binding energy is located at the vertex of the parabola. Neighboring nuclides are (in general) unstable. According to the systematic illustrated in Fig. 6.11, mean nucleon binding energies increase more-or-less with every step (Z) approaching stable nucleon configuration. Typically, the value of $\Delta\bar{E}_B = f(Z)$ increases exponentially and this is reflected by the exponential expression of a parabola, shown also in

---

7 It is important to note that the process of type $_z^*$K1 $\rightarrow$ $_{z+1}$K2 creates an isotope of a heavier chemical element. The basic requirement of β-transformation holds true – the absolute mass of K2 is less than the absolute mass of *K1 – yet it is a heavier element!

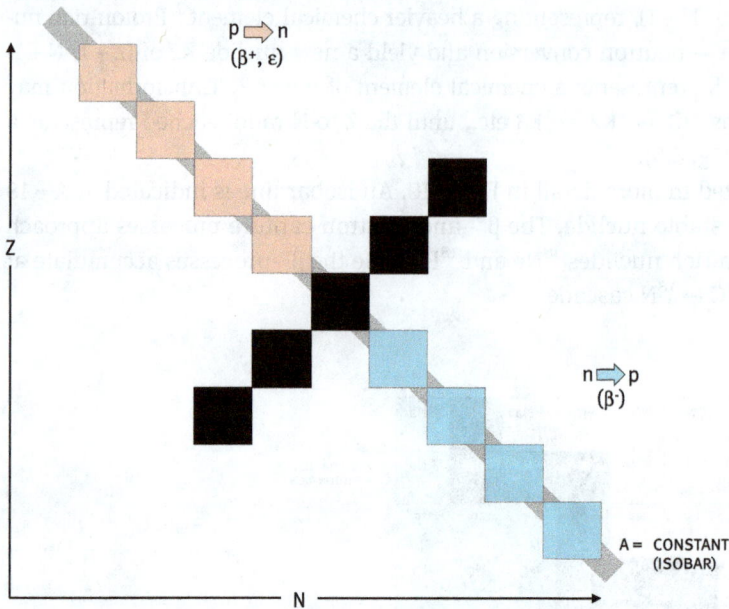

Fig. 6.11: Unstable nuclides undergoing β-transformation processes along an isobar line of the Chart of Nuclides.

Figs. 6.14 and 6.15. The blue nuclides from Fig. 6.11 (located on the right) shift towards the left-hand side of the parabola in Fig. 6.12. This is because they are of low Z compared to the red nuclides, which are of higher Z.

6.4.3 Turning one parabola of A = constant into the valley of β-stability

The two branches of β-processes (β⁻-transformation, and β⁺ and/or EC) accumulate at the same vertex of the parabola, but arrive from the left and right sides, respectively. Since the various terms of the WEIZSÄCKER equation all include a multiple of mass number A, the equation may be transformed for the value of Z which lies at the vertex of the parabola. The expression is $Z_A = f(A^{CONSTANT})$ and is obtained from eq. (3.1) with A = constant, and yields eq. (6.5). Z_A is the proton number with optimum mean nucleon binding energy.[8,9]

---

8 According to eq. (6.5), noninteger values may result for Z. In this case, the most stable nucleon configuration is next to this value.
9 Equation (6.5) uses the five main terms of the WEIZSÄCKER equation. If the equation were modified and extended, e.g. by the promiscuity term, this expression would slightly change.

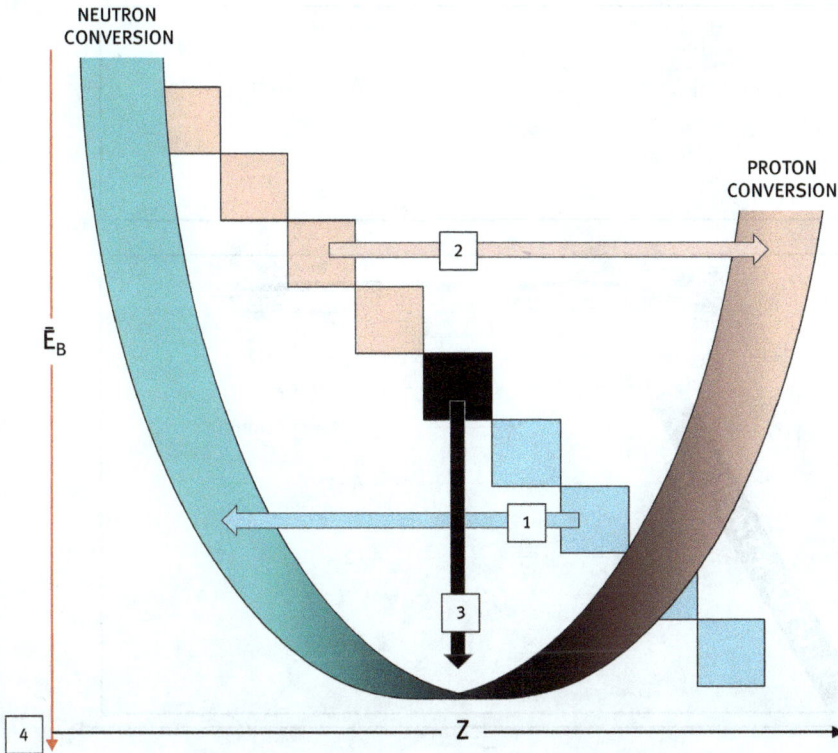

Fig. 6.12: Unstable nuclides undergoing β-transformation processes as (1) β⁻ route and (2) β⁺ or ε route along a coordinate system of mean nucleon binding energy $\bar{E}_B$ vs. Z at A = constant form a parabola. Note that the y-axis showing $\bar{E}_B$ is reversed – in contrast to how mean nucleon binding energy correlates with mass number. (4). The maximum mean nucleon binding energy is located at the vertex of the parabola (3). Interestingly (and unlike a normal parabola in algebra), the parabola here gets two "ends". The ends belong to the nucleon compositions that, according to the liquid drop model, are energetically not permitted. They reflect the proton or neutron separation energies forming the proton and neutron drop lines, as per Fig. 4.2.

$$Z_A = \frac{A}{2.0 + 0.0154A^{2/3}} \tag{6.5}$$

This is true for a single mass number A, and for the neighboring mass numbers as well. Each isobar line in the Chart of Nuclides thus has a maximum of mean nucleon binding energy for a specific value of Z.

Arranging these two-dimensional parabolas into a successive series creates a three-dimensional plot, shown in Fig. 6.13. Now a two-dimensional parabola has turned into a three-dimensional valley. Unstable nuclides are positioned along the hillsides, stable nuclides at the bottom of the valley. The latter is called the "valley of β-stability". The direction of the valley does not correspond to a straight line

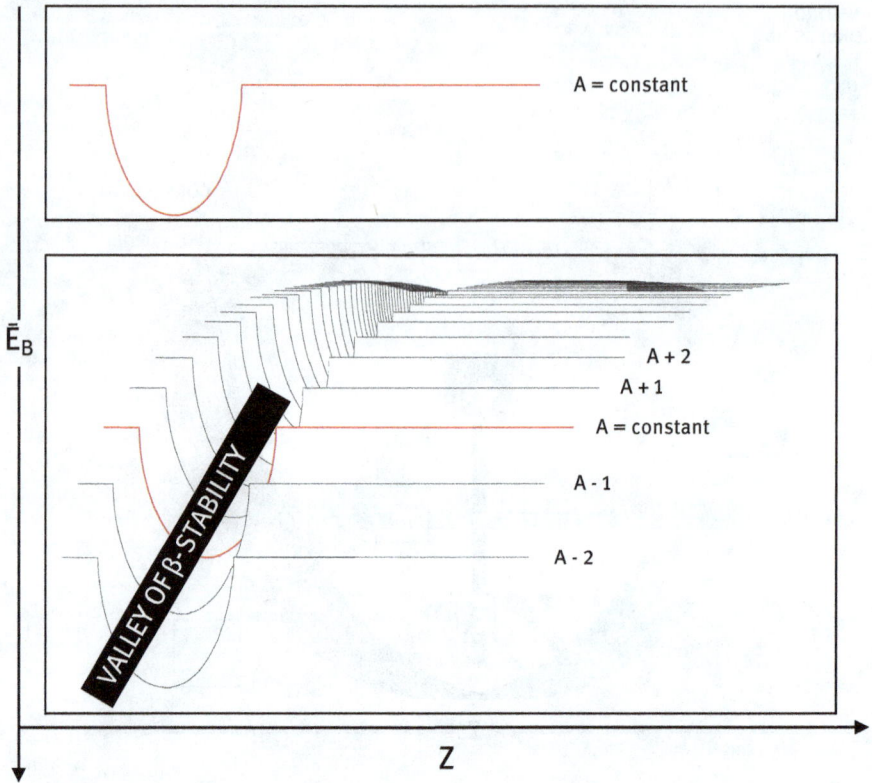

Fig. 6.13: The origin of the "valley of β-stability". Individual parabolas (as shown in Fig. 6.12) are arranged successively in order of increasing A, creating a "valley". Stable nuclides are positioned at the bottom of the valley, while unstable nuclides are along the hillsides.

(which would have been the isodiapher of (N − Z) = constant), but makes a sort of a soft turn (curve) towards higher Z-values.

The line of the valley represents the positioning of the stable nuclides within the Chart of Nuclides or *vice versa*. All the stable nuclides depicted in Fig. 3.3 of the Chart of Nuclide diagram lie in this "valley".

6.4.4 Details of the parabolas

Characteristic for the mathematics of a parabola, the two ascents become steeper and steeper. The x-axis, however, follows the linear scale of Z. This indicates that the differences in mean nucleon energy between successive transformations of $^*K1 \rightarrow K2$ are large at both "ends" of the parabola, and become less and less pronounced the

closer the transformation step is to the vertex. The sharper the ascent, the more unstable the nuclides. This is in good agreement with the LDM.

What about the complementary model of the nucleus of the atom: its shell
structure? Chapter 3 introduced the impact of specific occupancies of the various
nucleon shells. Ignoring magic numbers for a moment, *even* (= integer) nucleon
numbers are known to increase the mean nucleon binding energy. In part, this effect is covered by the WEIZSÄCKER equation too: it is the pairing term.[10] What are the
consequences in respect to β-processes? Starting from a composition of *K1 = (Z, N),
each step of nucleon conversion creates a new nuclide K2 of either (Z − 1),(N + 1) or
(Z + 1),(N − 1). There are 4 different version of this process, as reported in Tab. 6.2.
Consequently, any change in $\Delta\bar{E}_B$ caused by the coulomb term and the asymmetry
term is modified by the pairing term in a different way.

The impact of the pairing term is zero for (even, odd) and (odd, even) nuclides
*K1, because the new nuclide K2 is of the same category: (even, odd) turns into
(odd, even) and *vice versa*. This is the case for all isobars of odd mass numbers
A. Consequently, there is only one parabola for each mass number A, exemplified
in Fig. 6.14 for A = 95.

In contrast, an (odd, odd) nuclide *K1 turns into an (even, even) nuclide K2 and
an (even, even) *K1 turns into an (odd, odd) K2. In this case the pairing term mat
ters. The change in $\Delta\bar{E}_B$ caused by the coulomb term and the asymmetry term needs
correction as $\Delta\bar{E}_B^{(COULOMB + ASYM)} \pm \delta^{LDM}$. This happens for all isobars of even mass
number A. It must yield two separate curves as indicated in Fig. 6.15 for mass num-
ber A = 96.

Tab. 6.2: Changes of nucleon compositions in terms of odd or even nucleon
numbers relevant to the impact of the parity term of the Weizsäcker equation.

*K1	→	K2		$\Delta\delta^{LDM}$
(Z, N)	→	(Z − 1),(N + 1)	(Z + 1),(N − 1)	
(even, even)	→	(odd, odd)	(od, odd)	−
(odd, odd)	→	(even, even)	(even, even)	+
(even, odd)	→	(odd, even)	(odd, even)	0
(odd, even)	→	(even, odd)	(even, odd)	0

---

10 The pairing term was introduced to reflect the positive impact of nucleon composition of (even
Z, even N) nuclei in terms of a coefficient δ of positive sign. In contrast, the negative impact of (odd
Z, odd N) nucleon mixtures was addressed by a negative sign for δ^{LDM}. For (even, odd) and (odd,
even) compositions, the coefficient is $\delta^{LDM} = 0$.

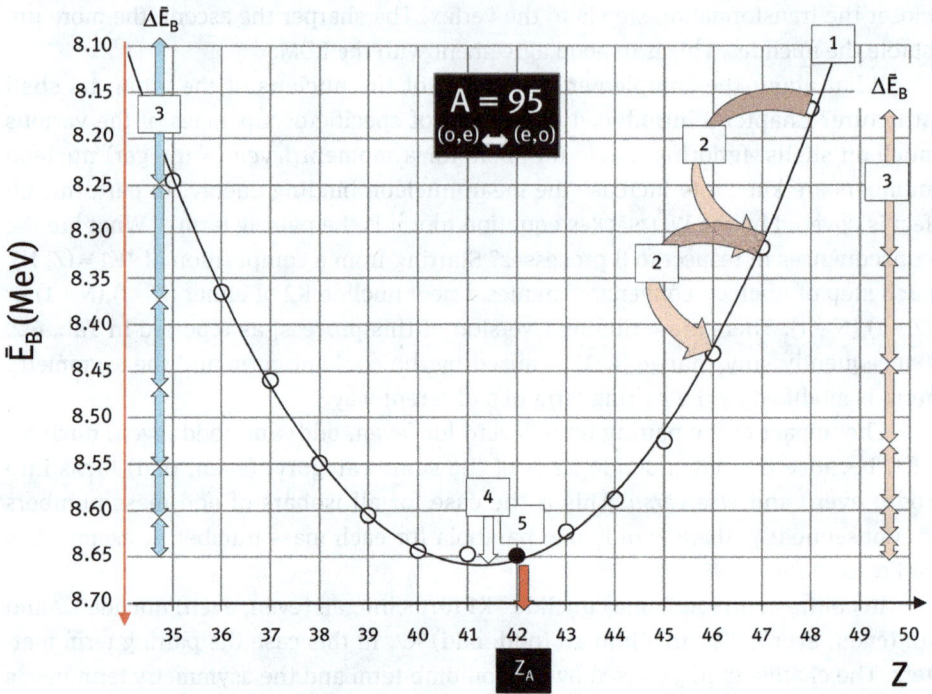

Fig. 6.14: β-parabola for mass number A = 95. Isobars of odd mass number A represent transformations of (even, odd) nuclides *K1 into (odd, even) nuclides K2 and *vice versa*. In the figure, (o,e) stands for (odd, even) and (e,o) for (even, odd). The impact of the pairing term is zero, so there is only one parabola (1). (2) The successive transformations of type β$^+$ or ε are shown at the right-hand side, and of type β$^-$ at the left-hand side (not shown). (3) Decreasing differences in mean nucleon binding energy are seen when approaching the vertex of the parabola. (4) Z_A is 40.937. (5) The only stable nuclide is ^{95}Mo (Z = 42) with the largest value of $\bar{E}_B$.

6.5 α-Emission, cluster emission, spontaneous fission processes (ΔA ≠ 0)

6.5.1 From β-transformation to α-emission

The β-transformations of nuclides at A = constant proceed along the line of isobars. Light- and medium-mass nuclides, but also many heavy nuclides, almost exclusively transform *via* β-processes and culminate within the valley of β-stability as stable nuclides of optimum Z_A. However, the Chart of Nuclides reveals a gap after mass number 208 and 209 for the heaviest stable nuclides available, namely ^{208}Pb and ^{209}Bi at magic numbers of 82 protons and/or 126 neutrons (see Fig. 6.5).

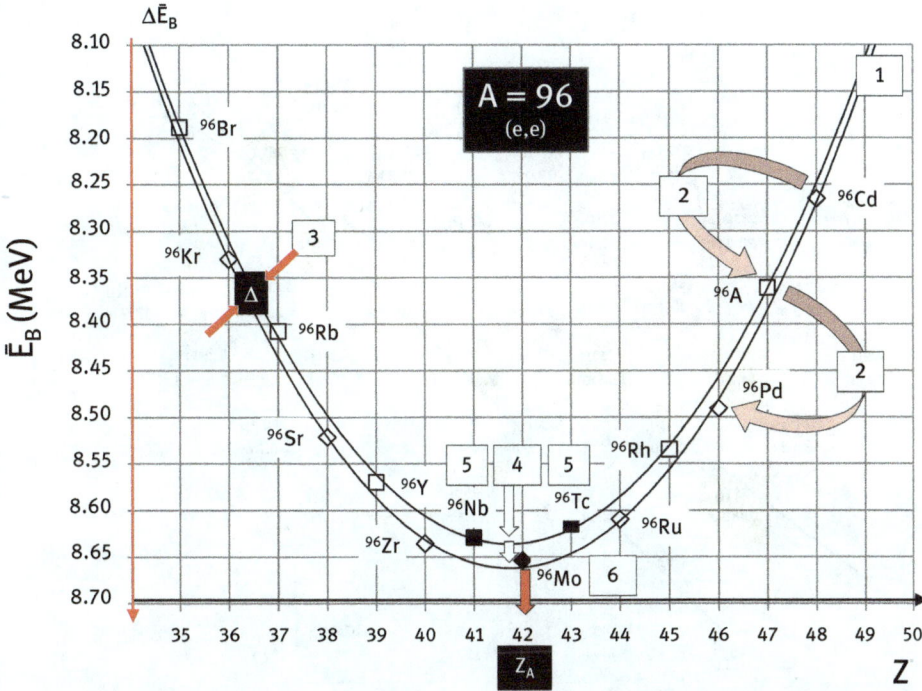

Fig. 6.15: β-Parabolas for mass number A = 96. Isobars of even mass number A represent transformations of (even, even) nuclides *K1 into (odd, odd) nuclides K2 and *vice versa*. In the figure, (e,e) stands for (even,even). The impact of the parity term of the Weizsäcker equation creates two parabolas, shifted in $\bar{E}_B$ by ± δ^{LDM}. (1) Two separate parabolas arise; the one for (even, even) nuclides is shown "below" (i.e., at higher values of $\bar{E}_B$). (2) The successive transformations of type β⁺ or ε are shown on the right-hand side, and of type β⁻ at the left-hand side (not shown). These alternate from (even, even) to (odd, odd) nuclides and so on. (3) The shift in energy between the two parabolas is Δ = 2δ. (4) Z_A is 41.328. (5) The most stable nuclide is ⁹⁶Mo (with the largest nucleon binding energy). The stability of the two (odd, odd) nuclides needs to be studied in detail. In this case, ⁹⁶Ru is also stable, while ⁹⁶Zr has a half-life of 3.9·10¹⁹ years (!)[11].

---

[11] This leads back to the question of the number of stable nuclides needed as an experimental database for developing models on the structure of nuclei of atoms. Beta transformations create stable nuclides for A = 1 – 209. Among these 210 mass numbers, there are 106 uneven numbers representing (e, o) or (o, e) nuclides. In these cases, there is only one stable nuclide (except for Z = 61, promethium). The other 104 mass numbers represent even values of A, and β-transformation switch between (e, e) and (o, o) nuclei. Here, there is at least one stable nuclide of (e, e) composition representing the ultimate value of $\bar{E}_B$ (except for Z = 42, technetium). This makes 105 + 103 = 208 stable nuclides. In 42 cases, there are two stable (o, o) nuclei, and for mass numbers 124 and 132 there are three, i.e. 208 + 42 + 2 × 2 = 254 stable nuclides. There are many (e, e) with very long half-lives, orders of magnitude larger than the age of the earth. The question is, how many of the remaining (o, o)

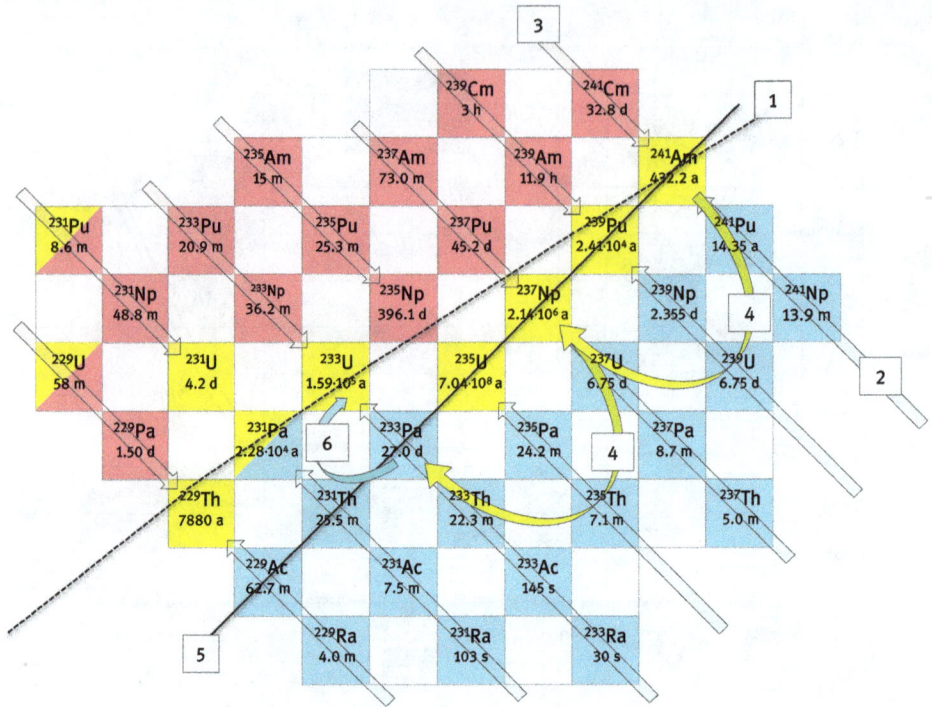

Fig. 6.16: Excerpt of the Chart of Nuclides for nuclides of mass number A = 229, 231, 233, 235, 237, 239 and 241 with half-lives indicated: The need for a new primary transformation pathway. α-emitting nuclides are color-coded in yellow. The dotted line represents the vertex of the β-parabolas (1), the arrows indicate the two arms of the parabola, driven by β⁻ (blue, (2)) and β⁺ and/or EC (red, (3)) primary transformations. The β-transformation definitely ends at the vertex of each parabola – but all the nuclides are (still) unstable (although of long half-life in the cases selected). The α-emission proceeds at $\Delta Z = -2$ and $\Delta A = -4$ and is indicated for successive transformations of ^{241}Am $\rightarrow$ ^{237}Np $\rightarrow$ ^{233}Th (4). These transformations are parallel to the isodiaphere line (5). ^{233}Pa continues by β⁻-transformation (6) along its β-parabola of A = 233 = constant to ^{233}U, which is at the vertex of that β-parabola. It stabilizes by α-emission.

Still, there are isobar lines for mass numbers of A > 208. These reach values of A = 252–262 for the heaviest lanthanide, lawrencium (Z =103) and approach 280 for the transactinide elements. Figure 6.16 depicts the situation for unstable nuclides of mass numbers A = 229, 231, 233, 235, 237, 239 and 241. The unstable nuclides representing these isobars follow the β-transformation character as described. The individual maximum of mean nucleon binding energy for those parabolas is reached for

---

nuclides are "really" stable? There are several candidates, such as ^{40}K, ^{50}V, ^{138}La, ^{176}Lu or ^{180}Ta (t½ = 1.28·10⁹ a, 1.4·10¹⁷ a, 1.05·10¹¹ a, 3.8·10¹⁰ a or >10¹⁵ a, respectively), which are not "really" stable. Only four are really stable: ^{2}H, ^{6}Li, ^{10}Be, ^{14}N. Thus, there are 258 "really" stable nuclides altogether.

the following nuclides: ^{229}Th, ^{231}Pa, ^{233}U, ^{235}U, ^{237}Np, ^{239}Pu, and ^{241}Am. However, the nuclides closest to the vertex of each parabola are not stable! And, there are no more options according to the β-transformation profile!

This is the moment stabilization of unstable nuclides cannot continue following the A = constant strategy. So, what should this unstable nuclide do? There are two classes of transformations where ^{A*}K2 < ^{A}K1: cluster emission (the most common is emission of α-particles) and spontaneous fission (see Fig. 6.7).

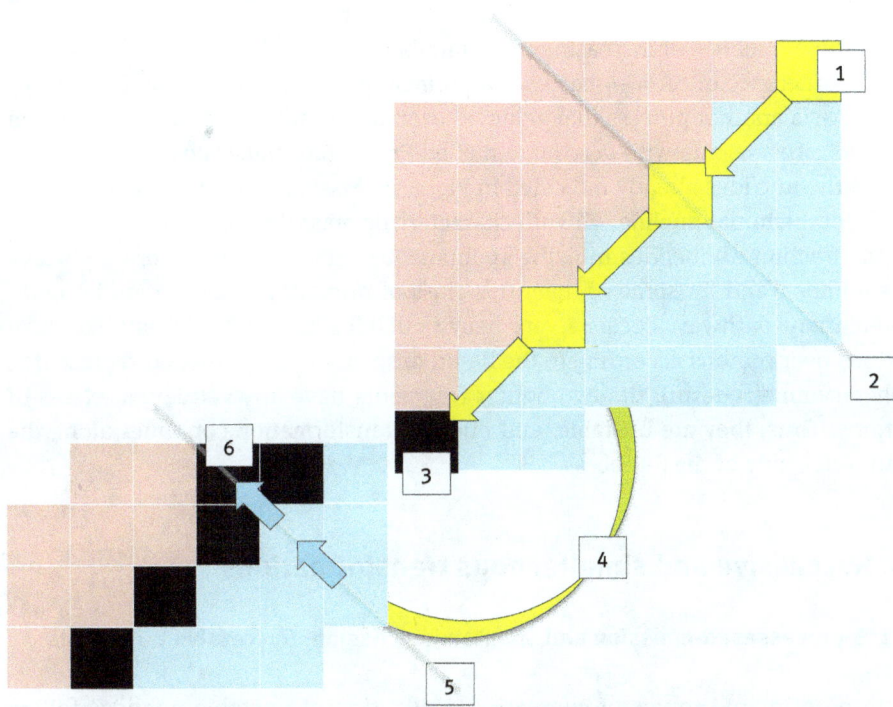

Fig. 6.17: The relation between β-transformation and α-emission processes. α-emission occurs at heavier mass number (1) and is most prominent at the vertex of isobar parabolas (2) of A > 209. It either ends at a stable nuclide (3) or at neutron-rich parts of isobar parabolas of lower mass number, and those unstable nuclides continue to stabilize *via* β-transformation within one more (4) step along the corresponding isobar parabolas (5) to finally terminates at the stable nuclide (6), either through a terminal α- or a β-transformation step (here shown for the latter version).

The α-emission process typically starts at the nuclide at the vertex of a parabola of A > 209.[12] Individual α-transformations or series of α-emissions proceed along a straight line paralleling the isodiaphere. This either produces the stable nucleon

---

12 But may also take place at "hillside" nuclides, see Chapter 9.

configuration directly, or an intermediate nuclide obtained preferably transforms *via* β-transformation processes.[13] The relationship between the two different classes of primary transformations is illustrated in Fig. 6.17. It shows how α-emission at heavier mass number is needed, and how it continues into β-transformation once lower mass numbers have been reached.

6.5.2 α-emission and spontaneous fission

For unstable nuclides of increasing mass number of about A > 234, α-emission with its mass difference of $\Delta A = -4$ becomes a promising choice of transformation. However, there is another process that achieves even larger differences in ΔA, as shown in Fig. 6.7. This occurs with very large nuclei. Figure 6.18 illustrates this option for some of the nuclides already indicated in Fig. 6.16. Fission yields two fragments, K2 and K3, of the initial nuclide *K1 with characteristic mass distributions.

The fraction of nuclides utilizing spontaneous fission increases for very heavy mass number, and for some of them this class of primary transformation becomes the dominant pathway. Because very heavy nuclides have a significant excess of neutrons over protons according to the liquid drop model (in particular represented by the asymmetry term), the two lighter fragments have an even larger excess of neutrons. Thus, they are unstable, and nuclide transformation continues along the neutron-rich side of the β-process.

6.6 Successive and simultaneous transformations

6.6.1 β-processes, α-emission and spontaneous fission: Successive transitions

The three principal routes of energetic stabilization of unstable nuclides follow their own specific characteristics. There are two strategies to minimize energy (i.e., mass): either by directly reducing the mass number A (α-transformation or sf) or by converting one sort of nucleon into the other (β-transformation). The three processes are summarized in Fig. 6.19.

---

13 A representative example was mentioned in the context of the natural decay chains, as per Fig. 5.23. The A = 4 n + 2 family includes the prominent nuclide ^{226}Ra. This was generated from ^{234}U and ^{230}Th *via* α-emission, continuing *via* α-emission to ^{222}Rn, then ^{218}Po and further to ^{214}Pb. Here the α-chain stops, and at A = 214 transformation is preferred following the β-characteristics "downhill" the neutron-rich side of the parabola at A = 214 = constant *via* ^{214}Pb (β⁻)→ ^{214}Bi (β⁻)→ ^{214}Po. The optimum value of $\bar{E}_{B(A=214)}$ is reached here for unstable ^{214}Po, and subsequent transformations continue *via* α-emission.

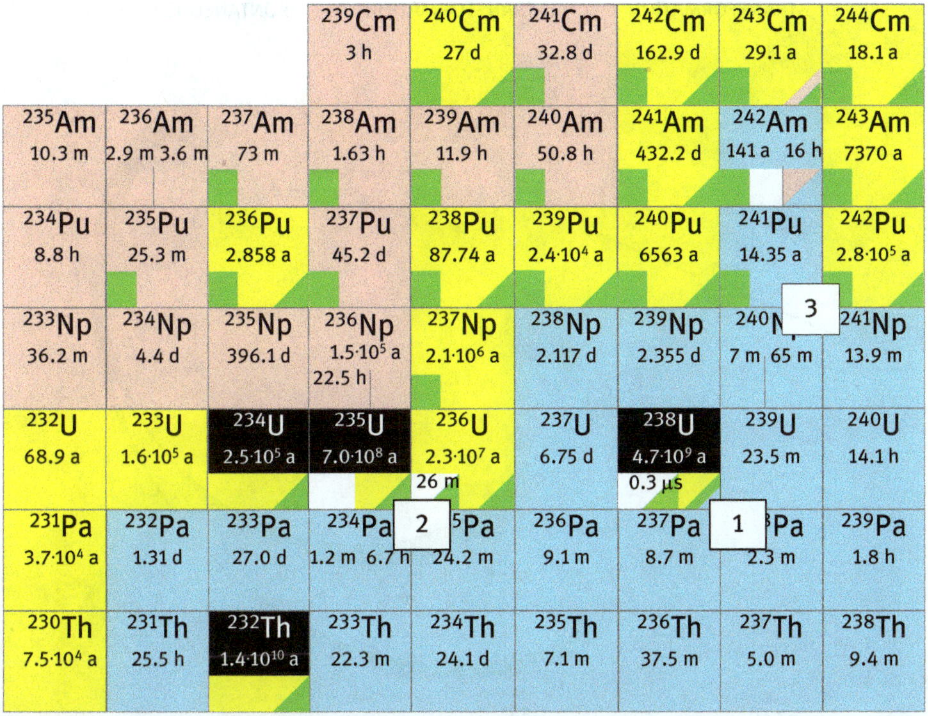

			^{239}Cm 3 h	^{240}Cm 27 d	^{241}Cm 32.8 d	^{242}Cm 162.9 d	^{243}Cm 29.1 a	^{244}Cm 18.1 a
^{235}Am 10.3 m	^{236}Am 2.9 m 3.6 m	^{237}Am 73 m	^{238}Am 1.63 h	^{239}Am 11.9 h	^{240}Am 50.8 h	^{241}Am 432.2 d	^{242}Am 141 a 16 h	^{243}Am 7370 a
^{234}Pu 8.8 h	^{235}Pu 25.3 m	^{236}Pu 2.858 a	^{237}Pu 45.2 d	^{238}Pu 87.74 a	^{239}Pu $2.4 \cdot 10^4$ a	^{240}Pu 6563 a	^{241}Pu 14.35 a	^{242}Pu $2.8 \cdot 10^5$ a
^{233}Np 36.2 m	^{234}Np 4.4 d	^{235}Np 396.1 d	^{236}Np $1.5 \cdot 10^5$ a 22.5 h	^{237}Np $2.1 \cdot 10^6$ a	^{238}Np 2.117 d	^{239}Np 2.355 d	^{240}Np 7 m 65 m (3)	^{241}Np 13.9 m
^{232}U 68.9 a	^{233}U $1.6 \cdot 10^5$ a	^{234}U $2.5 \cdot 10^5$ a	^{235}U $7.0 \cdot 10^8$ a	^{236}U $2.3 \cdot 10^7$ a 26 m	^{237}U 6.75 d	^{238}U $4.7 \cdot 10^9$ a 0.3 μs	^{239}U 23.5 m	^{240}U 14.1 h
^{231}Pa $3.7 \cdot 10^4$ a	^{232}Pa 1.31 d	^{233}Pa 27.0 d	^{234}Pa 1.2 m 6.7 h	^{235}Pa (2) 24.2 m	^{236}Pa 9.1 m	^{237}Pa 8.7 m	^{238}Pa (1) 2.3 m	^{239}Pa 1.8 h
^{230}Th $7.5 \cdot 10^4$ a	^{231}Th 25.5 h	^{232}Th $1.4 \cdot 10^{10}$ a	^{233}Th 22.3 m	^{234}Th 24.1 d	^{235}Th 7.1 m	^{236}Th 37.5 m	^{237}Th 5.0 m	^{238}Th 9.4 m

Fig. 6.18: Spontaneous fission for nuclides of mass number A = 231 to 241. Nuclides undergoing fission are color-coded in green (circles (1)). One of the nuclides undergoing fission is ^{238}U (2). It splits into two fragments of statistic mass distribution, with one fragment peaking around A = 140 (3), the other around A = 100 (d).

6.6.2 β-processes, α-emission and spontaneous fission: Simultaneous transitions

However, a given unstable nuclide may not only "choose" one of the three primary options – there are many radionuclides that can undergo two or even all three routes of transformation. In this case, ΔE is positive for all pathways, but the individual fractions are specific for each nuclide. For example, the prominent radionuclide uranium-235 shows a branch of α-emission and a branch of spontaneous fission. Spontaneous fission occurs rarely, in just 0.0014% of the transmissions. This means that given 1,000,000 atoms of ^{235}U, the majority (999,986) transform *via* α-emission, while only 14 undergo fission. The neutron-poor plutonium isotope 237 transforms basically *via* electron capture, but α-emission and spontaneous fission also occur. The neutron-rich plutonium isotope 241 transforms basically *via* β⁻ emission, but α-emission and spontaneous fission also occur, as shown in Fig. 6.20.

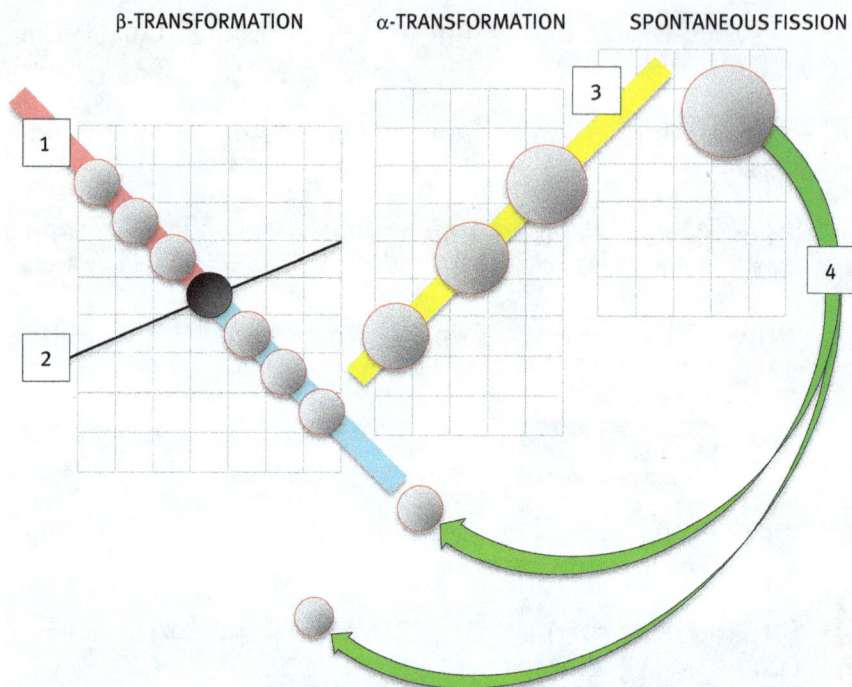

Fig. 6.19: Schematic comparison of the three main primary transformation processes of unstable nuclei. The mass number A remains constant for β-transformation and proceeds along an isobar line (1) and terminates at a stable nuclide positioned in the valley of β-stability (2). In the case of an α-transformation, mass numbers A are successively reduced by 4 along an isodiaphere (3). Because heavy nuclei start from a large excess of neutrons over protons (WEIZSÄCKER asymmetry term), transformation product nuclides are even more neutron-rich. Consequently, α-emission transits along β-transformation at the neutron-rich arm of the β-parabola. Thus, α-emission series overlap with β-transformation. In the case of spontaneous fission, a nuclide of very large mass number A splits into two fragments. It originates from a nuclide of very large excess of neutrons over protons, and the two fission product nuclides are even more neutron-rich and quite distant from the isodiaphere (4) of the initial nuclide. Each will undergo several steps of β-transformation until a stable nuclide is formed.

A similar idea of co-existing transformation routes for one nuclide holds true for the class of β-emissions. Nuclides such as ^{40}K or ^{64}Cu undergo all three subtypes – though at different abundances – as per Fig. 6.21.

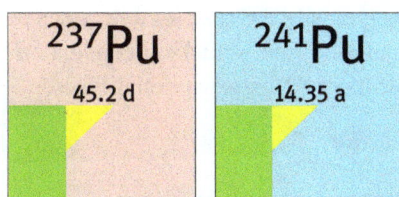

Fig. 6.20: Various transformation routes may apply to the same nuclide. The neutron-poor plutonium isotope 237 transforms basically *via* electron capture, but α-emission and spontaneous fission also occur. The neutron-rich plutonium isotope 241 transforms basically *via* β⁻-emission, but α-emission and spontaneous fission also occur. The colors are red for β⁺ and EC, blue for β⁻, yellow for α-emission, and green for spontaneous fission. The relative size of the different areas of one nuclide square indicates the probability of a certain transformation pathway.

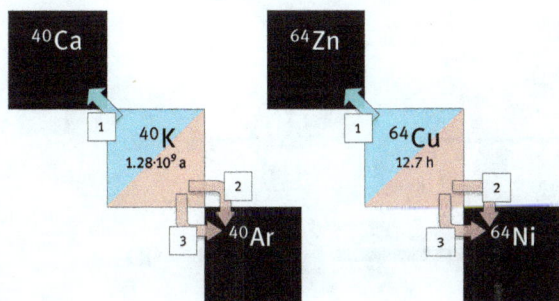

Fig. 6.21: Various transformation subtypes of β-emissions may apply to the same nuclide. ^{40}K or ^{64}Cu undergo all three subtypes, though at different abundances. For ^{40}K it is ≈89.3% β⁻-emission (1), ≈0.001% β⁺-emission (2) and ≈10.7% electron capture (3), while for ^{64}Cu it is ≈39.0% β⁻-emission (1), ≈17.9% β⁺-emission (2), and ≈43.1% electron capture (3).

6.7 Secondary transitions

6.7.1 Nuclear ground state and excited nuclear states

Chapter 3 introduced nuclear shell models applied to ground state levels of the nuclei. Primary transformation processes result in new numbers of protons and neutrons, i.e. the set of (Z, N) of the initial nuclide *K1 has changed.

However, not only the absolute number of each sort of nucleon matters due to the LDM and the corresponding WEIZSÄCKER equation, but their arrangements do too. Primary nuclear transformation does not necessarily result in proton and neutron shell occupancies that reflect the ground state levels of the corresponding nucleus as illustrated in Fig. 6.22. In contrast, the exothermic process of spontaneous primary transformation seen in unstable nuclides allows for higher energy levels in the outer shells of the nucleus to become populated. In fact, this is typically the case for all classes of primary transformation: spontaneous fission, α-emission and β-nucleon conversion. Direct, one-step processes from *K1 to the ground state of K2 are exceptional cases.

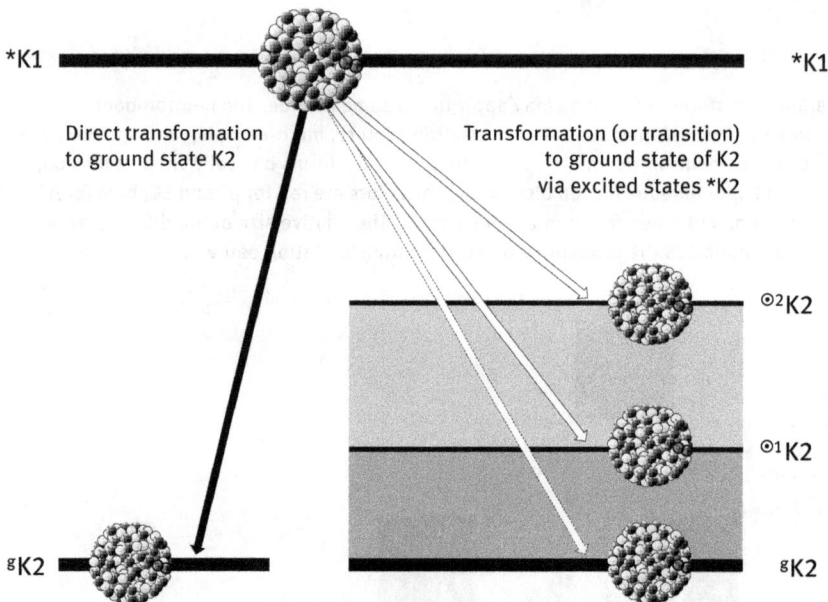

Fig. 6.22: Primary transformations (spontaneous fission, α-emission and β-nucleon conversion) tend to populate one or more high energy nucleon shells in the newly formed nucleus $^\odot$K2, rather than transforming directly and in one step to the ground state of K2.

There may even be several excited states $^{\odot i}$K2, depending on the number of sub-shells available and the energy difference between them. With more protons and neutrons per nucleus, the number of densely overlapping shell levels increases; this is paralleled by an increase in potential excited states. Each nuclear state is defined by the characteristic quantum parameters, such as nuclear spin and parity, as discussed in Chapter 7. Rather than the excited state energies of nuclide K2, it is the nuclear spin and parity of each nucleon which dictate why a primary transformation may terminate at a higher-energy nucleon shell. In other words, the unstable nucleus K1 prefers an "easy" pathway, not a "hindered" one.

What could a "hindrance" look like? This refers to changes in spin and parity between the transforming nucleus K1 and the energetic state of K2. If both the spin and parity values do not change, the transition is very easy – even called "super-allowed", cf. Figure 6.23. However, if the spin difference increases and the parity alternates, the transformation is considered "forbidden". This will be discussed further in Chapter 8.5.1.

Similar to electron excitation and de-excitation (i.e. relaxation) as discussed for the hydrogen emission spectrum, excited nucleons tend to leave high-energy shell

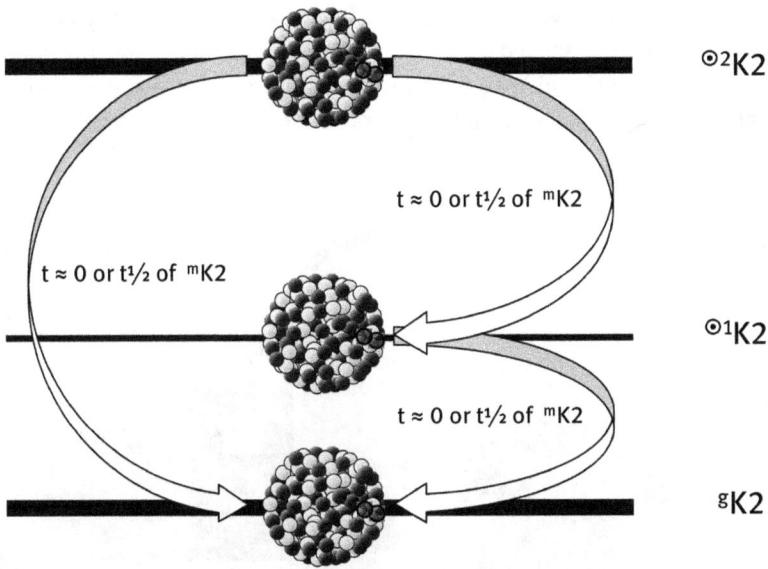

$$^{\odot 2}K2$$

t ≈ 0 or t½ of ^{m}K2

t ≈ 0 or t½ of ^{m}K2

$$^{\odot 1}K2$$

t ≈ 0 or t½ of ^{m}K2

$$^{g}K2$$

Fig. 6.23: Transitions between excited levels and between an excited and ground state. Transitions either proceed very fast (within about 10^{-12} s) or the excited state remains for a much longer period of time. This is indicated by having a "real" half-life.

levels for lower lying states in order to energetically relax. This corresponds to processes between individual excited levels of type $^{\odot 1}K2 \rightarrow {}^{\odot 2}K2$. More generally, this is written as $^{\odot i}K2 \rightarrow {}^{\odot (i-1)}K2$, but can also appear as $^{\odot i}K2 \rightarrow {}^{\odot (i-2)}K2$ and so on, like the hydrogen electron transitions. In parallel, transitions can proceed from excited levels to the ground state, i.e. $^{\odot i}K2 \rightarrow K2$, which represents the lowest energy available for the given nucleus.

This de-excitation stays within the nucleon configuration of K2, i.e. there are no changes in proton and neutron numbers. The most frequently found mechanism for this process is emission of energy in the form of electromagnetic radiation (γ-energy, photons); see Chapter 11 for details of the secondary nuclear transformation processes.

6.7.2 Unstable and metastable nuclides

The de-excitation processes in most cases proceed very quickly, with transitions being terminated within about 10^{-12} s post primary nuclear transition. However, some excited states may "survive" for longer periods of time, as per Fig. 6.24. These states thus exist for a much longer period of time than the usual transition speed. "Much" longer is a relative term, and in this case indicates factors of 100 to 1000 or more. This is known as "metastability". By definition, metastability is the prolonged duration of certain states[14] and applies to several systems in physics and chemistry. It is also known, for example, for electron excited states.[15] The definition of metastable nuclides actually refers – in analogy to chemistry – to *isomer forms* of the same nucleus. Those species are thus also referred to as metastable isomers ^{m}K2, or simply "isomers".[16]

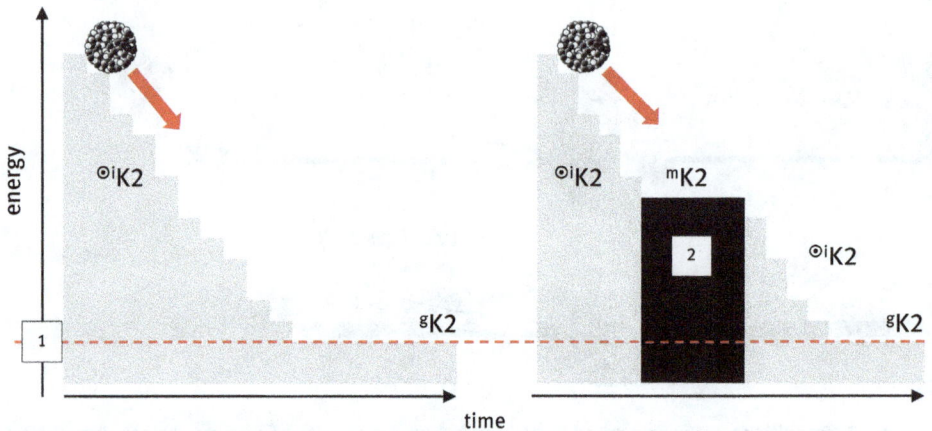

Fig. 6.24: Schematic illustration of metastable nuclei within successive transitions. (left) The excited nucleus relaxes step-by =step, thereby minimizing the energy of the nucleus. The dimension of the nucleus is depicted so that each step's energy difference is too small for the nucleus to remain there. It will immediately "fall down" the stairway to finally reach the energy of the ground state (1) of the nuclide ^{g}K2. (right) One step (2) in the stairway is sufficiently wide for the nucleus to remain for a longer duration in a metastable nucleon configuration ^{m}K2.

The excited level involved may be identified by its own half-life. The corresponding state of $^⊙$K2 is called "metastable". Metastable nuclide half-lives range from about 10^{-9} s to 10^{15} years (^{180m}Ta). The symbol of a metastable state of a nuclide becomes ^{m}K2, and in these cases the final ground state is denoted by ^{g}K. Within the isotope notation, this gives ^{99m}Tc for the metastable technetium-99 (t½ = 6.0 h) *versus* ^{99g}Tc for the ground-state technetium-99 (t½ = 2.1 · 10^5 a).

The Chart of Nuclides involves more than 3000 "boxes" identifying stable and unstable nucleon configurations. In addition, about 700 of these nuclides show one or

14 "Meta" refers to the virtual half-lives of the transitions between the other excited levels for the same nucleus.
15 Excited electrons relax within about 10^{-8} s. Metastable electron excited states have durations of about 10^{-3} s, i.e. several orders of magnitude higher. Within a cascade of subsequent electron transitions, this particular level will thus accumulate (or trap) electrons. This "population inversion" is very important for the design and use of lasers.
16 In the early years of nuclear and radiochemistry, when for example the successive daughters of the natural decay chains were analyzed chemically and spectroscopically, the nuclear parameters of metastable nuclides were detected – but not necessarily attributed to an isomer of the same nuclide.

more metastable isomer(s). Some of these metastable radionuclides are of interest in fundamental research, and others have important practical applications. Such an example is depicted in Fig. 11.24 for the ground and metastable states of technetium-99.[17]

6.8 Post-processes

β-processes, α-emission and spontaneous fission are primary transformations leading to either the ground state (K2) or an excited state (*K2) of the new nuclide. Excited levels relax *via* three subtypes of secondary transitions. Some primary and secondary transformations, however, induce additional inherent processes, which either continue as the final step of the nuclide's transformation or occur parallel to it. These post-effects are responsible for further radiative emissions beyond the basic features of primary and secondary transformations.

One may classify post-processes in two categories. Category I refers to phenomena originating inside the transforming atom, causing a "hole" in a certain electron shell. This "hole" turns into a "vacancy" which must be filled immediately. This proceeds via emission of new kinds of radiation: (a) electromagnetic radiation called X-rays, and (b) ejection of electrons from all shells of the atom, called AUGER and COSTER–KRONIG electrons.

Category II refers to phenomena which occur outside the transformed atom. It directly causes another kind of radiation, namely annihilation radiation (c). This is a logical consequence of the emission of a positron following primary β⁺-processes or the creation of a positron during pair-formation. The subsequent annihilation of that positron with a nearby electron in the surrounding matter creates exactly 511 keV of electromagnetic radiation.

Finally, electrons emitted during transformations travel through nearby condensed matter and interact with the atoms, both their nuclei and orbiting electrons.[18]

---

17 This isomer is used in nuclear medicine diagnosis by means of Single Photon Emission Computed Tomography (SPECT). Due to its availability from a ^{99}Mo/^{99m}Tc radionuclide generator and the success of radiopharmaceutical chemistry to develop various biologically relevant ^{99m}Tc-labeled molecules, ^{99m}Tc-SPECT imaging is performed in more than 100,000 patients worldwide every day (discussed in Chapter 13 of Volume II).

18 Independently, the electrons of surrounding atoms represent a target for almost all kinds of radiation emitted in transformations. When hitting an electron, radiation may release it and create a so-called ion pair, consisting of the positively charged ion and negatively charged "free" electron. Such processes represent the effect of radiation on condensed matter (in particular condensed biological "matter" such as humans) – and are the reason that radioactivity is commonly referred to as "ionizing radiation". They are not discussed in this textbook (Volume I: Introduction) because they do not yield a new kind of "radiation", but are covered in detail in Volume II: Modern Applications. Chapter 1 is on Radiation Measurement (and obviously, this relates to the interaction of radiation with well-defined condensed matter – the detector); Chapter 2 is on Radiation Dosimetry.

When traveling near the nucleus, beta-electrons of a certain initial kinetic energy can be slowed down or deflected, reducing their kinetic energy. The difference in energy is released as "bremsstrahlung", and represents another, new kind of radiation.

6.9 Outlook

The general principle of primary transformation of unstable nuclides is minimizing mass (which is energy). This is "organized" by conversion of nucleons (β-processes), emission of nucleons (α-emission), and spontaneous fission (sf).

In particular, β-processes and α-emission in most cases must be discussed in terms of the ground state and/or excited state(s) of the newly formed nucleus K2. This is the topic of secondary transition phenomena. This class of transformation is accompanied by one more sort of radiative emission, not directly belonging to the primary transformations: electromagnetic radiation, in particular photons. In parallel, secondary transformations "produce" electrons as well – but not *beta* electrons.

While the two subsequent pathways exclusively occur within the nucleus, there is finally a group of post-processes which contribute to the complete spectrum of radiative emissions of an unstable nuclide, originating from the shell of the transforming nuclides.

Specific discussion of these discrete processes needs to consider how:
- the individual pathways may be understood conceptually (in terms of theoretical and mathematical models),
- the value of ΔE corresponds to the velocity of the transformation (i.e. its half-life)
- the overall energy is disseminated among the products (particularly in terms of kinetic energy of the emitted particle "x" and/or "y").

These aspects are covered in the following chapters for the three types of β-processes, for α-emission, and for spontaneous fission.[19] The following chapters focus on the details of secondary transformations and of the two post-processes.

---

19 In contrast, some very special transformations are not treated here, mainly because they are almost irrelevant to nuclear and radiochemistry. One type is single or double proton emission, which is observed for very proton-rich nuclides close or beyond the proton drip line.

7 β-Transformations I: Elementary particles

Aim: For most unstable nuclides the dominant pathway of transformation is the β-process Its essence is turning either a neutron into a proton or *vice versa*, thus keeping the overall mass number constant. It involves either emission or capture of an electron, and there are three subtypes accordingly. These emissions and the secondary and post-processes they induce all contribute to a complex spectrum of radiative emissions.

Nucleon conversion processes are a consequence of processes of elementary particles not yet discussed: quarks, electron neutrinos and anti-electron neutrinos, as well as the corresponding mediators. Elementary particles are treated within the standard theory of particle physics. This chapter discusses the structure of quarks and leptons (forming the group of fermions) and the field quanta (mediators) allowing the interaction between fermions. These elementary particles form three families. In addition, each fermion has its antiparticle, thus extending and completing the exciting world of elementary particles.

Among the many different elementary particles, however, a rather limited number is needed in the context of β-transformations. These are the up quark and down quark, the electron and the anti-electron (the positron), the electron neutrino and the electron antineutrino, and finally the W bosons.

The characteristics of nucleon conversion also apply to nuclide transformations. This process is defined by quantum physics in a theory called "FERMI'S golden rule".

7.1 Elementary particles

7.1.1 The concept of elementary particles

Despite the general character of radionuclide transformation discussed in the context of the subatomic components of nucleons and electron introduced so far, the understanding of the β-process requires a more detailed, basically nuclear physical consideration.[1]

The three subatomic particles (electron, proton and neutron) were introduced in Chapter 1 as separate components of an atom located in its different compartments (the electrons in the shell, the two nucleons in the nucleus). In the present chapter

---

[1] This is different to the other two primary transitions. When accepting a "liquid drop" model of a nucleus of an atom, one would be willing to fancy the emission of an α-particle from a given nucleus, or even the fission of a large nucleus into two fragments. In contrast, when thinking about "subatomic particles" introduced so far, one should doubt the idea that one sort of a nucleon simply "turns" into the other one.

https://doi.org/10.1515/9783110742725-007

they appear as actors in a process of primary nuclide transformation. Whenever one sort of nucleon converts into the other one, the electron plays a significant role:
- for the β⁻- and β⁺-processes the electron is "created" and then emitted from the nuclide;
- in the electron capture process, an existing (inner shell) electron is captured by the nucleus of the atom to "combine" there with a proton.

In order to understand the processes responsible for these events it is not sufficient just to discuss the three subatomic particles introduced so far and their interaction – instead, it is necessary to go beyond the structure of nucleons. While the three particles until this stage have been classified as "subatomic", it does not necessarily mean that all of them are not further divisible. This is true for the electron, which is therefore classified as an "elementary particle". This is not true for the proton and the neutron! In contrast, the proton and the neutron are composed of other, subnucleon particles. This is the moment to consult the concept of elementary particles. Continuing in the tradition of the great Greek philosophers, elementary particles are the ones that are not further divisible. Note that this allegorizes the ultimate idea: Turning from visible matter to conceptual atoms, from experimentally proven atoms to evidence of subatomic particles, and finally postulating the elementary particles!

According to the development of the *standard theory of particle physics,* there are classes of elementary particles, arranged according to spin and electric charge. Spin may be integer or half-integer, making "fermions", and "bosons". Fermions all have spin ½. They involve elementary particles of integer electric charge (−1, 0, +1) and of noninteger charge (−⅓, +⅔), termed "leptons" and "quarks" respectively.[2]

The fermions interact and the mediators[3] allow for the interactions are called "field quanta". These are the photons, a group of "bosons" and a group of "gluons". Gluons mediate the strong interaction (strong in power, short in distance), attract nucleons and are responsible for the formation of nuclei of atoms. The photon is the field quantum mediating the electromagnetic interaction. Its range is longer, but the power is just 0.1 relative to the power of the strong interaction. The W and Z bosons are correlated with the weak interaction, of power just 10^{-12} compared to the strong interaction.

The elementary particles are summarized in Tab. 7.1. Each elementary particle is unique (analogous to the PAULI principle) in terms of charge and spin. Among all the elementary particles listed, the two most commonly known (at least in chemistry) are the electron and the photon. The electron is of electric charge −1 and spin ½. It is

---

2 For example, the Nobel Prize was awarded to MURRAY GELL-MANN "for his contributions and discoveries concerning the classification of elementary particles and their interactions".

3 This reminds us of the concept of an atom as suggested about 24 centuries ago, when PLATO (427–347 BC) added a fifth element, the ether, as the one allowing for interaction between the others. According to ARISTOTLE (384–322 BC) these were the four components all existing matter is built of (namely air, water, fire, and earth).

thus a fermion and a lepton. The photon has no charge and its spin is 1. Accordingly, it is a boson and belongs to the group of mediators. In Tab. 7.1 neither the neutron nor the proton appears – neither are elementary particles.

Tab. 7.1: Overview on the system of elementary particles showing electric charge and spin.

Elementary particle	Name	Symbol	El. charge	Spin ($\hbar$)	
Quarks	up	u (u_R, u_G, u_B)	$+\,^2/_3$	$^1/_2$	Fermi-on
	down	d (d_R, d_G, d_B)	$-\,^1/_3 + \,^2/_3$	$^1/_2$	
	charm	c (c_R, c_G, c_B)	$-\,^1/_3 + \,^2/_3$	$^1/_2$	
	strange	s (s_R, s_G, s_B)	$-\,^1/_3$	$^1/_2$	
	top	t (t_R, t_G, t_B)		$^1/_2$	
	bottom	b (b_R, b_G, b_B)		$^1/_2$	
Leptons	electron	e^-	-1	$^1/_2$	
	e^--neutrino	ν_e	0	$^1/_2$	
	muon	μ^-	-1	$^1/_2$	
	μ^--neutrino	ν_μ	0	$^1/_2$	
	tau	τ^-	-1	$^1/_2$	
	τ^--neutrino	ν_τ	0	$^1/_2$	
Mediators	photon	λ	$0 + 1$	1	Boson
(field quanta)	bosons	W^+	-1	1	
	gluons	W^-	0	1	
		Z^0	0	1	
		g_i ($i = 1 \ldots 8$)		1	

Instead, there are quarks, defined by noninteger spin and also noninteger electric charge ($+^2/_3$ or $-^1/_3$), and leptons, defined by half-integer spin and integer electric charge (0 or –1). Both the six quarks and the six leptons belong to the class of fermions (half-integer spin of $^1/_2$). In contrast, the last group of elementary particles is the mediators or field quanta, characterized by integer electric charge (0, +1, or –1) and integer spin. Due to the value of spin = 1, this group belongs to the class of bosons.

7.1.2 Families of elementary particles

Quarks and leptons are structured into three families, and – among other factors – arranged according to their mass (or energy). Figure 7.1 provides a sketch of this structure and of individual masses.

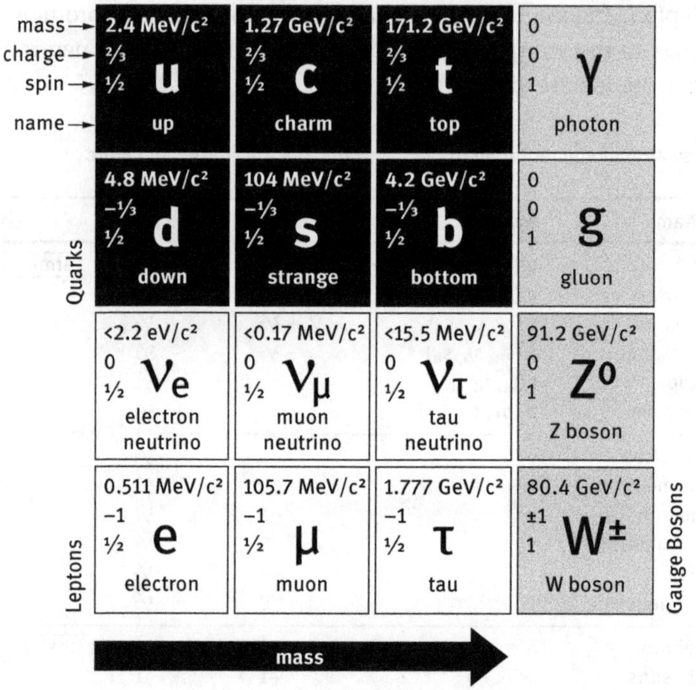

Fig. 7.1: Families of elementary particles and their field quanta. The first family includes the up quark and the down quark, the electron and the electron neutrino. The second family collects the charmed quark and strange quark, the muon and the muon neutrino etc.

7.1.3 Antimatter

The family of elementary particles thus lists 6 quarks, 6 leptons, and 12 bosons, representing altogether a world of 24 elementary particles. However, each elementary particle among the quarks and leptons has a "twin" that is identical in all parameters – except the charge. Those are called "anti"-particles, shown in Fig. 7.2. This arises from the criterion of symmetry in particle physics. Table 7.2 summarizes this for the leptons and their antileptons. The most prominent antiparticle is the positron – at least in the context of nuclear and radiochemistry. It has exactly all the properties of the electron (mass, spin), but its charge is +1 instead of –1.

MATTER

anti-MATTER

Fig. 7.2: Simplified representation of matter and antimatter existing like particles in a mirror. The difference here is just the charge, which is of opposite sign.

Tab. 7.2: Leptons and antileptons.

	Name	Symbol	El. charge	Spin ($\hbar$)
Leptons	electron	e^-	-1	$\frac{1}{2}$
	e^--neutrino	ν_e	0	
	muon	μ^-	-1	
	μ^--neutrino	ν_μ	0	
	tau	τ^-	-1	
	τ^--neutrino	ν_τ	0	
*Anti*leptons	positron	e^+	$+1$	$\frac{1}{2}$
	e^+-neutrino	ν_e	0	
	muon	μ^+	$+1$	
	μ^+-neutrino	ν_μ	0	
	tau	τ^+	$+1$	
	τ^+-neutrino	ν_τ	0	

7.2 Quarks

7.2.1 The group of quarks

Quarks[4] are fermions due to their value of spin ½. They are the elementary particles forming the matter of nucleons (and thereby the mass of nuclei and thus most of the mass of an atom). Properties of quarks thus should reproduce the different mass and charge of a proton and a neutron.

The group of quarks contains six different members, named "up", "down", "charm", "strange", "top", and "bottom". Quarks have electric charges of multiples of ⅓ of an integer value and of either positive or negative sign. One group is of +⅔ charge, the other of −⅓. Up, charm and top quarks appear to be identical due to the same spin and the same charge of +⅔ (but differ in mass), and the other three are identical due to the same spin and the same charge of −⅓ (but differ in mass). Quarks finally are understood to own a "color", be red, green or blue[5] – and thus each quark is finally unique in terms of quantum number "philosophy".

Although they differ in mass, the PAULI principle requires that every quantum physical object has a set of quantum mechanics parameters that makes it unique. This appears not to be the case. Quantum physics solves this problem by postulating one more property: the "flavor" of quarks, shown in Tab. 7.3. The combination of these properties creates a difference among quarks of the same charge.

Tab. 7.3: Parameters describing quantum mechanics properties of quarks expressed as "flavor".

Quark	Electric charge	Flavor quantum numbers				
		I_3 isospin	C charm	S strangeness	T topness (truth)	B′ bottomness (beauty)
u	+⅔	+½	0	0	0	0
d	−⅓	−½	0	0	0	0
c	+⅔	0	+1	0	0	0
s	−⅓	0	0	−1	0	0
t	+⅔	0	0	0	+1	0
b	−⅓	0	0	0	+1	−1

---

4 M GELL-MANN (1964) named quarks after JAMES Joyce's *Finnegan's Wake*: "Three quarks for master Mark". (R FEYNMAN suggested naming them "partons".)
5 Like the theory on quarks overall, the definition of flavor and color are of course abstract.

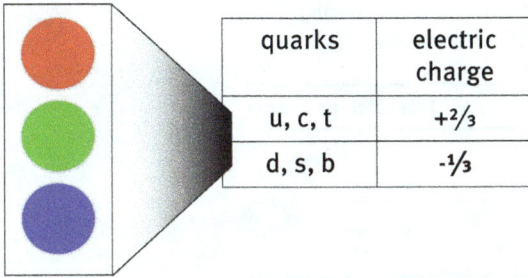

quarks	electric charge
u, c, t	$+2/3$
d, s, b	$-1/3$

Fig. 7.3: Concept of color in the quark's phenomenology.

7.2.2 Quarks and the composition of nucleons

The nucleons represent subatomic, but not elementary particles. The proton and the neutron consist of quarks, see Figure 7.4. However, among the many quarks (six "normal" quarks + six antiquarks) only two are relevant in the context of the β-transformation processes: the up quark and the down quark, i.e. those of the first family of elementary particles. The force the quarks interact in is the "weak interaction", mediated by bosons.

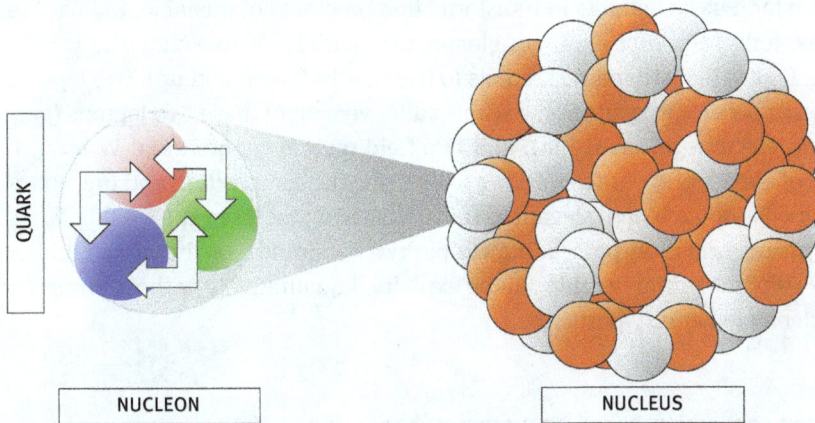

Fig. 7.4: Three quarks make one nucleon. The nucleons build the nucleus of an atom.

If related to nucleons, the "magic" number for quarks is 3: Each quark has an electric charge of a multiple of $1/3$, each quark exists at 3 colors, and 3 quarks make one nucleon. A proton is composed of two up quarks (2 times the electric charge of $+2/3$ makes a $+4/3$ charge) and one down quark (electric charge $-1/3$). The resulting total charge thus is $+3/3 = +1$. For the neutron, there is one up quark and two down quarks and their particular electric charges combine to the overall charge of 0 (see Tab. 7.4).

Tab. 7.4: Composition of nucleons. The three quarks are responsible for creating the different total electric charges of protons and neutrons.

Nucleon	Mixture of quarks			Total electric charge
Proton	u $+2/3$	+ u + $+2/3$	+ d + $-1/3$	+1
Neutron	u $+2/3$	+ d + $-1/3$	+ d + $-1/3$	0

7.3 Elementary particles relevant to β-transformations

7.3.1 The first family

Primary transformations proceeding at constant mass number A depend on internal nucleon conversion processes. Thus, the internal structure of nucleons is relevant. The variety of elementary particles, antiparticles and mediators creating the standard model in particle physics with all the different masses, spin, electric charge, color and flavor may appear to be confusing. In the context of nuclear and radiochemistry, however, there are two pieces of good news:

1. To understand basic features in transformation processes of unstable, radioactive nuclides, (only) the first family of elementary particles is relevant. The variety given in Figs. 7.1–7.3 therefore restricts to two quarks (down and up), two leptons (electron and electron neutrino), the antimatter version of these two leptons (positron and electron antineutrino), and two field quanta (photon and W boson). This set of elementary particles allows to describe the composition and the "internal" behavior of nucleons relevant to transformation processes at $\Delta A = 0$.

2. Primary transformation proceeding *via* pathways minimizing the mass number A ($\Delta A \neq 0$) of the initial nuclide can be explained qualitatively without straining subnucleon elementary particles.

7.3.2 Nucleon conversion based on metamorphoses of quarks

The essence of β-processes is now accessible by utilizing the concept of quarks. In all cases, only one particular quark of the three quarks of each nucleon is involved (the "actor"). The two other quarks just watch and are called "spectators".

- β⁻-process: Conversion of a neutron into a proton involves the conversion of one d-quark into one u-quark. This initial composition of $2 \times d + 1$ u ($= 2 \times -1/3 + 1 \times +2/3 = 0$) thus turns into $1 \times d + 2$ u ($= 1 \times -1/3 + 2 \times +2/3 = +1$). The mechanism is illustrated for the β⁻-process in Fig. 7.5.

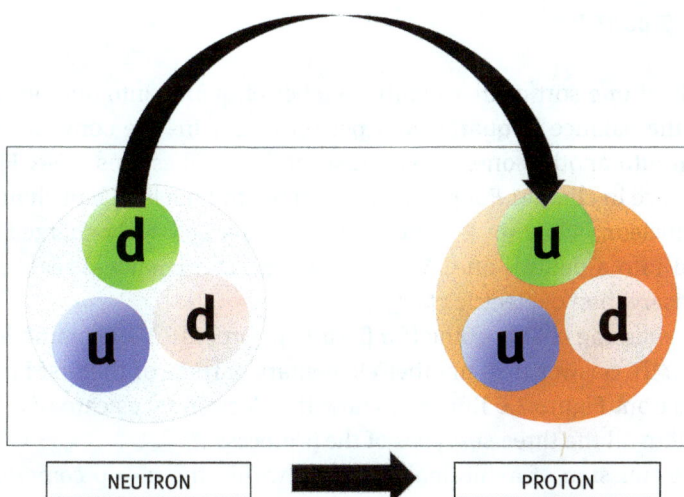

Fig. 7.5: Processes going on under the "surface" of nucleons, exemplified for the β⁻-process. The metamorphosis of a quark elementary particle (here one d-quark into a u-quark) explains the conversion of a neutron into a proton. The other down quark and the up quark of the neutron remain unchanged. This explains the balance in quarks, but provokes another question: What about the balance in charge? A neutral nucleon changes into a +1 charged nucleon . . . so what?

– β⁺- and EC processes: For the conversion of a proton into a neutron, the inverse occurs: a u-quark turns into a d-quark.

7.3.3 Balances of electric charge, spin and symmetry, and mass

As is already well established for conventional chemical reactions,[6] the transformation of unstable nuclei must also obey the laws of conservation. Within the processes of elementary particles, this refers for example to electric charge, mass, spin and symmetry. Here, some specific aspects should be addressed, which are needed to finally completely describe the metamorphoses of quarks and the consequences for the β-transformation phenomena. This refers to the impact of the electron, and leads to the role of the electron neutrino.

---

6 These are fundamental laws in chemistry. More than 200 years ago, LAVOISIER and PROUST stated the law of conservation of mass and the law of definite proportions, respectively. In chemical reactions and their stoichiometry, this refers to numbers of atoms (atoms cannot disappear or suddenly appear), to the corresponding values of mole and mass, for example, but also for electric charge.

7.3.4 Electrons vs. β-particles

The metamorphosis of one sort of first family member of quarks into another one perfectly explains the balance in quarks, and perfectly explains the conversion of one sort of nucleon into another one. However, several new questions arise. First, what about the balance in charge? For β⁻-processes, a neutral nucleon had changed into a +1 charged nucleon, while for β⁺- and EC processes a positively charged nucleon had changed into a neutral one. As with conventional chemical reactions, there must be a conservation of electric charge.

So, where is the missing charge going (for β⁻- and β⁺-processes) or coming from (for the EC process)? To answer this, another elementary particle of the first family is needed – the electron. Figure 7.6 illustrates how the electron takes care of a balance in charges within all the three subtypes of the β-process.

- β⁻-process: The emission of a "normal" electron within the n → p conversion satisfies the balance of electric charge: it is 0 → (+1) + (−1).
- β⁺-process: The p → n conversion requires the emission of the antimatter kind of electron, the +1 charged positron.[7] The balance of electric charge then is (+1) → (0) + (+1).
- EC process: The p → n conversion can occur through another pathway, electron capture (ε). Here, the proton captures a "normal" electron. The balance of electric charge is then (+1) + (−1) → (0).

Note that for all three subtypes of the β-transformation, an electron or positron is involved. While the terminology "electron" is obvious in the case of electron capture, it is more precise to call the electron released within the β⁻-process not an "electron", but a β⁻-particle. Similarly, the anti-electron ejected within the β⁺-transformation is called β⁺-particle, or just positron. This is illustrated in Fig. 7.7.

7.3.5 Mass

Mass balance should consider masses of the individual components plus the value of ΔE. Recall that energy is equivalent to mass and is much more relevant to nuclear processes than conventional chemical reactions. Let us consider the case of the conversion of a "free" neutron, i.e. one that is not part of an atom's nucleus. The mass of the neutron is a bit larger than the mass of the proton ($1.675 \cdot 10^{-24}$ vs. $1.673 \cdot 10^{-24}$ g or

---

7 This creation of the positron is a daily practice in many of the facilities related to Positron Emission Tomography. Every day, positron-emitting radionuclides are produced and used for human application at TBq radioactivity levels. Each transformation of a positron-emitting unstable radionuclide thus produces antimatter. This application is part of Chapter 12 of Volume II.

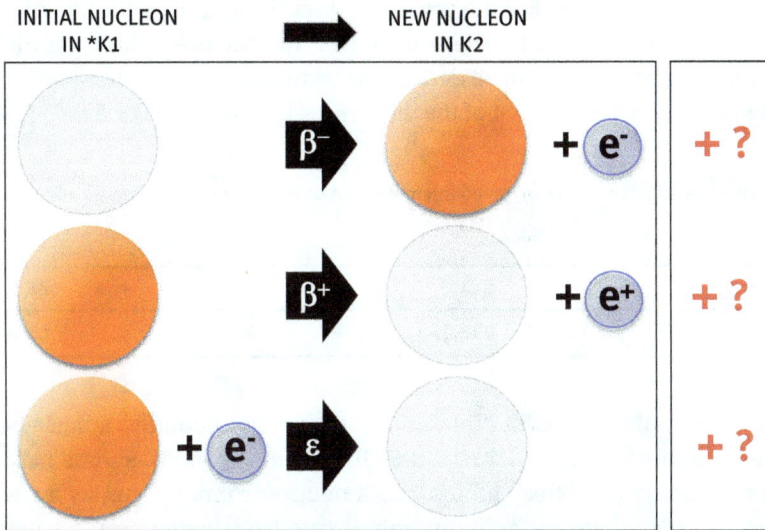

Fig. 7.6: Balances in electric charge for nucleon conversion representing the three subtypes of primary β-transformation. Gray and orange circles represent the neutron and the proton, respectively. In all cases, it is an electron which handles the balance in charge, though in different ways for the β⁻-process, the β⁺-process and the electron capture ε process. The question mark indicates that despite the balance in electric charge, something more needs to be addressed!

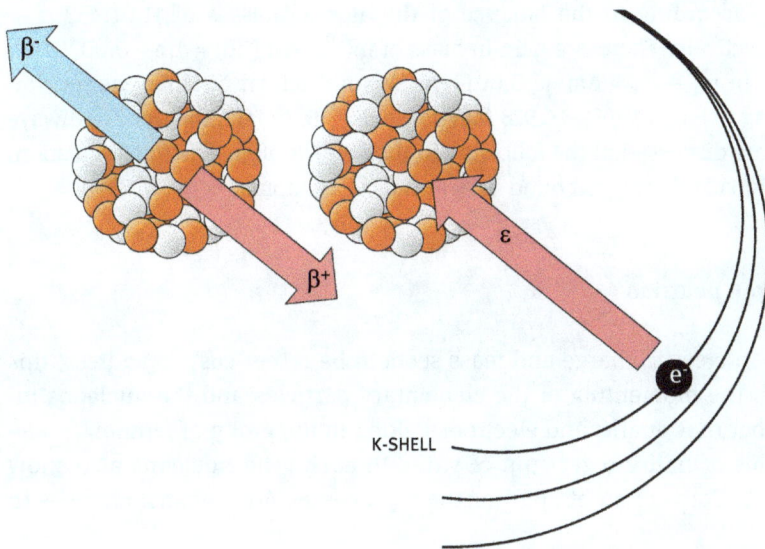

Fig. 7.7: Different mechanisms for β⁻- and β⁺-transformations (left) compared to electron capture (right). While for β⁻- and β⁺-transformations either a β⁻-particle or β⁺-particle is emitted, the ε subtype needs interaction with a "real" electron from the shell of the atom *K1.

1.00867 *vs.* 1.00728 u, respectively. For the unit energy see Tab. 7.5.).[8] With the total mass of the neutron, the proton and the electron known, the balance in mass for the three subtypes of β-transformations can be derived, as shown in Tab. 7.5. The value of Δm corresponds to the energy equivalent of the differences in mass $\Delta E = Q = \Delta m\ c^2$.

Tab. 7.5: Comparison of mass parameters involved in conversion of an unbound neutron into a proton.

	n	$\rightarrow$	p	+	e^-	+	ΔE
Energy (MeV)	939.565	=	938.272	+	0.511	+	0.78

Indeed, this conversion takes place for "free" neutrons (i.e. not bound in a nucleus of a nuclide) with a half-life of $t\frac{1}{2} = 611 \pm 1.0$ s (= 10.18 min). In contrast, one may believe that the conversion of a "free" proton into a neutron cannot occur, as mass increases for the p $\rightarrow$ n conversion. Actually, this is true for isolated protons; the β^+-process of unbound protons is impossible.

However, for bound nucleon systems, i.e. for nucleons located and interacting within a nucleus of an atom, it becomes possible. In this case the "increase" in mass for the p $\rightarrow$ n conversion is "overcompensated" by the balance of the overall mass of the two nuclides *K1 $\rightarrow$ K2 involved. For example, the transformation of the proton-rich and unstable isotope carbon-11 to stable boron-11 *via* the p $\rightarrow$ n conversion (β^+-subtype) occurs according to the balance of the atomic masses of 11.011433 u $\rightarrow$ 11.009305 u respectively. There is a gain in mass of $m(^{11}C) - m(^{11}B) = \Delta m = 0.002128$ u, despite the loss in $m_p - m_n = \Delta m = -0.00139$ u. Mean nucleon binding energy "improves" according to 6.676 MeV $\rightarrow$ 6.928 MeV for $^{11}C \rightarrow {}^{11}B$. In fact, the three pathways of β^+-processes are discussed in the following in the context of *nuclide* transformation, i.e. refer to neutrons and protons bound in a nucleus of an atom.

7.3.6 The electron neutrino

While balances in electric charge and mass seem to be rather easy to address, this is different with the momentum of the elementary particles and the nucleons involved. Remember that quarks and electrons belong to the group of fermions – elementary particles of half-integer spin of value ½ each. The nucleons also show values of spin ½. Conversion of spin for the β-processes now means: changes in

---

[8] From this balance, it is obvious that a neutron may convert into a proton because the mass balance $m_n - m_p$ is positive. In terms of the atomic mass unit u, the difference is $\Delta m = 0.00139$ u.

spin between the initial and the final states of the transformations should be zero or of integer value. So – here comes a problem!

Again let us consider the n → p conversion of the free neutron. The neutron's spin is ½, so the total spin of the left-hand side of the transformation shown in Fig. 7.5 is noninteger. Among the transformation products, the spin ½ of the proton and the spin ½ of the electron combine to an integer number.

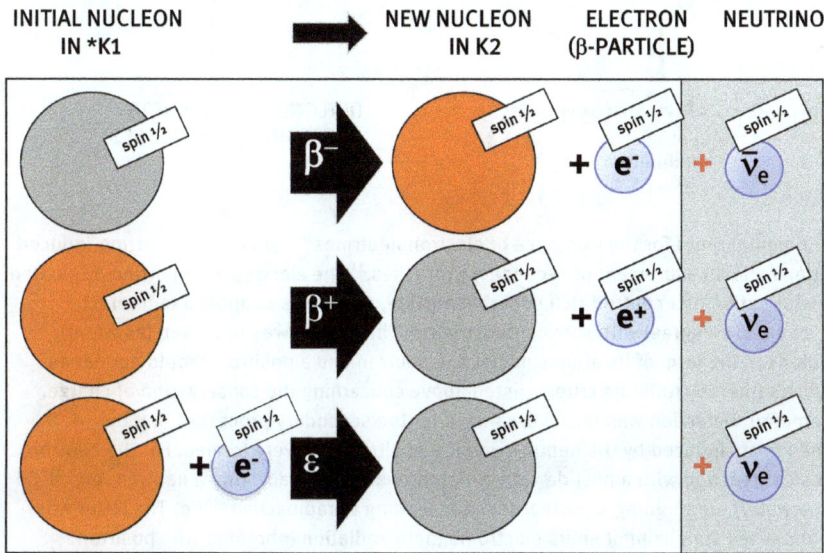

Fig. 7.8: The electron neutrino within the three subtypes of the β⁻-process. In all cases, an electron neutrino υe is emitted in the context of nucleon conversions. It guarantees the conservation of spin.

This problem was addressed first by W PAULI, who in 1930 postulated the existence of a hitherto unknown particle. It should have no electric charge (so he called it "neutron"),[9] (almost) no mass, both in order to not violate the balances in electric charge and mass achieved for the β-transformation so far. However, it should carry a half-integer spin. This electron neutrino then achieved the conversion of spin (see Fig. 7.8). The neutrino hypothesis fits perfectly with all three subtypes of the β-process.

---

9 When J CHADWICK later (in 1932) discovered the "real" neutron, this became confusing. E FERMI therefore introduced an alternative terminology by suggesting instead of "neutron" to use "neutrino" (which is Italian and means "little neutral one").

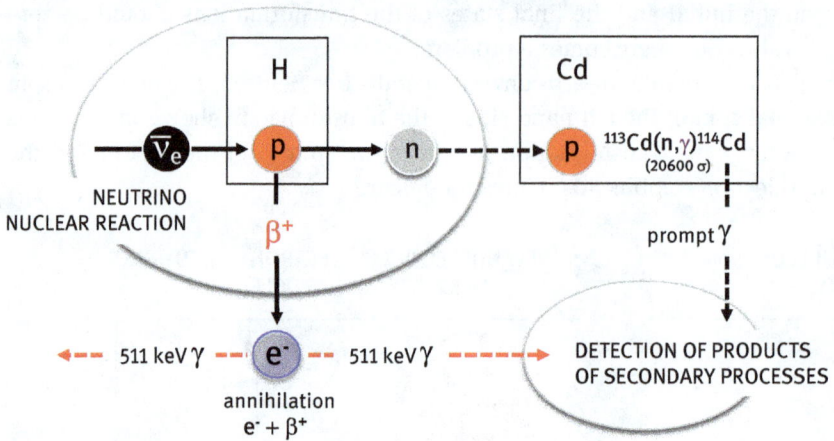

Fig. 7.9: Experimental proof for the existence of electron neutrinos[10] *via* electron neutrino-induced nuclear reactions and measurement of secondary processes.[11] The electron antineutrino originated from the transformation of a neutron-rich unstable nuclide, and it was supposed to interact (although to a very low degree) with surrounding matter. This matter was hydrogen (as water), providing protons in the form of its atomic nucleus. A neutron and a positron should appear as reaction products (thereby fulfilling criteria listed above concerning the conservation of charge, spin and symmetry). Detection was finally organized for the secondary processes, i.e. by measuring the effects induced by the neutron and the positron that were formed. For the neutron, this was a nuclear reaction with a nuclide with a high probability of capturing a neutron, e.g. ^{113}Cd (present in the water), undergoing a nuclear reaction leading to radioactive ^{114}Cd. The latter was radioactive, and when stabilizing it emits electromagnetic radiation (photons). The positron underwent another process, which was combination with an electron (from the water molecule), to convert the masses of the electron and the positron into two 511 keV photon quanta. These two different photon radiations served as indicators that the process postulated in eq. (7.1) was true.

7.3.7 Detection of the electron neutrino

The neutrino first appeared only as a theoretical concept. The experimental detection of neutrinos, however, remained (and remains) extremely difficult due to the extraordinarily small mass of this elementary particle. Historically, experimental

---

10 The other neutrinos (muon and tau) were detected much later, in 1962 and 2000, respectively. The discovery of the muon neutrino was honoured by the Nobel Prize in Physics in 1988 "for the neutrino beam method and the demonstration of the doublet structure of the leptons through the discovery of the muon neutrino" to LEON M LEDERMAN, MELVIN SCHWARTZ and JACK STEINBERGER.
11 Characteristics of nuclear reactions such as thermal neutron capture and secondary processes of nuclide transformations such as annihilation and photon emission will be discussed in following chapters.

verification was achieved utilizing nuclear reaction processes induced by electron antineutrinos *via* eq. (7.1) and post-processes illustrated in Fig. 7.9.[12]

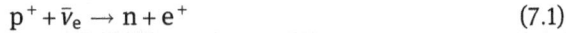

$$p^+ + \bar{v}_e \rightarrow n + e^+ \tag{7.1}$$

7.3.8 Symmetry

All electron neutrinos should carry a spin of ½ to account for a noninteger spin of the sum of the particles formed in β-transformations. In addition, it should exist as a particle in one case (β⁺-process) and as an antiparticle in another case (β⁻-process). Why? Whenever a new elementary particle is created, another complementary antiparticle should appear simultaneously, and *vice versa*.

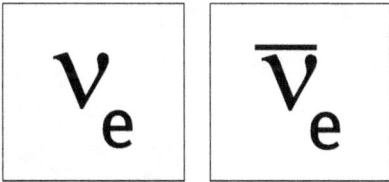

$$\nu_e \qquad \bar{\nu}_e$$

CHARGE = 0, SPIN = ½, MASS < 2 eV

Fig. 7.10: Electron neutrino (left) and electron antineutrino (right), and values of their relevant properties. While electric charge and spin are clearly defined, the mass of the neutrino is not known precisely.[13]

In the case of n → p conversion, the β⁻-electron formed is a "matter" particle. The electron neutrino should thus be an "anti-matter" particle, i.e. the electron antineutrino. This holds true for the β⁺-process as well. Here, antimatter is created (the positron, the β⁺-particle), thus a "normal matter" electron neutrino is needed. Figure 7.10 illustrates the two faces of the electron neutrino, i.e. the electron neutrino and the electron antineutrino, and lists the values of the relevant properties.

7.3.9 A more complete picture of the quark conversion mechanism

With electrons and electron neutrinos affiliated to the quark metamorphosis, all the laws of conservation now apply and yield a more complete picture, summarized in Fig. 7.11.

---

12 It thus was an indirect approach that finally proved the existence of neutrinos. In 1959 COWAN and REINES (Nobel Prize in Physics, REINES, 1995) investigated (i) the process of electron antineutrinos reacting with protons of the hydrogen nucleus, which was supposed to result (ii) in the formation of a neutron + positron pair, as per Fig. 7.9.

13 The experimental determination of the mass of the electron neutrino is an ongoing process. The current data indicate $mv_e < 2 \times 1.6 \cdot 10^{-19}$ kg·m²·s⁻² / c² = 3.2·10⁻¹⁹ kg·m²·s⁻² / (3·10⁸ m·s⁻¹)² = 3.56·10⁻³⁶ kg.

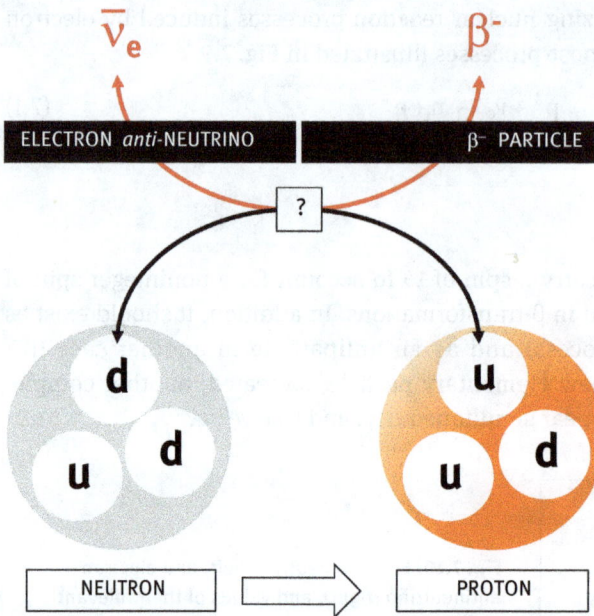

Fig. 7.11: Conservation of charge, mass, spin and elementary particle symmetry achieved throughout the β-processes. Exemplified for the β⁻-subtype induced by converting a down quark of a neutron into an up quark, creating a proton under simultaneous release of a β⁻-particle and an antimatter particle: the electron antineutrino. But how is this transformation "mediated"?

7.3.10 The mediators: FEYNMAN diagrams

Figure 7.12 illustrates the principal changes among the elementary particles involved (up quark, down quark, electron or anti-electron, electron neutrino or electron antineutrino). Yet, there is one question: why should one sort of quark turn into another one? A force is needed to manage this fundamental process.

The up quark, down quark, electron, anti-electron, electron neutrino and electron antineutrino all belong to the first family of elementary particles. This family is the subject of the weak interaction and the field quanta responsible for carrying interactions are the bosons. In the case of the β⁻-transformation, the boson in question is the W⁻ boson; for the β⁺-subtype it is the W⁺, and for electron capture the Z° boson. FEYNMAN[14] suggested graphical representations of this process (and many other processes in elementary particle physics). This is shown in Fig. 7.12 with a focus on the

---

14 The Nobel Prize in Physics, 1965, was awarded to SIN-ITIRO TOMONAGA, JULIAN SCHWINGER, and RICHARD P FEYNMAN "for their fundamental work in quantum electrodynamics, with deep-ploughing consequences for the physics of elementary particles".

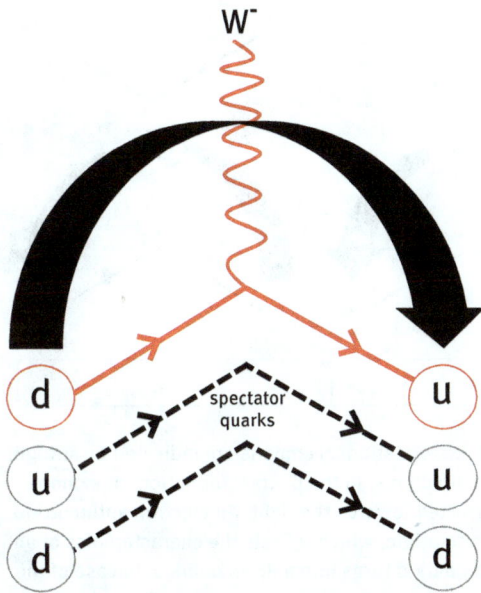

Fig. 7.12: FEYNMAN diagrams of β⁻-transformation. Fermions are indicated by straight lines, bosons by wavy lines, arrows indicate time directions. The down quark d (the "actor") turns into an up quark u. (The two other quarks are just "spectators".) At the turning point, the field quantum W⁻ comes into action, indicated by a wavy line.

three quarks and the mediator, and in Fig. 7.13 with a focus on the role of the β-particles / electrons and the electron neutrinos.

7.4 Quantum theory of β-transformation phenomena

7.4.1 Complete β-transformation balances

The resulting balances including the electron neutrino are listed in Tab. 7.6. For the n → p conversion, a β⁻-particle (rather than an ordinary electron) is emitted. It is a particle and in this case, the co-emitted electron neutrino is an antiparticle (i.e., the electron antineutrino). For the p → n conversion, a positron is emitted. It is an antiparticle (the anti-electron), so in this case the co-emitted neutrino is a normal elementary particle (i.e., the electron neutrino). For the electron capture process, the electron neutrino is a normal elementary particle, too. Note that in this case no β-particle is formed. All three nucleon conversions proceed inside the nucleus. For all three versions an electron neutrino or antineutrino is ejected. However, β-particles only appear for β⁻- and β⁺-transformations.

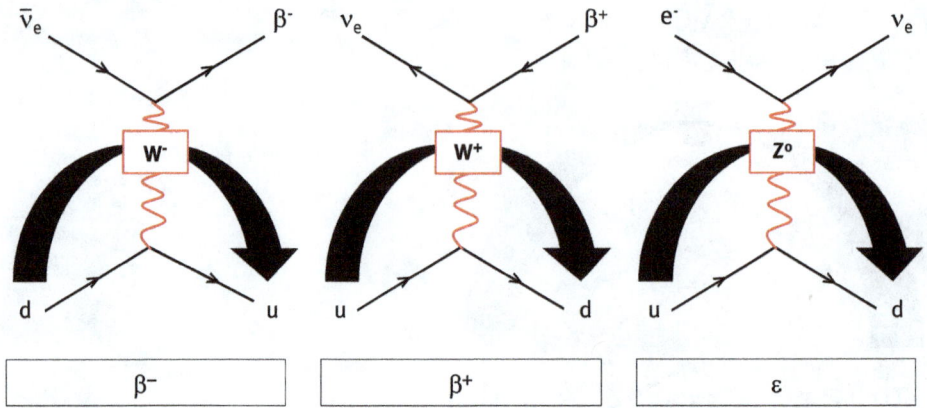

Fig. 7.13: FEYNMAN diagrams of all subtypes of β-transformations. Fermions are indicated by straight lines, bosons by wavy lines, arrows indicate time directions. In the β⁻-transformation for example, the β⁻-particle emitted is shown with an arrow directed towards the right. An electron antineutrino is co-emitted; its arrow points in the direction of the origin, which reflects the characteristics of an "anti"-particle. β⁺- and ε-transformations: the up quark d turns into a down quark u. In case of the β⁺-process, the field quantum W⁺ mediates the emission of the antimatter particle β⁺, and an electron neutrino is co-emitted. In case of the electron capture subtype, the field quantum Z° exclusively causes the emission of an electron neutrino.

Tab. 7.6: Balances in β-transformation. Relevant properties such as electric charge, mass, spin and symmetry of elementary particles are perfectly achieved by including electron neutrino and electron antineutrino.

Subtype	Nucleon level	Electron emission	Neutrino emission	+
β⁻	$n \rightarrow p^+$	$+\beta^-$	$+\bar{v}_e$	$\Delta E_{\beta-}$
β⁺	$p^+ \rightarrow n$	$+\beta^+$	$+v_e$	$\Delta E_{\beta+}$
ε	$p^+ + e^- \rightarrow n$	$-$	$+v_e$	ΔE_{ε}

7.4.2 Initial *vs*. final states and the matrix transition element

The process of nucleon transformation inside the nucleus of an atom is explained by quantum physics. The basic terminology is called "FERMI'S golden rule". It defines the probability P_{fi} of transition (per unit of time) between initial (*i*) and final (*f*) states from one energy eigenstate of a quantum system (here represented by the nuclide *K1) into another (the final nuclide K2). Figure 7.14 compares the phenomenological process and the quantum physical approach.

The transformation process is considered as the probability to convert from initial states into final nuclear states. Key parameters are phase space volumes, densities of energy states, probabilities of transition and the overlap of wave functions of

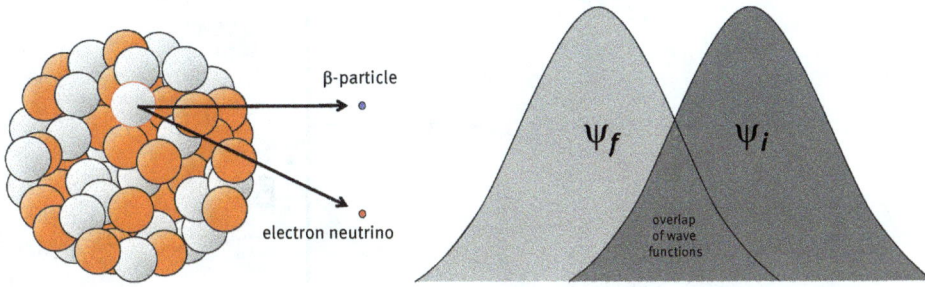

Fig. 7.14: The phenomenon of β-transformation (left). Its principal concept in quantum theory (right) showing wave functions Ψ_i and Ψ_f.

the two states. These states are expressed by a density, i.e. a number n of states per unit of energy (dn/dE). The mathematics relates the probability (P_{fi}) of transition (transition rate = transitions per unit time) to phase spaces *via* a matrix element $\{M_{fi}\}^2$. This matrix element considers the overlapping wave functions of the final and initial states Ψ_f and Ψ_i and includes the Hamiltonian operator $\hat{H}$ of the weak interaction. Suppose the overlap of the wave functions is large; then the probability of transition is high.

$$P_{fi} = \frac{2\pi}{h} \left| \psi_f \left| \hat{H}_{int} \right| \psi_i \right|^2 \frac{dn}{dE} \tag{7.2}$$

The idea of overlapping wave functions is illustrated in Fig. 7.15 for energies. The energy level of the final state may be higher than that of the initial states. However, transitions only occur to final states of lower energy.

Each state corresponds to a space phase volume with x, y and z coordinates within a Cartesian coordinate system of p_i. These are located on the surface of a sphere as illustrated in Fig. 7.16. Each energy of the β-particle and the electron neutrino is described by its corresponding impulse coordinates p_i and is:[15] $\Delta x \Delta y \Delta z$ $\Delta p_x \Delta p_y \Delta p_z = \hbar^3$. Energy and impulse of the β-particle correlate by $E^2 = p^2 c^2 + m^2 c^4$. For a given volume segment, only specific states are possible by $V \approx \Delta p_x \Delta p_y \Delta p_z$. The most relevant equations and their relationship are illustrated in Fig. 7.17. Equation (7.6) finally relates quantum physics to energies of nuclide transformations.

In the case of the β⁻- and β⁺-subtypes of β-transformation, the nuclide transformation energy Q_β is in part delivered simultaneously to the β-particle and the electron neutrino. These two elementary particles share the energy available. Consequently, β-particle energies show a continuous spectrum, and the same is true for the electron neutrino. The expression $Q_\beta - E_\beta^{max}$ refers to the impact of the electron neutrino. The

---

15 HEISENBERG uncertainty principle: $\Delta x \Delta p \geq \hbar$.

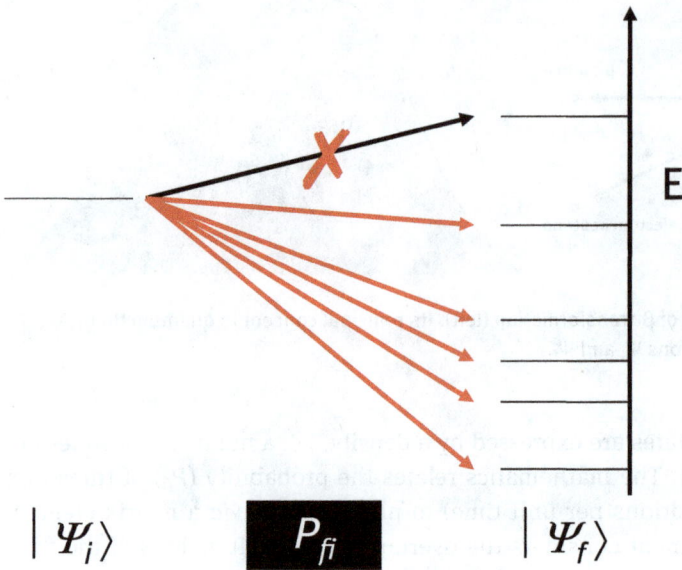

$$|\Psi_i\rangle \qquad \boxed{P_{fi}} \qquad |\Psi_f\rangle$$

Fig. 7.15: Overlapping wave functions for energy levels of initial and final states. Transition only occurs to final states of lower energy with a probability of P_{fi}.

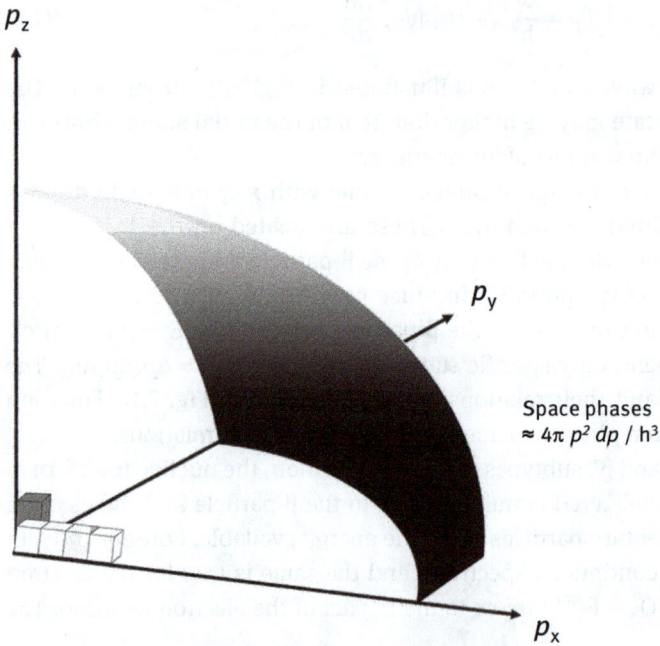

Space phases
$\approx 4\pi\, p^2\, dp\, /\, h^3$

Fig. 7.16: Space phase volumes within a Cartesian system of impulses p_i. For a given volume segment, specific states are characterized by $\Delta p_x \Delta p_y \Delta p_z$.

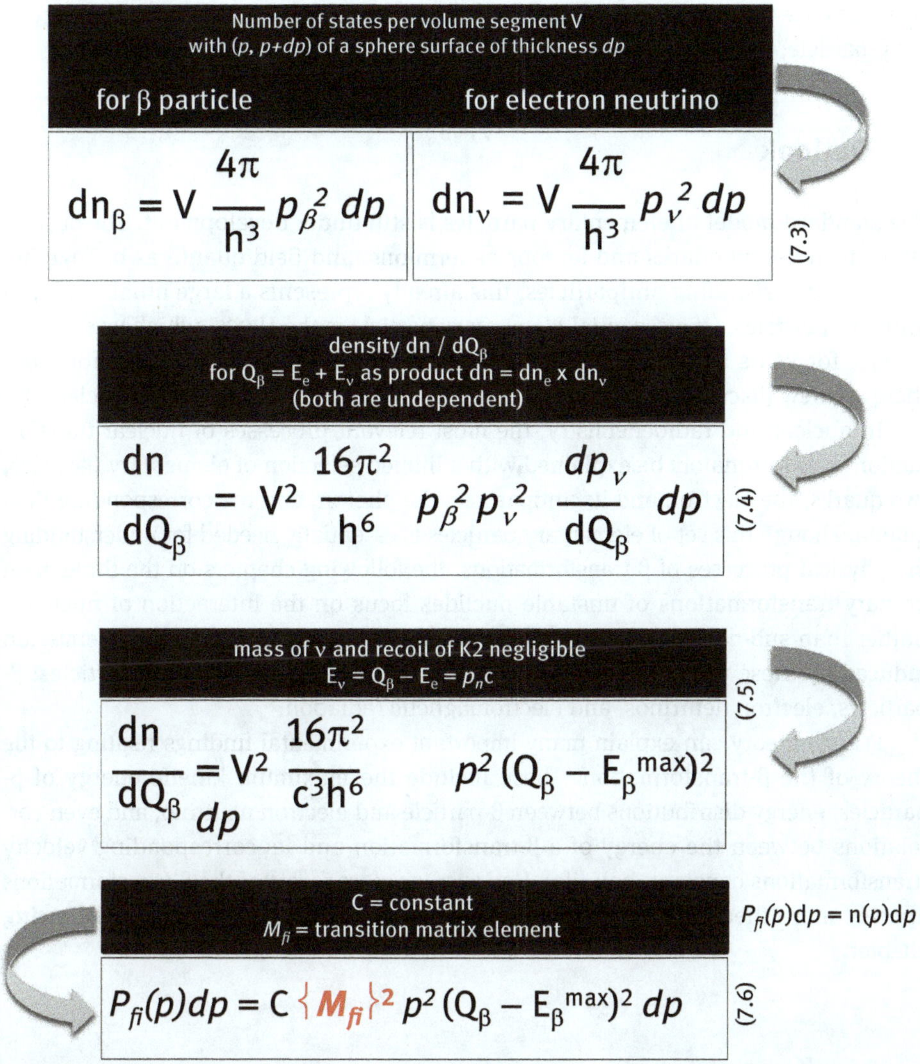

Number of states per volume segment V with $(p, p+dp)$ of a sphere surface of thickness dp	
for β particle	for electron neutrino
$$dn_\beta = V \frac{4\pi}{h^3} p_\beta^2 \, dp$$	$$dn_\nu = V \frac{4\pi}{h^3} p_\nu^2 \, dp$$

(7.3)

density dn / dQ_β for $Q_\beta = E_e + E_\nu$ as product $dn = dn_e \times dn_\nu$ (both are undependent)
$$\frac{dn}{dQ_\beta} = V^2 \frac{16\pi^2}{h^6} p_\beta^2 p_\nu^2 \frac{dp_\nu}{dQ_\beta} \, dp$$

(7.4)

mass of ν and recoil of K2 negligible $E_\nu = Q_\beta - E_e = p_n c$
$$\frac{dn}{dQ_\beta \, dp} = V^2 \frac{16\pi^2}{c^3 h^6} p^2 \left(Q_\beta - E_\beta{}^{max}\right)^2$$

(7.5)

C = constant M_{fi} = transition matrix element	$P_{fi}(p)dp = n(p)dp$
$$P_{fi}(p)dp = C \left\{ M_{fi} \right\}^2 p^2 \left(Q_\beta - E_\beta{}^{max}\right)^2 dp$$	

(7.6)

Fig. 7.17: FERMI's golden rule: Basic equations. Number of states for β-particle and electron neutrino per volume segments (7.3) combine to densities of states for both elementary particles (7.4). With negligible mass of the electron neutrino and very small recoil energy of K2, the density of states is expressed in terms of overall energy Q_β of the transformation relative to the maximum kinetic energy of the β-particle emitted (7.5). Probabilities of transition are finally described by the transition matrix element M_{fi}, while several numerical parameters are combined to a constant $C = V^2 / (2p^3 \, c^3 \, \hbar^7)$, eq. (7.6).

Q_β values are not totally represented by the maximum value of the kinetic energy of the β-particle; instead, a fraction of kinetic energy is left for the electron neutrino.

7.5 Outlook

The standard model of elementary particles is still under development. The current status is to define quarks and leptons as fermions, and field quanta as b. Together with the corresponding antiparticles, this already represents a large number of very different particles. One essential elementary particle is the HIGGS, which was sought eagerly for years and was discovered experimentally only in 2012.[16] In addition, there are new discussions towards an even richer world of elementary particles (!).

In nuclear and radiochemistry, the most relevant processes of nuclear transformations can fortunately be explained with a limited selection of elementary particles; two quarks, the electron and its antiparticle, the photon, and the corresponding field quanta. Though this set of elementary particles is essentially needed for understanding the physical processes of β-transformations, the following chapters on the three main primary transformations of unstable nuclides focus on the interaction of nucleons (rather than sub-nucleon elementary particles). Nevertheless, the radiative emission induced by those primary processes again refers to several elementary particles: β-particles, electron neutrinos, and electromagnetic radiation.

FERMI'S theory can explain many important experimental findings relating to the theory of the β-transformation. These include the maximum kinetic energy of β-particles, energy distributions between β-particle and electron neutrino, and even correlations between the energy of a β-transformation and its corresponding velocity (transformations constant, half-life). However, as more facets of the β-transformations appear, more theoretical concepts are needed. This is discussed in the following chapter.

---

16 Nobel Prize in Physics, 2013, jointly awarded to F ENGLERT and PW HIGGS "for the theoretical discovery of a mechanism that to our understanding of the origin of mass of subatomic particles, and which recently was confirmed through the discovery of the predicted fundamental particles . . . ".

8 β-Transformations II: β⁻-process, β⁺-process and electron capture

Aim: For most unstable nuclides, the dominant pathway of transformation is the β-process. Its essence is turning either a neutron into a proton or *vice versa*. It involves either ejection of an electron from the nucleus (then called the β-particle) or capture of an electron from the electron shell. The first subtype applies to neutron excess nuclides, and because a negatively charged β⁻-particle is emitted, it is called the β⁻-process. For neutron-deficient nuclides, β-transformations proceed either by emission of a positively charged β⁺-particle (the positron) and are called β⁺-processes. Alternatively or in parallel, a proton of a neutron-deficient nucleus captures an electron from (mainly) the K electron shell. This subtype is named electron capture (EC or ε).

This chapter discusses the individual characteristics of these subtypes. Topics covered include how the overall energy ΔE is disseminated among the partners, in particular in terms of kinetic energies of the β-particles emitted. The experimentally obtained correlation between the gain in energy ΔE for the whole transformation and the velocity of the transformation, i.e. its half-life, are presented. The theoretical and mathematical models to describe these and more effects are introduced.

Finally, prominent β-emitters are mentioned, which are important to specific aspects of nuclear and radiochemistry, to science and technology.

8.1 Phenomenon

The essence of the β-processes is turning either a neutron into a proton or *vice versa*. It is the main primary transformation and characterized by $\Delta A = 0$ and $\Delta Z = \pm 1$. As the total number of nucleons does not change for the nuclides *K1 and K2, the transformations proceed along the isobars. Nucleon binding energies improve, which is best expressed by the isobar parabola of $\bar{E}_B = f(Z)$, shown in Fig. 8.1.

The first subtype applies to neutron-excess nuclides, and because a negatively charged particle is emitted, it is called the β⁻-process. The second process, valid for neutron-deficient nuclides, proceeds by emission of a positively charged electron and is called the β⁺-process. Alternatively, or in parallel, a proton of a neutron-deficient nucleus captures an electron from the K electron shell, and this subtype is named electron capture (EC or ε). The β⁻- and β⁺-processes and electron capture show the β-particles electrons released and the electron captured either on the side of the transformation products or on the side of the initial nuclide, as shown in Fig. 8.2. The β⁻-process forms a nuclide K2 of (Z + 1) composition and the new nuclide needs an additional electron. In contrast, the β⁺-transformation forms a nuclide of (Z − 1) composition and one of the initial electrons of the nuclide K1 becomes obsolete.

https://doi.org/10.1515/9783110742725-008

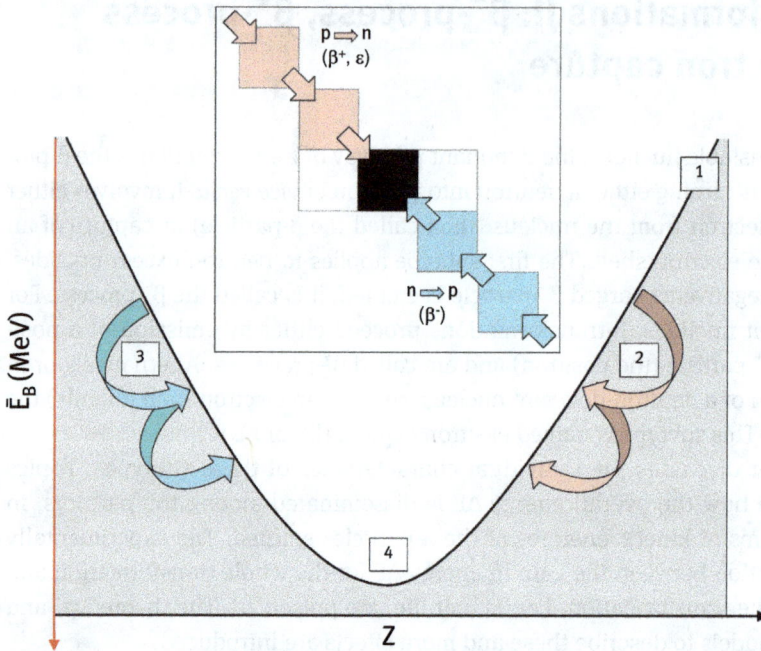

Fig. 8.1: β-Transformation along isobars and in terms of $\bar{E}_B = f(Z)$. Neutron-rich nuclides transform *via* the neutron → proton conversion (i.e. the β⁻-process). The new nuclide K2 has a composition of (Z + 1, N – 1). Proton-rich nuclides utilize proton → neutron conversion (i.e. the β⁺-process and electron capture) and yield a new nuclide K2 of (Z – 1, N + 1) composition. Transformation may continue along the line of the same isobar stepwise, unless the Z to N ratio represents a stable nuclide. A stable nuclide will have adequate mean nucleon binding energy according to the liquid drop and nuclear shell models. Along the parabola of A = constant (1) one single parabola; successive transformations of type β⁺ or ε at the right-hand side (2), and of type β⁻ at the left-hand side (3) improve mean nucleon binding energy and finally approach the vertex of the parabola (4) with the largest value of $\bar{E}_B$.

While the main goal is to optimize nucleon binding energy among the unstable nuclides involved, the β-processes are characterized by specific radioactive emissions. Electron neutrinos are released during the primary transformation steps of all subtypes. More relevant to nuclear and radiochemistry is the emission of the two kinds of β-particles within the β⁻- and β⁺-processes.

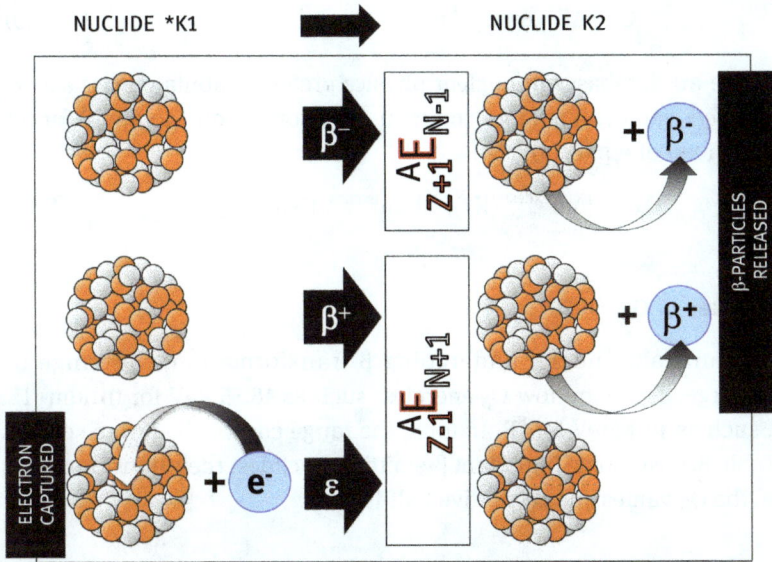

Fig. 8.2: Release of the β-particles and capture of the electron involved in the three β-transformation subtypes relative to the balance of the nucleons.

8.2 Energetics of β-transformations

8.2.1 Values of ΔE and Q

The three subtypes of β-transformation are all characterized by a balance of mass between the initial unstable nuclide *K1 and the transformation product nuclide K2 in terms of $m_{K2} < m_{*K1}$. The new nuclide *must* have lower mass in order to guarantee an exothermic transformation. The value of mass refers to the whole nuclide rather than the masses of nucleons or nuclei alone, and atomic mass data (in u) can be used as tabulated. As indicated in Chapter 2, mass defect as well as mass excess are analogous parameters used to quantify nuclear transformation processes. Accordingly, energy balance may be expressed in those units as well.

The value of ΔE is specified as the Q-value of the process. The three subtypes thus have individual values, i.e. Q_{β^-}, Q_{β^+}, and Q_ε. Suppose a given unstable nuclide is able to undergo two or all three subtypes of the β-transformation; each branch will thus be characterized by its individual amount of ΔE.

$$Q_{\beta^-} = \Delta E = \Delta mc^2 = [m_{*K1} - (m_{K2})]c^2 \qquad (8.1a)$$

$$Q_{\beta^+} = \Delta E = \Delta mc^2 = [m_{*K1} - (m_{K2} + 2m_\beta)]c^2 \qquad (8.2)$$

$$Q_\varepsilon = \Delta E = \Delta mc^2 = [m_{*K1} - m_{K2}]c^2 \tag{8.3}$$

Because state-of-the-art databases in nuclear physics prefer to tabulate mass *excess* instead of mass *defect* values, it is more common (and convenient) to apply for instance eq. (8.1b) instead of eq. (8.1a):[1]

$$Q_{\beta^-} = |\Delta m^{\text{excess}}(K2)| - |\Delta m^{\text{excess}}(K1)| \tag{8.1b}$$

8.2.2 Absolute values of Q_β

Among the many unstable nuclei undergoing β-transformation, the range of Q_β-values is very large. There are low Q_β-energies, such as 18.55 keV for tritium ^{3}H, and large ones, such as 14.1 MeV for ^{8}B. In total, the range covers about three orders of magnitude. Table 8.1 lists some important β-emitting nuclides, their transformation product nuclide, the Q_β-values and respective half-lives.

Tab. 8.1: Important nuclides *K1 undergoing β-transformation, their transformation product nuclide K2, the Q_β-values and the half-lives. (May I suggest indicating what a, m, d etc represent in the caption? Units of t1/2).

*K$_1$	K2	Q_β (keV)	t½	*K1	K2	Q_β (keV)	t½
		β⁻				β⁺	
^{3}H	^{3}He	18.6	12.323 a	^{11}C	^{11}B	1 982.4	20.38 m
^{14}C	^{14}N	156.5	5730 a	^{13}N	^{13}C	2 220.5	9.96 m
^{24}Na	^{24}Mg	5 515.5	14.96 h	^{15}O	^{15}N	2 754.2	2.03 m
^{32}P	^{32}S	1 710.5	14.26 d	^{18}F	^{18}O	1 655.2	109.7 m
^{35}S	^{35}Cl	167.2	87.5 d	^{22}Na	^{22}Ne	2 842.3	2.603 a
^{60}Co	^{60}Ni	2 823.1	5.272 a	^{68}Ga	^{68}Zn	2 921.1	67.7 m
^{89}Sr	^{89}Y	1 492.3	50.5 d	^{124}I	^{124}Te	295.1	4.15 d
^{90}Y	^{90}Zr	2 279.8	64.1 h			ε	
^{99}Mo	^{99}Tc	1 357.3	66.0 h	^{68}Ge	^{68}Ga	106	270.8 d
^{131}I	^{131}Xe	970.8	8.02 d	^{125}I	^{125}Te	185.8	59.41 d
^{137}Cs	^{137}Ba	1 175.6	30.17 a	^{165}Er	^{165}Ho	373.6	10.3 h
^{177}Lu	^{177}Hf	500.6	6.71 d	^{201}Tl	^{201}Hg	481.0	73.1 h

---

1 Use of mass Δm^{excess}: For the β⁻-transformation ^{131}I → ^{131}Xe, for example, this would give $Q_\beta = \Delta m^{\text{excess}}$ (^{131}Xe) − Δm^{excess} (^{131}I) = 88.414 MeV − 87.443 MeV = + 0.9701 MeV. For the β⁺-transformation ^{18}F → ^{18}O, for example, it is $Q_\beta = \Delta m^{\text{excess}}$ (^{18}O) − Δm^{excess} (^{18}F) = (− 0.783 MeV) − 0.873 MeV = + 0.0903 MeV.

8.2.3 Specific effects for β⁺-emission vs. electron capture

The way the Q-value is calculated, however, is in part modified according to the role of the electron captured and the β-particles emitted. The β⁻- and β⁺-particles do not belong to the mass m of the nuclide *K1 nor to K2. In contrast, the electron capture subtype starts at a state of nuclide *K1, collecting one *additional* electron on top of the initial electron shell configuration of the corresponding atom.

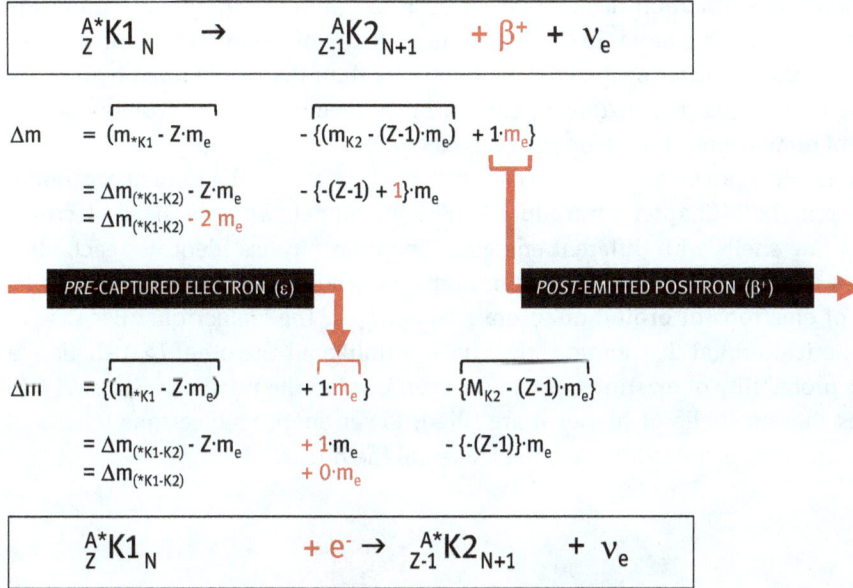

$$
{}^{A*}_{Z}\text{K1}_{N} \quad \rightarrow \quad {}^{A}_{Z-1}\text{K2}_{N+1} \quad + \beta^{+} + \nu_e
$$

$$
\Delta m = (m_{*K1} - Z \cdot m_e) \quad - \{(m_{K2} - (Z-1) \cdot m_e) + 1 \cdot m_e\}
$$

$$
= \Delta m_{(*K1-K2)} - Z \cdot m_e \quad - \{-(Z-1) + 1\} \cdot m_e
$$

$$
= \Delta m_{(*K1-K2)} - 2\, m_e
$$

PRE-CAPTURED ELECTRON (ε)   **POST-EMITTED POSITRON (β⁺)**

$$
\Delta m = \{(m_{*K1} - Z \cdot m_e) \quad + 1 \cdot m_e \} \quad - \{M_{K2} - (Z-1) \cdot m_e\}
$$

$$
= \Delta m_{(*K1-K2)} - Z \cdot m_e \quad + 1 \cdot m_e \quad - \{-(Z-1)\} \cdot m_e
$$

$$
= \Delta m_{(*K1-K2)} \quad + 0 \cdot m_e
$$

$$
{}^{A*}_{Z}\text{K1}_{N} \quad + e^- \rightarrow {}^{A*}_{Z-1}\text{K2}_{N+1} \quad + \nu_e
$$

Fig. 8.3: Mass balances for β⁺- and ε-transformation identifying the different impact of the positron emitted and the electron captured. For both subtypes, a proton-rich nuclide *K1 forms a nuclide K2 of (Z − 1) composition. The difference is that for β⁺, one more electron mass appears at the side of the products (which is the positron emitted). Likewise for ε, the mass of one electron captured counts on the other side of the balance, i.e. its mass must be added to that of nuclide *K1. Consequently, the balance in electron rest mass is either 2 or 0.

$$
\beta^- \qquad m_{K2} < m_{*K1} \qquad\qquad (8.4)
$$

$$
\beta^+ \qquad m_{K2} < m_{*K1} + 2m_e \qquad\qquad (8.5)
$$

$$
\varepsilon \qquad m_{K2} < m_{*K1} \qquad\qquad (8.6)
$$

Using the masses m of the nucleus and electrons m_e (calculated as Z times the rest mass of the electron m_e), values of Δm are calculated as illustrated in Fig. 8.3. Regardless of the difference in mass between the two nuclides, the β⁺-transformation requires an excess of that Δm plus $2 \cdot m_e$. The amount of energy which equals the mass of two electrons is $2 \cdot m_e \cdot c^2 = 2 \cdot 0.511\ \text{MeV} = 1.022\ \text{MeV}$. In contrast, electron capture (and β^--processes) are energetically satisfied by $m_{K2} < m_{*K1}$. This in particular discriminates between pathways of proton-rich unstable nuclides, i.e. β^+- and ε-transformation.[2]

8.2.4 The role of the electron shell of the unstable nuclide

The electron shell is not involved at all in the β^-- and β^+-transformations. In contrast, the ε-transformation needs an electron to be captured by one proton of the nucleus of an unstable atom. In Chapter 1 the structure of the atom was introduced, stating that the diameter of the nucleus is smaller than the whole atom by a factor of about 1000. Thus, the distance between nucleus and shell is enormous in the context of dimensions of nucleons or even electrons.

Nevertheless, an electron *can* be captured by the nucleus *K1 from its surrounding electron shell! Chapter 1 introduced the BOHR model, which assigns electrons into separate shells with different energies. Quantum physics identifies each electron by a characteristic set of quantum numbers. And here comes a very specific feature of electrons of orbital quantum number $l = 0$, the s-electrons: Because of their spherical orbital distribution, they have – unlike all the other $l > 0$ shells – a nonzero probability of existing close to (or even inside!) the nucleus.[3] Figure 8.4 illustrates that as shells of higher n are filled, lower-energy s-electrons (shells of lower n) are more attracted by (and get closer to) the nucleus.

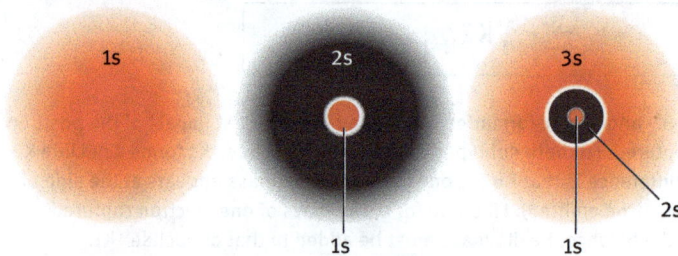

Fig. 8.4: Changing dimensions of s-electron orbitals for main quantum numbers $n = 1, 2, 3$.

---

2 Example: proton-rich ^{7}Be ($t\frac{1}{2} = 53.29$ d) transforms to stable ^{7}Li. Atomic masses are 7.016003 u and 7.016929 u, respectively; thus Δm is 0.000925 u = 0.863 MeV. (Utilizing mass excess values, this is $Q_\beta = \Delta m^{excess}$ (^{7}Be) – Δm^{excess} (^{7}Li) = 15.769 MeV – 14.907 MeV = + 0.862 MeV.) Consequently, the only pathway open for ^{7}Be is electron capture.
3 Note that this generally refers to the particle–wave duality of quantum physics.

8.2.5 Positron emission *vs.* electron capture

In the context of β-processes, proton-rich unstable nuclides have two options to improve their nucleon binding energy: β^+-emission and electron capture. The impact of energy was discussed above. The first criterion is energy in terms of mass balances. Electron capture needs a mass balance of $m_{*K1} > m_{*K2}$, while the β^+-option appears only at $m_{*K1} > m_{*K2} + 2\ m_e$ (see Fig. 8.3). At $\Delta E > 1.022$ MeV; the question remains, why should a given proton-rich unstable nuclide prefer one pathway to the other? A further consideration is how the ratio between electron capture and the β^+-process for similar proton number Z depends on the Q-value of the two subtypes. In this case, the β^+-process increases in relative frequency with increasing Q-values.

Finally, the spatial distribution of electrons is defined by the orbital quantum numbers l, giving rise to s, p, d, f (etc) orbitals. As mentioned, the spherical three-dimensional distribution of the s-orbital overlaps in part with the nucleus of the atom. Thus, there is a quantum mechanics probability (albeit very low) for s-orbital electrons to exist "inside" the nucleus. On a relative scale, this probability is most pronounced for K-shell electrons over L- or M-shell s-electrons. Electrons with $n > 1$ have higher energies and orbit at greater distances from the nucleus (see Fig. 8.4).

From this, several conclusions can be drawn:[4]

1. The probability of electron capture increases with decreasing distance of the K-shell to the nucleus. The distance between the nucleus and K-shell follows a function of $1/Z^2$, as per Chapter 1. The higher the element's proton number Z, the higher the probability of electron capture.
2. Consequently, electron capture dominates in the case of unstable proton-rich nuclides of heavy elements. Likewise, β^+-emission dominates for light elements.

The corresponding ratio between electron capture and the β^+-process thus depend on the proton number Z of $*$K1. A ratio of 1 indicates that both pathways are equally probable. In fact, most of the proton-rich unstable nuclides proceed *via* both pathways, though at different percentages. For medium and large Z nuclides, K-shell s-electron

---

4 Theoretically, there may even be a third conclusion: for a chemical element (in particular of high Z number), the chemical oxidation state may influence the quality of electron capture. For the element technetium, for example, oxidation states range from + VII (TcO_4^-, pertechnetate) to –I (carbonyl complexes), revealing almost all the intermediate oxidation states for individual chemical species. The distance of each electron shell (not just the valence shell) from the nucleus depends on the ratio between the positive and negative charge carriers (i.e., protons and electrons). With a decreasing number of outer electrons (oxidation state + VII → –I) the electron shell dimensions expand. Consequently, the probability of electron capture should decrease for a number of e.g. ^{99m}Tc species of oxidation state + VII → –I. This may – at least theoretically – even correlate with changes in the physical half-life of one and the same radionuclide. This would represent an exception from the general statement that processes of radionuclide transformation do not depend on chemistry.

capture starts to dominate. In contrast, the β^+-process dominates for proton-rich un-stable nuclides of relatively low Z.

Unstable proton-rich nuclides which preferentially undergo β^+-transformation are found among the second period of the Periodic Table of Elements. These include carbon (^{11}C, $t\frac{1}{2}$ = 20.38 min), nitrogen (^{13}N, $t\frac{1}{2}$ = 9.96 min), oxygen (^{15}O, $t\frac{1}{2}$ = 2.03 min), and fluorine (^{18}F, $t\frac{1}{2}$ = 109.7 min). Abundances of the β^+-subtype are 99.76 %, 99 %, 99.9 % and 96.7 % for ^{11}C, ^{13}N, ^{15}O and ^{18}F respectively. These are key nuclides in medically important molecular imaging and diagnostics *via* positron emission tomography (PET) and find extensive application in radiopharmaceutical chemistry. Nevertheless, some unstable nuclides of elements above Z = 20 also emit positrons at percentages which are relevant for practical applications. However, the emission percentage drops with increasing proton number Z: ^{68}Ga = 88.0 %, ^{73}Se = 65.0 %, ^{86}Y = 34.0 %, ^{89}Zr = 23.0 %, ^{90}Nb = 51.1 %, ^{124}I = 24.0 %, for example.

8.3 Kinetic energetics of β-transformation products

8.3.1 Kinetic energy and momentum

Kinetic energy and momentum are relevant properties of all the species formed in the primary nuclear transition, i.e. to the new nuclide K2, the β-particle and/or the electron neutrino emitted. For example, the transformation products for the β^--transformation are K2 + β^- + $\bar{\nu}_e$. The overall energy $\Delta E = Q_\beta$ - is shared among all the products. The question is: how?

Momentum[5] p is $m \cdot v$ (mass m, velocity v) and kinetic energy is $E = \frac{1}{2}m \cdot v^2$. Momentum is conserved for all the species *via* $m_i \cdot v_i = m_j \cdot v_j$. The mass distributions are very different for the three species formed. The rest mass for an electron and electron neutrino are $m_e = 9.1094 \cdot 10^{-31}$ kg and $3.56 \cdot 10^{-36}$ kg, respectively. The absolute masses m of K2 depend on the mass number of the nuclide. For example, this mass can be 10 u (for A = 10) or 100 u (for A = 100). In terms of absolute mass, according to $m = A \cdot u$, these values are $1.66054 \cdot 10^{-25}$ kg or $1.66054 \cdot 10^{-24}$ kg, respectively. The nuclide masses are thus about six orders of magnitude larger than the electron rest mass.

8.3.2 Distribution of kinetic energy between the β-particle and K2

Conservation laws refer to momenta of K2, the β-particle and the electron neutrino. The overall energy Q is disseminated between the species according to eqs. (8.7–8.9).

---

5 To discriminate from the symbol p (used for the proton), momenta are denoted by an italic *p*.

The higher the Q-value, the higher the maximum kinetic energy of the elementary particles emitted.

Because mass values are very low for the β-particle and the electron neutrino compared to K2, their kinetic energies are opposite, as $m_\beta \cdot E_\beta \approx m_{K2} \cdot E_{K2}$. In the case of β⁻- and β⁺-transformations, both elementary particles share this main fraction of energy Q_β, which is less than the total energy of the nuclear transformation (see eqs. (8.7) and (8.8) for the care of β⁻ transformation). This is different for electron capture, because no β-particle is emitted. In this case, the kinetic energy is shared between K2 and the electron neutrino only, following eq. (8.9).

$$Q_\beta = E_{K2} + E_{\beta^-} + E_{\bar{\nu}_e} \tag{8.7}$$

$$E_{\beta^-} + E_{\bar{\nu}_e} = Q_\beta - E_{K2} \tag{8.8}$$

$$E_{\nu_e} = Q_\varepsilon - E_{K2} \tag{8.9}$$

Momentum and kinetic energy, however, must consider the high speed of the low mass β-particle and neutrino. The β-particle and the electron neutrino have relatively high velocities and thus differences between rest mass m_β^0, i.e. at zero kinetic energy, and "real" mass m_β of the β-particle, at any velocity relatively close to the speed of light c, must be considered. For relativistic velocity, eqs. (8.10) and (8.11) apply.

$$E_\beta^2 = p^2 c^2 + \left(m_\beta^0 c^2 \right)^2 \tag{8.10}$$

$$E_\beta = \left(m_\beta - m_\beta^0 \right) c^2 \tag{8.11}$$

8.3.3 Distribution of kinetic energy: Recoil energy

Let us assume the β-particle is ejected from K2, i.e. the former *K1. The impulse of the β-particle causes a somehow opposite impulse to K2. This is referred to as "recoil energy" of K2. It is linked to:
(a) the Q-value,
(b) the mass of K2,
(c) the kinetic energy E_β of the emitted β-particle and the electron neutrino[6] (or the electron neutrino exclusively in case of electron capture).

In addition, recoil energy is influenced by the spatial arrangement of the two elementary particles during emission (see Fig. 8.5).The K2 recoil energy thus lies between

---

6 The impact of the electron neutrino on the recoil energy can be neglected due to the extremely low mass of the electron neutrino.

the theoretical maximum value and zero. The maximum kinetic energy $^{\text{RECOIL}}E_{K2}^{max}$ a recoil nucleus may receive is calculated with eq. (8.12).

$$^{\text{RECOIL}}E_{K2}^{max} = \left(\frac{E_{\beta}^{max}}{2c^2} + m_{\beta}^0\right)\frac{E_{\beta}^{max}}{m_{K2}} \tag{8.12}$$

For example, the β⁻-transformation of the carbon radioisotope ^{14}C into ^{14}N yields $^{\text{RECOIL}}E_{K2}^{max} = E^{max}(^{14}N) = 6.9$ eV. ($m_{K2} = 14$ u, $m_{\beta}^0 = 0.511$ keV, $E_{\beta}^{max} = 0.156$ MeV).

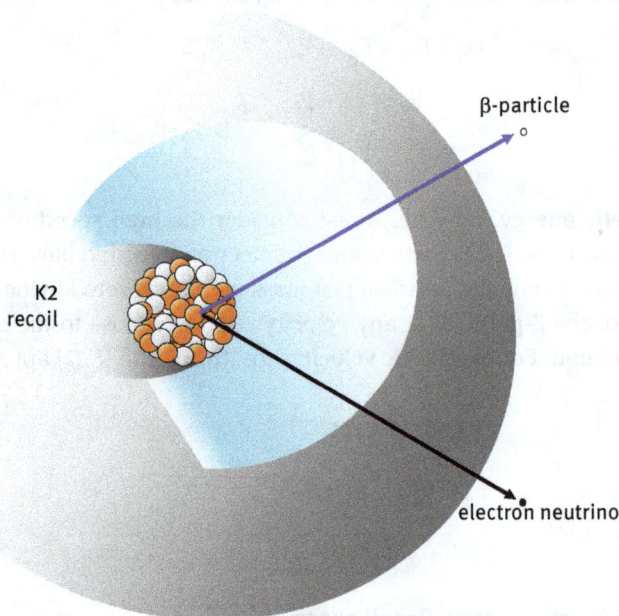

Fig. 8.5: Simplified two-dimensional illustration of spatial emission profiles of a β-particle and an electron neutrino that influence the recoil energy of K2. Absolute recoil energy depends on the mass of K2, the kinetic energy of the emitted β-particle and/or the electron neutrino and the angular distribution the two elementary particles during emission.

The recoil energy of K2 is higher when its mass number is low and the emitted β-particle's kinetic energy is high. For example, at mass number A ≈ 100 and a maximum kinetic energy for the emitted β-particle of 1 MeV, the value of $^{\text{RECOIL}}E_{K2}^{max}$ is about 10 eV.[7] Table 8.2 gives typical values for other mass numbers and β-particle energies.

---

[7] Recoil energies are, in many cases, below the electron binding energies defining chemical bonds between atoms, discussed in Chapter 1. However, in some cases, the energy is above that order of magnitude and thus may break chemical bonds. This is a special topic in nuclear and radiochemistry.

Tab. 8.2: Typical range in maximum K2 recoil energy $^{RECOIL}E_{K2}^{max}$ by mass number A and maximum kinetic energy of the emitted β-particle E_β^{max}.

	$^{RECOIL}E_{K2}^{max}$ (eV)		
E_β^{max} (MeV)	0.5	1.0	2.0
$A_{K2} = 50$	8	25	70
$A_{K2} = 100$	4	12	35
$A_{K2} = 200$	2	5	18

8.3.4 Distribution of kinetic energy: β-particle and electron neutrino

Since the recoil nucleus receives a very small amount of kinetic energy, the dominant fraction is left for the small particles emitted. For electron capture, all the remaining kinetic energy goes to the electron neutrino, as per eq. (8.9). Consequently, the electron neutrino receives a kinetic energy of a discrete energy value.

However, for the β⁻- and β⁺-transformations, this is different, as per eq. (8.8). How do the β-particles and the electron neutrinos share their fraction of kinetic energy? The answer is: "statistically". There are cases where the β-particle gets all the kinetic energy and nothing is left for the electron neutrino – or *vice versa*. In the first case, the kinetic energy of the β-particle is at maximum and denoted by E_β^{max}. Its value depends on the value of Q_β. In reality, there is a distribution between both the elementary particles, and consequently, kinetic energies observed for β-particles and for electron neutrinos show a continuous spectrum. The β-particle kinetic energy thus lies between the theoretical maximum value and zero. The typical maximum energy for β⁻- and β⁺-particles emitted from neutron-rich and neutron-poor unstable nuclides, respectively, ranges from about 20 keV to a few MeV.

Figure 8.6 shows a simplified course of kinetic energy for β⁻-particles emitted from unstable ³H and ¹⁴C. The y-axis shows the relative number of β-particles observed for a given energy. Maximum kinetic energies E_β^{max} are 18.591 keV and 156.476 keV, respectively, but the fraction of β-particles that reaches this energy is very low. Only a small fraction of the β⁻-particles shows this maximum energy or energies close to this value. More than ⅔ of the β-particles emitted have energies less than ⅔E_β^{max}.

Most β⁻-particles typically show an average or mean energy (E_β^{mean}. or $\bar{E}_\beta$) around ⅓E_β^{max}. However, both the very low energy and maximum kinetic energy of β-particles requires a deeper look (see Fig. 8.8).

The same applies to positrons emitted within the β⁺-subtype of β-transformation. Figure 8.7 shows profiles of the continuous spectra of the positrons emitted from four relevant nuclides used in medical diagnosis (PET).

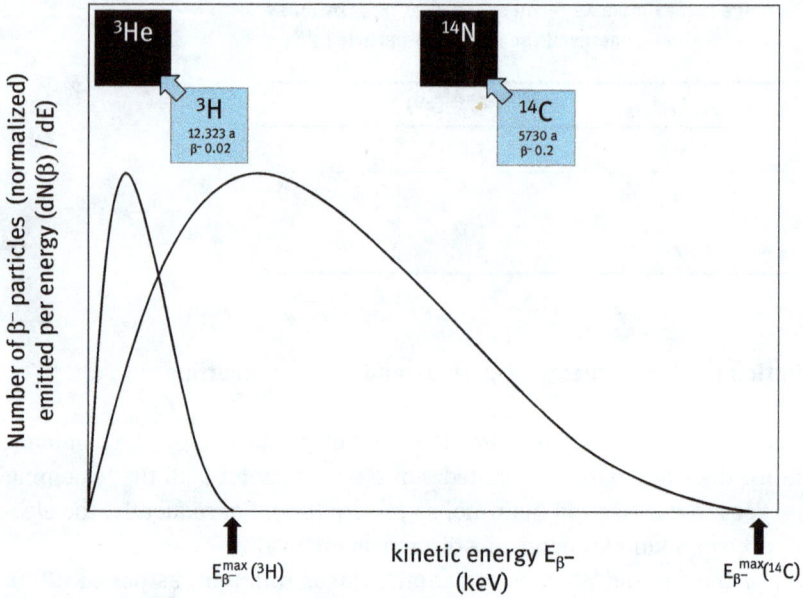

Fig. 8.6: Simplified β⁻-particle spectra of ^{3}H and ^{14}C.

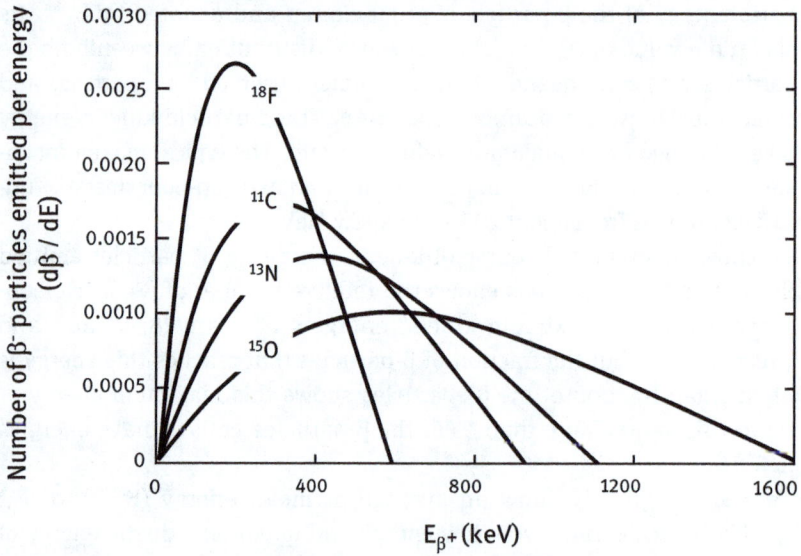

Fig. 8.7: Theoretical β⁺-particle energy spectra of ^{11}C, ^{13}N, ^{15}O and ^{18}F. Spectra are normalized to have the same area under the curve. Maximum values of kinetic energy differ depending on the Q-values, as per Tab. 8.1. With permission: CS Levin, EJ Hoffman, Calculation of positron range and its effect on the fundamental limit of positron emission tomography system spatial resolution, Phys. Med. Biol. 44 (1999) 781–799.

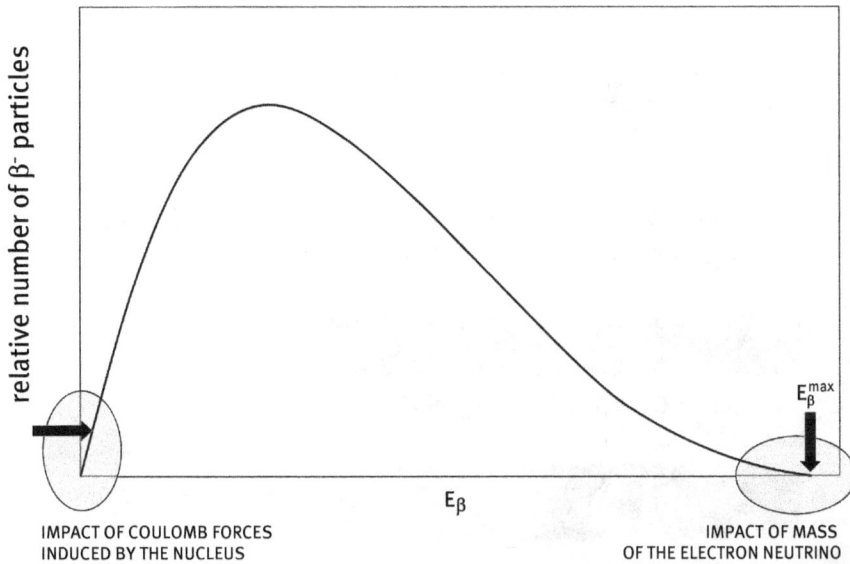

Fig. 8.8: Continuous spectrum of β-particle kinetic energy and areas needing additional comments.

8.3.5 Distribution of kinetic energy: Effects caused by coulomb interaction

This discussion refers to the low kinetic energy part of the theoretically continuous β-particle energy spectrum. The abundance of β⁻-particles sharing their kinetic energy with the electron neutrinos is explained by eq. (8.8). However, there is a special effect caused by the different coulomb interaction between the positively charged nucleus of the radionuclide and the negatively or positively charged β⁻- and β⁺-particles emitted, summarized in Fig. 8.9. It seems obvious that the negatively charged β⁻-particles "feel" some coulomb attraction caused by the nucleus – they are therefore a bit "slower" than theoretically predicted. In other words, the abundance of low-energy β⁻-electrons observed is higher than the theoretical line of the continuous energy spectrum of e.g. Fig. 8.8. In contrast, positively charged β⁺-particles (positrons) receive some coulomb repulsion, which increases their initial kinetic energy. Extremely speaking, there are almost no positrons at zero or close to zero theoretical kinetic energy, because they have obtained an additional fraction of kinetic energy through coulomb repulsion.

Quantum physics treats this phenomenon mathematically. It introduces a so-called Fermi correction term $F(Z, E_\beta)$ into the basic equation (7.6), shown in Fig. 8.10. It may be understood as a numerical factor being either >1 or <1, respectively, for β⁻- and β⁺-particles.

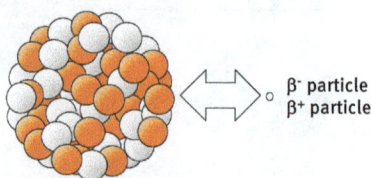

Fig. 8.9: β-particle attraction (β⁻) or repulsion (β⁺) influencing the energy distribution of the low-energy particles emitted in the course of β⁻- or β⁺-transformations.

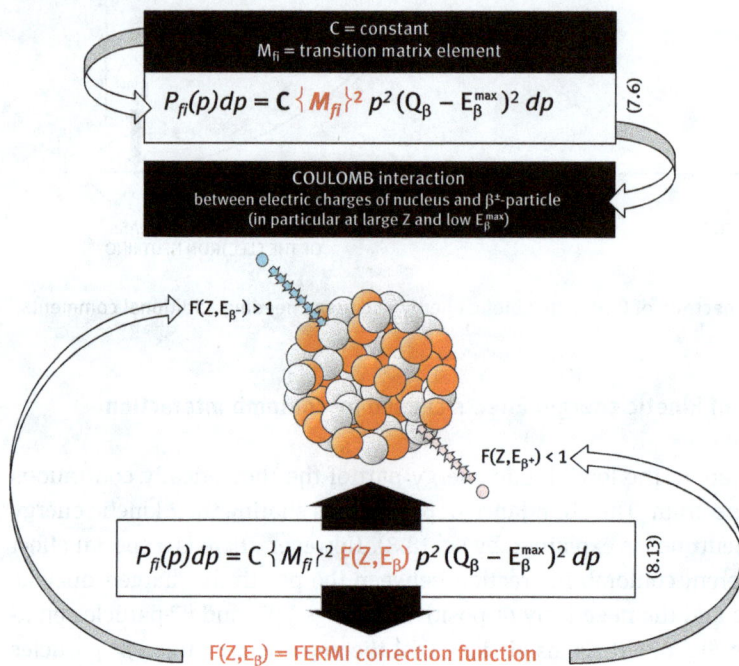

$$P_{fi}(p)dp = C\left\{M_{fi}\right\}^2 p^2 (Q_\beta - E_\beta^{max})^2\, dp \qquad (7.6)$$

COULOMB interaction
between electric charges of nucleus and β±-particle
(in particular at large Z and low E_β^{max})

$F(Z, E_{\beta^-}) > 1$

$F(Z, E_{\beta^+}) < 1$

$$P_{fi}(p)dp = C\left\{M_{fi}\right\}^2 F(Z,E_\beta)\, p^2 (Q_\beta - E_\beta^{max})^2\, dp \qquad (8.13)$$

$F(Z,E_\beta)$ = FERMI correction function

Fig. 8.10: Fermi correction function. The term $F(Z,E_\beta)$ "corrects" for the coulomb interaction between the positively charged nucleus and the negatively charged β⁻-particle or the positively charge β⁺-particle. The term F thus depends on the proton number Z (if high) of the initial unstable nuclide and the initial kinetic energy of the β-particle (if low), as per eq. (8.13).

Because the charge (+1 or −1) and mass of β-particles are always the same, this refers (a) to the small or large number of protons in the nucleus (Z), and (b) to the initial kinetic energy the β-particles obtain according to the Q-value. As a consequence of the latter effect, it will only modify the "early" part of the continuous β-particle spectrum, i.e. the lowest-energy β-particles.

8.3.6 Determination of maximum β-particle energy: The impact of the electron neutrino

This discussion refers to the high kinetic energy part of the theoretically continuous β-particle energy spectra, shown in Fig. 8.8. In the case of kinetic energies close to the maximum value E_β^{max}, the kinetic energy of the co-emitted electron neutrino (at least theoretically) modifies the spectrum. The very low, but definitely >0 mass of the electron neutrino must be subtracted from the measured endpoint energy of the β-particle. Figure 8.11 shows this effect for the spectrum of the β⁻-transformation of tritium.

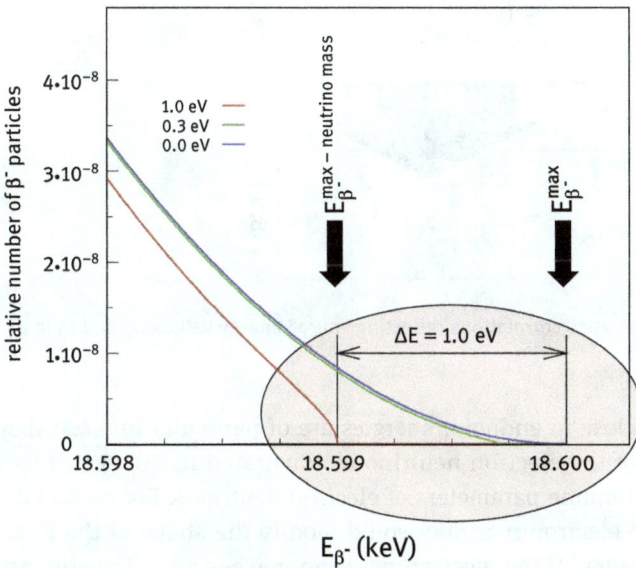

Fig. 8.11: Part of the energy spectrum of the β⁻-transformation of tritium lying very close to the maximum energy values. Detected within the Katrin experiment to determine the mass of the electron neutrino. Spectrum endpoint energies are corrected for three possible masses of the electron antineutrino: 0 eV, 0.3 eV and 1.0 eV. (https://de.wikipedia.org/wiki/Katrin).

8.3.7 Determination of maximum β-particle energy: The impact of the Q-value

Equation (8.13) may be arranged by isolating the term $(Q_\beta - E_\beta^{max})$ to yield eq. (8.16). It reflects a proportionality between the square root of $n(p) / F(Z,E_\beta) p^2$ and $(Q_\beta - E_\beta^{max})$. Other parameters of eq. (8.13), such as the constant C and the matrix transition element, become obsolete for this purpose. The term $n(p) / F(Z,E_\beta) p^2$ parallels the count rate of β-particles, which are measured experimentally. This results in a linear plot as illustrated in Fig. 8.12, otherwise known as a KURIE plot.[8] It is mainly relevant in deriving experimental values for E_β^{max} from experimentally measured continuous spectra of β-particles.[9]

$$n(p)dp = C\left\{M_{fi}\right\}^2 F(Z,E_\beta)\, p^2\, (Q_\beta - E_\beta^{max})^2\, dp \qquad (8.14)$$

$$\frac{n(p)dp}{F(Z,E_\beta)\, p^2\, dp} \approx (Q_\beta - E_\beta^{max})^2 \qquad (8.15)$$

$$\sqrt{\frac{n(p)}{F(Z,E_\beta)\, p^2}} \approx Q_\beta - E_\beta^{max} \qquad (8.16)$$

Fig. 8.12: Kurie plot: Quantum physics correlations reflecting "pure" energy values Q_β and E_β in a linear relationship.

Deviations from linearity close to endpoint energies are of particular interest; they may indicate the impact of the electron neutrino as illustrated in Fig. 8.11. KURIE plots are thus a tool to determine parameters of electron neutrinos. Figure 8.13 depicts how the mass of the electron neutrino would modify the shape of the KURIE plot at low β-particle energies. If the electron neutrino mass is zero, linearity applies. If the electron neutrino has a mass, linearity is lost and the difference between the line and the experimentally determined shape of the curve points to a specific mass of the electron neutrino.

---

8 . . . which has nothing to do with M CURIE.
9 . . . which would be rather difficult when extrapolating the exponential curve of e.g. Fig. 8.8 to a y-axis = zero value.

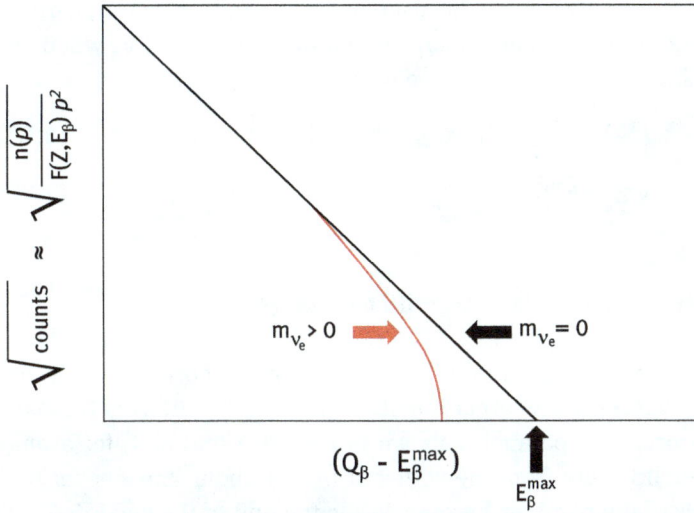

Fig. 8.13: Kurie plot: Graphical representation in terms of counts $\approx \{n(p) / F(Z,E_\beta)\, p^2\} = f(Q_\beta - E_\beta^{max})$. At β-particle energies approaching zero, the mass of the electron neutrino would modify the shape of the Kurie plot differently.

8.4 Velocities of β-transformations

8.4.1 Correlations between E_β^{max} and half-life

Historically, a huge number of β-emitters has been analyzed by measuring the ranges d of the emitted β-particles, e.g. in air.[10] The distance appeared to be proportional to the β-particle's energies, i.e. $E_\beta^{max} \sim d^{max}$. A systematic correlation between transformation constant λ and d_{air} was derived (SARGENT 1933). The double-logarithmic correlation is $\log \lambda = A \cdot \log d_{air} + B$, with $A^{SARGENT}$ and $B^{SARGENT}$ representing numerical factors. With $E_\beta \sim d$, eq. (8.18) is derived with analogue coefficients $a^{SARGENT}$ and $b^{SARGENT}$. In this version, it represents a correlation between maximum kinetic energy of the β-particles emitted and the transformation constant of the nuclide transformation *K1 → K2. The same is true for the Q_β-value when used instead of E_β^{max} because of

---

10 The β-particles emitted from a nuclide trespass into the matter surrounding the nuclide. This could be gas, liquid or solid matter such as metals, water etc. The electrons interact with the components of matter, which is atoms, and more specifically, nuclei and electron shells. Details of this "interaction of radiation and matter" are discussed in Chapter 1 of Volume II. The β-electrons will successively lose their kinetic energy until rest energy. The maximum traveling distance d_β^{max} (remember there is a continuous spectrum of β-particle energies) is determined by E_β^{max}. At values of, for example, 0.1 MeV and 1.0 MeV an electron travels about 0.1 m or 3.8 m in air, 0.1 mm or 4.3 mm in water, and 0.07 mm or 2.1 mm in aluminum, respectively.

$Q_\beta \sim E_\beta^{max}$. Similarly, the half-life $t\frac{1}{2}$ may be inserted instead of λ. However, correct mathematical treatment of this semi-empirical correlation is only possible when including quantum physical considerations – see below.

$$\log\lambda = A^{SARGENT}\log d_{air} + B^{SARGENT} \qquad (8.17)$$

$$\log\lambda = a^{SARGENT}\log E_\beta^{max} + b^{SARGENT} \qquad (8.18)$$

8.4.2 Correlations between Q-value and $\Delta\bar{E}_B$ with half-life

When comparing the many β-transforming nuclides, Q-values correlate with the half-life. For larger Q_β-values, transformation steps proceed faster. This fits perfectly with the β-transformation parabolas shown in Figs. 6.13 and 6.14, for example. The further the nuclides are from the vertex of the parabola, the steeper the hillside of the two arms of the parabola become. While the unit of the x-axis is $Z \pm 1$ and thus is linear, the y-axis representing mean nucleon binding energy is exponential. Figure 8.14 compares the "win" in mean nucleon binding energy $\Delta\bar{E}_B$ with the corresponding half-life of this transformation for all the unstable nuclides covered by both Figs. 6.13 and 6.14, i.e. for all β-transformations along the isobars of mass numbers 95 and 96.

Obviously, there is $t\frac{1}{2} = f(\Delta\bar{E}_B)$. Similarly, the Q-values are (in general) inversely proportional to the half-life or directly proportional to the transformation constant, eq. (8.19); see also Tab. 8.1. This also applies to E_β (eq. (8.20)), because the value of Q_β generally determines the value of E_β according to eqs. (8.7)–(8.9). The larger E_β^{max} (or Q_β), the faster the transformations proceed. Figure 8.15 correlates experimental data for the same nuclides as shown in Figs. 8.14, 6.13, and 6.14. Energies and velocity of transformation are proportional, but not linearly. Small changes in energy (Q_β or E_β^{max}) have an impressive impact on the half-life of the transformation, which is indicated by the exponent n in eq. (8.20).

$$Q_\beta \sim \frac{1}{t_{1/2}} \sim \lambda \qquad (8.19)$$

$$\lambda \sim \left(E_\beta^{max}\right)^n \qquad (8.20)$$

8.4.3 Half-lives of β-transitions and quantum physics: The logft-value

Quantum physics considerations for β-transitions already introduced can be extended to mathematically define these correlations between energetic and velocity of β-transformations. Remember that the transitions follow probabilities from an

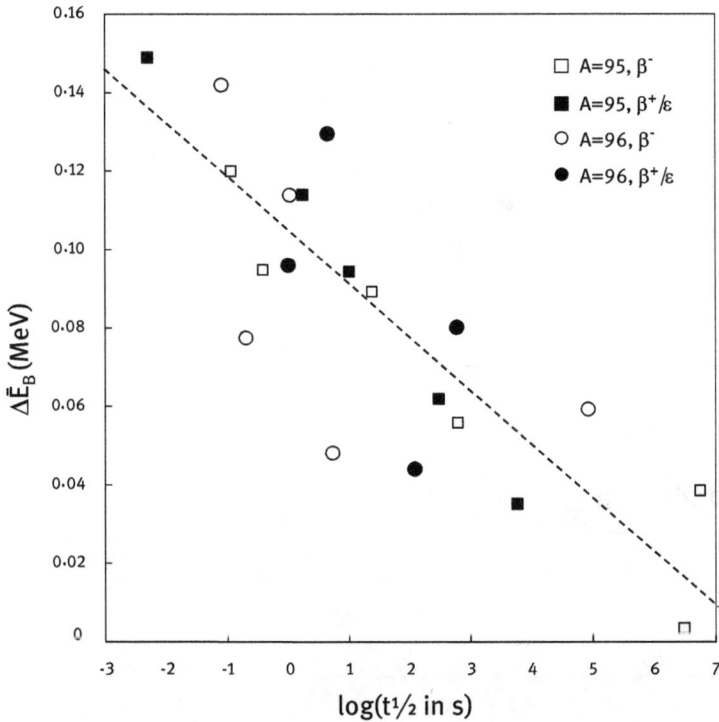

Fig. 8.14: Successive gain in $\Delta\bar{E}_B$ in β-transformation along parabolas for mass numbers A = 95 and 96 correlates with half-lives $t^{1/2}$. For values between the gain in successive mean nucleon binding energy $\Delta\bar{E}_B > 0.1$ MeV, half-lives are seconds or less, while for $\Delta\bar{E}_B < 0.1$ MeV, half-lives approach hours, days and years.

initial to several possible nuclear states, as shown in eq. (8.14). If transition rates for all individual states n are all taken together, integrals cover a range of $p = 0$ to p_{max} and result in eq. (8.21), shown in Fig. 8.16.

The right part of eq. (8.21) creates an integral representing the FERMI correction function $F = f(Z,E_β)$ and the term of overall transformation and β-particle kinetic energies $f(Q_β - E_β^{max})$. The latter becomes the "Fermi integral function", with f as a function of Z and $Q_β$. It reflects the impact the individual nuclear structure of the nucleus has on the transformation process. The constant C and the matrix transition element, both energy-independent, are not part of the integral, eq. (8.22a). Integrating $n(p)dp$ corresponds to the overall probability of transformation processes, i.e. represents all the individual transition states with their individual energetic levels and individual transition probabilities, eq. (8.22b). It gives $\lambda = f\,C\,\{M_{fi}\}^2$. When using half-life instead of transformation constant, it is $\ln 2 / t^{1/2} = f\,C\,\{M_{fi}\}^2$, eqs. (8.23a,b). Parameters of energy ($Q_β$ and $E_β^{max}$) are correlated with internal structures of the

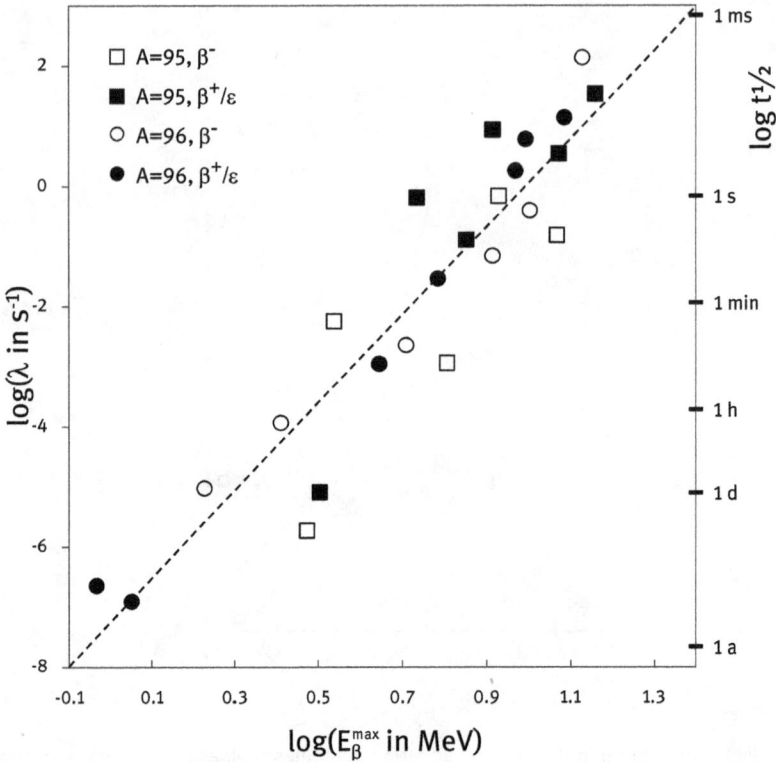

Fig. 8.15: Correlation of maximum kinetic energy of the emitted β-particles and the transformation constant (a) and half-life (b) of the β-transformation for nuclides of mass numbers A = 95 and 96. For changes in E_β^{max} of one order of magnitude (e.g. from 1 to 10 MeV), the transformation constant changes by seven orders of magnitude.

nuclides described by quantum physics and now bridge with transformation veloci-ties λ or t½.[11]

In addition, eq. (8.24) shows the product of f and t½. It is usually written as "ft" and gives ft = ln2 / C {M_{fi}}². Here, the value of the FERMI integral function f is multiplied by the half-life t½ (in seconds). Typically, it is expressed as logarithm log(ft). For each β-transformation of nuclides *K1 → K2, these values are tabulated. The data for selected unstable nuclides undergoing β-transformation are listed in Tab. 8.6.

---

11 This becomes the rationale of the semi-empiric correlations of type log(λ) *vs.* log(E_β^{max}) within the "historical" SARGENT plot.

$$\int_{p=0}^{p_{max}} n(p)\,dp = C\{M_{fi}\}^2 \int_{p=0}^{p_{max}} F(Z,E_\beta)p^2(Q_\beta - E_\beta)^2\,dp \qquad (8.21)$$

C and energy-independent $\{M_{fi}\}^2$ not part of the integral

$$\lambda = \int_{p=0}^{p_{max}} n(p)\,dp \qquad (8.22b)$$

$$\int_{p=0}^{p_{max}} F(Z,E_\beta)p^2(Q_\beta - E_\beta)^2\,dp \equiv f(Z,Q_\beta) \equiv f \qquad (8.22a)$$

$$\lambda = fC\{M_{fi}\}^2 \qquad (8.23a)$$

$$\lambda = \ln 2 / t_{1/2}$$

$$\ln 2 / t_{1/2} = fC\{M_{fi}\}^2 \qquad (8.23b)$$

$$ft = f \cdot t_{1/2}$$
with $t_{1/2}$ (in s)

$$ft = \ln 2 / C\{M_{fi}\}^2 \qquad (8.24)$$

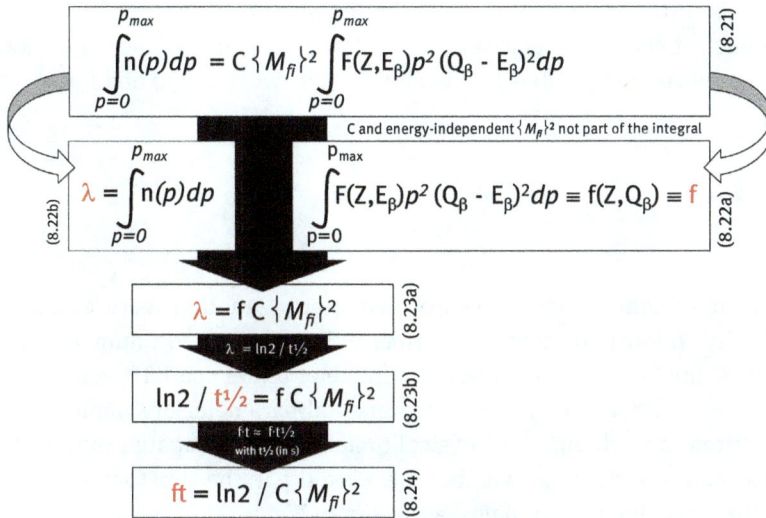

Fig. 8.16: ft-values: The Fermi integral function f linked to the half-life of β-transition processes. The ft-values thus combine quantum physical characteristics of transitions and the half-lives of the transition and are tabulated for each β-transformation.

8.5 Selection rules

Until this moment, probabilities of transitions have been discussed for β-transformations with respect to energetic levels of initial and final states. However, something more is needed to quantitatively understand β-transformation processes. In quantum physics, but also in physics and chemistry in general, there are so-called rules of transition, also called selection rules.[12] The fundamental point of view here is "symmetry". Symmetry belongs to the fundamental laws of quantum mechanics.

8.5.1 Spin

Symmetry specifically refers to electromagnetic aspects involved in angular and orbital moments (spin). The changes in quantum numbers between initial and final state thus concern orbital spin L and angular spin S and selection rules thus apply to

---

12 Rules like these also work in chemistry and are relevant for understanding, for example, electronic spectra (i.e. different selection rules apply to transitions of electrons between s-s, d-d and f-f atomic orbitals compared to p-s etc. transitions), or to make use of infrared and RAMAN spectroscopy (due to selection rules in vibrational states).

all objects defined by orbital momenta – including the photon.[13] Changes in overall spin of a certain nuclear state of a nucleus between the initial and the final state may according to spin-orbit coupling values J cover all values between $\langle J_i - J_f \rangle$ and $J_i + J_f$.

$$l_i - l_f \leq l \leq l_i + l_f \qquad (8.25)$$

8.5.2 Parity

For qualitatively understanding the impact of "symmetry" another feature needs consideration: Parity Π. Parity originally describes either even or odd numbers. In physics, it applies to the behavior of a physical parameter under spatial inversion.[14] In quantum physics, parity is realized as a fundamental parameter of symmetry in quantum states. It refers to changes of physical quantities under spatial inversion within a polar coordinate system, i.e. whether there is a flip in the sign (odd to even or *vice versa*) of the three polar coordinates, as shown in Fig. 8.17.

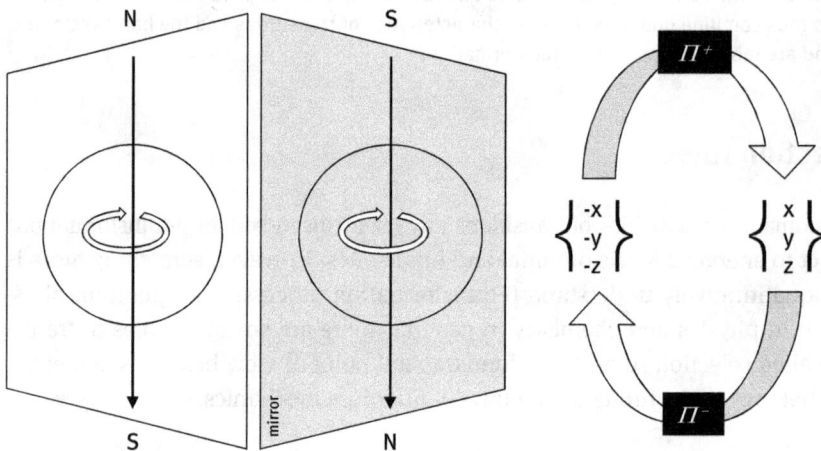

Fig. 8.17: Spatial inversion of a quantum physical object may occur as in a mirror (N = north, S = south). Parity symmetry would create a further inversion.

Inversion of an object or system in polar coordinate systems along a single point refers to three polar coordinates of different sign. Because these coordinates are

---

13 . . . which will become an important issue when discussing secondary transition processes in Chapter 11.

14 An analogy to chemistry may be found in chirality, when a given molecular structure appears as seen in a mirror and left-hand arrangements turn into right-hand ones. It remains the same molecule, but its constituents are arranged conformationally different in space – and create different chemical properties.

involved in wave functions, eigenvalues of the wave functions change, which is expressed by parity Π. The polar coordinates x, y and z are involved in wave functions Ψ. Mathematically, parity refers to how wave functions with corresponding eigenvalues and parity operators P change in the course of spatial inversion, as shown in Tab. 8.3. The process is thus either invariant (i.e. symmetric) or not. If not, it is called parity inversion (or transformation) or parity violation.[15] If the system is invariant to inversion and changes into itself, the parity is +. In the opposite case, the system is variant to inversion and changes, symmetry is violated and the parity is −.

$$\frac{\Pi_i}{\Pi_f} = (-1)^n \tag{8.26}$$

Tab. 8.3: Parity: Changes in wave functions of initial to final states follow eigenvalues corresponding to the parity operator P, parity Π, with different impact on symmetry.

Wave functions initial vs. final	Eigenvalues of operator P	Parity Π	System	Parity change	Symmetry
$\Psi_i(x,y,z) = \Psi_f(-x,-y,-z)$	+1	+	converts into itself	even	Inversion-invariance = symmetry
$\Psi_i(x,y,z) = -\Psi_f(-x,-y,-z)$	−1	−	changes sign	odd	Inversion-variance = violation

8.5.3 Selection rules

For nuclear transitions, symmetry in orbital and quantum spin and in parity influence the processes of nuclear transformation. Selection rules for transitions between initial and final states (beyond energy) are expressed mathematically by eqs. (8.25) and (8.26) for overall spin and for parity. Changes are $\Delta J = 0, 1, 2, 3, 4, \ldots$, and either $\Delta \Pi = +$ or −. Overall parameters combine both changes in terms of overall spin of the nuclides I_{fi}^{Π}. Note that spin states are defined according to the shell model of the harmonic oscillator involving spin-orbit coupling.

The impact of symmetry may qualitatively (maybe oversimplified) be compared to effects well established for chemical reactions (see Tab. 8.4). Of course, the most important parameter is energy, for example, the degree of exothermic character. The mechanism of the chemical reaction, however, may be influenced be some hindrances: steric, kinetic, etc. The lower the hindrance, the more favorably the reaction

---

15 While quantum field theory implies that parity violation contradicts fundamental laws in physics and cannot appear, this is true for strong force, electromagnetism and gravity – but not necessarily true for the weak interaction. In 1957 S E Wu et al. studied effects of ^{60}Co beta transformation and experimentally revealed violation of parity conservation.

proceeds.[16] For nuclear transitions, this is paralleled by the Q-value of the transformation. In addition, selection rules apply: whenever a change in overall spin or parity is zero or high (ΔJ and/or $\Delta \Pi$), β-transition occurs straightforwardly or is hindered. These transitions are either "allowed" or "forbidden". One may say that the more changes involved in overall spin and parity, the more difficult (more forbidden) the nuclear transformation is.

Tab. 8.4: Qualitative comparison of selection rules in nuclear sciences and in chemistry.

Criteria	Energy	Selection rules	Symmetry change
Nuclear transformation	Large Q_β = short $t^{1/2}$	Changes in ΔJ and/or $\Delta \Pi$	No = High probability of transformation Yes = Low (or no) probability of transformation
Analogy to chemistry	Exothermic character	Kinetic, steric or other hindrances	

Example: The β⁻-transformation of tritium ^{3}H, i.e. $_1$H$_2 \rightarrow _2$He$_1$, is of J^i_{Π}= s$^{1/2}$ from initial state to J^i_{Π}= = s$^{1/2}$ for final state (see Fig. 8.18). Parities are + for both the initial and final states. Consequently, transitions are of $\Delta J = 0$ and $\Delta \Pi = +$.[17]

8.5.4 Allowed and forbidden transformations

In the context of symmetry, changes in overall spin and parity decide whether the transition process is straightforward – or inhibited. The termini used are "allowed" and "forbidden" with internal gradations.

– "Allowed": Allowed transitions are either "super-allowed" or just "allowed". Super-allowed refers to the absence of changes in overall spin and parity, i.e. $\Delta J = 0$, and $\Delta \Pi = +$. They overlap with "allowed" transitions, which still remain $\Delta \Pi = +$, but may accept the lowest change in overall spin: $\Delta J = 1$.
– "Forbidden": The more changes there are in ΔJ, the more the transitions become forbidden. The lowest graduation is the retention of $\Delta \Pi = +$, but an increased change in spin: $\Delta J = 1$ or 2. This is "first forbidden". Further progress in

---

16 For example: "*similia similibus solvuntur*", similar substances will dissolve similar substances.
17 The same set of parameters is true for the transition of a free neutron to a proton. The more nucleons involved, the more complex the determinations of overall spin values of individual nuclear states. For ^{3}H the unpaired proton at 1s level contributes a spin of s$^{1/2}$, while the two paired neutrons at 1s level add spins of $+^{1/2}$ and $-^{1/2}$ to result in an overall spin of $^{1/2}$. For the product nucleus ^{3}He, the two paired protons combine with one unpaired neutron. The remaining overall spin is again $^{1/2}$.

prohibition arises due to changes in parity and / or more dramatic changes in overall spin, see Tab. 8.5. However, for weak interaction "forbidden" is not an absolute measure; it rather means "hindered", and first to second to third etc. hindrance reflects a gradual system.

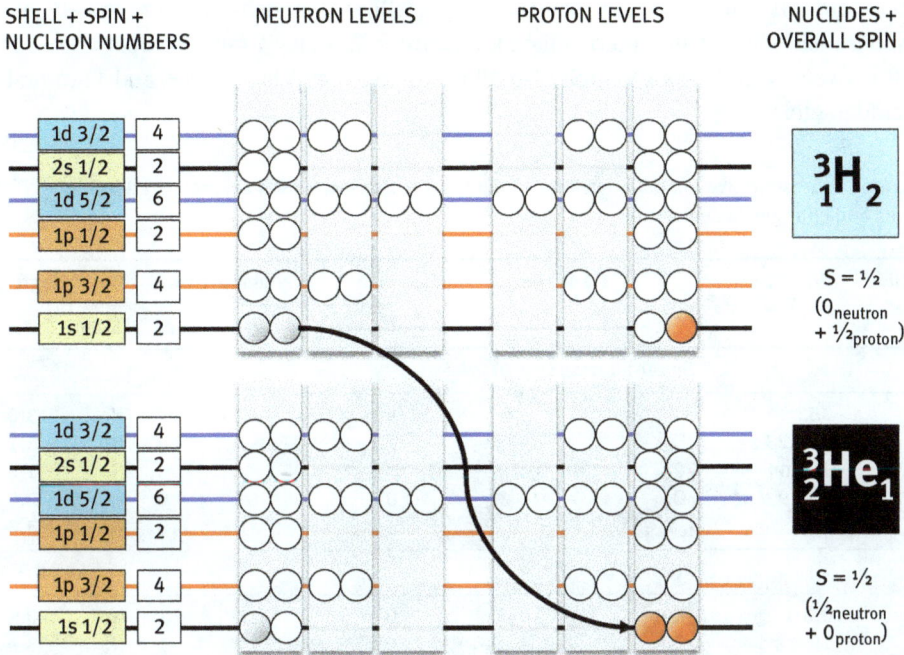

Fig. 8.18: Overall spin and parity values for the β^--transformation of tritium.

Tab. 8.5: Structure of allowed and forbidden transitions applied to β-transformation.

Transitions	Subgroup	Symmetry	
		ΔJ	$\Delta \Pi$
Allowed	Super-allowed	0	+ (no)
	Allowed	0, 1	+ (no)
Forbidden	First forbidden (n = 1)	1, 2	– (yes)
	Second forbidden (n = 2)	2, 3	+ (no)
	Third forbidden (n = 3)	4	+ (no)
	Fourth forbidden (n = 4)	4	– (yes)

Selection rules modify the relationship between Q-values and half-lives of β-transitions. The characteristics of nuclear structure are involved in the FERMI integral function f. Figure 8.19 represents a kind of statistic of logft-values for unstable nuclides undergoing β-transformation. Most of the β-transformations are characterized by low changes in symmetry and thus belong to super-allowed, allowed and first forbidden categories. For example, super-allowed transitions have lowest, allowed transitions low and intermediate logft-values. A logft < 4 reflects super-allowed, logft between 4 and about 9 applies to allowed, logft between about 6 and 12 to first forbidden etc.

Tab. 8.6: Representative unstable nuclides undergoing β-transformation characterized by logft-values and changes in symmetry.

Nuclide	$t_{1/2}$	Transition		I^π_i		I^π_f	ΔJ	$\Delta \Pi$	logft
		Super-allowed ($n = 0$)							
^{0}n	885 s	n	$\rightarrow$ p	$s_{1/2+}$	$\rightarrow$	$s_{1/2+}$	0	No	3.0
^{3}H	12.323 a	$_1H_2$	$\rightarrow$ $_2He_1$	$s_{1/2+}$	$\rightarrow$	$s_{1/2+}$	0	No	3.1
^{11}C	20 min	$_6C_5$	$\rightarrow$ $_5B_6$	$p_{3/2-}$	$\rightarrow$	$P_{3/2-}$	0	No	3.6
^{14}O	70.59 s	$_8O_6$	$\rightarrow$ $_7N_7$	0^+	$\rightarrow$	0^+	0	No	3.5
^{31}S	2.62 h	$_{16}S_{15}$	$\rightarrow$ $_{15}P_{16}$	$s_{1/2+}$	$\rightarrow$	$s_{1/2+}$	0	No	3.7
		Allowed ($n = 0$)							
^{6}He	0.8 s	$_2He_4$	$\rightarrow$ $_3Li_3$	0^+	$\rightarrow$	1^+	1	No	2.9
^{10}C	19.3 s	$_6C_4$	$\rightarrow$ $_5B_5$	0^+	$\rightarrow$	1^+	1	No	3.0
^{18}F	110 min	$_9F_9$	$\rightarrow$ $_8O_{10}$	1^+	$\rightarrow$	0^+	1	No	3.6
^{33}P	25.34 days	$_{15}P_{18}$	$\rightarrow$ $_{16}S_{17}$	$s_{1/2+}$	$\rightarrow$	$d_{3/2+}$	1	No	5.0
^{35}S	87.5 days	$_{16}S_{19}$	$\rightarrow$ $_{17}Cl_{18}$	$d_{3/2+}$	$\rightarrow$	$d_{3/2+}$	0	No	5.0
^{53}Fe	8.51 min	$_{26}Fe_{27}$	$\rightarrow$ $_{25}Mn_{28}$	$f_{7/2-}$	$\rightarrow$	$f_{7/2-}$	0	No	5.2
		First forbidden ($n = 1$)							
^{85}Kr	10.76 a	$_{36}Kr_{49}$	$\rightarrow$ $_{37}Rb_{48}$	$g_{9/2+}$	$\rightarrow$	$f_{5/2+}$	2	Yes	9.4
^{113}Cd	$9 \cdot 10^{15}$ a	$_{48}Cd_{65}$	$\rightarrow$ $_{49}In_{64}$	$h_{11/2-}$	$\rightarrow$	$g_{9/2+}$	1	Yes	9.2
^{133}I	20.8 h	$_{53}I_{80}$	$\rightarrow$ $_{54}Xe_{79}$	$g_{7/2+}$	$\rightarrow$	$h_{11/2-}$	2	Yes	10.0
		Second forbidden ($n = 2$)							
^{22}Na	2.603 a	$_{11}Na_{11}$	$\rightarrow$ $_{10}Ne_{12}$	3^+	$\rightarrow$	0^+	3	No	15.1
^{93}Zr	1.510^6 a	$_{40}Zr_{53}$	$\rightarrow$ $_{41}Nb_{52}$	$d_{5/2+}$	$\rightarrow$	$g_{9/2+}$	2	No	>12.8
^{99}Tc	2.110^5 a	$_{43}Tc_{56}$	$\rightarrow$ $_{44}Ru_{55}$	$g_{9/2+}$	$\rightarrow$	$d_{5/2+}$	2	No	12.3
^{137}Cs	30.17 a	$_{55}Cs_{82}$	$\rightarrow$ $_{56}Ba_{81}$	$g_{7/2+}$	$\rightarrow$	$d_{3/2+}$	2	No	12.1
		Fourth forbidden ($n = 4$)							
^{40}K	1.2810^9 a	$_{19}K_{21}$	$\rightarrow$ $_{18}Ar_{22}$	4	$\rightarrow$	0^+	4	Yes	21.0

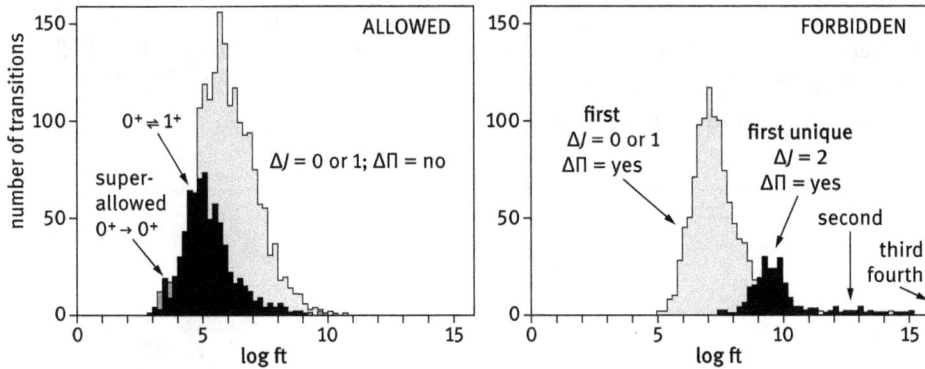

Fig. 8.19: logft-values for unstable nuclides undergoing β-transformation grouped according to allowed and forbidden transitions.

Thus, it is not only the half-life (in seconds), it is also the impact of symmetry according to quantum mechanics which characterizes the transformation. For example, the half-life of sodium-22 of 2,603 years is shorter than the half-life of tritium of 12,323 years, but ft values are much larger for ^{22}Na (logft = 15.1) than for ^{3}H (logft = 3.1). The transition of ^{3}H as illustrated in Fig. 8.18 is "super"-allowed, the one of ^{22}Na is "second" forbidden. The changes in ΔJ are responsible and these are 0 and 3, respectively (see Tab. 8.5). Note that there is no change in parity for both transitions. Table 8.6 summarizes some unstable nuclides undergoing β-transformation.

8.5.5 Symmetry and SARGENT graph

The correlation between energetic and velocities of β-transformations as shown in Fig. 8.15 was the SARGENT graph as derived from $\log \lambda = f$. The relationships of transformation constants of β^--emitting radioisotopes against corresponding logarithms of their maximum β^--energies reflected what might be expected: the larger the gain in total energy of the transformation is, the larger the maximum kinetic energy of the β^--particle emitted, and the faster the process proceeds. Thus, the principal relationship shown in Fig. 8.15 was not a surprise.

When considering the β^--emitters available within the natural decay chains that time,[18] however, the graph revealed strange empirical evidence: there is not just one line – but two! Due to the double-logarithmic version of eq. (8.18), they split by a significant factor of two orders of magnitude in λ for the same value of E_β^{max}, as seen in

---

18 Historically, some β^--emitters available within the natural decay chains have been analyzed in that way.

Fig. 8.20. For a given value of E_β^{max}, there may either be a corresponding value of λ or another one, larger by 2 to 3 orders of magnitude. For example, for a value of $E_\beta^{max} \approx$ 1 MeV ($\log E_\beta^{max} \approx 0$), the half-lives of ^{210}Bi (5.013 d = 4.33·10^5 s) and ^{214}Pb (26.8 min = 1.61·10^3 s) differ by a factor of ca. 270.

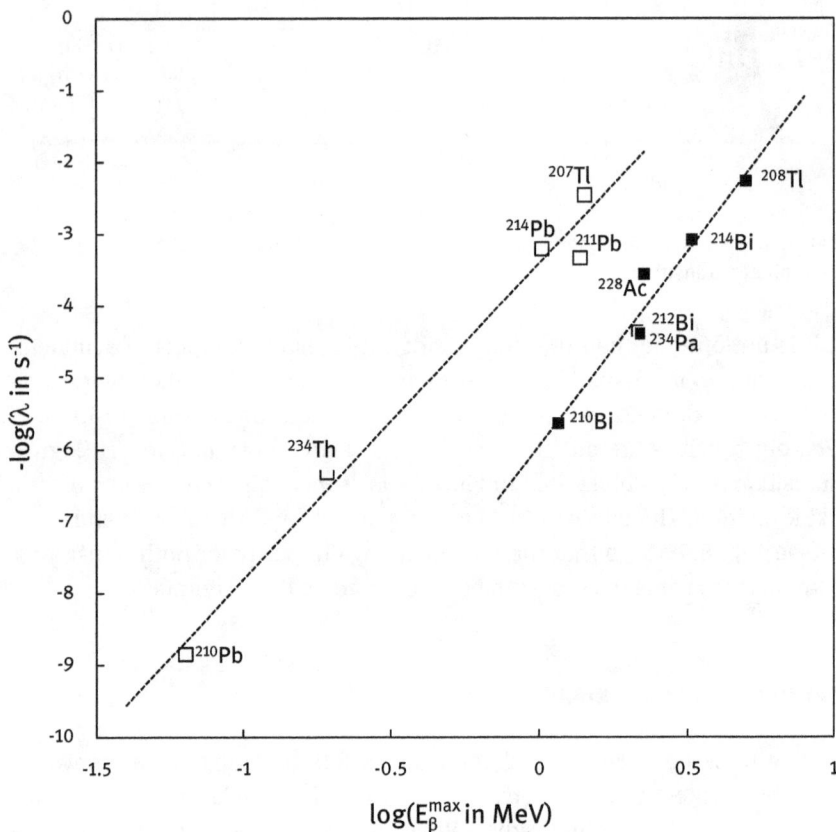

Fig. 8.20: β-emitting nuclides originally utilized to create the Sargent plot. The two lines correspond to allowed and forbidden (dotted line) transformations.

Quantum physics provides an explanation. The impact of wave functions and transitions expressed via $\{M_{fi}\}^2$ considers violation of symmetry. Whenever transitions are forbidden they proceed much slower than an allowed transition – despite the same balances in energy in Q_β-values or maximum kinetic β-particle energy. The unstable nuclides illustrated in Fig. 8.20 belong either to those with allowed transitions or to those with forbidden ones.

8.6 Excited states

Primary transformation processes of unstable nuclides *K1 do not necessarily yield the ground state of the newly formed nuclide K2 directly. Instead, energetically excited levels $^{\odot i}$K2 of the new nuclide may be populated. These energetically different states all belong to the new nuclide in terms of mass number A, proton number Z, and neutron number N, but a nucleon may occupy a higher-energetic nucleon shell. Their quantum mechanics characteristics differ from the ground state level. The number and the characteristics of potential excited levels depend on the shell structure, in particular in terms of the overall spin. Consequently, selection rules apply also to transitions within one and the same nucleus, but with different shells involved.

Figure 8.21 illustrates various excited levels for a nuclide of intermediate mass number, ^{60}Co. The initial state of ^{60}Co is of 4^+. The transformation product nuclide is stable ^{60}Ni with three individual excited nuclear states. Its highest-energetic excited state ($^{\odot 3}$K2) is of 4^+. The energetically lower excited level of $^{\odot 2}$K2 is of 2^+. Next, there is an excited level $^{\odot 1}$K2 of 2^+ again. Finally, there is the ground state ^{8}K2, which is of 0^+. Theoretically, there are four principle primary transformations, namely *K1 → $^{\odot}$K2, *K1 → $^{\odot 2}$K2, *K1 → $^{\odot 1}$K2 and *K1 → ^{8}K2 with the corresponding logft-values. The most probable transformations are those with the lowest logft-values. Accordingly, the frequencies of the experimentally observed transformations are 99.88 %, <0.002 %, 0.12 % and 0 % respectively for the four options of ΔJ = 1, 2, 2 and 5. The dominating transformation is *K1 → $^{\odot 1}$K2. And: there is no direct transformation into the ground state (ΔJ = 5).[19]

The more nucleons there are allocated in a nucleus and the closer the nucleon shells approach and overlap, the larger the number of excited states is. There are just a few excited states for light elements, but several for heavy elements. There are none for ^{3}H, but several for ^{131}I which are populated by individual electrons. Table 8.7 lists the excited levels and their energies, abundances, selection rules categories and logft-values for the β^--transformation of ^{131}I to ^{131}Xe. Accordingly, the "easiest" transformation is to the exited level no. 4 with logft = 6.64. This route occurs in 89.4 of 100 events of ^{131}I transformation. The next most probable transitions are those of logft = 6.86 and 6.98 with 7.36 % and 2.114 %, respectively – all three belonging to the allowed pathways of parity invariance. The three other transition change parity, and belong to first forbidden transitions and are of <1 % probability. The ground state of ^{131}Xe is not addressed at all.

---

19 So what about the ground state level of ^{60}Ni? It is not populated directly from ^{60}Co due to selection rules. This ground state is finally achieved by another mechanism: de-excitation along excited nucleon shell levels. The process is similar to those introduced for energetically excited electrons existing at higher-energetic electron shell orbitals. A further similarity is that the most common approach here is emission of electromagnetic radiation, i.e. photons. This is the topic of Chapter 11 on secondary transformation pathways.

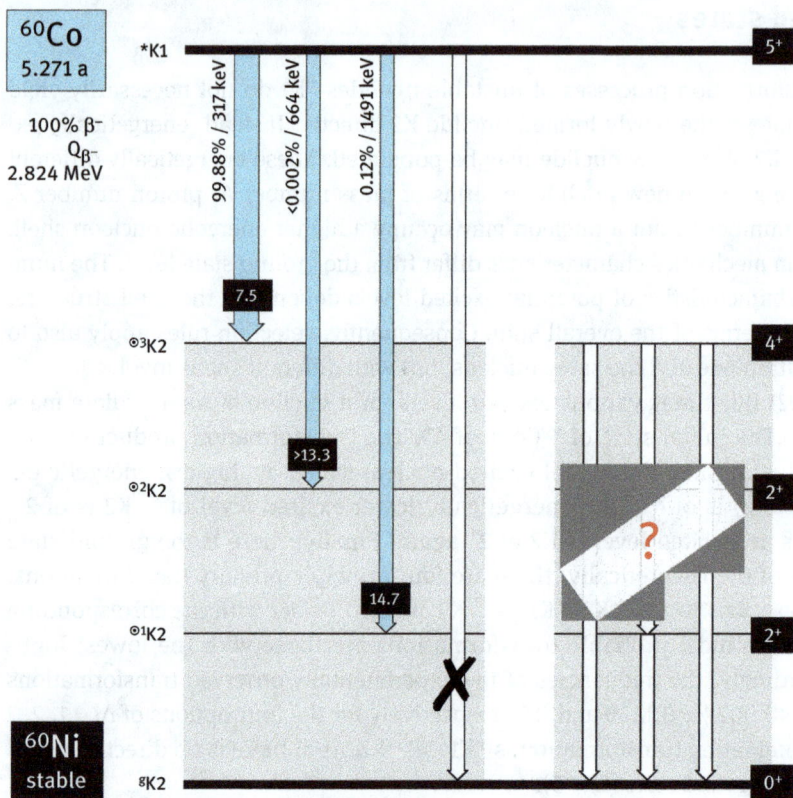

Fig. 8.21: Excited levels following the β⁻-transformation of ^{60}Co. Primary transition processes of ^{60}Co populate three separate excited levels $^{\odot i}$K2 in the product nucleus ^{60}Ni, indicated by gray color. The population of the levels follows symmetry considerations. The primary transformation populating the first excited state is the most abundant one because the changes in spin $\Delta J = 1$ are the lowest relative to all the other options. The direct primary transformation into the ground state is strongly forbidden and does not occur at all ($\Delta J = 5$). De-excitation from one excited state to an energetically lower one ("?") requires secondary transformation processes.

As a consequence, the β⁻-particles emitted in the course of this β⁻-transformation are of different kinetic energy depending on the specific excited state $^{\odot}$K2. The closer the excited levels are to the ground state, the higher the difference in energy between the initial state of ^{131}I and the corresponding excited states of ^{131}Xe. For ^{131}I, the dominating β⁻-particles show a maximum kinetic energy of 606.3 keV, but there are also others of higher (e.g. 806.9 keV) or lower (e.g. 247.9 keV) E_β^{max}. Thus, there is a whole group of β⁻-particles emitted that carry less than the 0.971 MeV of the Q_β-value.

Tab. 8.7: Excited levels in the β^--transformation of ^{131}I ($t\frac{1}{2}$ = 8.0233 d, 7/2$^+$, Q_β = 0.971 MeV) to ^{131}Xe (stable).[20]

Level	E_β^{max}	Abundance	ΔJ	$\Delta\Pi$	Category	logft
	(keV)	(%)				
8	247.9	2.114	5/2	+	Allowed	6.98
7	303.9	0.643	7/2	–	1st forbidden	7.79
6	333.8	7.36	7/2	+	Allowed	6.86
4	606.3	89.4	5/2	+	Allowed	6.64
3	629.7	0.053	9/2	–	1st forbidden	9.8
2	806.9	0.396	11/2	–	1st forbidden	10.03

8.7 Examples and applications

The type β-transformations comprise most of the unstable nuclides. Among these nuclides are many that have become valuable for research and technology, for industrial and medical applications. Table 8.8 gives a selection of relevant nuclides.

All the three subtypes of transformations provide different kinds of emitted radiation that is of benefit for specific applications. In some cases, the β^--particle is the main tool. In many cases, the secondary effect is more important for practical applications than the primary transformation. The secondary effect of interest is the emission of low- or high-energetic photons resulting from de-excitation of excited nucleon levels populated within the primary transformation process, discussed in Chapter 11. In the β^+-process, the positrons emitted usually serve as a source of photons, which are created in the course of annihilation of that positron after combining with an electron. This effect belongs to the class of post-transformation processes to be discussed in Chapter 12.

All these applications will also be discussed separately in Chapter 11 according to the various directions of utilization of radionuclides and their emissions in research and technology. Here, the most relevant nuclides undergoing β-transformation are only mentioned with respect to their β-transformation details.

8.7.1 Important applications of β^--particle emitters based on the β^--electron

In the case of low-energy and relatively long-lived nuclides, β^--particles are of value for several purposes. Well-known examples are tritium and ^{14}C. Tritium or ^{14}C-labeled analogue molecules are used to investigate chemical reaction mechanisms and to analyze biological processes of relevant organic molecules (hydrocarbons) *in vitro*.

---

20 Nuclear data are from BNM-LNHB/CEA-Table de Radionucléides.

Tab. 8.8: Selected unstable nuclides undergoing β-transformation relevant for applications in research and technology.

Nuclide	t½	E^{max} of β⁻- or β⁺-particles (MeV)	Main photon emission (MeV)	Application
β⁻-subtype				
^{3}H	12.323 a	0.02	–	Chemistry, *in vitro* molecular biology and assays, dating
^{14}C	5730 a	0.2	–	
^{32}P	14.26 d	1.7	–	*In vitro* molecular biology and assays
^{35}S	87.5 d	0.2	–	
^{60}Co	5.272 a	0.3, 1.5	1.332	External radiation therapy
^{89}Sr	50.5 d	1.5	–	Nuclear medicine therapy
^{90}Y	64.1	2.3	–	
^{99}Mo	66 h	1.2	–	Parent of ^{99m}Tc generator
^{137}Cs	30.17 a	0.5, 1.2	0.662	Calibration source
^{153}Sm	1.93 d	0.7, 0.8	0.103	Nuclear medicine therapy
^{177}Lu	6.71 d	0.5	0.208	
^{186}Re	89.25 h	1.1	0.137	
^{188}Re	16.98 h	2.1	0.155	
β⁺-subtype				
^{11}C	20.38 min	1.0	–	Nuclear medicine diagnosis
^{13}N	9.96 min	1.2	–	
^{15}O	2.03 min	1.7	–	
^{18}F	109.7 min	0.6	–	
^{26}Al	$7.16 \cdot 10^5$ a	1.2	1.809, 1.130	Astrophysics
^{68}Ga	67.6 min	1.9	–	Nuclear medicine diagnosis
^{124}I	4.15 d	2.1	0.603, 1.619	
ε-subtype				
^{57}Co	271.79 d	–	0.122, 0.136	MÖSSBAUER spectroscopy
^{123}I	13.2 h	–	0.159	Nuclear medicine diagnosis *via* SPECT
^{125}I	59.41 d	–	0.035	*In vitro* molecular biology and assays
^{165}Er	10.3 h	–	–	Nuclear medicine therapy

These nuclides do exclusively emit beta particles and can be detected by liquid scintillation. This deserves separate discussion; see Chapter 12 of Volume II.

In tumor therapy, β^--particles are employed to induce dense ionization of water
within or around a tumor cell. Here, a key nuclide is ^{90}Y. It transforms almost
completely (99.983 %) into the ground state of ^{90}Zr according to the logft-value of
8.0 of the 2^- to 0^+ transition, shown in Fig. 8.22. Consequently, the β-particles emitted carry almost the complete amount of energy of the transition (Q_{β^-} = 2.280 MeV)
and the E_β^{max} value is identical, namely 2.280 MeV. Since there are two almost un-
populated excited levels ($^{\odot 2}$K2 with logft = 11.1 and $1.4 \cdot 10^{-6}$ %, $^{\odot 1}$K2 with logft =
9.4 and 0.017 %), there is hardly any accompanying photon emission due to de-
excitation. Thus, ^{90}Y is an almost "pure" β^--emitter of high-energy β^--particles of
long range in tissue (maximum range = 12 mm, mean range = 5 mm) and is preferred
for treatment of tumors of larger dimension.

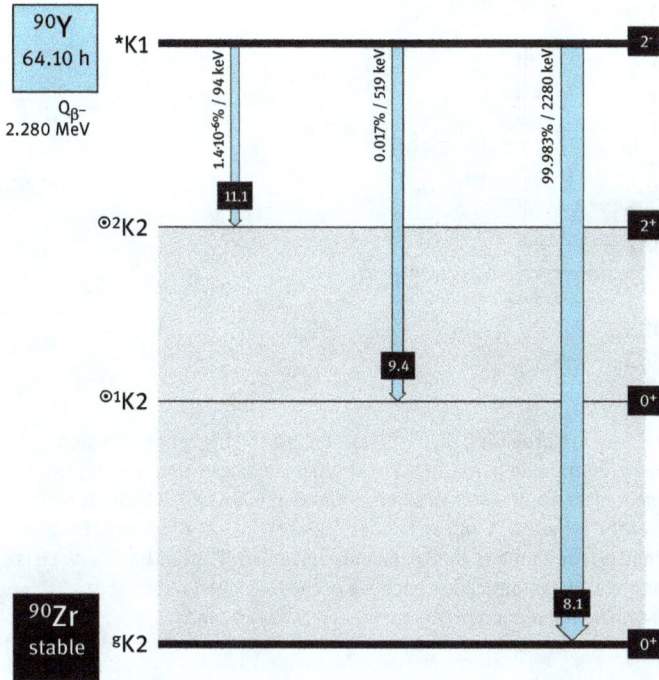

Fig. 8.22: β^--transformation of ^{90}Y to ^{90}Zr. Direct transformation to the ground state is the
dominant strategy (99.983 %) due to the logft-value of 8.0. There is negligible population of two
excited states.

Once a biological carrier system has delivered the radionuclide close to the tumor
cell or even inside the cell close to the cell nucleus, ionization caused by the emitted

β⁻-particles[21] leads to double-strand breaks of the tumor DNA and selective death of the malignant cell. This approach may be tuned to large or small tumors according to the kinetic energy of the β⁻-particles, which is proportional to their range in tissue, as shown in Fig. 8.23.

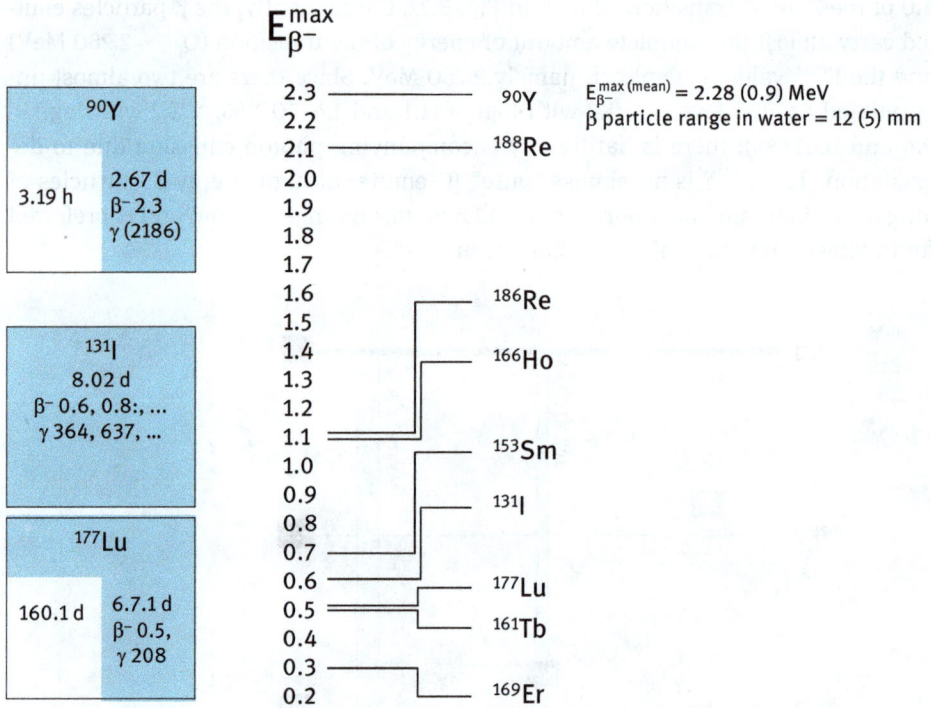

Fig. 8.23: Medically relevant β⁻-emitters providing β⁻-particles of short to long range in water. β⁻-emitters such as ^{90}Y, ^{131}I and ^{177}Lu provide β⁻-particles of varying maximum kinetic energies. These β⁻-particles induce dense ionization of water within or around a tumor cell, which ultimately leads to double-strand breaks of the tumor DNA and selective death of the malignant cell. By selecting the appropriate β⁻-emitter, the range in tissue may be controlled. The lanthanide ^{177}Lu is among the very useful trivalent metallic radionuclides because of the low mean range of its β⁻-particles in tissue of 0.67 mm. The dimension of one tumor cell is 10–20 μm.[22]

---

21 The effects induced by the various kinds of emissions when interacting with matter are discussed in Chapter 1 of Volume II, and nuclear medicine therapy is discussed in Chapter 14 of Volume II.

22 Remember that ranges of β⁻-electrons are not discrete due to the continuous energy spectrum of the electrons. Mean ranges thus correspond to the energy of most of the electrons initially emitted.

8.7.2 Important applications of β⁻-particle emitters based on the photon

Another important neutron-rich unstable nuclide undergoing β⁻-transformation is ¹³¹I, which decays to its primary transformation product, ¹³¹Xe. Initially ¹³¹Xe is generated in various excited nuclear states, which de-excite under emission of electromagnetic radiation. According to the dominant population of one state, the photons emitted from this level approaching the ground state of ¹³¹Xe either directly or *via* another excited state create intense photon emissions at 364 keV and 284 keV energy, respectively. Thus, the β⁻-transformation process is accompanied by photon emission which means that ¹³¹I is not a "pure" β⁻-emitter like ⁹⁰Y. While the β⁻-particles are used for therapeutic purposes, the additional photon emission is useful for nuclear medicine diagnosis.

Another important β⁻-emitter is ⁹⁹Mo. This is not because this radionuclide is of particular direct value; instead, it serves as the parent nuclide of the ⁹⁹Mo/⁹⁹ᵐTc radionuclide generator. Figure 8.24 shows the scheme of β⁻-emissions of ⁹⁹Mo.

Similar to ¹³¹I, there are many excited levels in the product nuclide ⁹⁹Tc, populated to very different degrees following the logft-values, and there is no direct pathway to the ground state through β⁻-emission. The dominating transformation leads to an excited level with logft = 7.1 and shows 82.4 % branching.

Other uses include the application of relatively long-lived nuclides for calibrating radiation detectors on the basis of prominent γ-quanta (e.g. ¹³⁷Cs). In the case of intense and high-energetic photons ($E_\gamma > 1$ MeV), the corresponding radionuclides, e.g. ⁶⁰Co, are applied in cancer treatment by means of external irradiation ("γ-knife"). Other applications are in the field of astrophysics. Photon radiation emitted from unstable nuclides (²⁶Al and others) of extra-stellar objects is detected and used to investigate fundamental data about the past, present and future of our universe. The same is true for our own planet; see Chapter 5, Volume II on "dating".

8.7.3 Important applications of β⁺-particle emitters based on annihilation

β⁺-electron emitters (i.e. proton-rich unstable nuclides of low mass number A) are of interest in most cases not because of the radiation caused by the primary nuclear transformation (which is the positron), but because of post-effects the positron undergoes once released from the initial nucleus. This is the subsequent annihilation of the positron when combined with a normal electron. This process transforms the mass of the two elementary particles into electromagnetic emission in the form of two 511 keV photon quanta, see Chapter 12. These photons are detected by Positron Emission Tomography (PET). The nuclides providing intense and almost exclusive positron emission, such as and ¹⁸F, are labeled to biologically relevant molecules and thus are

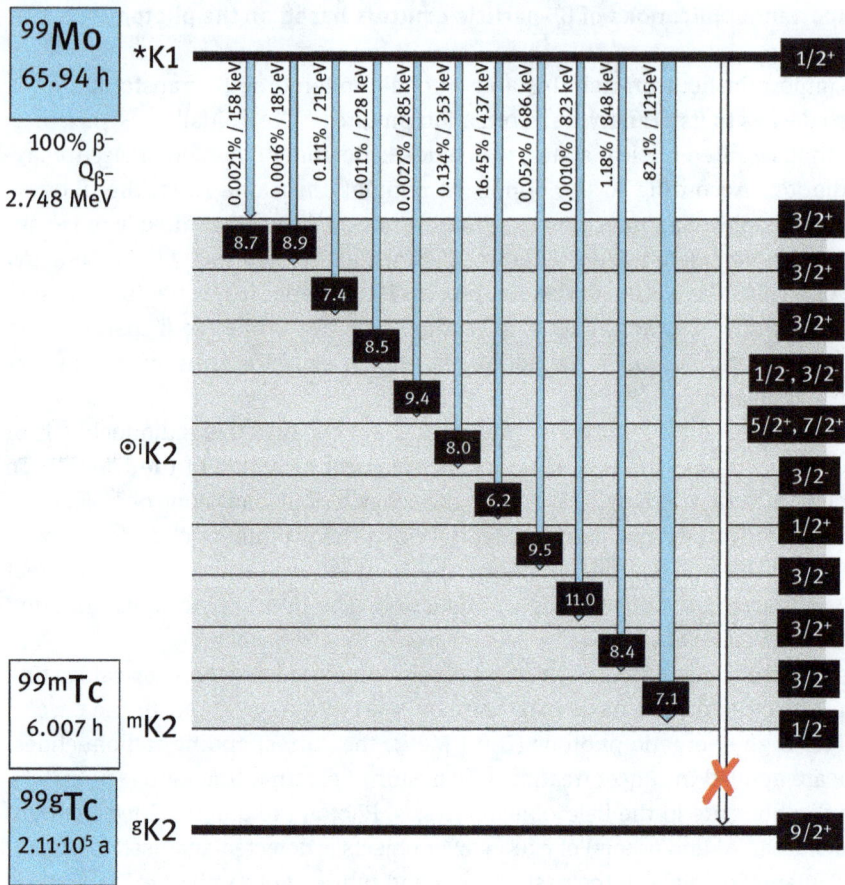

Fig. 8.24: β⁻-transformation of ^{99}Mo to ^{99g}Tc and ^{99m}Tc. Direct transformation to the ground state is forbidden. There are many individual excited levels available, but one preferred pathway (82.1 %) with logft of 7.1. Unlike the processes shown for ^{131}I, this excited state exhibits its own half-life of 6.0 h and thus represents the metastable nuclide ^{99m}Tc.

valuable tools for molecular imaging in nuclear medicine.[23] Despite their promising nuclear data, they are particularly relevant in radiopharmaceutical chemistry and molecular imaging, because these nuclides represent isotopes of "organic" atoms, i.e. those typically constituting organic and biologically essential molecules.

---

23 Nuclear medicine diagnosis is discussed in Chapter 13 of Volume II.

8.8 Outlook

8.8.1 De-excitation

In many cases, nucleon transformation does not cover the complete process of $^*K1 \rightarrow {}^gK2$; but rather the route $^*K1 \rightarrow {}^\odot K2$. This de-excitation procedure requires transitions:
(a) between excited states and
(b) from exited states to the ground state.

Similar to the excited electrons populating electron shells of higher energy, there must by a process of de-excitation towards the ground state. The most common approach for the excited nucleus of an atom is emission of electromagnetic radiation, i.e. photons. Interestingly, this applies to most nuclides $^\odot K2$ providing excited states, this means that it does not matter whether the nuclide K2 itself is stable or not (see the example given in Fig. 8.23 with stable ^{60}Ni). This belongs to the class of "secondary" nuclear transformation in terms of $^*K2 \rightarrow K2$ and is the topic of Chapter 11.

8.8.2 Inverse β-transition

The β-process itself shows some special features: inverse β-transition and double beta transformation. Nuclear sciences make use of the inverse process of proton-to-neutron conversion. Instead of the emission of an electron neutrino in the course of the process $p \rightarrow n + \beta^+ + v_e$, the proton may capture an electron antineutrino. This also creates a neutron (similar to the process of electron capture), but exclusively releases a positron, summarized in eq. (8.27). The product can be easily and sensitively detected, and thus serves as a measure of whether the electron neutrino was there or not. This was briefly mentioned in the context of the (indirect) experimental detection of the electron neutrino itself.

$$\bar{v}_e + p \rightarrow n + \beta^+ \tag{8.27}$$

8.8.3 Double β-transformation

Despite the standard procedure of converting an excess neutron into a proton (or a free neutron into a proton), there are two situations in which two "bound" neutrons *simultaneously* transform into two protons. The mechanism is either according to a real "double" β-process, i.e. emitting two β⁻-particles and two electron antineutrinos (see eq. (8.28) and Fig. 8.25), or a "neutrino-less" process (eq. (8.29)). While the second version is a matter of intense research, the first has been experimentally observed for about 35 unstable neutron-rich nuclides, such as ^{48}Ca, ^{76}Ge, ^{82}Se, ^{96}Zr, ^{100}Mo, ^{116}Cd, ^{128}Te, ^{130}Te, ^{136}Xe, and ^{150}Nd.

$$2\bar{\nu}_e\beta\beta \qquad {}^{A^*}_{Z}K1 \;\rightarrow\; {}^{A}_{Z+2}K2 + 2\beta^- + 2\bar{\nu}_e \qquad (8.28)$$

$$0\bar{\nu}_e\beta\beta \qquad {}^{A^*}_{Z}K1 \;\rightarrow\; {}^{A}_{Z+2}K2 + 2\beta^- \qquad (8.29)$$

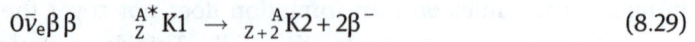

The rationale is, that whenever the "normal", single β⁻-transformation process is not energetically possible or symmetry-wise forbidden, the double process becomes an option. Its probability, however, is small. The nuclides mentioned thus may show very large half-lives approaching 10^{20} years.

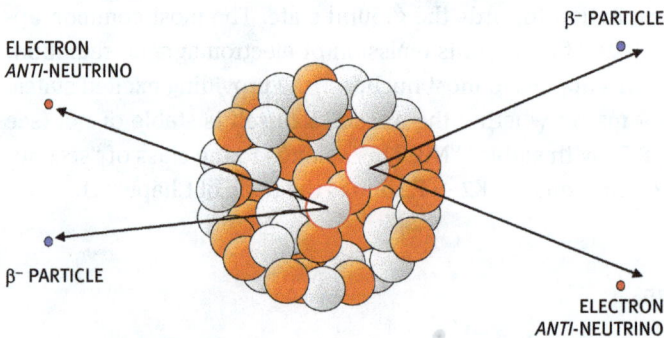

Fig. 8.25: Mechanism of a double β⁻-transformation process. Kinetic energy in this case is distributed among all four emission products.

Two examples are shown in Fig. 8.26 for the double β-transformation process in ^{82}Se and ^{96}Zr. The latter nuclide was illustrated within the β-parabola of the A = 96 isobar chain in Fig. 6.14. Its expected transformation product was $_{40}$Zr$_{56}$ → $_{41}$Nb$_{55}$. Nevertheless, ^{96}Zr is not the nuclide at the vertex of the parabola, so there should be a pathway to transform to the final and stable nuclide, which is ^{96}Mo ($\bar{E}_B = 8653.988$ keV). However, mean nucleon binding energies of the two nuclides are $\bar{E}_B = 8635.328$ and $\bar{E}_B = 8628.997$ keV, respectively, for ^{96}Zr and ^{96}Nb so there is (energetically) no good reason to transform. The alternative is indeed the double β⁻-route, although the half-life of this transformation is $3.9 \cdot 10^{19}$ years.

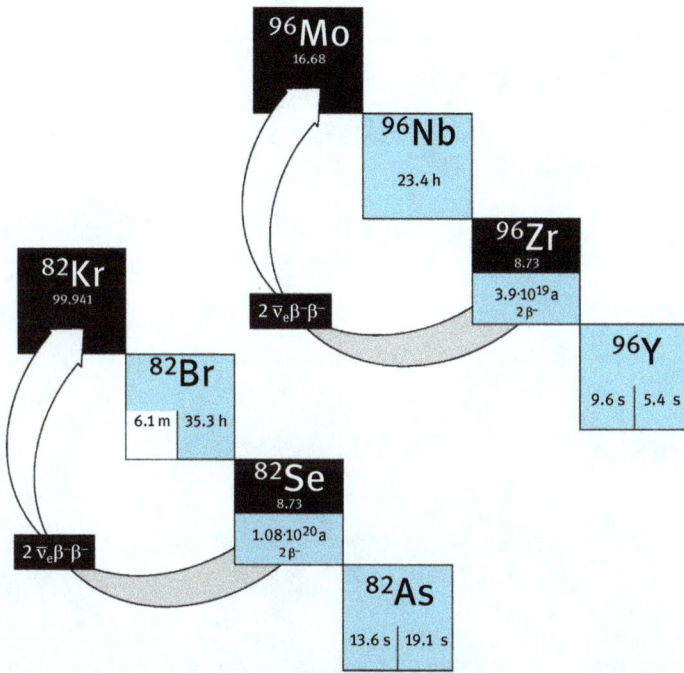

Fig. 8.26: Examples of double β⁻-transformation processes. "Normal" β⁻-transformation from ^{82}Se to ^{82}Br and from ^{96}Zr to ^{96}Nb is "problematic" due to balances in mean nucleon binding energy and masses of the nuclides involved. Atomic masses (in u) for the nuclides involved are 81.9166995, 81.9168018, 81.9134812 for ^{82}Se, ^{82}Br, ^{82}Kr and 95.9082776, 95.9081016, 95.9046748 for ^{96}Zr, ^{96}Nb, ^{96}Mo, respectively.

9 α-Emission

Aim: For mass numbers 1 to 209, the β-process transforms unstable nuclides into stable final product nuclides, forming the valley of β-stability. Whenever heavier nuclides (Z > 83, A > 210) behave according to the β-transformation profile, the most stable nuclide of the A = constant parabola does not necessarily represent a stable product.

Consequently, another option for stabilizing unstable nucleon compositions is needed, and the second most "popular" one (after β-transitions) is α-emission.

The essence of α-emission is the emission of a cluster of four nucleons, two protons and two neutrons – the nucleus of the ^{4}He atom, the α-particle. The new nuclide K2 thus will be of (A – 4), (Z – 2) and (N – 2). According to the liquid drop model, mean nucleon binding energy improves for heavy nuclides in the direction of a lower mass (A – 4) position until medium mass numbers are reached.

This chapter discusses:
(i)   how the overall energy is disseminated among the new nuclide K2 and the α-particle, in particular in terms of kinetic energies of the α-particles emitted,
(ii) the correlation between the nuclear transformation energies and the velocity of the transformation, i.e. its half-life, and
(iii) the corresponding theoretical and mathematical models. The main feature is the quantum mechanics tunneling effect.

Finally, prominent α-emitters are introduced, which are important to specific aspects of science, technology and medicine.

9.1 Introduction

For all mass numbers from 1 to 209, β-processes yield one definite stable nuclide or two – depending on the (even, even) or (even, odd) nucleon composition. This paradigm does not continue after A > 209. For example, let us consider the mechanism of a β-transformation of the A = 226 isobar. The radium isotope 226 represents the most stable nucleon composition. ^{226}Ra shows by far the longest half-life of this isobar, namely 1600 years. The neighbors at (Z + i) are of much lower stability, with half-lives the range of hours (29 hours for ^{226}Ac), minutes (31 min and 1.8 min for ^{226}Th and ^{226}Pa) and decrease further down to milliseconds (280 ms for ^{226}U and 31 ms for ^{226}Np). For the (Z – i) arm of the parabola, ^{226}Fr and ^{226}Rn show half-lives of 48 s and 7.4 min, respectively. In the present case, the nuclide at the vertex of the isobar parabola of $\bar{E}_B = f(Z)$ is ^{226}Ra, yet it is not stable, as shown in Fig. 9.1. Thus, β-transformation has done its best to build the most stable nuclide of the A = 226 isobar, but was not able to create a stable nucleon configuration. Consequently, ^{226}Ra must transform into a more stable nucleon configuration by a different mode to the β-transformation.

https://doi.org/10.1515/9783110742725-009

This is the ultimate reason to consider another type of stabilization process – in this case, the α-emission. In the following, the process of α-emission is explained for the naturally occurring transformation chains between mass numbers A = 210 and 238. Of course, this also applies to any other radionuclide undergoing α-transformation.

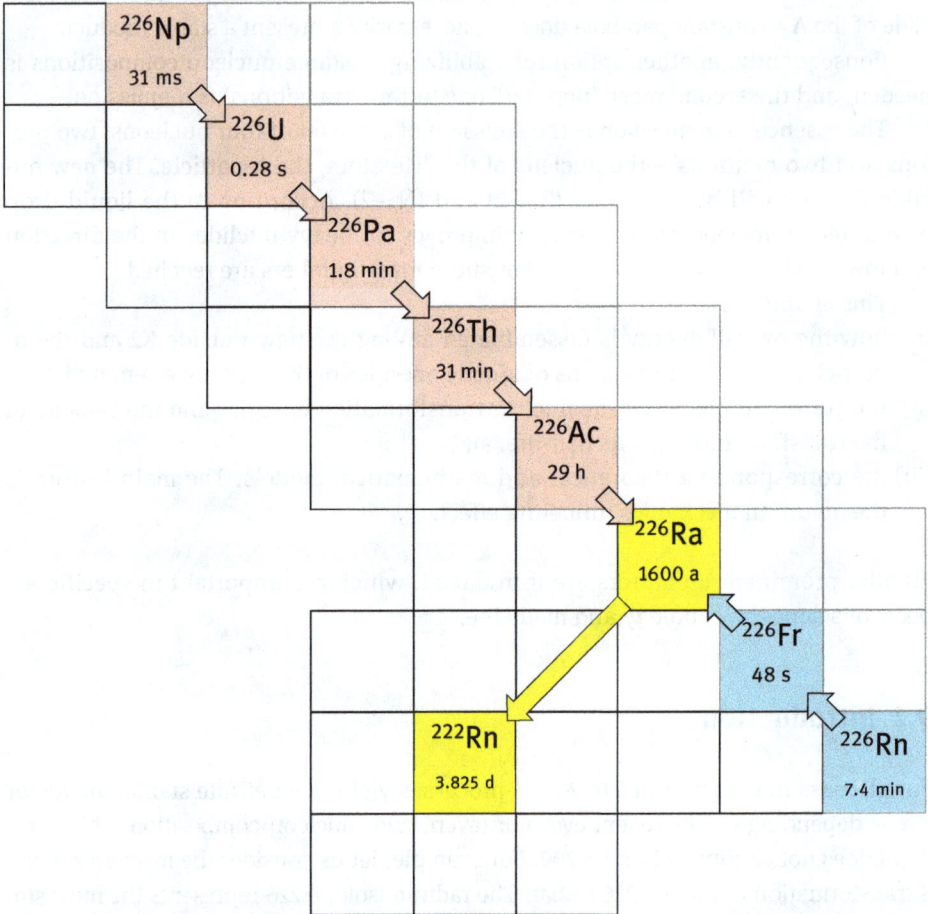

Fig. 9.1: β-transformation processes along the isobar A = 226. The most stable, but not really stable, only just-stable and longest-lived nuclide is ^{226}Ra. This unstable nuclide of optimum mean nucleon binding energy along the isobar transforms through α-emission to ^{222}Rn, thereby switching to a new, lower isobar. In the Karlsruhe Chart of Nuclides, α-emitting radionuclides are indicated by yellow color.

9.2 Mass balances in α-transformations

Emission of an α-particle immediately reduces the mass of the unstable nuclide *K1 and changes both proton and neutron numbers – it is a primary transformation. The reason the small component is the α-particle and not any other mixture of nucleons lies in its very high "internal" stability. The mean nucleon binding energy of the ^{4}He nucleus is 7.052 MeV, and the nucleus is further stabilized due to a double magic nucleon shell configuration ($Z = 2$, $N = 2$).

The α-transformation thus balances mass between the initial unstable nuclide *K1 and the transformation product nuclide K2 in a clear way: the mass number of the new nucleus is lowered to $(A - 4)$. At the same time, the proton number decreases by $(Z - 2)$, and the newly formed nuclide K2 belongs to a chemical element located not as a neighbor, but at $(Z - 2)$ position, as per Fig. 9.2. All α-emissions follow an isodiaphere line.

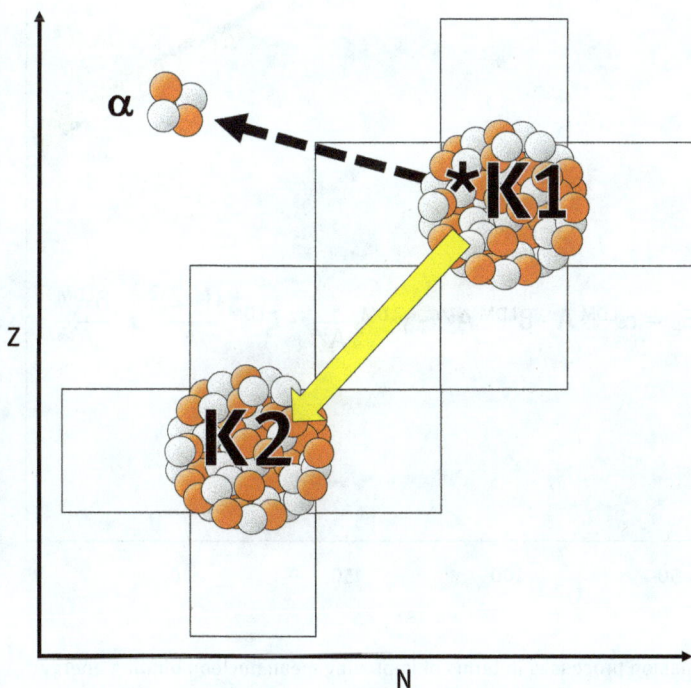

$$^{*}K1 \rightarrow K2 + \alpha + \Delta E \tag{9.1}$$

$$^{A}_{Z}{}^{*}K1_{N} \rightarrow {}^{A-4}_{Z-2}K2_{N-2} + {}^{4}_{2}\alpha_{2} + \Delta E \tag{9.2}$$

Fig. 9.2: Principal scheme of α-emission within the Chart of Nuclides. The initial unstable nuclide *K1 emits an α-particle, and the new nuclide K2 is of $(Z - 2)$, $(N - 2)$ composition. α-emission follows an isodiaphere line.

The "motivation" to transform *via* α-emission is the same as for any other process of radioactive transformation, namely to increase mean nucleon binding energy. This is directly achieved for large values of A when shifting 4 mass units to the left, as per Fig. 9.3. Within the LDM, the main impact originates from the decrease of the value of the coulomb term of the WEIZSÄCKER equation, which is $-\gamma^{LDM}$ $Z^2 / A^{\frac{1}{3}}$. In contrast, the asymmetry term distributes less than the value of (N – Z), for *K1 is identical with that of (N – 2) – (Z – 2) for K2 after α-emission. Also, the sign of the parity terms does not change within the course of the α-emission: any (even, even)-nucleus will remain (even, even) after transformation, any (even, odd) or (odd, even) composition will remain the same, etc.

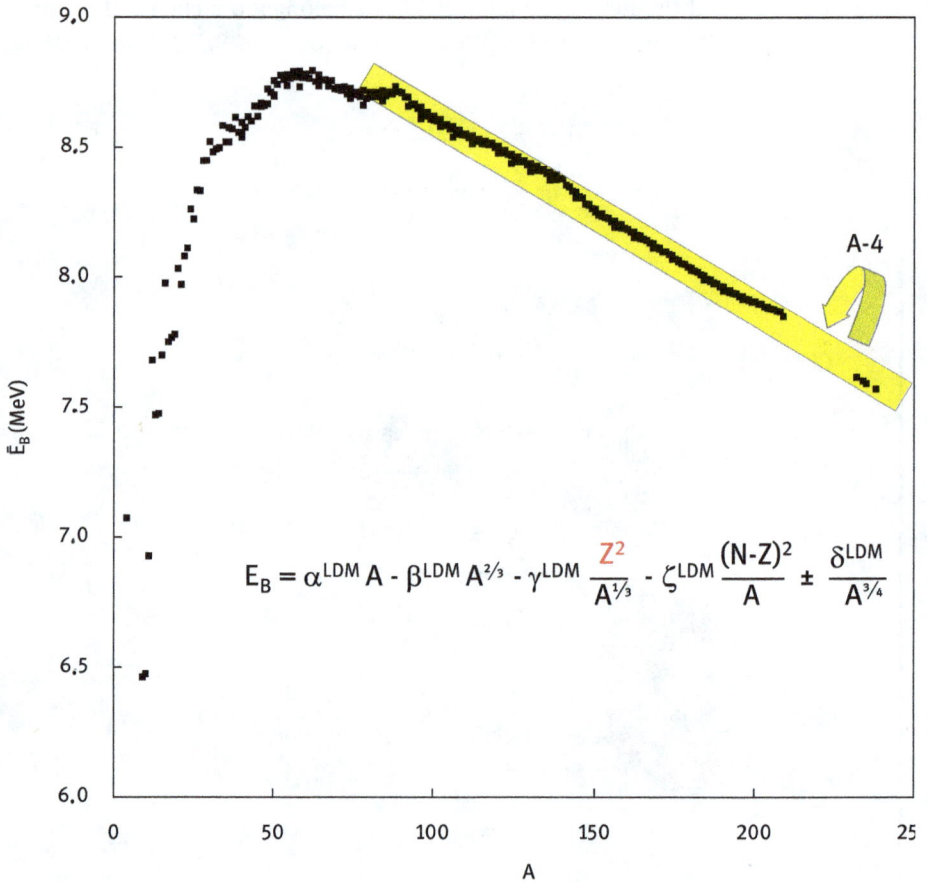

$$E_B = \alpha^{LDM} A - \beta^{LDM} A^{\frac{2}{3}} - \gamma^{LDM} \frac{Z^2}{A^{\frac{1}{3}}} - \zeta^{LDM} \frac{(N-Z)^2}{A} \pm \frac{\delta^{LDM}}{A^{\frac{3}{4}}}$$

Fig. 9.3: Directions of α-emission processes in terms of improving mean nucleon binding energy according to the LDM. The line gives the polynomial of the Weizsäcker equation, and the squares represent the mean nucleon binding energy of stable nuclides.

9.3 Pathways of α-emission

9.3.1 From β-transformations to α-processes

Figure 9.1 introduced the scenario of nuclide stabilization at mass numbers A > 209 along an isobar line, i.e. for β-transformation pathways. In these cases, the local maximum in $\tilde{E}_B$ for A = constant does not represent a stable nucleon configuration, and consequently, another transformation pathway is needed.

9.3.2 Unstable α-transformation products

However, the emission of one α-particle again does not necessarily generate a stable nucleon mixture. This effect can be explained following the example given in Fig. 9.1. ^{226}Ra is a very neutron-rich nuclide. The excess of neutrons is 138 − 88 = 50, the ratio between neutrons and protons is 138:88 = 1.568. Whenever ^{226}Ra starts to transform by α-emission it must follow an isodiaphere line, forming a product of (Z − 2) and (N − 2) composition. Thereby, the absolute excess of neutrons remains constant, but the relative excess of neutrons and the ratio of neutron-to-proton number further increases. The transformation product is ^{222}Rn with 86 protons and 136 neutrons; the neutron excess remains the same (50), while the ratio increases (136:86 = 1.5814). Consequently, the new nuclide is not expected to be stable. Again, this single α-emission transformation is not able to produce a stable transformation product, although there definitely is a reduction in mass: the mass of ^{222}Rn is 222.017578 u and is thus less than the mass of ^{226}Ra of 226.025410 u by the mass of the α-particle. This is accompanied by a gain in mean nucleon binding energy of $\tilde{E}_B = 7.662$ MeV for ^{2226}Ra and 7.695 MeV for ^{222}Rn.

So what? The transformation may continue *via* the next α-emission. This is exactly the case for ^{226}Ra, as introduced for the natural chain of transformation of the 4n + 2 series. ^{226}Ra originates from ^{230}Th by α-emission, and ^{226}Ra itself continues to form daughters by successive α-emission as ^{226}Ra → α → ^{222}Rn → α → ^{218}Po → α → ^{214}Pb, shown in Fig. 5.23. With this nuclide, the α-chain finally terminates; the excess of neutrons has reached a high level.

9.3.3 From α-transformations to β-processi

Now, here comes the "teamwork" of α- and β⁻-transformations: for ^{214}Pb, the β⁻-process becomes the only pathway to further stabilize the nucleus. For this naturally occurring decay chain, this β⁻-transformation now represents the continuation of the preceding α-emission step(s). It happens along the neutron-rich arm of the isobar parabola at A = 214, until a new, local maximum in $\tilde{E}_B$ is reached for this particular isobar. This new maximum could represent a stable nuclide; if not, a situation occurs like that explained above for transformations along the isobar A = 226, and another α-emission follows, as per Fig. 9.4.

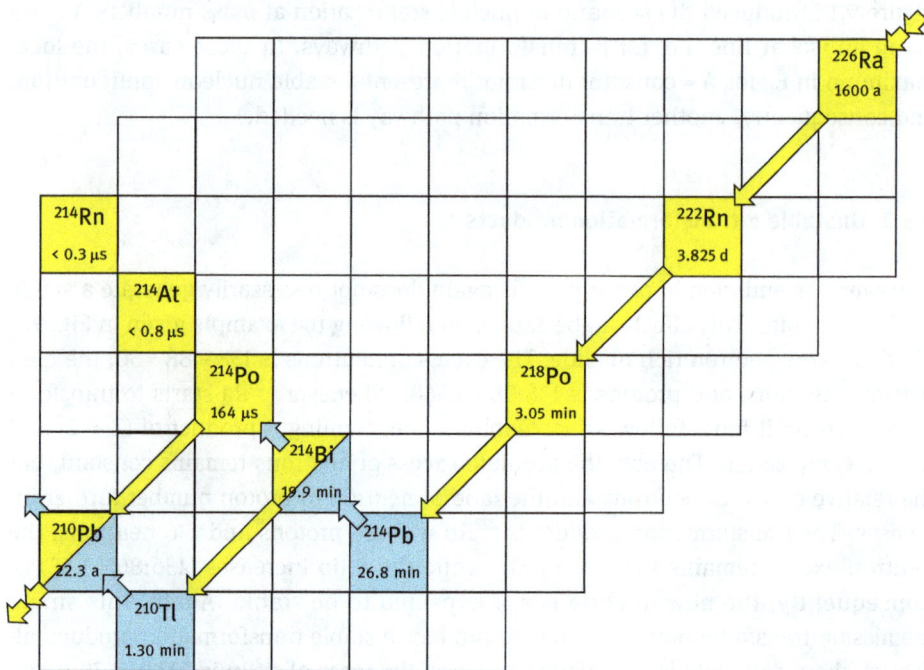

Fig. 9.4: Continuation of the naturally occurring ^{238}U transformation chain subsequent to the α-emissions from ^{226}Ra. The direct chain of α-emission stops at ^{214}Pb to continue along the isobar 214 via β⁻-processes.

9.3.4 Simultaneous β- and α-emission

Following the transformation of ^{214}Pb, the transformation chain consequently continues *via* α-emission at ^{214}Bi and ^{214}Po – but only as an additional option in parallel to a β-transformation. Pure α-emission continues at ^{214}At and ^{214}Rn. The two processes thus may appear in a subsidiary fashion, as per Fig. 9.5. It looks like a zigzag of the two primary types of transformation, i.e. α and β⁻, but drives nucleon composition and mean nucleon binding energy towards the ultimate goal – the stable nuclide.[1]

Thus, α- and β-transformation not only may alternate from one transformation step to the subsequent one – they may appear simultaneously for one and the same

---

1 Nature has realized this within the naturally occurring chains of transformation.

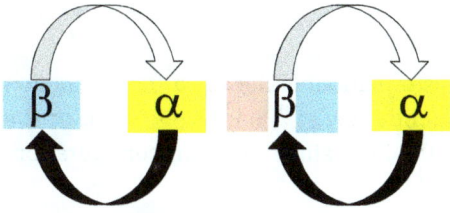

Fig. 9.5: Alternating processes of α- and β⁻-transformations in naturally occurring transformation chains (a) and for other chains (b), until a stable nucleus is finally formed through one of the primary transformation processes.

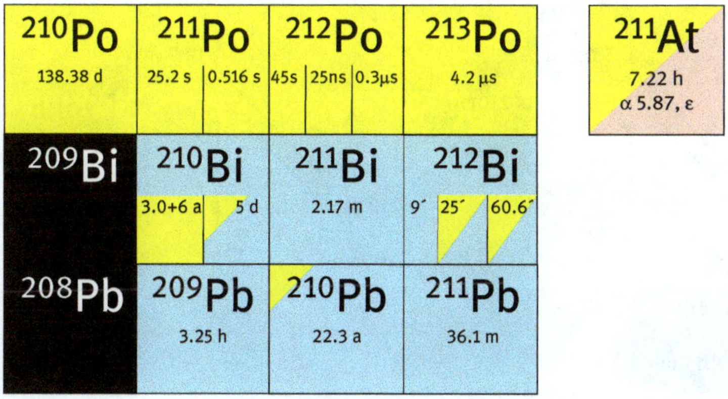

Fig. 9.6: Notation of parallel options of primary transformations for one and the same nuclide. The size of the color-coded area qualitatively indicates the proportions between the different branches of the transformation. ^{212}Bi: 35.93% α-emission (Q_α = 6.207 MeV, main α-energy 6.167 MeV) + 64.07% β⁻-emission (Q_β = 2.252 MeV, main β⁻-energy 2.252 MeV). ^{211}At: 41.78% α-emission (Q_α = 5.982 MeV, main α-energy 5.87 MeV) + 58.2% electron capture (ε) emission (Q_ε = 785 keV).

nuclide![2] This indicates that Q values are positive for different primary transformation options. In this case, each pathway gets its individual absolute value according to the different balances in mass, and the notations are Q_α and Q_β, respectively. The branching of the two transformations is specific. Figure 9.6 shows examples of a nuclide generated in a naturally occurring chain of transformations utilizing β⁻-transformation in parallel to α-emission (^{212}Bi) and an example of an artificially produced radionuclide performing both α- and electron capture transformations (^{211}At). Prominent examples are listed in Tab. 9.1.

---

2 This simultaneous transformation is the reason for the branched chains within the four naturally occurring chains of transformation.

9.3.5 The final stable nuclide

Within the naturally occurring chains of transformation, either the α-transformation itself finally creates the ultimate stable nuclide, e.g. ^{210}Po *via* α → ^{206}Pb, or the β⁻-emission along an A < 210 isobar creates the final stable nucleon configuration, e.g. ^{206}Tl *via* β⁻ → ^{206}Pb, as shown in Fig. 9.7.

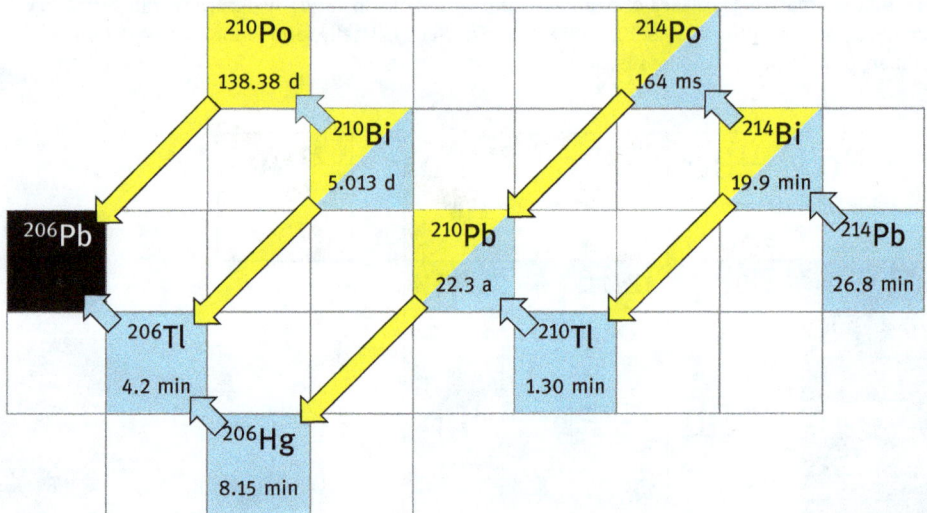

Fig. 9.7: The terminal stable nuclide ^{206}Pb within the naturally occurring ^{238}U chain. Either the α-transformation forms a stable nuclide, i.e. ^{210}Po *via* α → ^{206}Pb, or the β⁻-transformation of ^{206}Tl *via* β⁻ → ^{206}Pb does so.

9.4 Energetics

The α-transformation process occurs spontaneously and is nonreversible like all the other primary transformation pathways. Thus, α-transformation is exothermic, and the loss in mass (lower mass for the sum of the transformation products ($m_{K2} + m_\alpha$) relative to m_{*K1}) corresponds to energy ΔE, as per eq. (9.3). The corresponding energy may also be derived using the mass defect Δm^{defect} or Δm^{excess} values of the three components involved, as per eq. (9.4). The value of ΔE may specifically equal the Q_α-value of the process.

$$\Delta E = Q_\alpha = \Delta mc^2 = [m_{*K1} - (m_{K2} + m_\alpha)]c^2 \tag{9.3}$$

$$Q_\alpha = \Delta m^{\text{defect}}(*K1) - [\Delta m^{\text{defect}}(K2) + \Delta m^{\text{defect}}(\alpha)] \tag{9.4}$$

9.4.1 Values of Q_α

The new nuclide K2 definitely *is* lighter in mass, guaranteeing an exothermic transformation. The absolute value of Q_α basically depends on the masses m of the two nuclides and their difference, and involves the mass of the α-particle emitted. According to the WEIZSÄCKER equation (illustrated in Fig. 9.3) and eq. (9.3), α-emission is possible for $Q_\alpha > 0$. This is the case at about A > 135. However, in this range of mass number, unstable nuclides in most cases prefer β-transformation. Therefore, tendencies in $Q_\alpha = f(A)$ are best illustrated for those unstable nuclides not undergoing β-transformation. In general, there is an increase in Q_α with increasing mass number A, which is explained by the LDM, as per Fig. 9.8.

In addition, the nuclear shell model influences the individual values of Q_α. For some α-transformations, Q_α-values are larger than predicted by the LDM. This applies if the protons and neutrons of the formed nuclides K2 are magic numbers of N = 82 and Z = 82 or N = 126. The corresponding mass numbers are A ≈ 144 – 148 (N = 82), A ≈ 184 – 192 (Z = 82), and overlapping A ≈ 210 – 219 and 206 – 218 for Z = 82 and N = 126. It reflects the additional gain in nucleon binding energy whenever a transformation starts from an unstable nuclide *K1 of a proton and/or a neutron number of 82 + 2 or neutron number 126 + 2. This effect is explained in Fig. 9.9.

The α-emission is thus the dominant pathway to stabilize unstable nuclides
- among the naturally occurring decay chains, i.e. between A = 210 and 238,
- for many artificially produced heavy nuclides of Z > 83,
- for a few lighter neutron-deficient nuclides only, such as $_{72}$Hf – $_{83}$Bi (but extremely short-lived),
- for very neutron-deficient nuclides around $_{52}$Te – $_{55}$Cs (but extremely short-lived with half-lives of seconds to fractions of seconds),
- for a limited number of lanthanide nuclides (with long half-lives).

Figure 9.10 illustrates the regions these α-emitters populate within the Chart of Nuclides, and Tab. 9.1 list several important α-emitters.

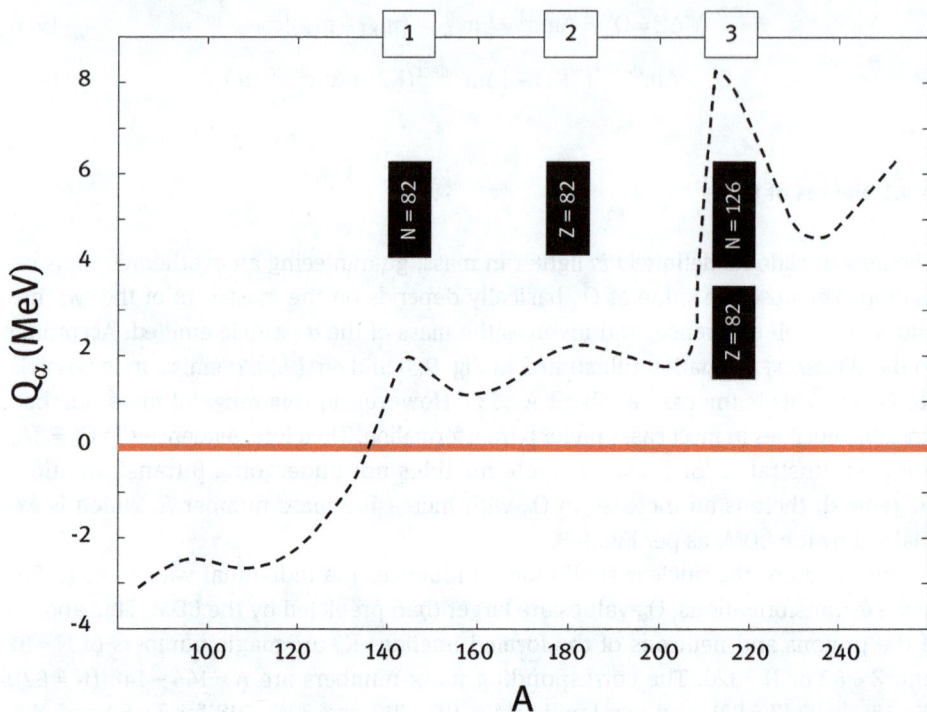

Fig. 9.8: Values of Q_α for unstable nuclides not undergoing β-transformation. There is a general increase in Q_α with increasing mass number. At the same time, several mass numbers show a significantly increased Q_α-value. This reflects the impact of magic numbers of protons and neutrons. (a) $N = 82$, (b) $Z = 82$, (c) $Z = 82$ or $N = 126$. The corresponding mass numbers are $A \approx 144{-}148$ ($N = 82$), $A \approx 184{-}192$ ($Z = 82$), and overlapping $A \approx 210{-}219$ and $206{-}218$ for $N = 126$ and $Z = 82$.

9.4.2 Values of Q_α

Among the many unstable nuclei undergoing α-transformation, the range of Q_α-values is very large. There are low Q_α-energies of 1.8 MeV (^{144}Nd) and large ones such as 11.6 MeV (^{212m}Po). This covers a range of about one order of magnitude (which is less than for the β-process). While maximum energies are quite similar for emitted α- and β-particles, the very low values of Q_β seen, e.g. the β-emitters ^{3}H (0.0186 MeV) or ^{14}C (0.1565 MeV) do not appear in the case of α-transformations. Table 9.1 lists some important α-emitters, their transformation product nuclide K2, the E_α-values and the half-lives.

A = 144-148

A = 210-219

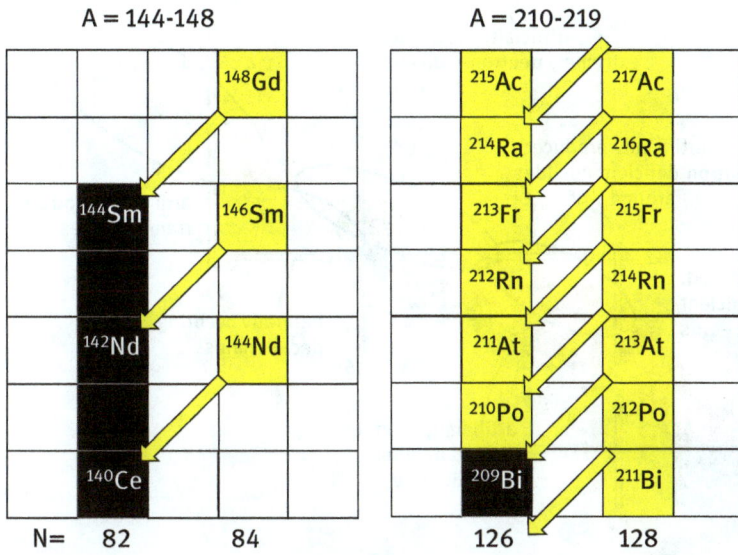

N=	82	84

A = 184-192 ...

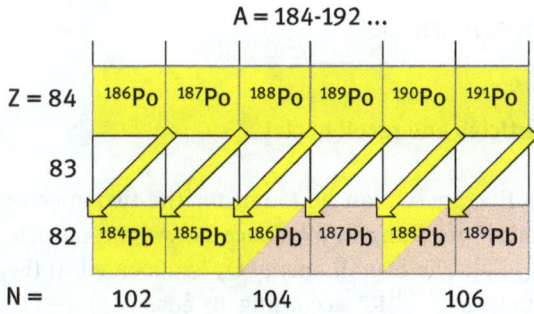

A = 206-218

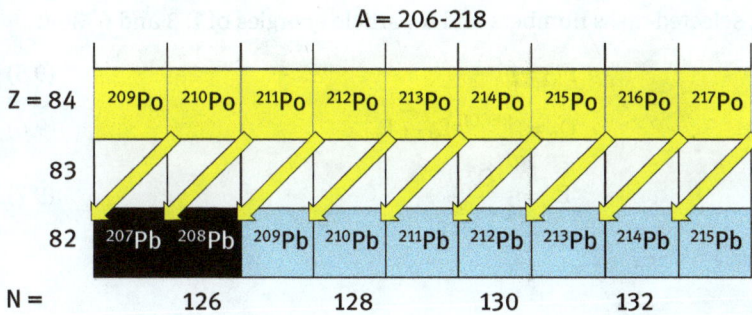

Fig. 9.9: Impact of the shell model of the nucleus on Q_α-energy. Whenever an α-emission creates a nuclide K2 of magic nucleon number, the mean nucleon binding energy increases. The corresponding mass numbers representing large Q_α-values are thus around A ≈ 144–148 (N = 82), A ≈ 184–192 (Z = 82), and A ≈ 210–219 and 206–218 for N = 126 and Z = 82, respectively.

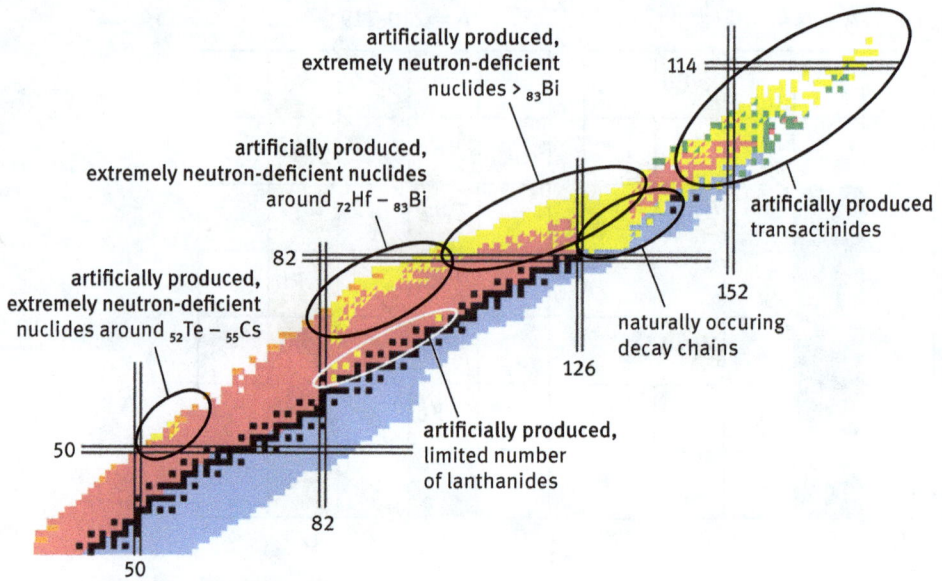

Fig. 9.10: Occurrence of α-emitters within the Chart of Nuclides.

9.4.3 Kinetic energy of emitted α-particles and recoil nuclei

Impulse $p = m \cdot v$ and kinetic energy $E = \frac{1}{2}m \cdot v^2$ can be found for the two species formed in the primary nuclear transition, resulting in a balance between α-particle emission and K2 recoil of $m_\alpha \cdot E_\alpha = m_{K2} \cdot E_{K2}$. The overall energy Q_α is allocated to the emitted α-particle and the recoil nucleus $^{RECOIL}K2$ according to eqs. (9.5)–(9.7).[3] Table 9.1 gives the corresponding numbers for the α-emission of ^{238}U. Table 9.3 gives typical values for selected mass numbers and α-particle energies of 1, 3 and 6 MeV.

$$p_{K2} = p_\alpha \tag{9.5}$$

$$Q_\alpha = {}^{RECOIL}E_{K2} + E_\alpha \tag{9.6}$$

$$E_\alpha = \frac{Q_\alpha}{1 + \frac{m_\alpha}{m_{K2}}} \tag{9.7}$$

---

3 This is different to the β⁻- and β⁺-subtypes of the β-process, but similar to electron capture.

Tab. 9.1: Important nuclides undergoing α-transformation, their proton number, the transformation product nuclide, main* kinetic energy E_α, half-lives, and (if applicable) parallel primary transformations (sf = spontaneous fission, see Chapter 10).

Z	*K1	K2	E_α (MeV)	$t_{1/2}$	Other primary transformation
64	^{148}Gd	^{144}Sm	3.183	74.6 a	–
65	^{149}Tb	^{145}Eu	3.97	4.1 h	ε, β⁺
83	^{212}Bi	^{208}Tl	6.051	60 min	β⁻
83	^{213}Bi	^{209}Tl	6.623	45.59 min	β⁻
84	^{210}Po	^{206}Pb	5.304	138.38 d	–
85	^{211}At	^{207}Bi	5.867	7.22 h	ε
88	^{223}Ra	^{219}Rn	5.716	11.43 d	–
88	^{224}Ra	^{220}Rn	5.685	3.66 d	–
88	^{226}Ra	^{222}Rn	4.784	1600 a	–
89	^{225}Ac	^{221}Fr	5.830	10.0 d	–
90	^{229}Th	^{225}Ra	4.845	7880 a	–
90	^{232}Th	^{228}Ra	4.013	$1.405\cdot10^{10}$ a	sf
92	^{233}U	^{229}Th	4.824	$1.592\cdot10^{5}$ a	–
92	^{235}U	^{231}Th	4.398	$7.038\cdot10^{8}$ a	sf
92	^{238}U	^{234}Th	4.198	$4.468\cdot10^{9}$ a	sf, β⁻
93	^{237}Np	^{233}Pa	4.790	$2.144\cdot10^{6}$ a	sf
94	^{238}Pu	^{234}U	5.499	87.74 a	sf
94	^{239}Pu	^{235}U	5.157	$2.411\cdot10^{6}$ a	sf
95	^{241}Am	^{237}Np	5.486	432.2 a	sf
96	^{245}Cm	^{241}Pu	5.361	8500 a	sf
97	^{247}Bk	^{243}Am	5.531	1380 a	–
98	^{252}Cf	^{248}Cm	6.118	2.645 a	sf

*Main α-emission: Similar to β-transformation, there may be several emissions for the same nuclide due to the population of excited nuclear states of the product nucleus. The table shows the most abundant one, which is not necessarily the most energetic one (if primary transformation prefers an excited state of the product nuclide).

Several conclusions can be drawn from the overall trend:
1. Kinetic energy is distributed between α-particle and K2 directly and depends only on the mass number of K2 (the mass of the α-particle is always the same).[4]
2. Values of kinetic energies of the α-particle are higher if the Q_α-value is high and the mass number of the nuclide is low, as per eq. (9.7). Its value is nuclide-specific and representative, like a fingerprint.

---

4 For the naturally occurring transformation chains, for example, most of the α-emissions occur at A > 208, and the parent nuclides are ^{232}Th, ^{235}Ac and ^{238}U. Consequently, the ratio between masses of the α-particle and the recoiled nucleus K2 are of 4 to 204–238 and thus relatively similar, i.e. 1 to ≈55 ± 5. This ratio of m_{K2}/m_α is completely different compared to the balances of the β-processes: Masses of nuclides for A = 10 or A = 100 were about six orders of magnitude larger than the rest mass of the β-particle.

3. Consequently, kinetic energies of the α-particle are discrete. Figure 9.11 shows an α-spectrum of the nuclide ^{148}Gd ($t^{1/2}$ = 74.6 a) with the discrete energy of the 3.183 MeV α-particle emitted.
4. Compared to the (maximum) kinetic energies of β⁻- and β⁺-particles, (discrete) kinetic energies of α-particles are relatively high. Kinetic energies of α-particles emitted from the most relevant α-emitters are in the range of 3–6 MeV, although extreme values of ca. 2 and ca. 12 MeV can be reached.
5. The recoil energy of the product K2 is $^{RECOIL}E_{K2} = Q_\alpha - E_\alpha$, as per eq. (9.6).
6. The recoil energy of K2 is higher if: the kinetic energy of the α-particle emitted is high and the mass number of the nuclide is low. Recoil energy may typically range from ca. 10 to 100 keV, as per Tab. 9.3.

Tab. 9.2: Values of Q_α and kinetic and recoil energies of the transformation products of the ^{238}U α-emission process. Mass excess data are used for determination of Q_α. Kinetic energies of the α-particles according to eq. (9.7) are obtained by simply using mass numbers.

			MeV
238U	Q_α	$= \Delta E = \Delta m^{excess}_{*K1} - (\Delta m^{excess}_{K2} + \Delta m^{excess}_{\alpha})$ $= 47.3091 - (40.6140 + 2.4249)$ MeV	$= 4.270$
	E_α	$= Q_\alpha / (1 + m_\alpha / m_{K2}) = 4.270$ MeV $/ 1 + 4/234 = Q_\alpha$ $(/ m_{K2} / m_{*K1}) = 4.270$ MeV $(234/238)$	$= 4.198$
234Th	$^{RECOIL}E$ (^{234}Th)	$= Q_\alpha - E_\alpha = 4.270$ MeV $- 4.198$ MeV	$= 0.072$

9.5 Velocities of α-transformation

9.5.1 Values of Q_α and velocities of transformation

Similar to the many β-transforming nuclides, Q_α-values correlate with the half-life of the radionuclide, as per eq. (9.8). The larger the Q_α-value, the larger the gain in energy a nuclide "wins" in terms of mean nucleon binding energy $\Delta \bar{E}_B$ when transforming. In these cases, the transformation steps proceed quickly. The same principle applies to small Q_α-values and/or transformations with special nucleon shell configurations involved. The α-transformation thus shows – similar to β-transformation – half-lives covering milliseconds to billions of years.

Similar to the SARGENT graph for β-emissions, there is a plot for α-emissions (GEIGER / NUTTALL, back in 1911, 1912). Here, d_α is the range of an α-particle in air, and $A^{GEIGER/NUTTALL}$ and B $^{GEIGER/NUTTALL}$ are constants of a linear graph when plotted in the double-logarithmic version. With a given (exponential) proportionality between the energy of an α-particle and its range in air (the higher the energy, the greater its

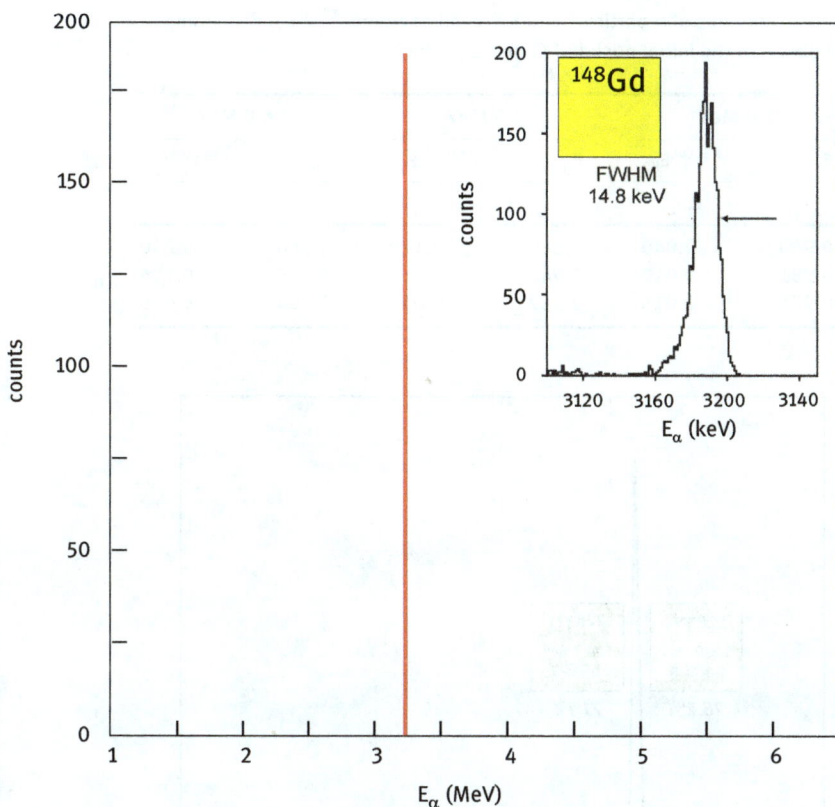

Fig. 9.11: Experimental α-spectrum of a ^{148}Gd sample over a broad range of energy (0–8 MeV) and (inset) in the range of 3.10 to 3.25 MeV. The discrete energy of the 3.183 MeV α-particle emitted is shown. According to the resolution of the α-spectrometer applied and the thickness of the ^{148}Gd source, this peak has a full width at half maximum (FWHM) of 14.8 keV.[5]

distance; and d_α (in cm) $\approx 0.3\, E_\alpha^{2/3}$), eq. (9.9) turns into eq. (9.10). Small changes in energy cause a significant impact on the half-life of the transformation.

$$Q_\alpha\,(\sim E_\alpha) \sim \frac{1}{t_{1/2}} \sim \lambda \tag{9.8}$$

$$\log \lambda = A^{\text{GEIGER/NUTTALL}} \log d_\alpha + B^{\text{GEIGER/NUTTALL}} \tag{9.9}$$

$$\log \lambda = a^{\text{GEIGER/NUTTALL}} \log E_\alpha + b^{\text{GEIGER/NUTTALL}} \tag{9.10}$$

---

5 This should not be confused with the initial discrete energy of this emission.

Tab. 9.3: Kinetic energies of α-particles and K2 recoil energies $^{RECOIL}E_{K2}$ depending on mass number A of the initial nuclide *K1 and Q_α-values.

Q_α	1.0 MeV		3.0 MeV		6.0 MeV	
A_{*K1}	E_α	$^{RECOIL}E_{K2}$	E_α	$^{RECOIL}E_{K2}$	E_α	$^{RECOIL}E_{K2}$
A			(MeV)			
100	0.960	0.040	2.880	0.120	5.760	0.240
210	0.981	0.019	2.943	0.057	5.886	0.114
250	0.984	0.016	2.952	0.048	5.904	0.096

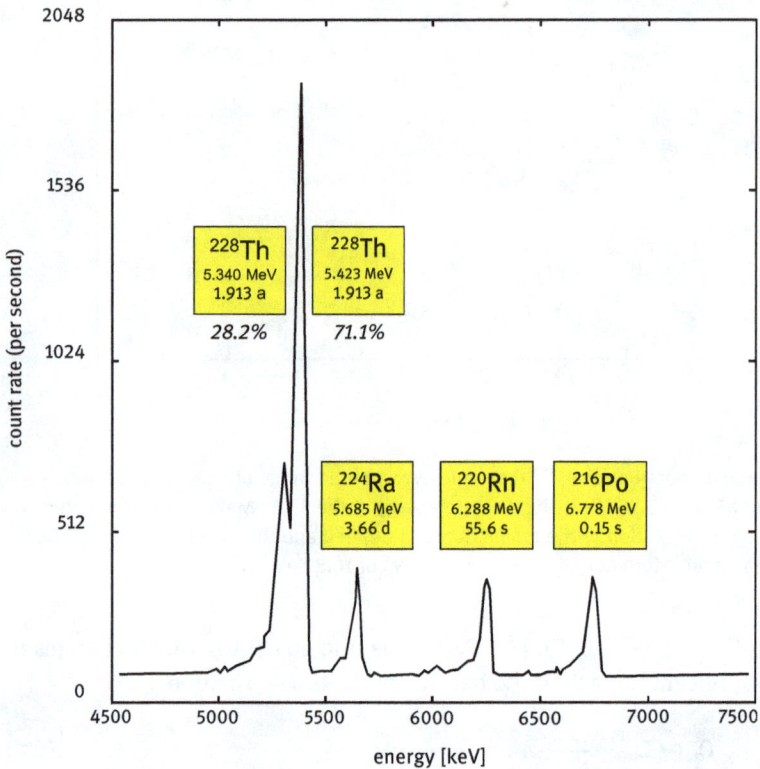

Fig. 9.12: Experimental α-spectrum including the four α-emitting members successively formed in the ^{232}Th chain. The higher the energy, the shorter the half-life. ^{228}Th ($t^{1}/_{2}$ = 1.913 a, E_α = 5.340 and 5.423 MeV),[6] ^{224}Ra ($t^{1}/_{2}$ = 3.66 d, E_α = 5.685 MeV), ^{220}Rn ($t^{1}/_{2}$ = 55.6 s, E_α = 6.288 MeV) and ^{216}Po ($t^{1}/_{2}$ = 0.15 s, E_α = 6.778 MeV).

---

6 The two different α-emissions originating from ^{228}Th represent (similar to β-transformation) an excited nuclear state and the ground state.

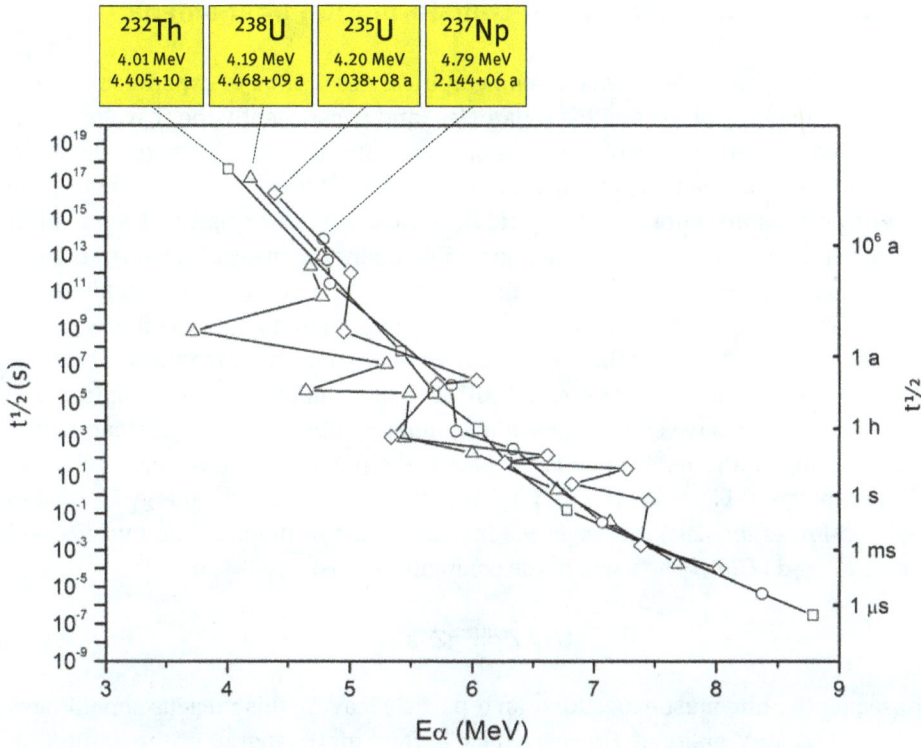

^{232}Th	^{238}U	^{235}U	^{237}Np
4.01 MeV	4.19 MeV	4.20 MeV	4.79 MeV
4.405+10 a	4.468+09 a	7.038+08 a	2.144+06 a

Fig. 9.13: Correlation between kinetic energies of the emitted α-particles and the half-life of the transformation for nuclides present within the naturally occurring transformation chains. The four chains show more or less comparable values for the constant a$^{\text{GEIGER/NUTTALL}}$ in eq. (9.10), but have different values for the constant b$^{\text{GEIGER/NUTTALL}}$. Again, this hints at the influence of shell effects. The (4 n + 0) = ^{232}Th and (4n + 2) = ^{238}U chains represent (even, even) nuclei in contrast to the (4n + 1) = ^{237}Np and (4n + 3) = ^{235}U chains.

The relationship materializes when, for example, correlating the α-particle energies of four members of the naturally occurring ^{232}Th transformation chain with their half-lives. Figure 9.12 shows an experimental α-spectrum similar to Fig. 9.11: ^{228}Th → ^{224}Ra → ^{220}Rn → ^{216}Po. It reflects the relationship between α-particle energy and half-life: the higher the energy of the α-particle, the shorter the half-life. This correlation is shown in Fig. 9.13 for selected α-emitting nuclides generated within the naturally occurring transformation chains.

9.6 Quantum mechanics of α-transformation phenomena

Equations such as (9.10) provide a semi-empirical description of experimental data, i.e. how the energy of the emitted α-particles (and consequently the Q_α-value itself) corresponds to the velocity of the transformation. In this case, the larger the differences in mass between *K1 and K2, and thus the greater the gain in energy is, the faster the transformation proceeds. However, correct mathematical treatment of this correlation is only possible when including quantum physical considerations.

The basic consideration refers to the potential well of a nucleus of an atom. For protons and other positively charged particles about to join a nucleus this well represents a kind of "barrier" that requires a given amount of energy from the positively charged particle so that it may "fall" into the well and finally profit from the strong interaction between nucleons at maximum proximity. Let us consider the nucleus ^{238}U again and realize two facts. Fact 1: the potential well for the nucleus of ^{238}U, as shown in Fig. 9.14. The protons inside induce a potential energy U^C (due to coulomb forces) of ≈28.5 MeV over 9.3 fm, the radius of the uranium nucleus (see eqs. (2.2a) and (2.2b)). The value of the coulomb force is:[7]

$$U^C = Z^{nucleus} Z^\alpha k^C \frac{e^2}{r} \tag{9.11}$$

Consequently, one must expect that an α-particle leaving this nucleus should have at least 28.5 MeV energy.[8] However, fact 2: the kinetic energy of the emitted α-particle is 4.198 MeV (only!), as calculated in Tab. 9.2. This is not because it is a simplified calculation or deduced from a wrong equation – it precisely corresponds to the experimentally measured kinetic energy of the α-particle as released from ^{238}U. Both facts are absolutely correct. So, what now?

---

7 Coulomb's constant $k^C = 8.9876 \cdot 10^9$ N m^2 C^{-2}, elementary charge unit e = 1.602.10^{-19} C, 1 MeV = $1.602 \cdot 10^{-13}$ Nm.

8 Interestingly, this is also true for the reverse process: the α-particle entering the nucleus of an atom. In this context, it gives an explanation for the RUTHERFORD scattering experiments. The highest energy α-particles he directed towards a gold foil were those of the short-lived ^{214}Po. This nuclide was generated *in situ* from a longer-lived ^{224}Ra source as a member of the ^{232}Th chain. Its half-life was $t\frac{1}{2}$ = 164 μs and E_α = 7.687 MeV (remember: short half-life = high α-energy, as per eq. (9.8) and Fig. 9.13). The effect that the α-particles (only) *scattered* at the nucleus of the gold atom, i.e. did not *enter* the nucleus, fits well with the idea of a higher potential energy barrier. This, in a more general context, is one of the characteristics of charged particle induced nuclear reaction, discussed in Chapter 13.

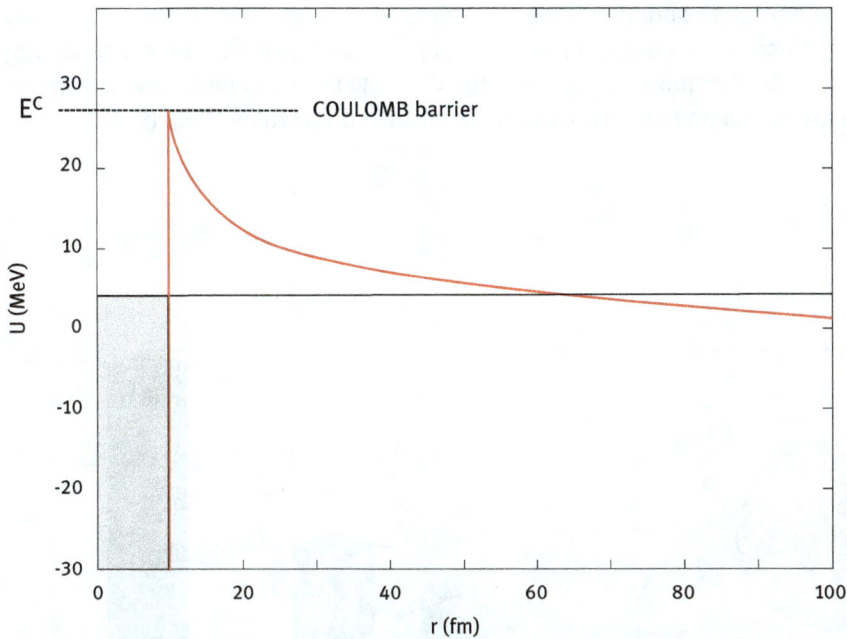

Fig. 9.14: One-dimensional potential well showing the potential energy $U = f(r)$ of the nucleus of ^{238}U with the height of the coulomb barrier E^C. The energy at the radius of the uranium nucleus of 9.3 fm is ca. 28.5 MeV. With permission: W Loveland, DJ Morrissey, GT Seaborg, Modern Nuclear Chemistry, Wiley Interscience, 2006, J Wiley & Sons, Inc., Hoboken, New Jersey.

9.6.1 Mechanism of the tunnel effect

To reconcile these two (seemingly incompatible) facts, we turn to the concept of "tunneling". Recall that wave-particle duality can be used to describe the α-particle as a wave with a given wavelength[9] and amplitude, representing the energy and mass of the corresponding particle arriving at the barrier. Wave functions and eigenstates apply. The wave may penetrate the wall and arrive at the other side of the well. What in the real world was an impenetrable barrier is about to become penetrable.[10]

The energetics of α-particle emission may look as schematically drawn in Fig. 9.15. The α-particle has left the potential well (which holds the nucleons together according

---

9 DE BROGLIE wavelength = $[\hbar/2mE]^{1/2}$.

10 The wave properties of matter also explain the inverse process, i.e. that there is a certain probability for a positively charged particle to penetrate *into* a nucleus even if the velocity (energy) of the particle does not suffice to overcome the electric repulsion from the nucleus.

to the strong force at bounded states according to nucleus shell structure) and has "tunneled" through the potential wall. It becomes a "free" particle. The kinetic energy of the α-particle after tunneling through the COULOMB barrier is much lower than the height of the barrier and exactly of the value calculated according to eq. (9.7).

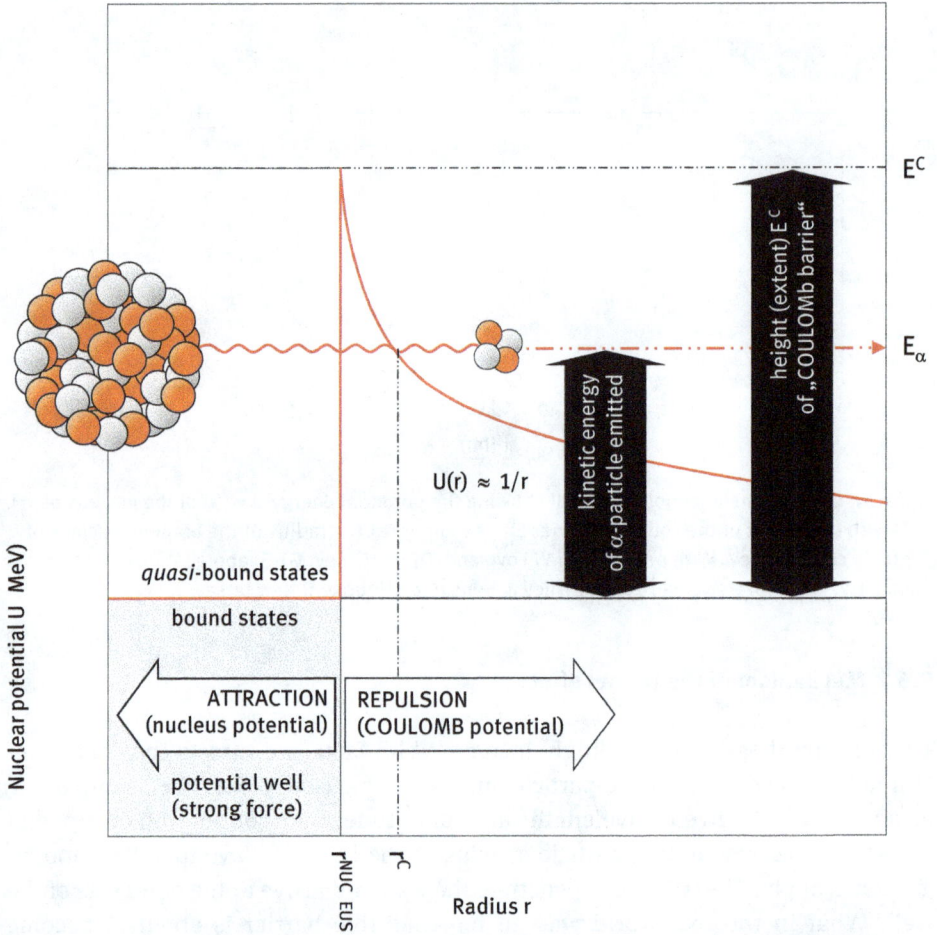

Fig. 9.15: Concept of an α-particle tunneling through a coulomb barrier of a nucleus. E^C gives the amount of the potential energy due to the coulomb forces, E_α the kinetic energy of the α-particle after tunneling at a virtual radius r^C. The smaller r^C is, the lower the kinetic energy of the α-particle emitted.

9.6.2 Mathematics of the tunnel effect

A quantum mechanical model (GAMOW 1928, CONDON and GURNEY 1928, 1929) considers an initial and a final state of transformation, with a corresponding probability p_{fi} of transition (per unit of time) from one energy eigenstate of a quantum system (nuclide $*K1$) to another (nuclide K2).[11]

For α-emissions, the model should allow for three relevant steps:

1. Prior to the emission, the α-particle must be pre-formed as such inside the homogeneous ensemble of the many individual nucleons within the nucleus. This may happen with a given probability due to the special stability of a ($Z = 2$, $N = 2$) cluster of double magic shell characteristics, accompanied by a gain in mean nucleon binding energy.[12]
2. If the cluster was formed anywhere within the nucleus, this cluster must be present close to the surface of the nucleus. This includes an anticipated sort of "transport" or "diffusion".
3. Only following its formation and virtual transport can the tunneling of the particle through the barrier be discussed.

Figure 9.16 illustrates the mathematical factors involved in α-particle tunneling. Although there is evidence for the processes of pre-formation of an α-particle (1) and its diffusion to the surface of the nucleus (2), the mathematical model collects these two steps into the frequency factor f (also called "reduced transition probability"). This describes how often an α-cluster appears at the surface of the nucleus (and "knocks on the door"). Once it has appeared at the surface, it gets a chance to leave the nucleus *via* the tunneling effect. The probability for this step is defined by the penetrability factor P (also known as "transition factor" or "penetrability").

The two components to be mathematically handled are thus the frequency factor f and the penetrability factor P. Obviously, both factors determine the overall efficacy of α-emission. The overall approach is to define the total velocity of an α-emission as a product of type $\lambda = f \cdot P$. The individual expressions for the two factors are summarized in Fig. 9.17.

Despite the fact that the concept is based on quantum mechanics, most of the parameters involved in eqs. (9.12) and (9.13) reflect basic parameters of the nucleus of an atom. Parameters used are:

---

[11] This is quite similar to β-transformations. However, the β-transformation is just based on the conversion of one sort of nucleon into the other one.

[12] In fact, *in situ* formation of an α-cluster within a nucleus is known for light nuclides, where it explains the existence of halo-nuclei. Here, internal clusters are formed with the remaining nucleons occupying "outer shells". This yields radii for those nuclei that are larger than predicted by charge or mass radii according to eqs. (2.2a) and (2.2b), e.g. for ^{6}He and ^{8}He nuclei with one α-cluster each and two and four neutrons, respectively, existing at outer spheres.

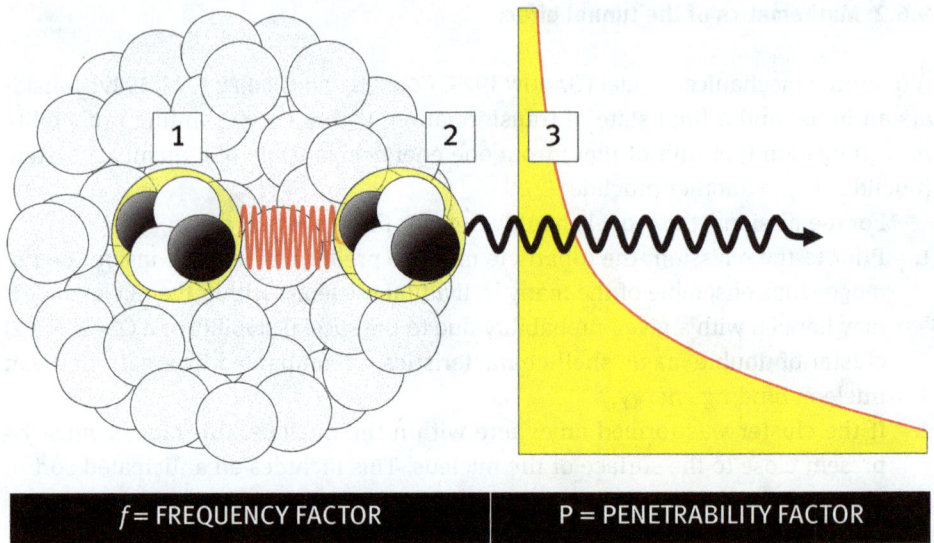

f = FREQUENCY FACTOR

How often does an a-particle
appear at the surface of the nucleus?

P = PENETRABILITY FACTOR

What is its chance
to penetrate the coulomb barrier?

Fig. 9.16: The steps involved in the α-emission process. (1) Formation of an α-cluster inside the nucleus, (2) delivery of this α-cluster towards the surface of the nucleus, (3) release of the α-particle and tunneling through the coulomb barrier of the nucleus.

E^C = energy of the potential wall (typically between 28–30 MeV);
$m´$ = reduced mass ($m´ = m_\alpha m_{K2} / (m_\alpha + m_{K2})$, i.e. for heavy elements $m´ \approx m_{K2}$);
Z_{K2} = proton number of K2;
Q_α = overall energy of the α-transformation;
r_{K2} = radius of K2.

The exponent in eq. (9.13) is expressed as 2G, forming eq. (9.14). G is the GAMOW factor and its values are on the order of G = 30–60. With this dimension, it becomes clear that the probability of penetrating (tunneling) a potential well is in fact extremely low. The same is true of the frequency factor. Depending on the proton number, it is about $Z_{K2}^{-4/3}$, which makes $1.9 \cdot 10^{-20}$ s^{-1} for Z_{K2} between 58 and 98.

9.7 Excited states

Primary transformation processes of unstable nuclides *K1 do not necessarily yield the ground state of the newly formed nuclide K2 directly, but may populate energetically enriched levels of the new nuclide K2. This holds true for the α-transformation as well, though to a lesser extent than for β-transitions. The number and the characteristics of

FREQUENCY FACTOR	PENETRABILITY FACTOR

$$f = \frac{[2(E^C + Q_\alpha) / m\,]^{\frac{1}{2}}}{2r_{K2}}$$ (9.12)

$$P \approx \exp\left[-\frac{2}{\hbar}\int_{R_{K2}}^{r_\alpha}\sqrt{2m'\,(\frac{2Z_{K2}e^2}{r_{K2}} - Q_\alpha)}\,dr\right]$$ (9.13)

$$P = e^{-2G}$$
G = GAMOV factor
(2G ca. 60 -120) (9.14)

$$\lambda = f\,P$$ (9.15a)

$$t^{1/2} = \ln 2 / f\,P$$ (9.15b)

$$t^{1/2} = \ln 2 / \frac{[2(E^C + Q_\alpha) / m'\,]^{\frac{1}{2}}}{2r_{K2}}\,e^{-2G}$$ (9.16)

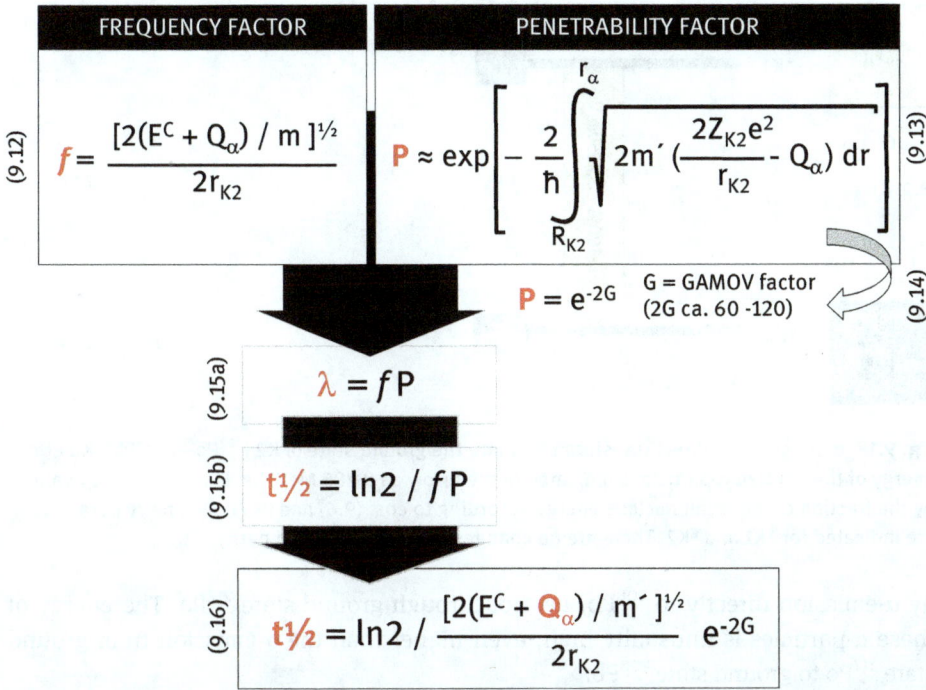

Fig. 9.17: Half-lives of α-emissions correlated with quantum mechanical parameters as well as with Q_α– values in eq. (9.16). Key parameters are the frequency factor f and the penetrability factor P. The exponent in eq. (9.13) is called the GAMOW factor and yields eq. (9.14).

potential exited levels depend on the shell structure, in particular in terms of overall spin.

Figures 9.18 and 9.19 show two α-emitting nuclides. In Fig. 9.18 the α-emission directly and exclusively addresses the ground state of K2. This is illustrated by ^{212}Po, the last α-transformation step in the 4n + 0 natural transformation chain. Figure 9.19 shows an alternative route: a transformation cascade through several excited states (^{226}Ra.)

In the context of excited states, it is interesting to note that α-emission (like in β-transformation) may not only result in the formation of an excited state; it may even start from an excited state. Figure 9.20 gives an example for ^{212}Bi. In addition to de-excitation *via* emission of electromagnetic radiation, some excited states following a primary β-transformation are subject to a subsequent second primary transformation, thereby forming a nuclide K3 instead of de-exciting $^{\odot i}$K2 to ^{g}K2. This nuclide undergoes α-emission to form stable ^{208}Tl (33.93%) and β⁻-emission to form unstable ^{212}Po (64.07%). The ground state of ^{212}Po behaves as illustrated in Fig. 9.18. However, several exited states of ^{212}Po are accessible via the β⁻-transition of ^{212}Bi. A very low percentage (compared to the main de-excitation *via* electromagnetic radiation) de-excite

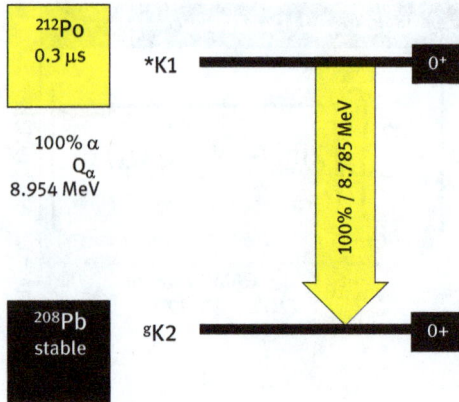

Fig. 9.18: α-emission as direct transformation into the ground state of K2 (^{212}Po → ^{208}Pb). Kinetic energy of the emitted α-particle is indicated in the arrow as 8.785 MeV, i.e. less than the Q_α-value by the fraction of the recoil nucleus energy according to eqs. (9.6) and (9.7). Symmetry parameters are indicated for *K1 and ^{g}K2. There are no changes in overall spin and parity.

by α-emission directly to ^{208}Pb, i.e. not through ground state ^{212}Po. The energy of these α-particles is unusually high, even higher than the α-emission from ground state ^{212}Po to ground state ^{208}Pb.

9.8 Example applications

9.8.1 Applications of α-emitters based on the biological action of the α-particle

The α-particle emitted travels a given distance until its kinetic energy approaches zero. The α-particles typically emitted in a range of 4 to 7 MeV focus their therapeutic action[13] on a short distance of 30 to 80 μm. This causes a high density of ionization within a short distance. Due to the higher mass of the α-particle relative to β⁻-particles, their linear energy transfer (LET) is high. Consequently, ionization effects can be well focused, resulting in irreparable tumor DNA double-strand breaks close to a tumor cell. The corresponding LET values for ^{211}At, for example, are 97 keV/μm.[14] Several α-emitters are thus attracting interest in tumor therapy.

Another therapeutic effect in tumor treatment caused by α-particles is achieved via neutron-induced nuclear reaction on ^{10}B. Once a biological carrier system has delivered this stable isotope of boron close to the tumor cell (or even inside it, near

---

13 Utilization of therapeutically relevant emissions is a topic in Chapter 14 of Volume II.
14 In contrast, β⁻-emitters show longer ranges. For example, ^{90}Y (high β⁻-particle energy and mean range in tissue = 3.9 mm) or ^{177}Lu (low β⁻-particle energy and mean range = 0.67 mm), as per Fig. 8.25. The LET for e.g. ^{90}Y is 0.22 keV/μm.

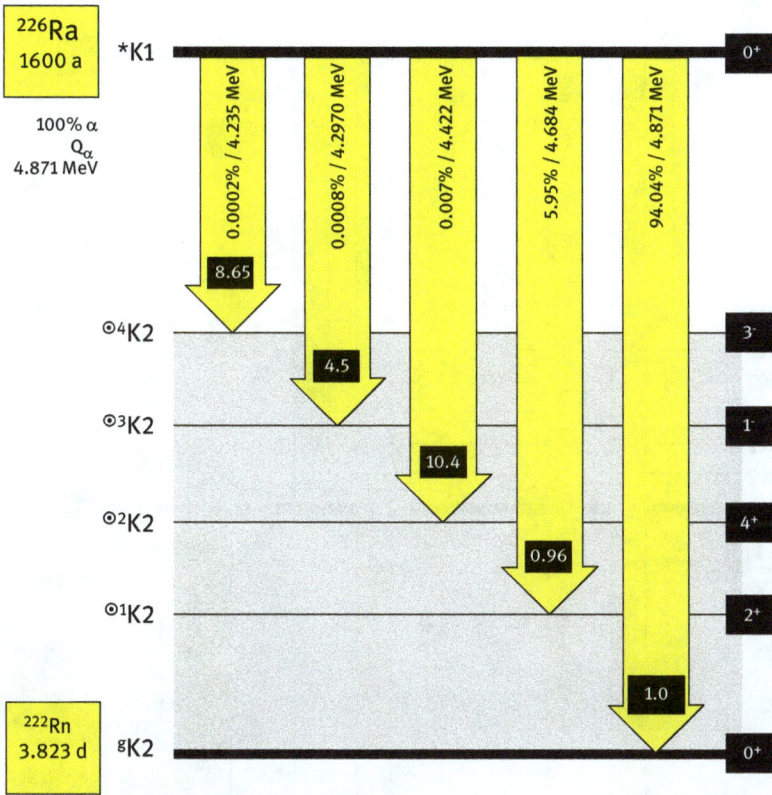

Fig. 9.19: α-emission populating several excited states of K2 (^{226}Ra → ^{222}Rn). Symmetry parameters are indicated for *K1, $^{\odot i}$K2 and ^{g}K2. Each arrow shows the logft-values for the transformation, the corresponding abundance of the separate transformations, and the corresponding energy of the emitted α-particle. Kinetic energy of the emitted α-particle is maximum for *K1 → ^{g}K2 (4.871 MeV), which is also the most probable transformation (94.04%). The next most probable one is *K1 → $^{\odot 1}$K2 (5.95%) with less kinetic energy for the α-particle (4.684 MeV).

the cell's nucleus), the nuclear reaction converts the ^{10}B nucleus into an α-particle and a ^{7}Li nucleus (shown in Fig. 9.21). The kinetic energy of both particles (α and ^{7}Li) induces ionization along their paths. Again, this leads to double-strand breaks of the tumor DNA and selective death of the malignant cell. This process is called Boron Neutron Capture Therapy (BNCT). A commonly used carrier molecule is ^{10}B-phenylalanin, a modified amino acid that accumulates in proliferating tumor tissue to a higher degree than in normal tissue.

$$n + {}^{10}B = \alpha + {}^{7}Li \qquad (9.17)$$

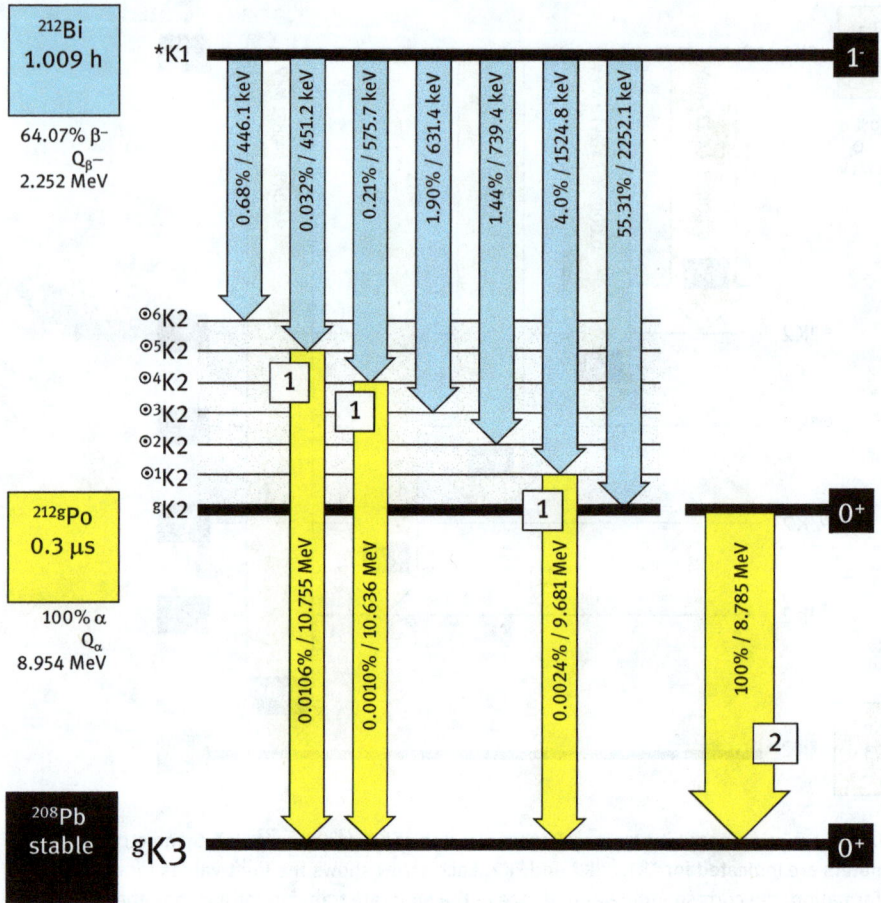

Fig. 9.20: α-emission starting from excited nuclear states $^{\odot i}$K2. Three α-emissions originate (1) from excited nuclear states of ^{212}Po as populated in initial β⁻-transformations ^{212}Bi → ^{212}Po. Correspondingly, energies of these α-emissions are larger (E$_\alpha$ > 9 and > 10 MeV, respectively) than the one shown in Fig. 9.18(2) (E$_{\alpha 0}$ = 8.785 MeV).

Tab. 9.4: Select nuclides undergo α-transformation that is relevant for application in therapeutic nuclear medicine.

Nuclide	t½	Production
^{225}Ac	10 d	^{233}U-chain, ^{229}Th-chain, ^{226}Ra(p,2n)^{225}Ac,
^{224}Ra	3.66 d	^{228}Th
^{223}Ra	11.4 d	^{227}Ac-chain, ^{227}Th-chain, ^{226}Ra(n,γ)^{227}Ac
^{213}Bi	45.6 m	^{225}Ac-chain, ^{225}Ac/^{213}Bi-generator
^{212}Bi	60 m	^{224}Ra-chain, ^{212}Bi/^{212}Pb-Generator
^{211}At	7.2 h	^{209}Bi(α,2n)^{211}At
^{149}Tb	4.1 h	Ta(p,spall), ^{152}Gd(p,4n)^{149}Tb

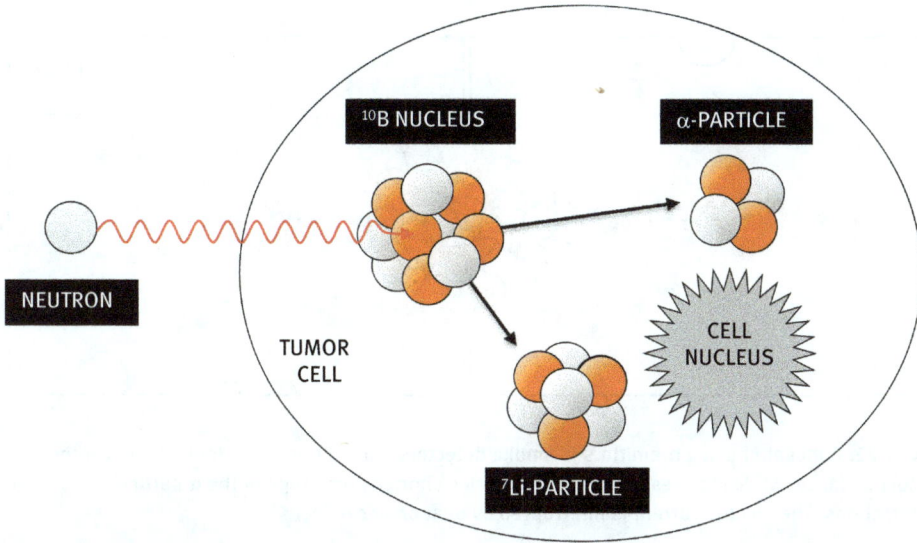

Fig. 9.21: Concept of BNCT. A compound containing ^{10}B is irradiated by neutrons. The nuclear reaction creates two products: an α-particle and a ^{7}Li-particle.

9.8.2 Applications of α-emitters in research and technology

Since MARIE CURIE produced pure samples of ^{226}Ra and other isotopes, long-lived nuclides present in the naturally occurring transformation chains have not only become an object of research but also a tool for providing α-radiation. The α-particles were and are utilized e.g. as projectiles for scattering studies on atoms (see RUTHERFORD'S irradiations of gold foil) and as projectiles for nuclear reactions (see I. CURIE / P. JULIOT'S first man-made synthesis of unstable isotopes and the current production pathways of artificial radionuclides). This will be discussed in Chapter 13.

Furthermore, the emission of α-particles has found application in smoke detectors. The concept is to permanently measure the ionization induced by the interaction of α-particles with air to register a constant electric current between two electrodes. The conductivity of air is low, but measurable (<10 pA). If dust or smoke particles are in the air, they will adsorb the ions formed. Consequently, the electric current is disturbed, which induces an alarm signal (see Fig. 9.22). To guarantee a long shelf-life of these systems, long-lived α-emitting radionuclides are needed. Typical sources include ^{226}Ra and ^{241}Am. In order to provide a similar α-emission intensity, a higher radioactivity is needed for longer-lived nuclides. Thus, typical activities are in the range of 0.5 MBq and 50 kBq for ^{226}Ra ($t_{1/2}$ = 1600 a) and ^{241}Am ($t_{1/2}$ = 432.2 a), respectively.

While smoke detectors these days are often constructed without radioactive material, there is another application which is rather unknown to most of us: batteries. Have you ever wondered how satellites and space probes manage to send signals to

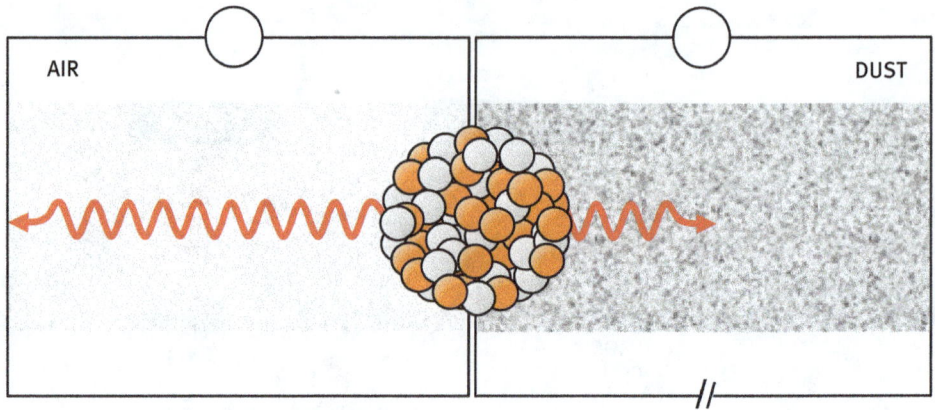

Fig. 9.22: Concept of using α-emitters in smoke detectors. Pure air provides ionization over the whole distance, while the presence of dust particles shortens the range of the α-particle interactions. The electric current is interrupted by dust or smoke.

Earth, even after many years? Remember Pioneer 10, the first man-made object to pass Jupiter and leave our solar system. Launched in 1972, it communicated with earth until 2003. How is this possible when solar energy is insufficient to generate electric power on board?

The answer lies in "radioisotope thermoelectric generators (RTG)", a type of "nuclear battery". For example, the α-emission of long-lived ^{238}Po ($t^{1/2}$ = 87.7 a) interacts with surrounding matter to create heat, which is transformed into electric power. A few grams are sufficient to provide several Watts of power. (Alternatively, the β-emitter ^{90}Y can be used instead.) ^{238}Po was used to generate energy for missions such as Pioneer 10 and 11 (in deep space) and Viking 1 and 2 (to Mars). Lunokhod, a lunar rover, utilized ^{210}Po ($t^{1/2}$ = 138.4 d) as a power source. In addition to ^{238}Po, ^{244}Cm ($t^{1/2}$ = 18.1 a) and ^{241}Am ($t^{1/2}$ = 432.2 a) can also be used, depending on the mission duration.

A simpler version of the RTG is the RHU, or "radioisotope heater unit (RHU)". These units provide only heat, not electric power, and are used in cold, remote sites such as outer space, Antarctica and other places on earth.

9.8.3 Applications of α-emitters to generate neutrons

After samples of ^{226}Ra and other long-lived α-emitters were used as tools to induce nuclear reactions, a new interest in neutrons arose. Since then, α-sources have been shown to convert α-particles into high-energy neutrons. Equation (9.18) gives one of the classical reactions representing an α/^{9}Be-neutron source and Fig. 9.23 illustrates the concept. The nuclear reaction products are stable ^{12}C and the "free" neutron. The mono-energetic α-particles provided by each α-emitting radionuclide generate mono-energetic neutrons (in the MeV energy region) according to the balance in mass of ^{9}Be and the kinetic energy of the α-particle emitted from the nuclide.[15]

$$\alpha + {}^9\text{Be} = {}^{12}\text{C} + \text{n} \tag{9.18}$$

The α-emitting radionuclides should ideally transform exclusively by α-emission and should be of long half-life.

The corresponding light-element nuclear reaction partners should have the following properties:

– mono-isotopic (i.e. of 100% natural abundance), to provide a highly efficient α/n-conversion and to avoid additional nuclear reactions;
– allow for the formation of a stable reaction product nuclide (such as ^{12}C in the case of ^{9}Be), and;
– have a high nuclear reaction yield.

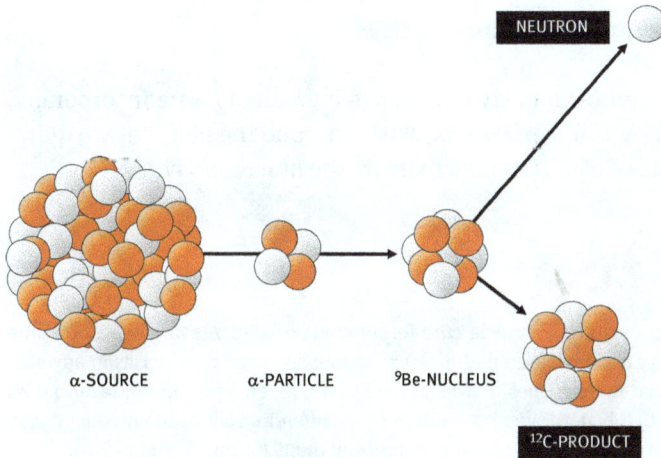

Fig. 9.23: Neutron sources based on α-emitting radionuclides. Illustration of the α-induced nuclear reaction on ^{9}Be, forming a stable nuclear reaction product ^{12}C.

^{9}Be qualifies as a target for the α/n conversion process. Naturally occurring beryllium consists of 100% ^{9}Be (^{9}Be = "mono-isotopic" among the stable isotopes), it has a high nuclear cross-section for α-induced nuclear reaction, and it forms stable ^{12}C.

---

15 The idea is to dilute the α-source (remember the short range of α-particles) with light chemical elements, allowing an α-induced nuclear reaction that will form neutrons. However, the α-particle still loses kinetic energy before entering the ^{9}Be nucleus. Consequently, the kinetic energy of emitted neutrons is often less than the theoretical maximum. Finally, the neutron energy becomes somewhat continuous.

Table 9.5 lists four of the most frequently used α-emitters. Depending on the activity of the nuclides selected, neutron fluxes of about 10^7 per second are permanently generated. The α/n conversion sources are convenient systems that allow for basic experiments in nuclear chemistry and are routinely applied in teaching and training.

Tab. 9.5: Relevant α-emitting nuclides applied in neutron sources.

α-Emitting nuclide	t½	E_α (MeV)
^{226}Ra	1600 a	4.784, 4.601
^{241}Am	432 a	5.486, 5.443
^{210}Po	138 d	5.304
^{240}Pu	6563 a	5.168, 5.124

9.8.4 Applications of α-emitters in nuclear medicine

Historically, it is fascinating how quickly radioactive ingredients were incorporated into cosmetics and everyday commercial products — despite the fact that we didn't know (or care to know) the effects of radioactivity on the human body.[16]

---

16 At that time, radium was considered a miracle cure for all kinds of ailments and aches. Radium rays were associated with energy and increased performance, so industry used these terms in advertising and sales-promotions. At that time, crèmes, powders, toothpaste, water, beer, butter, razor blades and much more simply sold much better if you put radium or radiation in front of the product name. This ended when the hazards of radioactivity became obvious from the 1920s on, cf. Figure 9.24.

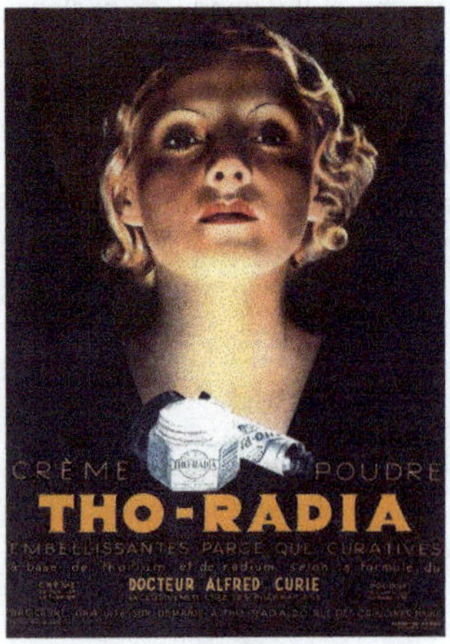

Fig. 9.24: Advertisement of cosmetic products containing α-emitting radioisotopes in the beginning of the nineteenth century.

A century later, the α-particle is being used more responsibly, as a prominent tool in nuclear medicine therapy. As previously mentioned, the concept is to deliver α-particle emitting radionuclides to tumor tissue, where the low-range (less then 1 mm) and high LET (linear energy transfer) radiation destroys tumor cell DNA and thus erases the tumor.[17] Only in only rare cases does the radionuclide itself show a clinically useful accumulation in the tumor – this is limited to ^{223}Ra, which (like calcium) is an alkaline earth metal and mimics calcium pathways in the organism. In particular, it accumulates in metastatic bone cancer cells, and is a registered drug to treat patients with prostate cancer (Alpharadin™, Xofigo^R, radium-223-chloride, Bayer HealthCare and Algeta ASA).

In contrast, other α-emitters relevant to nuclear medicine, such as ^{225}Ac and ^{211}At, must be chemically linked to other molecules. These molecules act as delivery vectors, bringing the radioisotope to well-defined biological target structures overexpressed in certain tumor cells. Such structures are the focus of radiopharmaceutical chemistry, overlapping the fields of pharmacy, medicinal chemistry, tumor biology, and medicine, cf. Table 9.4.[18]

9.9 Outlook

9.9.1 Hindrance factors

Similar to β-transformations, nuclear shell structure also plays a significant role for α-transformation, in addition to balances in mass and energy. So far, α-transitions have been discussed without explicitly considering shell effects. However, Fig. 9.13 already unveiled a difference between α-transitions along (even, even) nuclides for the 4n + 0 and 4n + 2 natural decay chains and the 4n + 1 and 4n + 3 analogue ones. For similar Q_α-values of ca. 6 MeV, for example, half-lives of the (even, even) nuclide chains are orders of magnitude shorter than for the ^{235}U chain.

This effect can be explained by a sort of hindrance for certain transitions. The idea is to consider α-transformations between (even, even) nuclides as proceeding "smoothly", while other (Z, N)-configurations appear somewhat "hindered", and thus may only proceed with a sort of "delay". In addition, their half-lives will be longer.[19] Consequently, a hindrance factor $\acute{H}$ is defined as the ratio between the

---

17 For the principal effects of radiation on tissue, see Chapter 2 on "Radiation Dosimetry" in Volume II of this textbook: Modern Applications of Nuclear- and Radiochemistry.

18 For further applications of α-particle emitting radionuclide in the context of radiopharmaceutical chemistry, nuclear medicine and tumor therapy, see Volume II, Chapters 13 on "Life Sciences: Therapy".

19 This qualitatively reminds us of the role of allowed and forbidden transitions for β-transformations, but hindrance factors discussed in α-transformation processes do not belong to selection rules.

"real" (experimentally measured) and "expected" (i.e., from (even, even) nuclide α-transformations) half-lives, as per eq. (9.19).

$$\acute{H} = \log \frac{t_{1/2}^{\text{experimental}}}{t_{1/2}^{\text{expected}}} \tag{9.19}$$

More recent models thus refer to correlations of type $t_{1/2} = f(E_\alpha)$, which consider nuclear shell effects — particularly for (magic) proton numbers Z. One approach uses E_α-values instead of Q_α because according to eq. (9.6) for a large range of A (190 ± 50), there are relatively constant relations of type $E_\alpha = 0.973 \pm 0.003\, Q_\alpha$, as per eq. (9.20) (TAAGEPARA, NURMIA 1961).

$$\log t_{1/2} = C_1 \left(Z_{K2} E_\alpha^{-1/2} - Z_{K2}^{2/3} \right) - C_2 \tag{9.20}$$

The point is to look for Z and to introduce constants C_1 and C_2 reflecting shell effects. When plotting many of the α-emitting nuclides according to eq. (9.20) in terms of $(Z_{K2} E_\alpha^{-1/2} - Z_{K2}^{2/3})$ as x-axis *vs.* log $t_{1/2}$ as y-axis, straight lines appear for different sets of nuclides. When selecting nuclides such as ^{215}Rn, ^{226}Ra, ^{238}U, ^{232}Th, values for C_1 and C_2 fit to 1.61 (MeV)$^{1/2}$ and 28.9, respectively, and create a straight line. This is defined as = 0.

Other nuclides follow different linear trendlines, which are parallel to the zero hindrance line towards longer half-lives. Hindrance factors of $0 < \acute{H} < 1.2$ indicate minor hindrance (reflecting nucleon numbers far from magic), while larger values of $\acute{H}$ correspond to nuclides with nucleon configurations close to magic numbers.

This may be further structured by another approach (DASGUPTA-SCHUBERT, REYES, 2007) according to the four possible combinations of nucleons, i.e. (even, even), (even, odd), (odd, even) and (odd, odd) (as per eq. (9.21); the coefficients a, b and c are listed below).[20]

$$\log t_{1/2} = a + b \left(A^{1/6} Z_{K1}^{1/2} \right) + c \left(Q_\alpha Z^{-1/2} \right)^{(t_{1/2} \text{ in seconds})} \tag{9.21}$$

$(Z,N)_{*K1}$	a	b	c
(even, even)	−25.31	−1.1629	1.5864
(even, odd)	−26.65	−1.0859	1.5848
(odd, even)	−25.68	−1.1423	1.5920
(odd, odd)	−29.48	−1.113	1.6971

---

[20] For more details, see: A Vértes, S Nagy, Z Klencsár, RG Lovas, F Rösch (eds.), Handbook of Nuclear Chemistry, second edition, Springer 2010, Volume I, Chapter 2: Basic properties of the atomic nucleus, T Fynes.

9.9.2 Other "particle" emissions

The emission of an α-particle is a common means of stabilizing a heavy unstable nuclide. The concept of separating a relatively small cluster (of increased mean nucleon binding energy and/or stability due to magic nucleon numbers) may hold true also for smaller or larger clusters, as per Fig. 9.25.

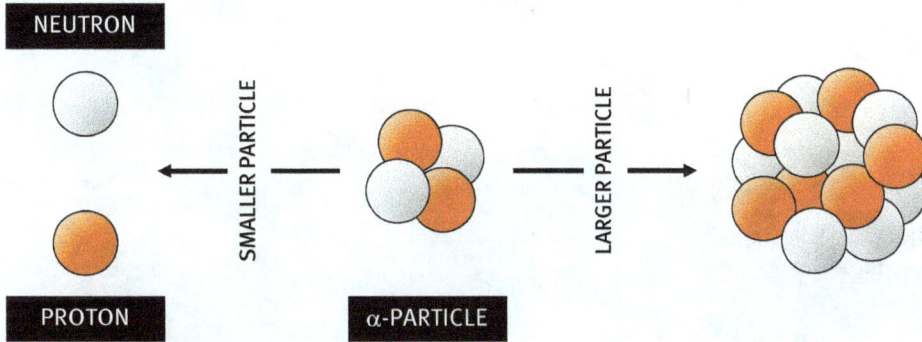

Fig. 9.25: Directions for emission of clusters smaller or larger than the α-particle. (left) Either one neutron or one proton, or (right) a ^{14}C cluster, for example.

9.9.2.1 Heavier clusters than α-particles

The emission of an α-particle is common because this unique particle represents a configuration of nucleons stabilized by double magic nucleon numbers. The next heaviest particle of this magic configuration would be the $Z = 8$, $N = 8$ nucleus, ^{16}O. However, other smaller nuclei have been observed as emitted clusters, namely ^{12}C and ^{14}C. The symmetric ^{12}C ($Z = 6$, $N = 6$) may be considered as a "triple" α-particle. The ^{14}C is a cluster nucleus containing the magic neutron number $N = 8$.

Emissions of those clusters have been observed in naturally occurring unstable nuclides and belong to the spontaneous primary transformations. Emission of ^{14}C has been observed for more nuclides than ^{12}C. However, in all cases, cluster emission is a very low-abundance process relative to the dominant α-emission. For the same *K1, the ratio of ^{14}C- to α-emission is $\approx 10^{-10}$:1. Larger particles such as ^{18}O and ^{24}Ne can also be emitted in a few cases, but with even lower abundance. Table 9.6 gives some of the nuclides involved in large cluster transformations. Figure 9.26 indicates the direction of the ^{14}C particle emission from ^{222}Rn.

Emission of clusters larger than the α-particle result in nuclides K2 of even more reduced mass number, proton number and neutron number. From an energy point of view, all these cluster emissions are thermodynamically allowed, provided they improve the nucleon binding energy (similar to the tendency shown for α-particle in Fig. 9.3) according to an even increased ΔA. Cluster emission becomes an

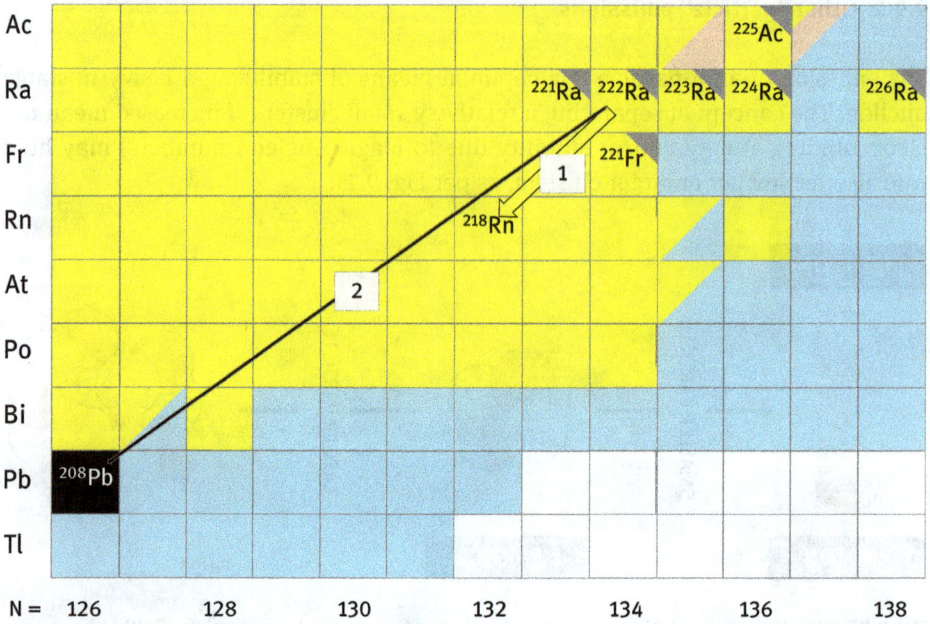

Fig. 9.26: ^{14}C cluster emission from ^{222}Ra shown within the Chart of Nuclides. The Chart of Nuclides identifies nuclides known to emit clusters with a gray triangle at the upper right corner of the nuclide's square. Note that the dominant pathway for ^{222}Ra to stabilize is *via* α-emission to ^{218}Rn (1). ^{14}C-emission from ^{222}Ra yields double magic ^{208}Pb (2). Only one out of about 10^{10} ^{222}Ra atoms utilizes ^{14}C-emission instead of α-emission.

option for A > 220 and Z > 87 nuclides. Nevertheless, pre-formation of such large clusters in heavy nuclei is very improbable. Consequently, only the creation of transformation product nuclides with stabilized nucleon shells increases the overall probability of cluster transformation.

9.9.2.2 Smaller than ^4He: From α-particles to single nucleons

The ability to emit clusters smaller than the ^4He nucleus is limited; for $Z = 2$ this could be the nucleus of another isotope of helium, ^3He, or nuclei of hydrogen isotopes for $Z = 1$, i.e. ^3H or ^2H. However, this does not happen, due to unpaired nucleon configuration (among other factors).

Several nuclides emit particles representing a single nucleon, i.e. either a neutron or a proton. Emission of a single proton shifts an unstable nuclide along an isotone line to an $(A - 1)$ and $(Z - 1)$ nucleus. Emission of a single neutron proceeds along an isotope line, yielding an $(A - 1)$ and $(N - 1)$ nucleus. However, this kind of emission is exclusively observed for "exotic" unstable nuclides far from the valley of β-stability, i.e. those with an enormous excess of either neutrons or protons.

Tab. 9.6: Cluster transformation processes showing the three clusters ^{14}C, ^{18}O and ^{24}Ne and nuclides involved.

Z_{K1}	*K1	Cluster	Z_{K2}	ΔA	K2	
87	^{221}Fr	$^{14}_{6}C_8$	81	−14	^{207}Ti	double
88	^{222}Ra		82		^{208}Pb	magic
88	^{232}Ra		82		^{218}Pb	magic Z
88	^{226}Ra		82		^{212}Pb	magic Z
89	^{225}Ac		83		^{211}Bi	
90	^{226}Th	$^{18}_{8}C_{10}$	82	−18	^{208}Pb	double magic
90	^{226}Th	$^{24}_{10}Ne_{14}$	80	−24	^{202}Hg	
90	^{231}Pa		83		^{207}Bi	
90	^{232}U		84		^{208}Po	

In the case of neutron or proton excess, β-transformation is the preferred stabilization pathway of unstable nuclides. Alternatively, in the case of an extreme excess of one type of nucleon, an excess nucleon may be emitted directly. This happens when the binding energy of that particular "last" nucleon is less than the Q_β-value (for example, the neutron of the corresponding $n \to p + \beta^- + \bar{v}_e$ pathway). Equations (9.22) and (9.23) describe the principal processes for proton emission and neutron emission, respectively. Figure 9.27 illustrates the pathways of these transformations.[21]

$$^AK1 \to {}^{A-1}K2 + p + \Delta E \tag{9.22}$$

$$^AK1 \to {}^{A-1}K2 + n + \Delta E \tag{9.23}$$

Figure 9.28 identifies the area of p-emitting and n-emitting nuclides within the Chart of Nuclides. Proton-emitting nuclides are very close to the proton drip line of the Chart of Nuclides. There are several such nuclides; however, all have very short half-lives of microseconds to nanoseconds (for most of these, the precise value is not yet determined). Only a few nuclides around the lanthanide group show half-lives of milliseconds, such as ^{151}Lu transforming to ^{150}Yb with $t_{1/2} \approx 81$ ms.

Neutron emission is a process that happens mainly in the context of stabilization of extremely neutron-rich "fragments" created within fission processes. The nuclides involved here are of mass numbers around A = 100 and 140, which parallel the frequent mass numbers of fission processes (see Chapter 10).

---

21 These transformations do not require "pre-formation" of a cluster such as the α-particle. Also (in the case of proton emission), the coulomb barrier is lower compared to α-emission.

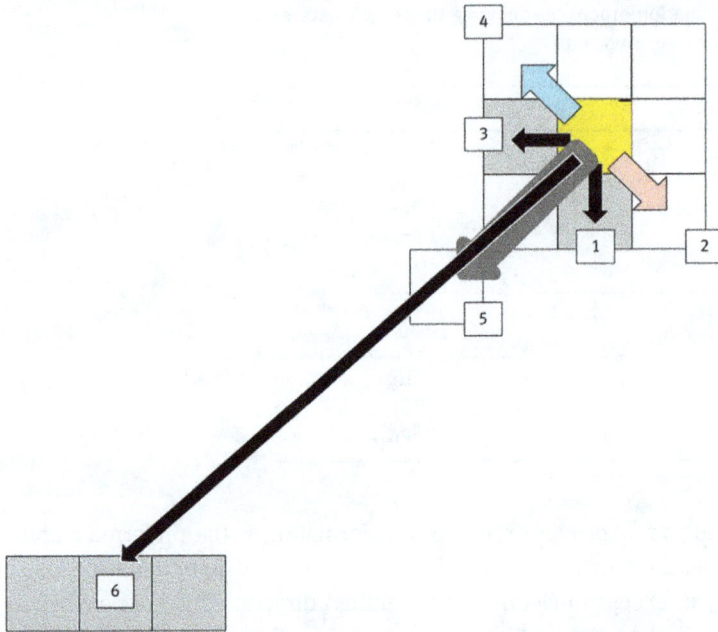

Fig. 9.27: Pathways of proton and neutron emission compared to β-transformations, α-emission and cluster emission. Proton emission (1) is an alternative to β^+- or electron capture transformation (2) of proton-rich nuclides, while neutron emission (3) is an alternative to β^--transformation (4) of neutron-rich nuclides. While β-transformation follows the isobar line, emission of protons proceeds along isotones and emission of a neutron takes place along isotopes. The dominant α-emission is shown with (5) along an isodiaphere relative to the much less abundant emission of carbon or larger clusters (6).

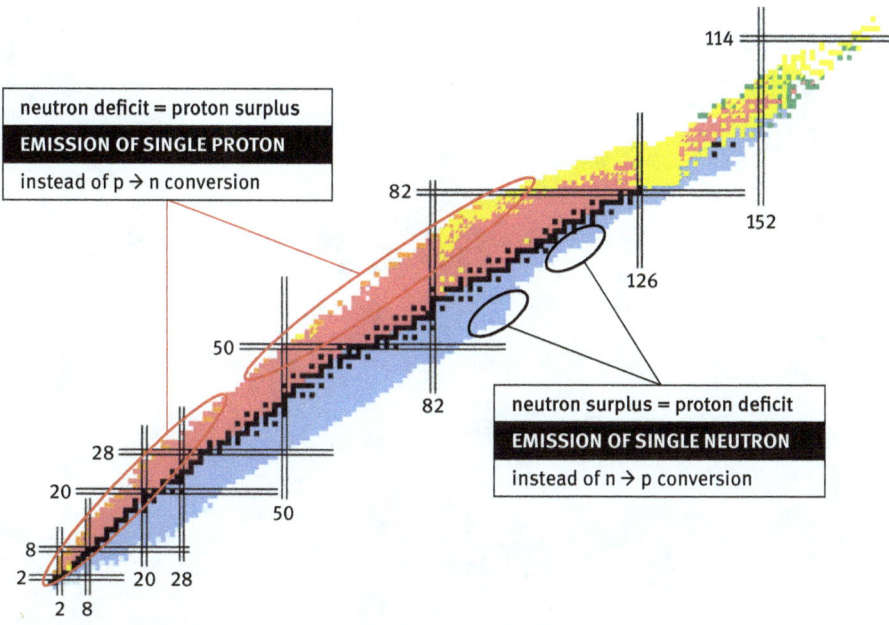

neutron deficit = proton surplus

EMISSION OF SINGLE PROTON

instead of p → n conversion

neutron surplus = proton deficit

EMISSION OF SINGLE NEUTRON

instead of n → p conversion

114

82

152

126

50

82

28

20

50

8

2

20 28

2 8

Fig. 9.28: Area of extreme p-rich and n-rich nuclides within the Chart of Nuclides undergoing p- or n-emission, respectively.

10 Spontaneous fission

Aim: In addition to β-processes and α-emission, there is a third option for naturally occurring unstable nuclides to stabilize: spontaneous fission. It becomes an option for very heavy unstable nuclides around A ≥ 232 and Z ≥ 90 (starting from ^{232}Th, ^{235}U and ^{238}U). Typically, it accompanies α-emission. For unstable nuclides of A > 250, the fraction of spontaneous fission relative to α-emission increases. For many of the artificially produced transactinide nuclides, spontaneous fission becomes the dominant route of transformation.

Fission is, by definition, the division of a nucleus into two or more parts with intermediate mass fragments, usually accompanied by the emission of neutrons, gamma radiation and (rarely) small charged nuclear fragments. Consequently, there is a completely different mechanism involved in the fission process compared to β-transformation and α-emission.

The most notheworthy impact of "fission" arises from "induced fission", representing a giant source of nuclear energy. Induced fission, however, will be discussed in Chapter 13 on "nuclear reactions". Nevertheless, many features of the fission of unstable heavy nuclides apply to both the spontaneous and the induced variant, and are introduced in this chapter.

This chapter discusses:
(i)   the formation of primary fission fragments,
(ii) the fate of these fragments,
(iii) some of the principal theoretical models and mathematical explanations of fission.

10.1 Introduction

Spontaneous fission (sf) represents one more option for naturally occurring unstable nuclides to stabilize – in addition to the β-processes and to α- and cluster emission. The number of unstable nuclides "naturally" transforming *via* the fission process is much lower compared to the two other routes. Spontaneous fission is most relevant from an academic point of view and for research on super-heavy chemical elements; its practical usage is quite limited. The rather dramatic impact of "fission" arises from "induced fission", representing an enormous source of nuclear energy. Induced fission, however, belongs to the chapter "nuclear reactions". Nevertheless, many features of the fission of unstable heavy nuclides apply to both the spontaneous and the induced variant.

What does "fission" mean? Fission is the concept of dividing a very large nucleus *K1 into two medium-large fragments K2 and K3 (*via* "binary fission", eq. (10.1)), or, more rarely, into three fragments K2, K3 and K4 (*via* "ternary fission", eq. (10.2)). The process is exothermic and the transformation energy gained is $\Delta E = Q_{sf}$.

https://doi.org/10.1515/9783110742725-010

$$*K1 \rightarrow K2 + K3 + \Delta E \tag{10.1}$$

$$*K1 \rightarrow K2 + K3 + K4 + \Delta E \tag{10.2}$$

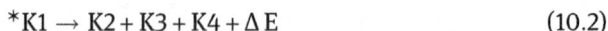

The process of fission, however, is not "just" the division of a certain large nucleus. It involves (a) several steps prior to the moment fission occurs and (b) interesting successive follow-up steps, until the initial fission *fragments* have finally turned into the individual fission product *nuclides*. During these successive transformations, neutrons, electromagnetic radiation, β^--electrons and electron neutrinos are also emitted. The process of nuclear fission is thus much more complex than that of the β- and α-transformation processes. In the following, spontaneous fission is explained with the focus on binary asymmetric fission of the uranium isotope ^{238}U.

10.2 Occurrence of spontaneous fission

Spontaneous fission belongs – like β-transformation and α-emission – to the naturally occurring primary modes of radioactive transformation processes. While β-transformation starts at A > 2 and most relevant α-emitters appear at A > 209, spontaneous fission starts to occur for the heavy unstable nuclides around A ≥ 232 (e.g. ^{232}Th, ^{235}U, ^{238}U), i.e. isotopes of the chemical elements of Z ≥ 90. Among these elements, only the isotopes of thorium (Z = 90) and uranium (Z = 92)[1] show sufficiently long half-lives to naturally exist on earth. ^{232}Th, ^{235}U and ^{238}U undergo spontaneous fission, though at an extremely low percentage relative to the preferred α-emission.[2]

Beyond thorium and uranium, all the other unstable nuclides which undergo sf are man-made, with progressively shorter half-lives. Yet, for many of the artificially produced transactinide nuclides of Z > 103, sf becomes an important route of transformation and constitutes a significant aspect of the synthesis and (in)stability of super-heavy elements.

Although more than 100 artificially produced nuclides of A ≥ 232 are known to undergo spontaneous fission, this number is low compared to the two other modes of primary transformation. Figure 10.1 locates corresponding regions of naturally occurring and artificially produced nuclides undergoing spontaneous fission within the Chart of Nuclides.

---

1 These nuclides are the parents of the three naturally occurring transformation chains.
2 Spontaneous nuclear fission was discovered by Soviet physicists PETRZHAK and FLEROV in 1940, carefully registering transformation events in uranium compounds. This was slightly later than the discovery of induced fission in 1938 by HAHN, STRASSMANN and MEITNER.

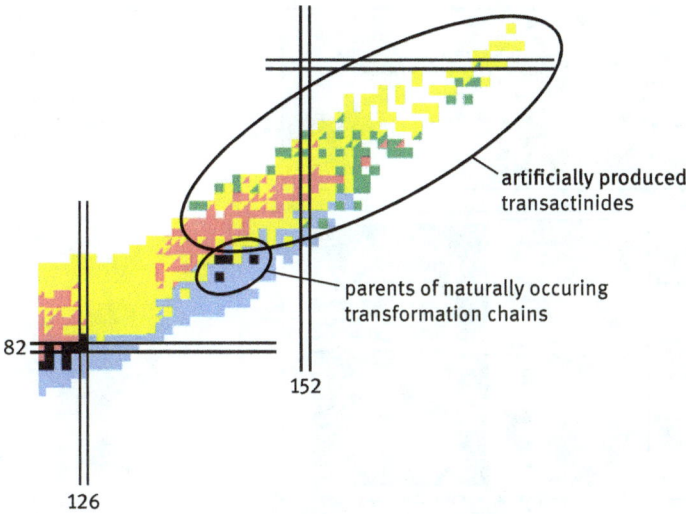

Fig. 10.1: Occurrence of nuclides within the Chart of Nuclides partially undergoing spontaneous fission.

10.2.1 Mass balances in fission

The relations in mass between the initial nuclide and the fission fragments are different compared to α-emission. Suppose α-emission is written as $^*K1 \rightarrow K2 + K3 + \Delta E$, where K2 represents the new nuclide and K3 the α-particle. The mass ratio of K2:K3 here is typically (>200):4, i.e. >50:1. In contrast, spontaneous fission forms K2 and K3 as two separate medium-large nuclei with masses of comparable order of magnitude. For a symmetrical division into two fragments of identical mass, the ratio between the product fragment masses would be 1:1. In this case, balances would ideally be $^AK2 \approx {}^AK3 \approx \frac{1}{2} {}^AK1$. Indeed, this happens for very heavy nuclides of $Z \geq 100$. For isotopes of thorium and uranium (and other nuclides), however, binary fission is asymmetric. In most cases, a larger (K2) and a smaller (K3) fragment are formed with varying mass distributions. For ^{238}U the two different primary fission fragments have masses of about 138 ± 15 and about 100 ± 15, i.e. at a ratio of ca. 1.4:1. Figure 10.2 schematically illustrates the difference between α-emission and fission concerning distributions of mass. Figure 10.3 illustrates the distribution of mass numbers between the two different primary fission fragments for ^{238}U. The most relevant fragments created have proton numbers of $^{Z2}K2 + {}^{Z1-Z2}K3 = {}^{Z1}K1$, with $^{Z2}K2$ and $^{Z3}K3$ ranging from ca. 47 to 70 and 31 to 51, respectively. Primary fission fragments represent much lighter chemical elements compared to the initial nuclide.

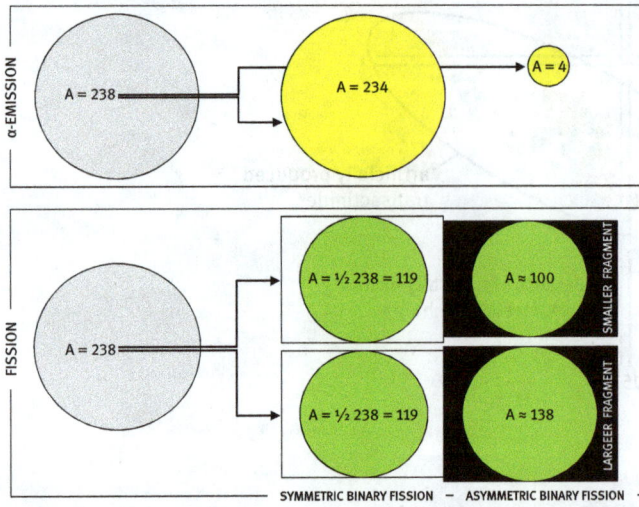

Fig. 10.2: Mass scale of unstable nuclides undergoing α-emission and spontaneous fission. ^{238}U is chosen as an example. Radii are proportional to $A^{⅓}$, as are the dimensions of the circles. The nucleus undergoing spontaneous fission is depicted as a sphere. Note that this is not correct, because spontaneous division of a large nuclide requires a nonspherical, i.e. deformed, nucleus.

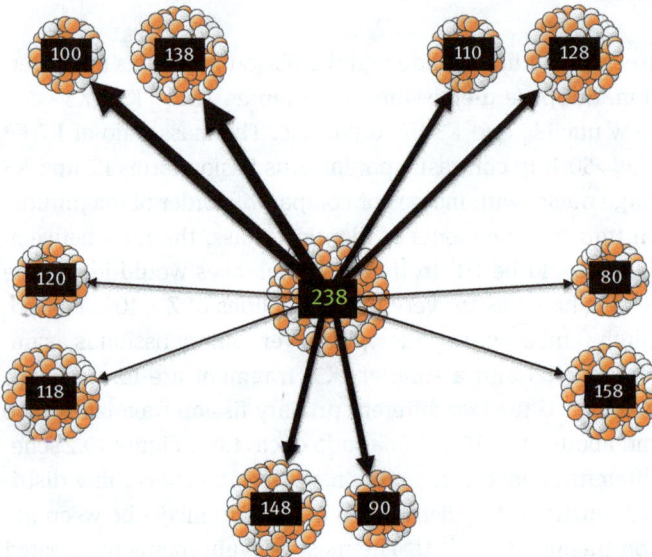

Fig. 10.3: Spontaneous fission of ^{238}U yields asymmetric mass distributions between the two primary fragments with combinations of $^{A}K1 = {}^{A}K2 + {}^{A}K3$. The most commonly occurring pairs are $^{A}K2 + {}^{A}K3 \approx$ 138 + ≈100, occurring in about 8% of all sf events. When the mass number of A2 shifts to larger (e.g. 145 or 150) or lower (e.g. 130 or 125) values, the value of A3 adjusts correspondingly. The probability of forming these pairs, however, decreases, as indicated by the thickness of the lines.

10.2.2 Simultaneous fission, β- and α-emission

Figure 9.6 already introduced the feature of parallel transformations, showing branches for β- and α-emission. The same applies to spontaneous fission. It indicates that the ΔE values are positive for various transformation options. Each pathway has its individual absolute value of Q_β, Q_α, and Q_{sf}, according to the different balances in mass between *K1 and transformation products. Examples of additional transformations include in almost all cases α-emission. In most cases, α-emission even represents the dominant transformation pathway. The relation between sf and α-emission starts with a very small fraction of sf relative to α-emission (<0.1%) for the nuclides of Z = 90 and 92 (^{232}Th, ^{235}U, ^{238}U) and then turns into a larger fraction, e.g. ^{248}Cm (ca. 8%) or ^{252}Cf (ca. 3%). Other parallel pathways are β⁻-emission (^{235}U), electron capture (e.g. ^{248}Fm), and the extremely rare cluster emission. Figure 10.4 shows examples of nuclides undergoing spontaneous fission and other types of transformation.

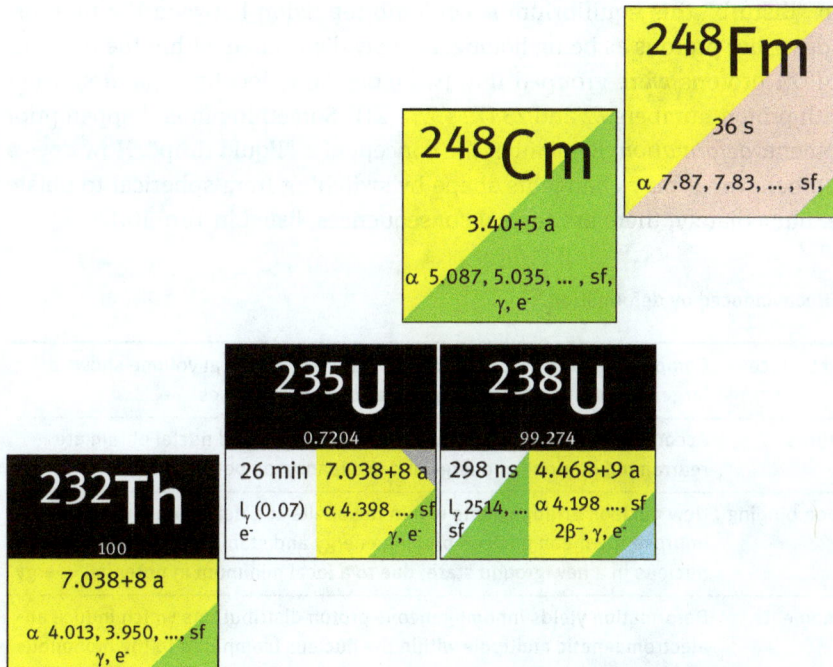

Fig. 10.4: Examples of nuclides simultaneously undergoing spontaneous fission and other types of transformation. The different colors indicate the various transformation routes: spontaneous fission = green, α-emission = yellow, β⁻-emission = blue, β⁺-emission and electron capture ε = red, cluster emission = gray. White fields refer to metastable isomers to be discussed in Chapter 11. The size of each triangular area symbolizes the relative abundance of that pathway.

10.3 Pathways of fission

Fission (spontaneous and induced) is by definition "the division of a nucleus into two or more parts with masses of equal order of magnitude, usually accompanied by the emission of neutrons, gamma radiation and, rarely, small charged nuclear fragments". Consequently, there are many individual steps to discuss, both at the initiation of the initial division and during the processes related to the fission fragments created. A key point is "deformation": if there is no deformation, there is no fission.

10.3.1 Deformation

Why should a nucleus divide? According to the liquid drop model, strong forces keep nucleons closely together as an ensemble, and there is a mass defect, i.e. a gain in energy, compared to the same number of unbound nucleons. The only force which can "disturb" this equilibrium is coulomb repulsion between the protons. The LDM describes protons as being homogeneously distributed within the nucleus. But what if $Z1$ protons were grouped into two (somehow) locally separated semi-centers with proton numbers $Z2$ and $Z3$ ($Z2 + Z3 = Z1$)? Something must happen prior to this moment: *deformation*. Remember the concept of a "liquid drop". How does a liquid drop behave? It may change its shape by switching from spherical to oblate or prolate. Immediately, there are several consequences, listed in Tab. 10.1.

Tab. 10.1: Effects induced by deformation.

Surface and surface tension	Compared to a sphere, a deformed sphere of identical volume shows a larger surface area. The surface tension also increases.
Shell structures	According to the shell model, the shells of deformed nuclei obtain are rearranged in terms of energy and occupancy, as per Fig. 3.23.
Mean nucleon binding energy	New nucleon arrangements within the shells of a deformed nucleus may improve the mean nucleon binding energy and stabilize the deformed nucleus in a new ground state, due to a local minimum in potential energy.
Multipole moments	Deformation yields inhomogeneous proton distributions which induce an electromagnetic multipole within the nucleus (compared to the monopole of a spherical nucleus).

10.3.2 From geometry to energy: surface area

The general theories describing the stability of a nucleus of an atom are the liquid drop and the nuclear shell model.[3] According to the WEIZSÄCKER equation, surface energy and coulomb repulsion play a role. For a sphere S, the following characteristics are also known:

-   volume is $^SV = 4/3\,\pi r^3$,
-   the relationship between mass number and volume and size is $^SA_V \approx r^{1/3}$
-   surface area is $^SS = 4\pi r^2$,
-   the relationship between the mass of a nucleus (reflecting the overall number of nucleons) and the surface and radius of a sphere is $^SA_S \approx r^{2/3}$.

In the context of the LDM, nuclei may transform (deform) into an ellipsoid. For an "ellipsoid of rotation", the three individual radii a, b and c turn into $a = r_{long}$ and $b = c = r_{short}$ (see Fig. 10.5). For a rotational prolate ellipsoid E, the volume is $^EV = 4\pi/3\,(r_{long}) \cdot (r_{short})^2$, as per eq. (10.3). The surface of an ellipsoid is expressed through eq. (10.4).

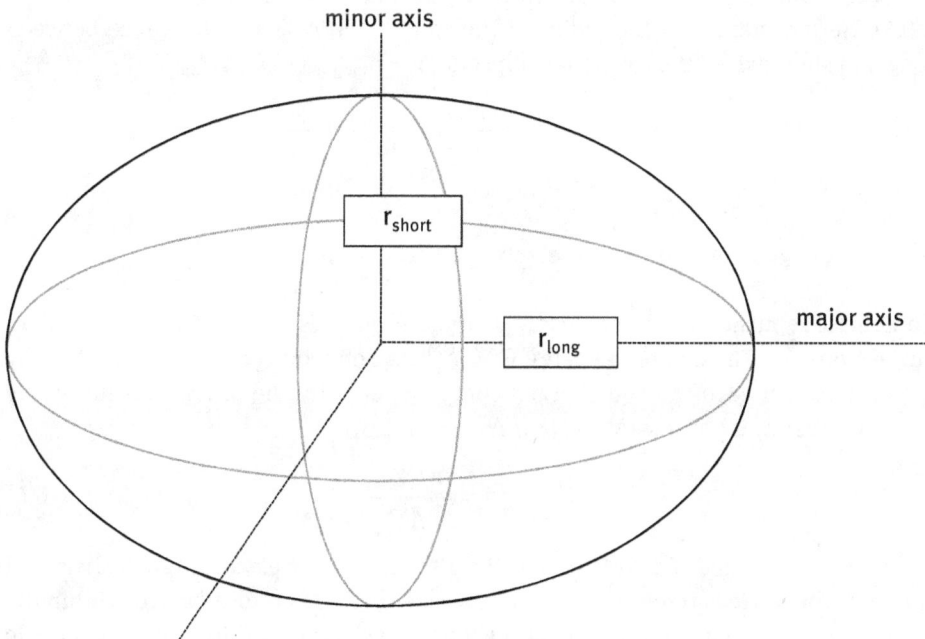

Fig. 10.5: Geometry of a prolate ellipsoid of rotation with two axes of r_{long} and r_{short}.

---

3 These two models are applied to unstable nuclei as well, but in particular for very heavy nuclei some of the essential inputs in both models may not apply – at least in detail.

Suppose this ellipsoid has the same volume as a sphere containing the same number of nucleons; then the surface of the two bodies (as per eq. (10.4)) will differ depending on the ratio between the two axes of the rotational ellipsoid. The difference increases the longer the deformation continues.[4]

$$^{E}V = \frac{4}{3}\pi\,(abc) = \frac{4}{3}\pi\,(r_{long})(r_{short})^2 \qquad (10.3)$$

$$^{E}S \approx 4\pi\left[\frac{(ab)^{1.6} + (ac)^{1.6} + (bc)^{1.6}}{3}\right]^{0.625} \qquad (10.4)$$

10.3.3 Deformation parameter

The radius r along an axis of rotation becomes smaller for prolate deformation and larger for oblate deformation. The increasing ratio of r_{lomg}/r_{short} from 1 (spherical) to >1 (prolate) or <1 (oblate) (as per Fig. 10.6) is described by the degree of *eccentricity* ε. This may be expressed by a deformation parameter $\varepsilon^{DEFORMATION}$, which involves the changes in radii r between the deformed shape and the spherical one, following eq. (10.5). Here r_0 stands for the radius of the sphere and Δr is the difference between long and short axis of the rotational ellipsoid, $\Delta r = r_{(long\,axis)} - r_{(short\,axis)}$.

$$\varepsilon^{DEFORMATION} \approx \frac{r_{(long\,axis)} - r_{(short\,axis)}}{r_0} \qquad (10.5)$$

10.3.4 From geometry to energy: electromagnetic forces

Coulombic repulsion of the positively charged protons is a force acting opposite to surface energy. The coulomb energy $E^{COULOMB}$ is defined according to eq. (9.11) by the radius r between the two charged components. With the radius substituted by mass number A, it becomes $V^C \approx Z_1 Z_2 / A^{1/3}$ (eq. (10.6)).[5]

$$E^{COULOMB} = \frac{0.72 Z_1 Z_2}{A^{1/3}} \qquad (10.6)$$

Another consequence of deformation is the alteration in the electromagnetic field of a nucleus. The nucleus represents a dynamic system, where protons "move" within the "liquid drop" nucleus. In a spherical nucleus, the electric field induces a monopole

---

4 For $r_{long}:r_{short} = 2:1$ or 4:1, the surface of a prolate ellipsoid relative to that of the sphere increases by factors of 1.077 and 1.279, respectively. For a rotational oblate ellipsoid ($r_{long}:r_{short} = 1:2$ or 1:4), it is even more extreme: 1.095 and 1.428, respectively.
5 The corresponding coulomb term of the WEIZSÄCKER equation was $- \gamma^{LDM} Z^2 / A^{1/3}$.

magnetic field. Within a rotational ellipsoid-shaped nucleus, a nonuniform distribution of electric charge appears. The deformed nucleus gets an electric quadrupole Q with quadrupole moments >0 (prolate) or <0 (oblate). The absolute value of the ellipsoid quadrupole ^{E}Q relative to the initial monopole value of a sphere ^{S}Q depends on the orbital moment I of the corresponding nucleus according to eq. (10.7) and Fig. 10.6.

$$^E Q = {}^S Q \left[\frac{-I(I+1)}{(I+1)(2I+3)} \right]$$

(10.7)

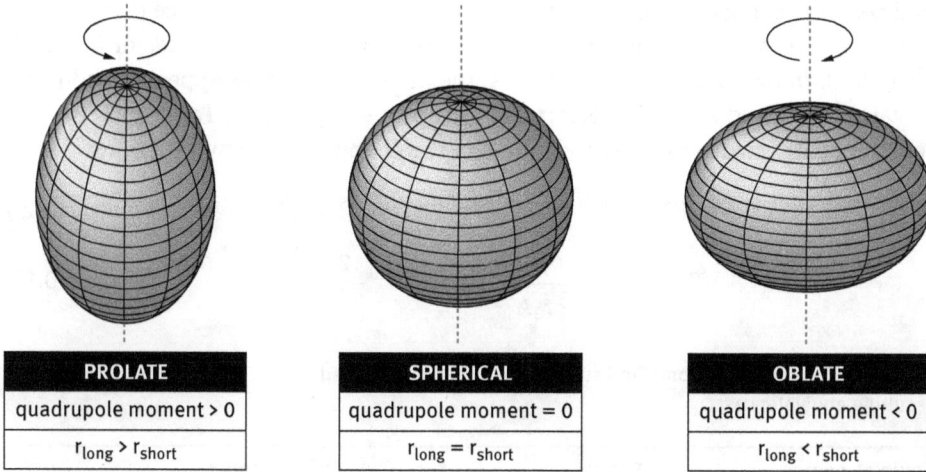

PROLATE	SPHERICAL	OBLATE
quadrupole moment > 0	quadrupole moment = 0	quadrupole moment < 0
$r_{long} > r_{short}$	$r_{long} = r_{short}$	$r_{long} < r_{short}$

Fig. 10.6: Ellipsoids of rotation along one selected axis, forming either prolate or oblate nuclei depending on the ratio between r_{long} and r_{short}.

10.3.5 Surface energy *vs.* coulomb energy

Surface energy $E^{SURFACE}$ is a force keeping a "liquid drop" body of nucleons together.[6] Increasing deformation may be expressed by a quadrupole stretching factor $^{qs}\alpha$. Upon deformation, the surface energy $\Delta E^{SURFACE}$ is positively affected according to eq. (10.8), while for coulomb energy $\Delta E^{COULOMB}$ the opposite occurs, as per eq. (10.9). The changes in surface and coulomb energy when deforming from a spherical to an ellipsoidal nucleus are expressed through eqs. (10.10) and (10.11).

$$\Delta E^{SURFACE} = {}^E E^{SURFACE} - {}^S E^{SURFACE}$$

(10.8)

$$\Delta E^{COULOMB} = {}^S E^{COULOMB} - {}^E E^{COULOMB}$$

(10.9)

---

6 Similar to the surface tension of a drop of water or even more extreme: mercury.

$$E_E^{SURFACE} = {}^S E^{SURFACE} \left(1 + \frac{2}{5} {}^{qs}\alpha^2\right) \qquad (10.10)$$

$$E_E^{COULOMB} = {}^S E^{COULOMB} \left(1 + \frac{1}{5} {}^{qs}\alpha^2\right) \qquad (10.11)$$

While these equations give a relationship between energies of deformed nuclei relative to the spherical one, individual energies can also be calculated directly following the WEIZSÄCKER equation. Here, analogue coefficients (in units of MeV) for surface and coulomb energy are denoted by $\alpha^{SURFACE}$ and $\alpha^{COULOMB}$. For surface energies, the radius r_0 and the mass number A are linked with a surface tension factor $S^{TENSION}$, eq. (10.12). In the case of coulomb forces, the number of protons appears in addition to values of r_0 and A and the electric charge unit e, eq. (10.13). Table 10.2 lists the resulting expressions for the ratio between the changes in the two energies.

$$^S E^{SURFACE} = 4\pi r_0^2 S^{TENSION} A^{2/3} = \alpha^{SURFACE} A^{2/3} \qquad (10.12)$$

$$^S E^{COULOMB} = \frac{3}{5} \frac{Z^2 e^2}{r_0 A^{1/3}} = \alpha^{COULOMB} \frac{Z}{A^{1/3}} \qquad (10.13)$$

Tab. 10.2: Changes in coulomb and surface energies for deformed nuclei. Individual equations and resulting equivalences.

Coulomb energy	Ratios	Surface energy	
$\gamma^{LDM} \frac{Z^2}{A^{1/3}} \approx \left[\alpha^{COULOMB} \frac{Z^2}{A^{1/3}}\right]$	$\frac{\gamma^{LDM} \frac{Z^2}{A^{1/3}}}{\beta^{LDM} A^{2/3}} \approx \frac{\left[\alpha^{COULOMB} \frac{Z^2}{A^{2/3}}\right]}{\left[\alpha^{SURFACE} A^{2/3}\right]}$	$\beta^{LDM} A^{2/3} \approx \left[\alpha^{SURFACE} A^{2/3}\right]$	
	$\approx \frac{Z^2}{A}$		(10.14)
	$\left|\Delta E^{SURFACE}\right| = \left|\Delta E^{COULOMB}\right|$		(10.15)
	$\frac{^S E^{SURFACE}}{^S E^{COULOMB}} = \frac{2}{1}$		(10.16)

There will exist a degree of stretching for which the stabilizing effect of deformation on surface energy $\Delta E^{SURFACE}$ is identical to its negative effect on coulomb energy $\Delta E^{COULOMB}$, eq. (10.15). Here, a quotient of type Z^2/A appears, eq. (10.14), indicating that the density of protons per volume (reflected by A) plays a significant role. Table 10.3 list values of Z^2/A for select nuclides transforming (at least in part) by spontaneous fission. According to eqs. (10.8)–(10.11), this corresponds to $^S E^{SURFACE}/^S E^{COULOMB} = 2/1$, eq. (10.16).

10.3.6 Fissility parameter

Equations (10.15) and (10.16) express an equilibrium between the changes in coulomb energy and the changes in surface energy for a ratio of $^S E^{SURFACE} / ^S E^{COULOMB} = 2$. When the ratio between the two contributions is changing, the value x is called the *fissility parameter*, shown in eqs. (10.17)–(10.19).

$$\frac{^S E^{COULOMB}}{2 ^S E^{SURFACE}} = x \tag{10.17}$$

$$x = \frac{\alpha^{COULOMB}}{2\alpha^{SURFACE}} \frac{Z^2}{A} \tag{10.18}$$

$$x = \frac{(Z^2/A)}{(Z^2/A)^{critical}} \tag{10.19}$$

Equation (10.18) involves basic parameters of nuclides, namely the mass number A, proton number Z and the two coefficients α_C and α_S. The value of x is thus directly derived.[7] The interesting point is to define a "critical" value for Z^2/A of a given nucleus. Given $\alpha_C / 2\alpha_S$ equals the reciprocal of Z^2/A, i.e. $\alpha_C / 2\alpha_S = (Z^2/A)^{-1}$, eq. (10.18) turns into (10.19). This term is called $(Z^2/A)^{critical}$. Thus, it is the proton number Z per mass number A of a nucleus, somehow a "proton density" per volume, which is responsible for an eventual fission process. For fissile nuclides, $x < 1$. For ^{235}U, ^{238}U and ^{252}Cf, for example, the values are 0.749, 0.740 and 0.793, respectively.

10.3.7 From geometrical deformation to nuclide division

At a *transition state* (or so-called *saddle point*[8]), the nucleus may switch to a new geometry by forming a neck between two semi-centers of the ellipsoid, shown in Fig. 10.7. From now on, coulomb repulsion may take over the process in an irreversible way, and within a very short period of time the phase of division starts. This moment is called the *scission point*.

---

7 With $\alpha_C = \beta^{LDM} = 15.94$ MeV and $\alpha_S = \gamma^{LDM} = 0.665$ MeV, the quotient $\alpha_C / 2\alpha_S$ gives a proportionality factor of 0.0203, and the fissility parameter x is now directly calculated by simply introducing specific numbers for Z and A of a candidate nucleus.

8 . . . which is about 10^{-20} s prior to the fission event, see below.

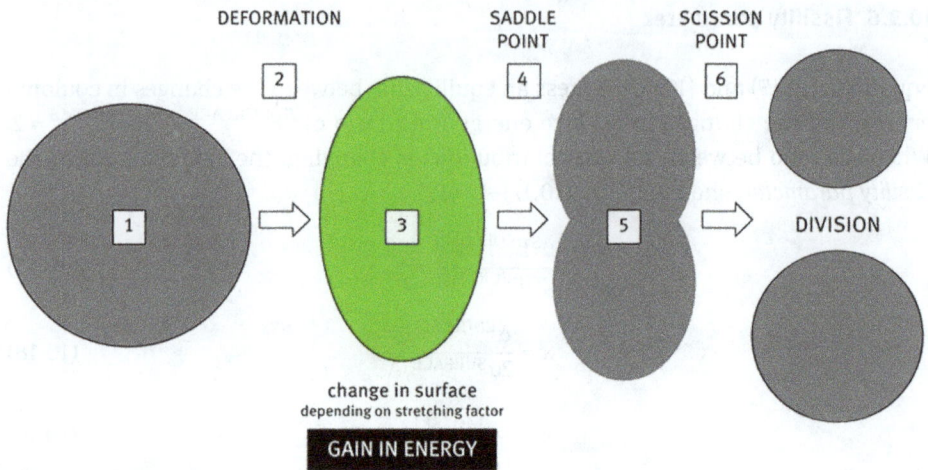

Fig. 10.7: Deformation (2) of a large spherical nucleus (1) to prolate shape (3) and subsequent necking (5) of the deformed nucleus. At a saddle point (4), coulomb repulsion between the two internal centers of high proton charges of the two fragments may succeed over surface tension, and at a scission point (6), the process may continue towards division to form two separate (primary) fragments.

10.3.8 Neck and fragments

The energetics of spontaneous fission according to the LDM explain the general aspects of this type of transformation. As already seen for β- and α-transformations, nucleon shell models overlay the LDM. The stability (nucleon binding energy) is enhanced for nuclei containing paired nucleons and magic nucleon numbers. Simply speaking, whenever there is a chance to collect a certain number of nucleons, why not prefer those reflecting magic numbers? And in fact, this is the strategy! What about the remaining nucleons? They would form the neck, as shown in Fig. 10.8. When fission takes place, the neck ruptures. The neck position is statistically determined, and the two parts of the neck (belonging to either of the two fragments) combine with the corresponding fragment. This will add a certain number of protons and neutrons to the two "pre-formed" fragments. Once the neck has divided, two separate fragments of high proton number ($Z \geq 28$ or ≥ 50) are repelled through coulomb forces.

10.3.9 Mass distributions and magic numbers

For the example of the heavy fragment A_{K2}, because the magic numbers involved are $Z = 50$ and/or $N = 82$, this would ideally correspond to a fragment of $Z = 50$ plus $N = 82$ yielding $A = 132$, with a double magic nucleon number. In parallel, nucleon numbers may alter as the corresponding "piece" of the neck is added, though at

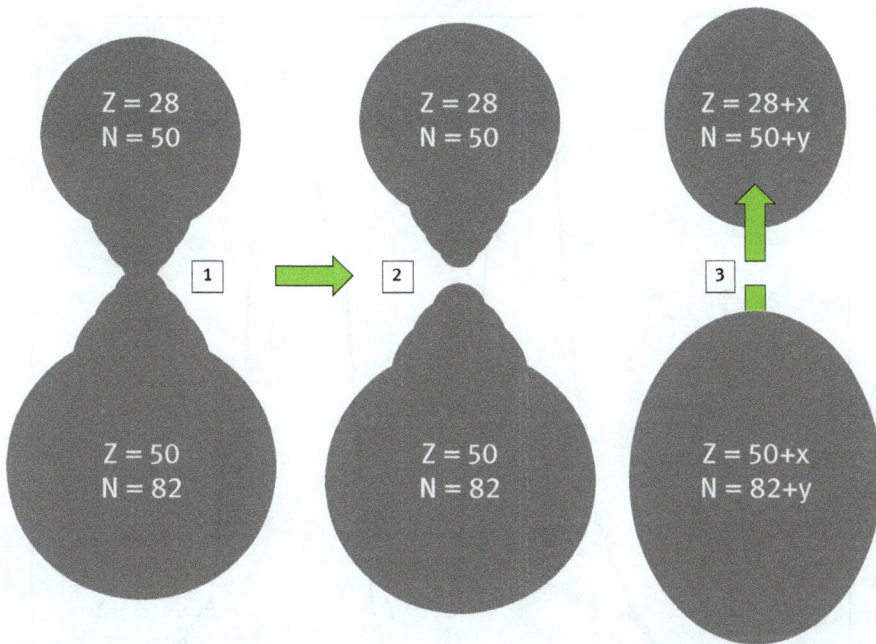

Fig. 10.8: The fate of the neck in the fission mechanism. A deformed nucleus starting to accumulate double magic nucleon numbers into two semi-centers of $Z = 28 + N = 50$ and $Z = 50 + N = 82$, while the remaining nucleons form the neck between them (1). The neck divides in a statistical pattern, and the two remaining parts of the neck combine with the corresponding fragment to add a certain number of x protons and y neutrons to the fragment (2). Once the neck has divided, two separate fragments of high proton number ($Z \geq 28$ or ≥ 50) remain and are repelled through coulomb forces (3).

least one sort of nucleon would try to "conserve" its magic value. This would explain a mass distribution of the larger fragment of $Z = 50$ (fixed) and $N = 79, 78$, etc. or $N = 81, 82$, etc., but also $N = 82$ (fixed) and $Z = 51, 50$, etc. or $Z = 53, 54$, etc. The distribution of the resulting primary fission fragments thus shows a high abundance of certain combination of mass numbers of the two fragments, while other combinations are less abundant. However, the sum of the nucleons within the initial fission step remains constant. Figure 10.9 gives the profile of mass distributions of primary fragments in the spontaneous fission of ^{252}Cf.

10.3.10 Paired nucleons

The effect of nucleons existing pairwise also improves overall nucleon binding energy. This holds true both for neutrons and protons. Consequently, even (versus odd) proton or neutron mass numbers should be more common in the initial fission

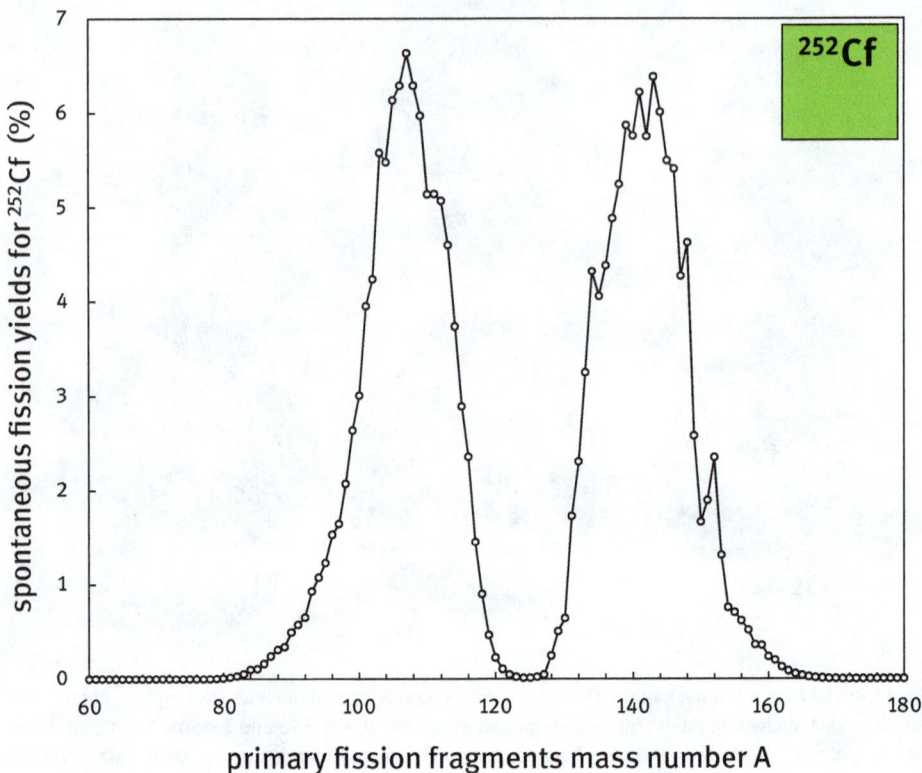

Fig. 10.9: Mass distributions of primary fragments in the spontaneous fission of ^{252}Cf (linear scale).[9]

fragment distribution. This is seen in Fig. 10.9 and is particularly obvious when the number of protons is considered, as in Fig. 10.10.

10.3.11 Asymmetric *vs.* symmetric fission

With increasing mass of the nuclides and with increasing neutron number among isotopes of the same element, spontaneous fission tends to turn from *asymmetric* into *symmetric* fission. Here, the two primarily formed fission fragments are of similar or identical mass number. This means that the initial nucleus divides into two more-or-less equal fragments of $A_{K2} \approx A_{K3} \approx \frac{1}{2}A_{K1}$, as per Fig. 10.11. Figure 10.12

---

9 Note that for sf of ^{252}Cf, the yields of individual fragment masses are given in percent relative to the fraction of ^{252}Cf undergoing sf. However, the sf branch of ^{252}Cf is just 3% relative to the dominant α-transformation pathway.

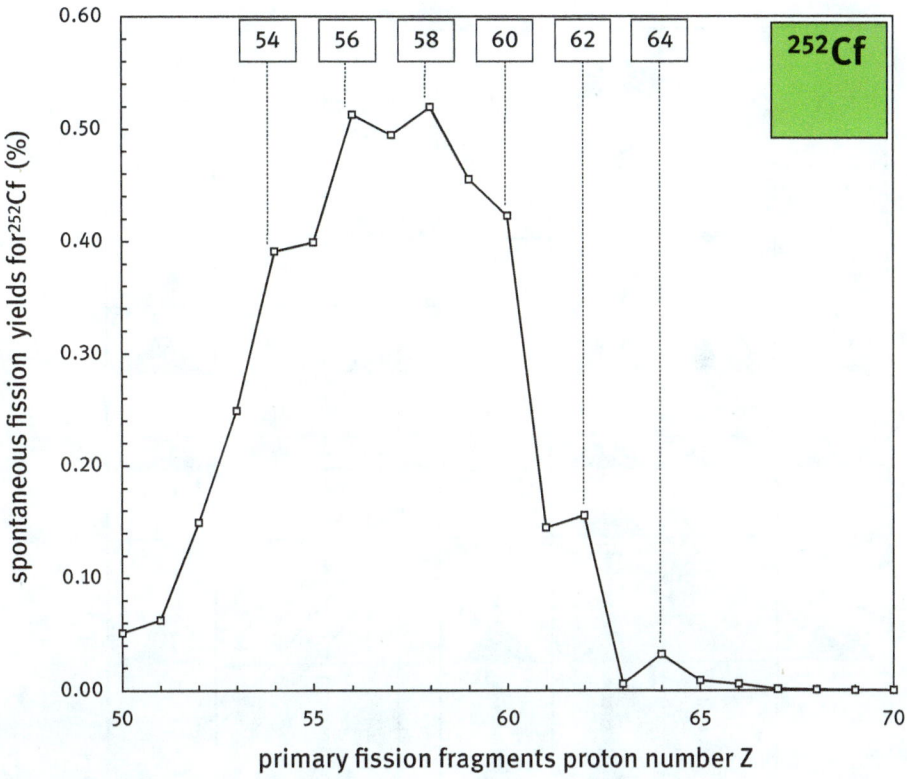

Fig. 10.10: Distribution of proton number Z of the two primary fission fragments in the spontaneous fission of ^{252}Cf.

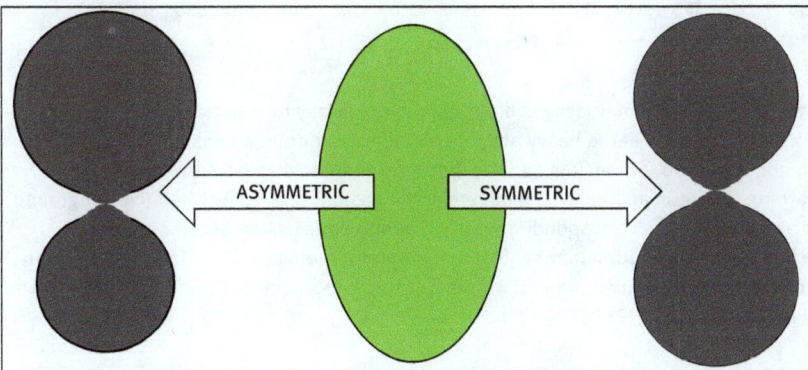

Fig. 10.11: Asymmetric *vs.* symmetric fission.

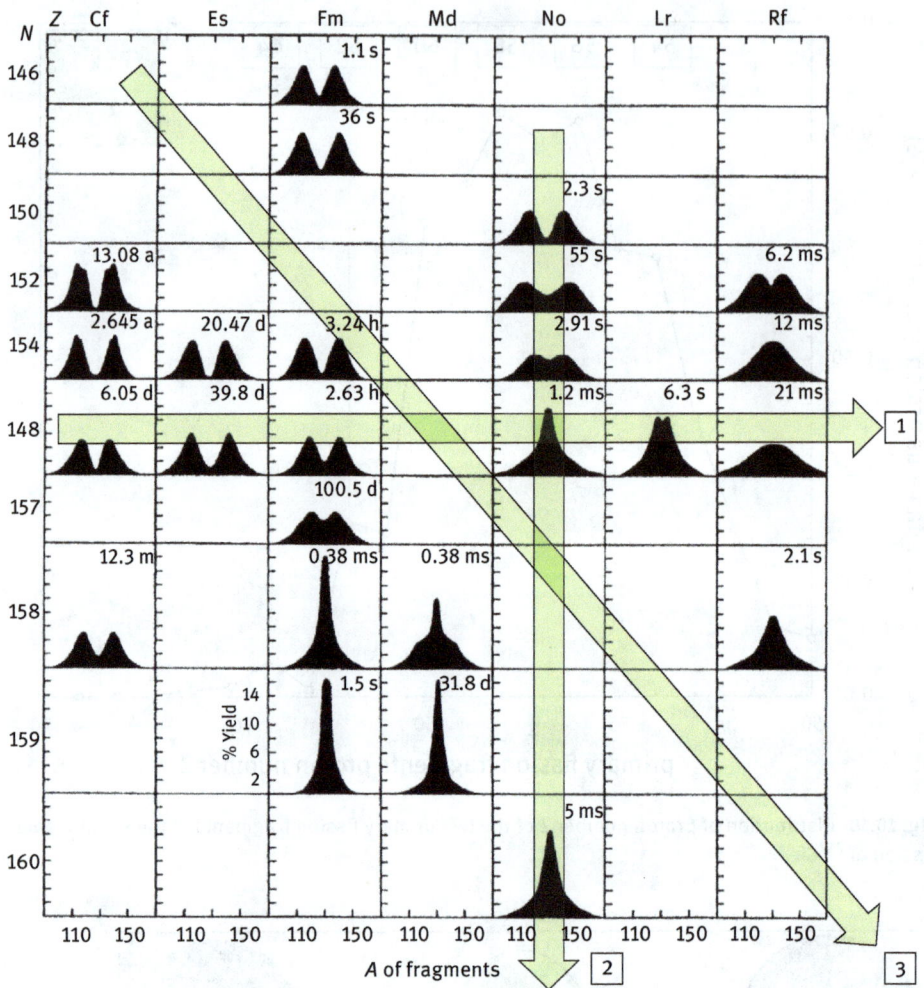

Fig. 10.12: Asymmetric and symmetric mass distributions of initially formed fragments in binary spontaneous fission. Isotopes of the heavy actinide elements californium, einsteinium, fermium, mendelevium, nobelium and lawrencium and the first transactinide element rutherfordium are arranged according to proton and neutron numbers of the fissioning nuclides. Half-lives of ground state isomers are added in the corresponding boxes. Spontaneous fission becomes more symmetric (1) for increasing neutron number at constant proton number, (2) for increasing proton number at constant neutron number, and (3) with increasing mass number (from: DC Hoffmann and MR Lane, Radiochimica Acta 70/71 (1995) 135).

shows the occurrence of symmetric and asymmetric fission profiles for nuclides of the very heavy elements of $Z = 94 - 104$.

The neck model explains this trend in mass distribution. For less heavy nuclides the two fragments attract magic compositions of type ($Z = 50$, $N = 82$) and ($Z = 28$, $N = 50$).

Increasing the nucleon numbers does not change this pattern; thus, it is the neck that increases in mass and size. However, as even more nucleons become available (starting at element $Z = 100$), a mass distribution of type two times $Z = 50$ becomes possible. This turns into binary symmetric fission. At the same time, the neck shrinks, because the two fragments accumulate almost all the nucleons available to obtain shell occupancies approaching magic numbers.

This can be summarized by three general tendencies. Asymmetric fission becomes more symmetric:
1. for increasing neutron number at constant proton number,
2. for increasing proton number at constant neutron number,
3. with increasing mass number from $A = 246$ ($_{100}Fm_{146}$) or 250 ($_{98}Cf_{152}$) to $A = 262$ ($_{102}No_{160}$ or $_{104}Rf_{158}$).

10.3.12 Binary *vs.* ternary fission

The mechanism of spontaneous fission discussed so far refers to the formation of two fission fragments. However, "ternary" fission[10] also occurs in parallel. "Ternary" is the division of the initial nucleus into three primary fragments, as per Fig. 10.13. Mechanically, this refers to the ternary fission of a neck that splits in two places instead of one. Consequently, a minor component remains (let us denote it as K4), which is not attributed to the two essential primary fragments, following eq. (10.2). There are two questions: What is the new fragment A_{K4}? And what is the ratio between binary and ternary fission?

Fragment distribution in ternary spontaneous fission: For the new primary fragment, one would expect a statistical distribution of mass numbers theoretically ranging from 1 (the proton or the neutron) to ≤ 40, because the pool of nucleons available is limited to those found in the neck. Whenever the impact of the shell model holds true, this spectrum of fragments should represent a few fragments of particular stability. Indeed, the majority of the lightest fragments are α-particles (generated in about 90% of all events). It may vary down to fragments of hydrogen or up to fragments of carbon or even argon. The mass distribution of the two larger primary fragments is thus shifted to somewhat lower values. When the binary fragments are, for example, $^{binary}A_{K2} \approx 135 \pm 15$ and $^{binary}A_{K3} \approx 95 \pm 15$, then the ternary fragments are $^{ternary}A_{K2} \approx {}^{binary}A_{K2} - 4$ or $^{ternary}A_{K3} \approx {}^{binary}A_{K3} - 4$. This is shifted in a few cases, according to the somewhat smaller or larger mass number of $^{ternary}A_{K4}$. Experiments have demonstrated that this small fragment is released perpendicular to the direction of the two large primary fragments (thereby proving the postulated

---

10 GREEN, LIVESEY, TSIEN et al. discovered ternary nuclear fission in 1946.

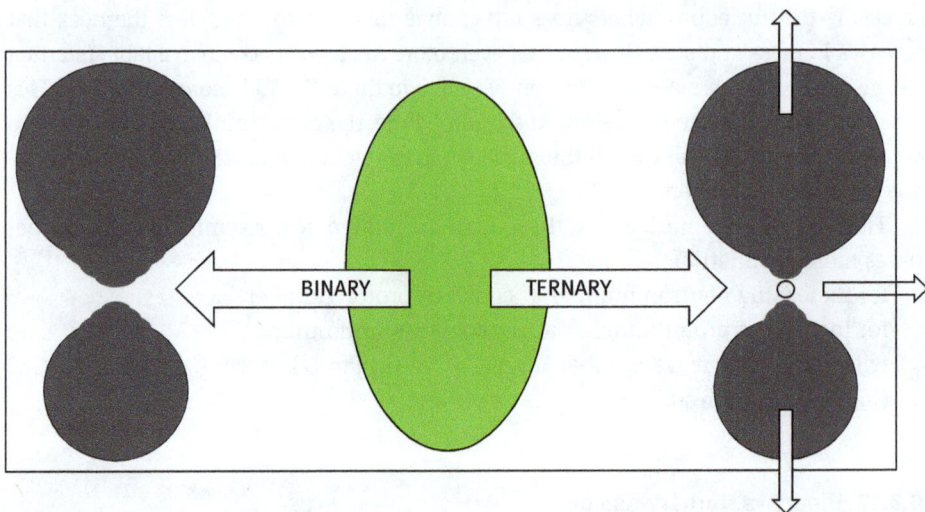

Fig. 10.13: Illustration of the asymmetric binary and ternary divisions of a fissile nucleus.

mechanism of ternary fission). While the recoil energies of fragments K2 and K3 are similar to those for binary fission, the kinetic energy of each α-particle released as third primary fragment in ternary fission is large, reaching about 16 MeV.[11]

Ratio of ternary fission relative to binary fission: Because of energetic consider-ations, ternary fission is a rare event relative to binary fission. A typical ratio be-tween binary and ternary fission is ca. $1:10^{-5}$ or $1:10^{-6}$.

10.4 Energetics

10.4.1 Trends

The motivation to transform *via* spontaneous fission is the same as for any other pro-cess of exothermic radioactive transformation, namely to increase the mean nucleon binding energy. Along the curve of mean nucleon binding energy according to the WEIZSÄCKER equation, any shift from a nucleus of very high mass number A to inter-mediate mass nuclei increases the mean nucleon binding energy, as per Fig. 6.7. In this context, fission is similar in principle to α-emission, but the gain in mean nu-cleon binding energy is much higher than in the case of $\Delta A = 4$.

---

[11] This is about 10 MeV larger than observed for α-particles released during α-transformation of unstable nuclides.

10.4.2 Absolute scale

Fission of heavy nuclei into two smaller fragments belongs to a process of utmost importance as a source of energy. The absolute value of the energy of the spontaneous fission process follows eq. (10.1) for a binary asymmetric spontaneous fission. Equations (10.20) and (10.21) utilize real masses or mass defect or mass excess values, respectively:

$$\Delta E = \Delta m\, c^2 = Q_{sf} = [m_{*K1} - (m_{K2} + m_{K3})]c^2 \tag{10.20}$$

$$Q_{sf} = \Delta m^{defect}(*K1) - [\Delta m^{defect}(K2) + \Delta m^{defect}(K3)] \tag{10.21a}$$

$$Q_{sf} = \Delta m^{excess}(*K1) - [\Delta m^{excess}(K2) + \Delta m^{excess}(K3)] \tag{10.21b}$$

Each of the two primary fission fragments shows an increase in $\bar{E}_B$ relative to the very heavy initial nucleus. Compared to α-emission, the gain in energy in terms of $\Delta\bar{E}_B = \bar{E}_B(K2) - \bar{E}_B(*K1)$ and $\Delta\bar{E}_B = \bar{E}_B(K3) - \bar{E}_B(*K1)$ is much higher.

For α-emission, let us consider an $A = 238$ nuclide emitting an α-particle. $\bar{E}_B$ values are ≈7.570 MeV for ^{238}U and ≈7.597 MeV for ^{234}Th, yielding $\Delta\bar{E}_B \approx 0.047$ MeV. For fission, let us again take the same nuclide of $A = 238$. In the hypothetical case of symmetric fission, for each of the $A = 119$ product nuclei $\bar{E}_B$ is ca. 8.35 ± 0.10 MeV.[12] The difference is $\Delta\bar{E}_B \approx 0.78 \pm 0.10$ MeV, which is larger than in the case of α-emission by a factor of almost 20.

When calculating the overall (instead of mean) nucleon binding energy, each of the 238 nucleons thus improves by ca. 0.8 MeV, yielding an overall nucleon binding energy of ca. 190 MeV per one nuclide of $A = 238$! It is among the largest sources of energy in the universe and the essence of "nuclear power".

10.4.3 Fission barrier

Simple mathematics would state that the energy gained by dividing one large nucleus into two identical pieces should be the same as that needed to recombine them into the original nucleus.

According to eqs. (9.11) and (10.6), $E^{COULOMB}$ energy for ^{238}U and the two hypothetical primary fragments of $A/2 = 119$ and $Z/2 = 46$ is about 197 MeV. This is about 7 MeV larger than the energy *released* in the fission process. This value is called the "fission barrier" H_{fb}, described in eq. (10.22). The fission barrier is the difference in energy between the corresponding nuclear state of the nucleus prior to fission and

---

12 There are many unstable nuclides of $A = 119$ formed along the isobar . . . .

the maximum potential energy needed to fission.[13] Hence, H_{fb} is >0 (and typically about 5–8 MeV), which would make spontaneous fission energetically impossible.

$$H_{fb} = E^{COULOMB} Q_{sf} \tag{10.22}$$

Spontaneous fission should require additional energy to surpass the fission barrier – but this extra energy is not available, except from external sources.[14] Thus, spontaneous fission may (in part) be understood by the quantum mechanical tunneling effect. Figure 10.14 illustrates the concept of spontaneous fission in the context of changes of potential energy depending on the degree of deformation in terms of tunneling. Thus, spontaneous fission may take place without this extra energy only because of quantum mechanics tunneling the barrier – similar to α-emission.

10.4.4 Kinetic energies of primary fission fragments and neutrons

These strongly deformed and extremely neutron-rich primary fragments observe extreme coulomb repulsion, thereby obtaining a defined kinetic energy and a high velocity. "Flying" away at that velocity, the fragments release single excess neutrons (about 2–3 so called "prompt" neutrons, see below) and transfer kinetic energy.[15]

10.5 Velocities of spontaneous fission

Spontaneous fission may be (similar to α-emission) described quantum mechanically in terms of the frequency f (of pre-forming two potential fragments) and a penetration factor P (transmitting through the fission barrier). The transformation constant is the product of the two components of type fP. Again, both overall energy and internal nucleon structure combine. Large values of fP correspond to large values of the transformation constant λ^{sf}, yielding a short half-life $t_{1/2}^{sf}$ for spontaneous fission. Half-lives of spontaneous fission cover a giant range of more than 10^9 years up to microseconds. While half-lives $t_{1/2}^{sf}$ of nuclides belonging to the light

---

13 The height of the fission barrier depends on several parameters and is significantly determined by the grade of deformation, the size of the neck, and the kind of fission (symmetric or asymmetric).

14 Here comes the relevance of induced fission. Whatever tool is employed to transfer energy (particles with kinetic energies, electromagnetic radiation etc.) to a potential candidate nucleus for fission, this imported energy may help to overcome the fission barrier. Consequently, the fission process will become possible or (if it was already allowed through the tunneling effect) will be facilitated significantly; see Chapter 13 on nuclear reactions.

15 Maximum neutron kinetic energies E_n are about 10 MeV, corresponding to a velocity of ca. 10^7 m/s (which is about 1/30 of the speed of light in vacuum). Due to spatial distributions, adsorption and interaction, only a few neutrons are released with the maximum kinetic energy. Most of the emitted neutrons show kinetic energies of ca. 0.2 to 2 MeV.

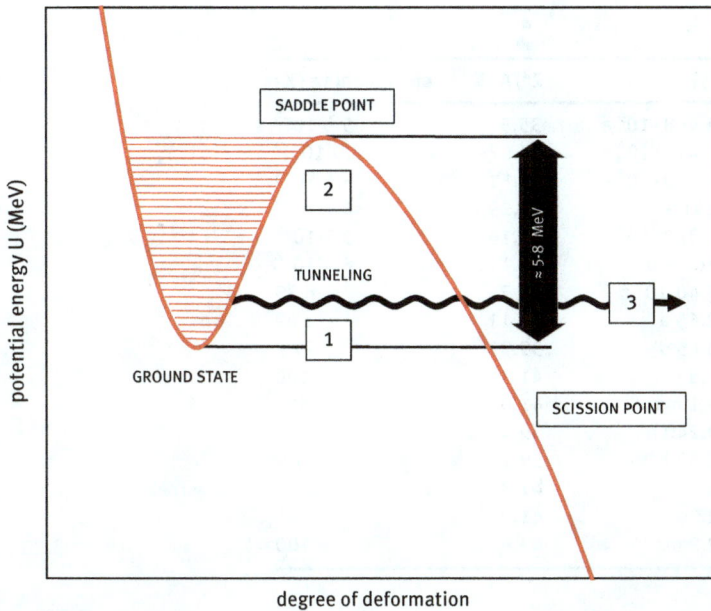

Fig. 10.14: Change of potential energy of the nucleus depending on the degree of deformation. The nucleus moves from the ground state of the deformed nucleus (1) either to surpass the fission barrier (2) at the saddle point and further to the scission point or by quantum mechanics tunneling (3) through the barrier.

actinide elements, for example, are $1.405 \cdot 10^{10}$ years for ^{232}Th and $7.038 \cdot 10^{8}$ years for ^{235}U, isotopes of heavy actinide or of transactinide elements $(Z \geq 104)^{16}$ show half-lives in the range of seconds and fractions of a second, given in Tab. 10.3.

Tab. 10.3: Prominent nuclides transforming (at least in part) by spontaneous fission. Proton number, half-life, quotient Z^2/A and fraction of spontaneous fission relative to α-emission. Branching data from National Nuclear Data Center, Brookhaven National Laboratory, USA / www.nndc.bnl.gov/chart.

Nuclide	Z	t½	Z^2/A	sf-Branching (%)
^{232}Th	90	$1.405 \cdot 10^{10}$ a	34.91	$1.1 \cdot 10^{-9}$
^{233}U	92	$1.592 \cdot 10^{5}$ a	36.33	$<6.0 \cdot 10^{-11}$
^{235}U	92	$7.038 \cdot 10^{8}$ a	36.02	$7.0 \cdot 10^{-9}$

---

16 Spontaneous fission is a dominant transformation pathway for super-heavy nuclides, and values of t½^{sf} approach microseconds for the heaviest elements identified so far. For details, see Chapter 10 of Volume II on the "Chemistry of transactinides".

Tab. 10.3 (continued)

Nuclide	Z	t½	Z²/A	sf-Branching (%)
^{238}U	92	$4.468 \cdot 10^9$ a	35.56	$5.5 \cdot 10^{-5}$
^{239}Pu	94	$2.411 \cdot 10^4$ a	36.97	$3.0 \cdot 10^{-10}$
^{242}Pu	94	$3.750 \cdot 10^5$ a	36.51	$5.5 \cdot 10^{-4}$
^{242}Am	95	141 a	37.29	$<4.7 \cdot 10^{-9}$
^{243}Am	95	$7.370 \cdot 10^3$ a	37.14	$3.7 \cdot 10^{-9}$
^{242}Cm	96	162.9 d	38.1	$6.2 \cdot 10^{-6}$
^{248}Cm	96	$3.40 \cdot 10^5$ a	37.2	8.39
^{252}Cf	98	2.65 a	38.11	3.09
^{254}Cf	98	60.5 d	39.37	99.7
^{242}Fm	100	0.8 ms	41.3	100
^{244}Fm	100	3.1 ms	41.0	>97
^{254}Fm	100	3.240 h	39.4	0.06
^{256}Fm	100	2.63 h	39.1	91.9
^{252}No	102	2.3 s	41.3	29.3
^{263}Db	105	27 s	41.9	41
^{258}Sg	106	2.9 ms	43.6	100

Figure 10.15 shows the half-lives of many nuclides undergoing spontaneous fission in terms of the quotient Z^2/A. One may further discriminate between nuclides of even or odd nucleon numbers (or even or odd mass numbers). For similar values of Z^2/A, even nucleon number configurations result in much shorter half-lives by many orders of magnitude. For one and the same heavy chemical element (Z = constant), half-lives of the isotopes change depending on the Z^2/A value. The larger the abundance of protons per overall nucleon number is, the faster the spontaneous fission proceeds. This is illustrated in Fig. 10.16 for even proton number elements.

10.6 Follow-up processes of initial fission

Complete understanding of spontaneous (and induced) nuclear fission needs to consider processes immediately following the division of the initial nucleus. In the case of binary fission, initial fission created two *primary* fission fragments. Several other processes immediately follow and these may be grouped by their character and by the order they appear over time, as per Tab. 10.4.

10.6.1 Extreme excess of neutrons in the initially formed fission fragments

Nuclides of increasing mass number have an excess of neutrons over protons. If the excess of neutrons is expressed as the ratio N/Z, the WEIZSÄCKER equation shows values of N/Z = 1.55 ± 0.05 for heavy nuclides of proton number Z = 90–100 and mass

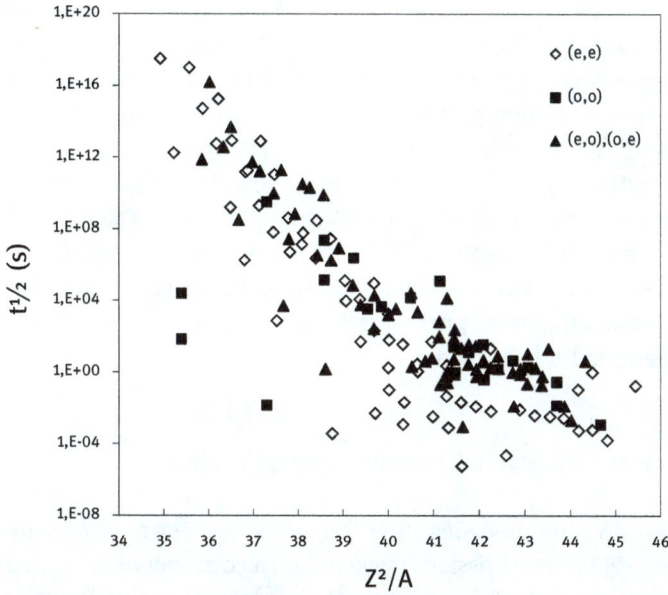

Fig. 10.15: Half-lives of nuclides undergoing spontaneous fission depending on the quotient Z^2/A. (Even, even), (odd, odd) and (even, odd) or (odd, even) nucleon compositions are identified.

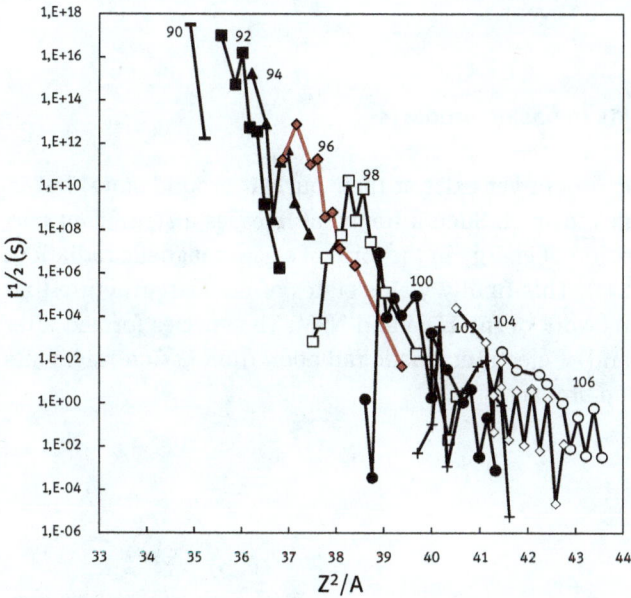

Fig. 10.16: Half-lives of nuclides undergoing spontaneous fission depending on the quotient Z^2/A along a constant proton number Z, representing isotopes of one and the same element. Z = 90 (Th), Z = 92 (U), Z = 94 (Pu), Z = 96 (Cm), red line, Z = 98 (Cf), Z = 100 (Fm), Z = 102 (No), Z = 104 (Rf), Z = 106 (Sg).

numbers of A = 230–250. For the two initial fragments of the spontaneous fission of ^{238}U, for example of A = 138 and A = 100, these ratios are ≈1.47 and ≈1.34 only, respectively.[17] However, the large excess of neutrons characteristic for the very heavy nuclide is distributed between the two primary fragments. Consequently, the two primary fission fragments ^{A2}K2 and ^{A3}K3 also have an excess of neutrons much too large for their intermediate mass. Primary fission fragments are thus not ensembles of nucleons arranged to satisfy the LDM or NSM. The terminology "fragment" rather than "nucleus" or "nuclide" is preferred for this reason and approaches are required to get rid of the excess of neutrons. There are two approaches for handling the excess in the *primary* fission fragments, namely separation of neutrons or successive transformation of a neutron into a proton.

10.6.2 From primary to secondary fission fragments: Prompt neutrons

Within a sequence of post-division cascades, first individual *prompt* neutrons are ejected, turning an ^{A2}K2 or ^{A3}K3 *primary fission fragment* into a corresponding $^{(A2)-1}$K2 or $^{(A3)-1}$K3 *secondary fission fragment*. This process takes place at about 10^{-14} s post division. The number of neutrons ejected is different for each mass number of the many primary fragments and depends on the individual excess of neutrons a primary fragment had when it incorporated a part of the neck. On average, there are two to three prompt neutrons per fission event.

10.6.3 From fission fragments to fission products

Secondary fission fragments do not yet exist at their nuclear ground state, but in highly enriched nuclear states instead. Such a fragment releases a significant part of its energy directly by emission of energy in the form of electromagnetic radiation, namely γ-rays (see Chapter 11). This finally yields states of nuclear structure that can be discussed in the framework of the LDM and NSM. The species formed after the separation of neutrons and of electromagnetic radiation from fission fragments are now called primary fission *products*.

---

17 Values for nuclides such as $^{235}_{92}$U$_{143}$, $^{238}_{92}$U$_{146}$, $^{251}_{98}$Cf$_{153}$ are 1.55, 1.59 and 1.56, those for $^{131}_{53}$I$_{78}$, $^{137}_{55}$CS$_{80}$ are 1.47 and 1.49, and those for $^{90}_{38}$Sr$_{52}$, $^{99}_{42}$Mo$_{55}$ are 1.37 and 1.36.

Tab. 10.4: Properties of initially formed fission fragments and the subsequent processes.

Primary fission fragments . . .	Consequences	Mass number A	Emission of
Have a neutron excess and are (according to their mass number A and excess of neutrons over protons) far from the optimum nucleon binding energy of the corresponding isobar, leading them to . . .	Relax a neutron by neutron separation	Lowering mass number A	*Prompt* neutrons
	Eject a neutron whenever $Q_\beta > \delta\bar{E}_B$ of the "last" neutron	Lowering mass number A	*Delayed* neutron
	Transform a neutron into a proton	Along A = constant	β^--particles + photons etc.
Have an excess of energy, and . . .	Release energy	Along A = constant	Electromagnetic radiation

10.6.4 From primary fission products to secondary fission products

The primary fission products created are now well characterized by nuclide termi-nology, namely mass number, proton number and neutron number. They have nu-cleon shell occupancies as described by the nuclear shell model. But these products are still neutron-rich. Because they are of mass number <200, further stabilization occurs *via* β^--transformation. The transformation optimizes the ratio of neutrons-to-protons, thereby increasing the mean nucleon binding energy along an isobar. It occurs at the left side of the β-transformation isobar parabola according to Fig. 6.11 (see Fig. 10.17).

10.6.5 Delayed neutrons

Within the course of subsequent, chain-like β^--transformations along an isobar line, an excess neutron may not be converted into a proton, but directly released from the nucleus.[18] A special feature is that the unstable fission product formed in a preceding β^--transformation step may not immediately eject that neutron, but "rest" for a certain period of time. This is the reason for calling such a neutron "de-layed". Delayed neutrons are ejected later than prompt neutrons.[19] An extreme

---

18 This happens if the binding energy of the "last" neutron is less than the Q_β-value of the corre-sponding $n \rightarrow p + \beta^- + \bar{\nu}_e$ pathway, as per eq. (9.21).
19 This effect becomes particularly relevant in the context of induced nuclear fission, as it is the tool to actively control the regime of a nuclear reactor – see Chapter 11 of Volume II.

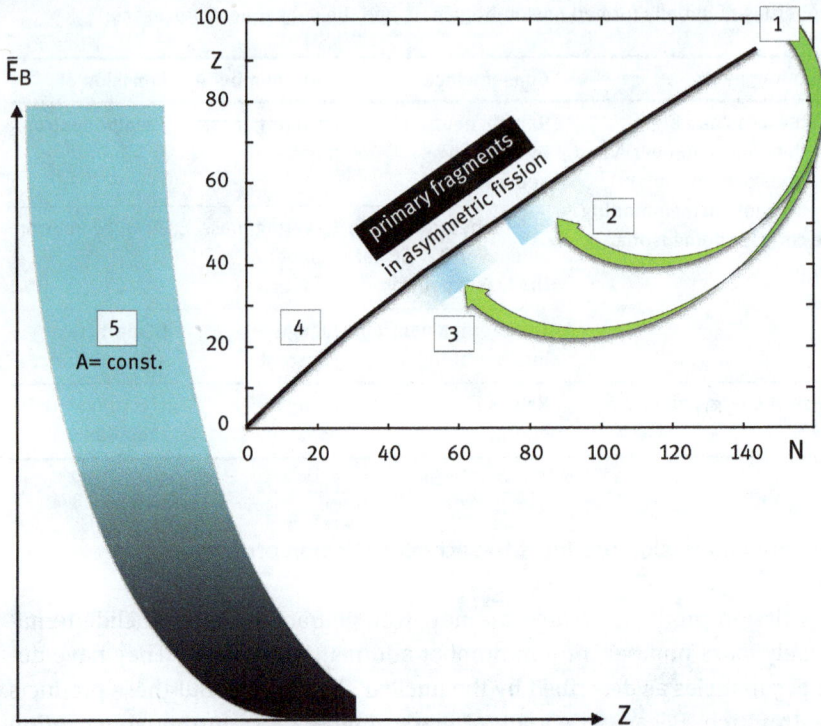

Fig. 10.17: Excess of neutrons in primary fission product leads to subsequent β^--transformation processes. (1) A heavy nuclide divides, forming large (2) and small (3) primary products. These have an excess of neutrons and are more or less far away from the line of β-stability (4). The excess of neutrons over protons is managed by converting a neutron into a proton, thereby optimizing values of $\bar{E}_B$ along an isobar parabola (5).

example is ^{87}Br, which transforms *via* β^--transformation *and* releases a neutron with a half-life of 55.7 s.

10.6.6 From secondary fission products to stable product nuclides

According to the characteristics of the β^--transformation, a stable nuclide is finally reached, representing the valley of β-stability. Thus, the initial event of spontaneous nuclear fission does not form stable nuclides; however, due to a cascade of follow-up transformations, all the initially formed fragments are converted into stable[20] nuclides.

---

20 There is just one exception – technetium. No stable isotopes of this element of Z = 43 exist.

10.6.7 Time scale

The successive steps involved in follow-up processes of spontaneous fission are characterized by individual velocities listed in Tab. 10.5 and illustrated in Fig. 10.18.

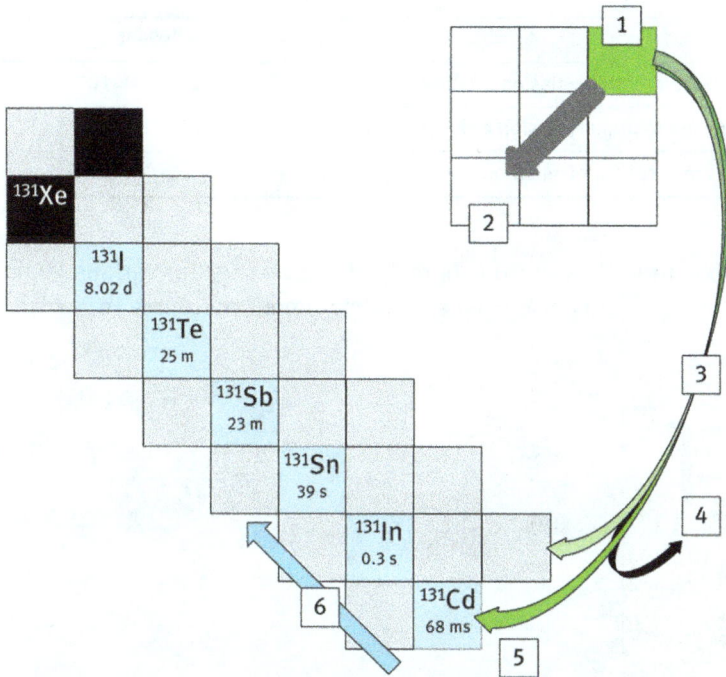

Fig. 10.18: Successive steps following initial fission events for binary fission. The division of a fissile nuclide (1) (in parallel to α-emission (2)) creates a larger primary fragment (3) plus a smaller primary fragment (not shown). This happens within ca. 10^{-14} s. These primary fragments de-excite within a period of about 10^{-11} s and separate some *prompt* neutrons at high kinetic energy (4). Only at this stage do primary nuclear products appear (5), which next continue to stabilize along an isobar line by successive β^--transformation (6), thereby forming secondary products as exemplified for mass number $A = 131$. The transformation along this isobar includes e.g. ^{131}I, and terminates at the stable nuclide ^{131}Xe.

10.7 Example applications

Spontaneous fission has a limited number of applications. Nevertheless, it is of practical relevance because high-energy neutrons are released at a constant flux. A radioisotope spontaneously transforming through fission thus represents a neutron "generator". A neutron generator may consist of ^{252}Cf; for example, a 2,645 years half-life nuclide undergoing spontaneous fission by ca. 3%, as per Fig. 10.19. It may

Tab. 10.5: Orders of magnitude of events *leading* to fission and *following* fission.

Process		Time
	Before fission event	
Necking	–	$\approx 10^{-20}$ s
Division	After fission event	
Primary fission fragments to secondary fission fragments	+	$\approx 10^{-14}$ s
Secondary fission fragments to primary fission products	+	$\approx 10^{-11}$ s
Primary fission products to secondary fission products	+	$> 10^{-3}$ s

be installed in a laboratory as a neutron source and can be used for various kinds of research and training. ^{252}Cf is also used as a primary neutron source in nuclear reactors.

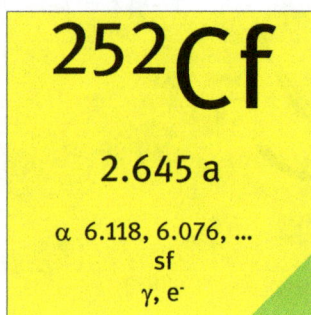

^{252}Cf

2.645 a

α 6.118, 6.076, ...
sf
γ, e⁻

Fig. 10.19: Transformation properties of the spontaneous fission-based neutron source ^{252}Cf.

10.8 Outlook

10.8.1 Isomers in spontaneous fission

Deformation of an unstable nuclide *K1 may yield a new nuclear ground state. For this degree of deformation, higher energy states may exist, and the nucleus may exist at an excited state.[21] From this excited state, de-excitation may proceed "backwards" to the ground state, but also "forwards" to spontaneous fission as first detected by S M POLIKANOV et al. in 1962.

---

[21] Conceived of as "super-deformed" due to an extreme ratio of about 2:1 between the long and short axes of the rotational ellipsoid.

These excited states represent *isomers* of the same nuclide, and their process of spontaneous fission is characterized by their own parameters of transformation. Similar to the processes described above, they experience tunneling from the excited state through the fission barrier to the scission point, as per Fig. 10.20. Because of differences between the ground and excited states energies, as well as the corresponding energies of the fission barriers, the half-lives of spontaneous fission isomers are different from the half-lives of the ground states. For example, the half-life of ground state ^{238}U is $4.468 \cdot 10^9$ years, while the half-life of its spontaneous fission isomer is 128 ns.[22]

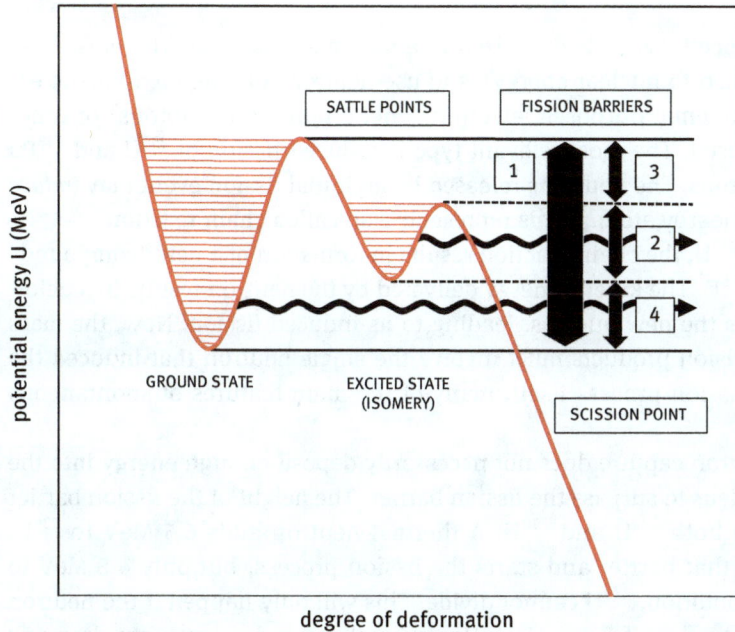

Fig. 10.20: Concept of isomers in spontaneous fission. The excited state may remain for a given period of time as a fission isomer. It proceeds further towards the scission point by tunneling through the corresponding fission barrier (2), which is lower than the fission barrier for the ground state (1). As this saves energy (3 and 4), spontaneous fission starting from a fission isomer state proceeds faster compared to the ground state situation.

---

22 The existence of these fission isomers is indicated by white fields within the nuclide boxes of the Chart of Nuclides, as per Fig. 10.4.

10.8.2 Spontaneous and induced fission

Spontaneous fission is a naturally occurring phenomenon, though only for a limited number of naturally occurring unstable nuclides, and for these few nuclides with a very low branching relative to α-emission. However, the fission of a "fissile" nuclide may be significantly facilitated by delivering various kinds of energy to the "candidate" nucleus. Consequently, this process is called *induced* fission. Induced fission suggests that a fissile nucleus receives *external* activation energy to achieve the conformation needed for nuclear division. It covers very important fission pathways of large nuclides, driven by artificial nuclear reactions involving ^{232}Th, ^{235}U and ^{239}Pu.[23]

In reality, induced fission is organized in a man-made fashion. The most common version (applied to nuclear energy) is to use neutrons of a defined kinetic energy. For more academic purposes, electromagnetic radiation (photons) or other vehicles are also used. The most relevant type is induced fission of ^{235}U and ^{238}Pu with thermal neutrons. The neutrons released by an initial fission event may *induce* fission at another nearby atom.[24] This represents a so-called chain reaction.

In the case of ^{235}U, the chain reaction results in formation of a new "compound" nucleus, namely ^{236}U. The kinetic energy delivered by the neutron results in a deformation and excites the new nucleus, leading to its induced fission. Now, the mass balance of the division products must involve the single neutron that induced the process. For the fission process itself, many of the main features of spontaneous fission apply.

However, neutron capture does not necessarily deposit enough energy into the newly formed nucleus to surpass the fission barrier. The height of the fission barrier is $H_{fb} \approx 6$ MeV for both ^{235}U and ^{238}U. A thermal neutron adds 6.5 MeV to ^{235}U, which overcomes that barrier and starts the fission process, but only 4.8 MeV to ^{238}U. Under this condition, ^{238}U cannot divide. This will only happen if the neutron delivers a kinetic energy of $E_n > 1$ MeV. However, these and many more dramatic aspects deserve very special consideration. Induced fission is covered in Chapter 13 in the context of "nuclear reactions" and Chapter 11 of Volume II in the context of "nuclear energy".

---

23 Induced fission was first experimentally realized by German chemists HAHN (Nobel Prize in Chemistry to HAHN in 1944 for "his discovery of the fission of heavy nuclei") and STRASSMANN and immediately accompanied by the theory of MEITNER and FRISCH.

24 However, several parameters can practically prevent this chain process: the kinetic energy of the neutrons released, their number (flux or density in terms of neutrons per second and cm^2) as well as the number of ^{235}U nuclides present within the range of the neutron initially emitted. There is evidence that, due to specific local circumstances, such a process indeed happened in the uranium mines of Oklo, Gabon (Africa) about $2 \cdot 10^9$ years ago.

11 Secondary transformations

Aim: β-processes, α-emission and spontaneous fission are primary transformations that do not necessarily lead to the ground state of the new nuclide K2 being formed. Instead, individual excited states $^{\ominus i}$K2 are populated in many cases, defined by nucleon shell occupancies different from the ground state. Similar to excited electrons of an atomic shell, which return to the ground state shell according to the nominal electron shell profile of the atom in its corresponding chemical state under emission of electromagnetic radiation, protons or neutrons populating energetically excited states of a nucleus "fall" towards lower energy levels until they finally may reach the ground state shell occupancy of the nucleus K2.

Because the values of Z, N and/or A do not change as they do during primary transformations, these transformations are "secondary" ones. They may be classified as "transitions" in the same nucleus, rather than transformations. Their main characteristic is the amount of energy ΔE between individual nuclear states, which in most cases is emitted as electromagnetic, i.e. γ-radiation. These γ-rays are the most commonly used means of identifying radioactive transformations and the nuclides involved. For several thousand unstable nuclides, their γ-spectra are in a way the most obvious "fingerprint" of radioactive transitions. This chapter introduces the origin and the various manifestations of these photons.

In addition, de-excitation may proceed without emission of electromagnetic radiation. Instead, the difference in energy between two nuclear states is used either to release a shell electron (internal conversion) or to create a pair of one electron and one anti-electron (pair formation).

11.1 From primary to secondary transformations

The *primary* processes of stabilization of an unstable nuclide do not always represent the total process of transformations of type *K1 → K2. For most of the radioactive nuclides, the initial primary event is followed by *secondary* processes addressing the fact that typically enriched levels of $^{\ominus i}$K2 are formed in the initial primary step – which need to de-excite. This de-excitation occurs mainly *via* photon emission, but there are two alternative de-excitation modes inducing different kinds of radiative emissions. Namely, they are internal conversion electrons and electron + positron pairs. These processes are responsible for even more radiative emissions not constituting the basic features of primary transformations.

The simultaneous emission of photons or conversion electrons or electron + positron pairs in secondary transformations as a follow-up phenomenon to the primary transformations with the emission of beta electrons, alpha particles or neutrons broadens the spectrum of individual radioactive emissions. Actually, Chapter 12 will

https://doi.org/10.1515/9783110742725-011

introduce even more of new modes of emissions. The whole spectrum of these different emissions thus must be considered in order to characterize a specific unstable nuclide and to understand what is called "radioactivity".

Unstable nuclei convert into stable ones by minimizing their absolute mass, which is mirrored in terms of nucleon binding energy. Primary processes modify the proton, neutron and/or overall mass numbers of the nuclides involved. Most of the primary processes are accompanied by the emission of small components x, i.e. beta particle, electron neutrino and electron antineutrino, alpha particle or neutron. Figure 11.1 illustrates the main directions of transformation of an unstable nuclide within the Chart of Nuclides and their respective major emissions. In contrast, the defining feature of secondary transformations is that the values of Z, A and N do not change, as per Table 6.3.

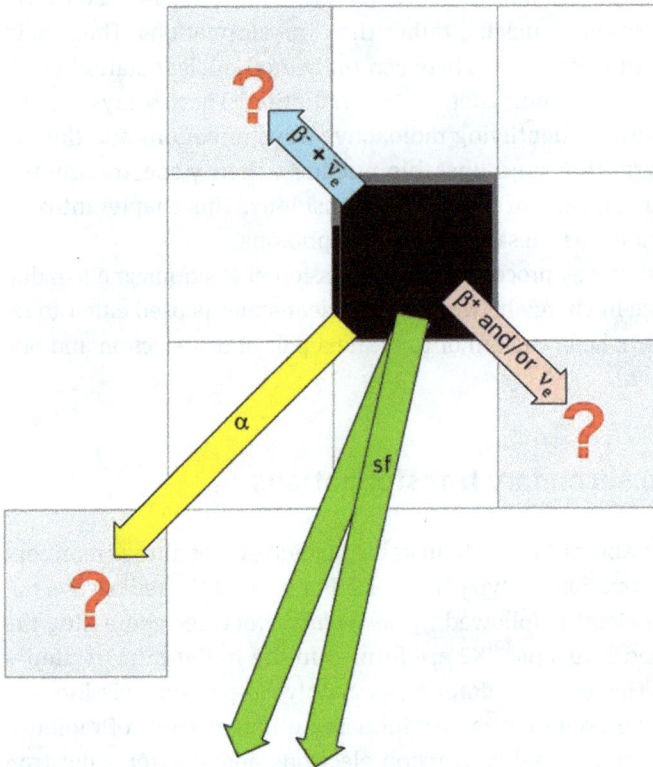

Fig. 11.1: Primary transformations of unstable nuclides indicating the three major emissions relevant to naturally occurring nuclides. The illustration indicates the changes of nuclide positions within the Chart of Nuclides due to changes in nucleon compositions.

11.1.1 From primary to secondary transformations

In many cases, a primary transformation of *K1 does not lead (and cannot lead, see below) to the ground state of the new nuclide K2, eq. (11.1), shown in Fig. 6.22. Instead, individual excited states $^{\odot i}$K2 are populated[1] as shown in eq. (11.2). Consequently, the newly formed nucleus $^{\odot i}$K2 may exist in an "excited" state for shorter or longer periods of time (see below), and the excited nucleons de-excite to levels of lower energy according to the shell model of the nucleus.

Excited electrons of an atomic shell return to the ground state shell according to the nominal electron shell profile of the atom in its corresponding chemical state under emission of electromagnetic radiation (see Chapter 1). In a similar fashion, the nucleons "move" towards lower-energy nuclear levels. The transitions may proceed from a higher-energy nuclear state to a lower-energy excited nuclear state (eq. (11.3)) and/or from a given excited state to the final ground state shell occupancy of the nucleus ^{g}K2 (eq. (11.4)). Secondary transformations proceed exothermically, and similar to conventional chemical reactions, exothermic processes yield an excess of energy ΔE. The corresponding emissions are denoted here as y. In most cases, y is electromagnetic energy (or photons), typically referred to as "γ-radiation".

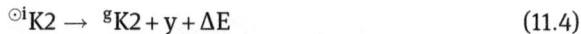

$$^*\text{K1} \rightarrow {}^g\text{K2} + x + \Delta E \tag{11.1}$$

$$^*\text{K1} \rightarrow {}^{\odot i}\text{K2} + x + \Delta E \tag{11.2}$$

$$^{\odot i}\text{K2} \rightarrow {}^{\odot j}\text{K2} + y + \Delta E \tag{11.3}$$

$$^{\odot i}\text{K2} \rightarrow {}^g\text{K2} + y + \Delta E \tag{11.4}$$

Secondary transitions exclusively start from excited levels of nuclei – see also Fig. 11.2. Suppose a primary transformation of *K1 leads directly to the ground state ^{g}K2 (eq. (11.1)), meaning no excited states of K2 are populated; in this case, no secondary process is needed at all. Thus, γ-emission in radioactive transformation processes is not a singular process, i.e. does not belong to the group of primary β-particle- or α-emissions.

---

1 In general, the larger the nucleus, the larger the number of possible excited states. These are defined by nucleon shell occupancies different to the ground state, and, consequently, by individual quantum mechanics characteristics such as overall spin numbers, parity, but also by energy.

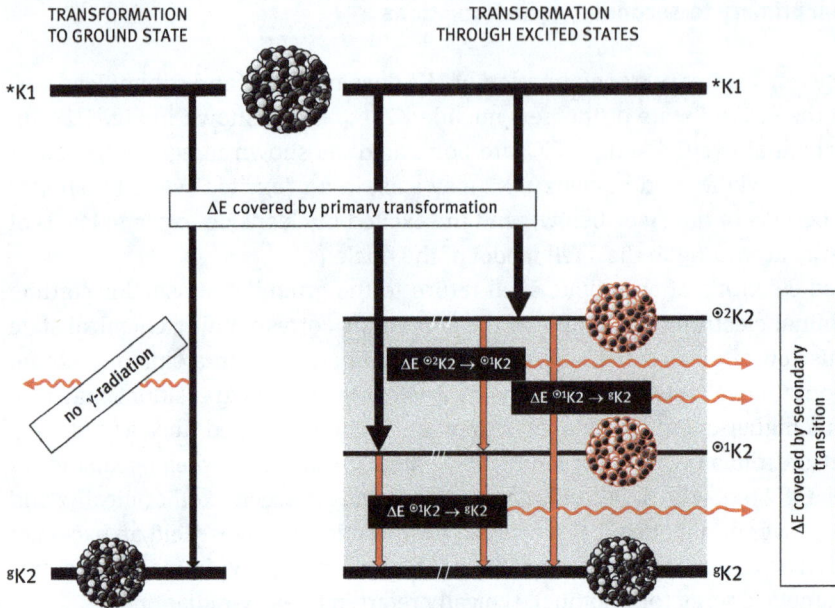

Fig. 11.2: Relation between primary and secondary transformations. Suppose a primary transformation of *K1 leads directly to the ground state of ^{g}K2; then, there is no γ-emission at all (left). Secondary transitions cover the transition from an excited level of $^{\odot i}$K2 to lower energy or ground state levels of *one and the same* nuclide K2. The difference in energy ΔE between these nuclear levels of K2 is typically emitted as γ-radiation, but also by other modes (right).

11.1.2 Options for secondary emissions

Altogether, there are three principal routes through which the difference in energy ΔE between different excited levels (or one exited level and the ground state) of a given nucleus can be expressed, as shown in Tab. 11.1.

Tab. 11.1: Three sorts of secondary processes transferring specific values of ΔE.

Transformation of ΔE into	Results in	Results in	Range of ΔE
Electromagnetic radiation	Emission of a photon	Of energy equivalent to ΔE	Any
Binding energy of (mainly) K- and L-shell electrons of K2	Ejection of one K- or L-shell "conversion" electron from the nuclide	Of energy equivalent to ΔE minus the electron binding energy	Any
Matter = "pair formation"	Creation of an electron + positron pair	Of energy equivalent to the mass of 2 electrons	Only ≥1.022 keV

The most common secondary emission is electromagnetic radiation (or photons) as γ-quanta, which are very valuable for the detection of radioactive transformation. The second option is the conversion of ΔE into the release of an already existing electron of that nuclide from its inner electron shell, creating a "conversion electron". The third option is the transformation of ΔE directly into matter, according to $E = mc^2$. It creates a pair of particles, representing matter (electron) and antimatter (positron). These two electrons existed in neither K2 nor in *K1; they are simply created.[2] Figure 11.3 illustrates the options.

PRIMARY TRANSFORMATION   SECONDARY TRANSFORMATION

| *K1 ⟶ ⊙iK2 + x | ⊙iK2 ⟶ ⊙i-1K2 or gK2 + y |

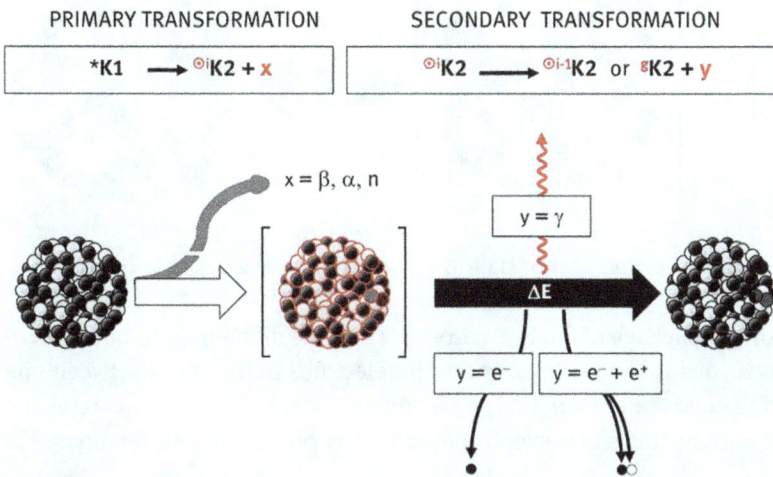

x = β, α, n

y = γ

ΔE

y = e⁻   y = e⁻ + e⁺

Fig. 11.3: Illustration of the successive character of primary and secondary transitions and summary of the three options of secondary processes. Secondary processes can proceed via emission of photons, release of internal conversion electrons and/or creation (formation) of an electron + positron pair. Note that all three pathways exclusively originate from excited states of ⊙K2 formed by primary nuclear transformations (or nuclear reaction processes). They proceed within one and the same nuclide and may occur simultaneously.

---

2 The opposite effect will be discussed in Chapter 12, namely the metamorphosis of matter into electromagnetic energy: matter (electron) and antimatter (positron) combine to 1.022 MeV energy. This is called "annihilation".

11.2 Photon emission

11.2.1 Similarities in photon emission from shell and nucleus of an atom

Similar to electron transitions in the shell of an atom, each nuclear transition has its individual and discrete amount of energy, Fig. 11.4. The difference in energy released in terms of electromagnetic radiation may be defined as $^{\odot i}\Delta E = Q_\gamma$.

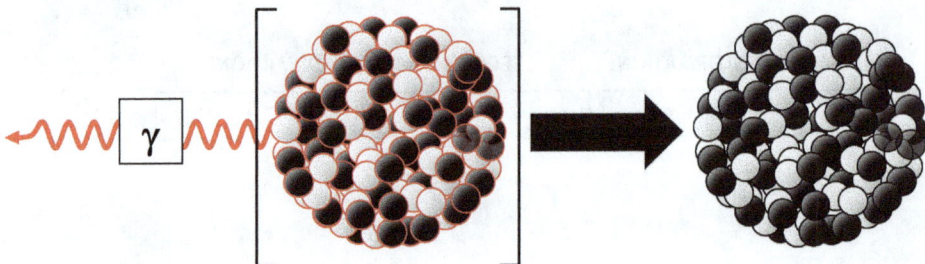

Fig. 11.4: De-excitation of an excited nuclear state to a ground state level *via* emission of a photon.

The photon is thus a package of energy, carrying away the differences in binding energy (here for nucleons in the nucleus, earlier for electrons in the shell) between one discrete excited nuclear state to another, lower energy one. Figure 11.5 illustrates the de-excitation of nuclear states of nucleons populated by primary transformations.

11.2.2 The γ-emission

The primary transformation of unstable nuclei *K1 into more stable or really stable nuclei, i.e. the change in nucleon composition, is a more or less "invisible" process. In contrast, the simultaneous ejection of beta electrons or alpha particles during primary transformations,[3] and now photons emitted during secondary transformation processes, is directly measurable. This is summarized in the well-known picture of the emissions α, β and γ, Fig. 11.6. The emissions can be detected quantitatively and used to identify the kind of radionuclide and its absolute activity. However, γ-emissions[4] are those most commonly used in nuclear and radiochemistry.

---

3 Of course, there are several more emissions already introduced, such as the positron and electron-(anti)neutrino (emitted in the course of β+- and electron capture transformations), clusters larger than the α-particle, and protons and neutrons as ejected from extremely proton-rich or neutron-rich artificially produced nuclides close to the corresponding nucleon drip lines.
4 Historically, these radiations are called α, β and γ, and have been identified in order of discovery from naturally occurring unstable nuclides. Following the Greek alphabet, the first radiation characterized was α-emission. The next generation of emission analyzed this way was named β. The

NUCLEON SHELLS
transition of nucleons
to lower-energetic levels of the nucleus
following a preceding primary nuclear transformation

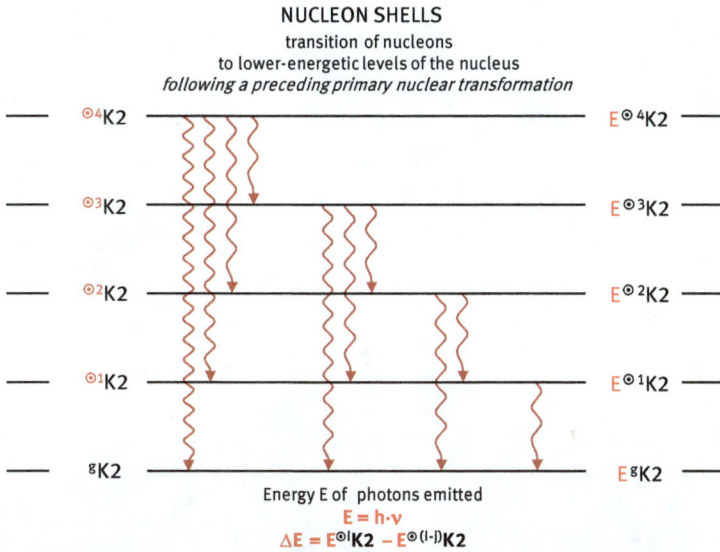

Energy E of photons emitted
$$E = h \cdot \nu$$
$$\Delta E = E^{\odot I}K2 - E^{\odot (I-i)}K2$$

Fig. 11.5: Transition of nucleons (i.e. protons or neutrons) from excited nuclear states to lower energy excited ones or to the ground state of the nucleus. Each nuclear state represents a characteristic profile of nucleon shell occupancies. Following a primary nuclear transformation, a nucleon may populate a higher energy (sub)shell and relax to a lower energy shell in a de-excitation cascade, finally reaching the initial ground state. The individual differences in binding energy are released as photons. This is comparable to the behaviour of excited electrons within electron shells, as in the hydrogen emission spectrum.

According to the standard model, the photon[5] is an elementary particle. It belongs to the field quanta and is the mediator of the electromagnetic force, discussed in Chapter 7. Unlike other mediators, its mass is zero; still, it obeys the wave-particle duality of quantum mechanics. Because the mass is zero, it cannot "decay" to a lower mass (i.e., lower energy) state, making it stable against transformation (unlike the unbound neutron). It has zero charge and therefore no antiparticle. Its spin is 1, which makes it a boson and its parity is −1.

---

final one categorized this way was the γ-emission, identified in 1900 by P VILLARD when measuring the different types of radiation emitted from radium. The term "γ-rays" was introduced by E RUTH-ERFORD in 1903 (in fact, RUTHERFORD named them all: α, β and γ emissions). Photons are symbolized by the Greek letter γ in nuclear sciences. In other sciences such as chemistry, the expression hν is used.

5 φῶς (Greek) = phôs = light.

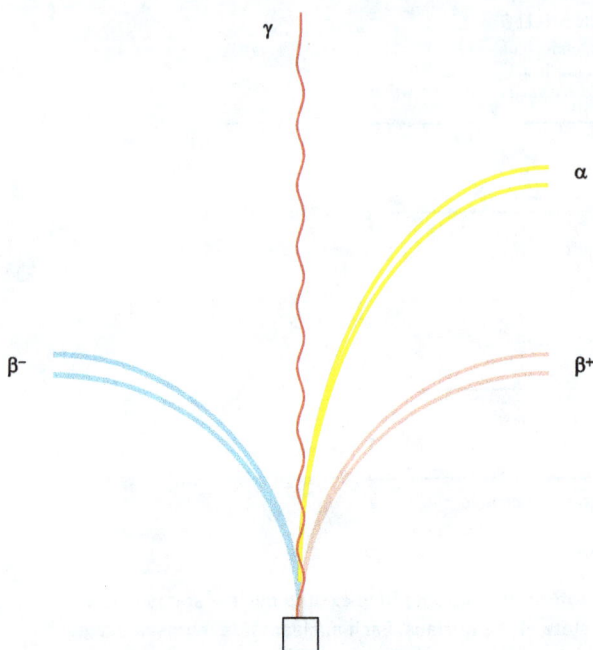

Fig. 11.6: Most relevant "radioactive" emissions (or "radiative" emissions) in nuclear and radiochemistry. They are schematically categorized in terms of deviation in an electromagnetic field, reflecting the charge and mass of each emission. For γ-emission, there is no charge, no mass and thus no deviation. In contrast, α-particles are positively charged and have a relatively heavy mass, while β⁻-electrons are negatively charged and about 4000× less massive than the α-particle.

11.2.3 Photons and X-rays

The photon is the quantum of electromagnetic radiation; this is true for all spectral regions, including visible, ultraviolet and infrared light, radio waves, microwaves, and those waves originating from nuclear transformations. Photons travel with the maximum velocity of the universe, which is one of the fundamental constants in science: the speed of light $c = 299,792,458$ m/s, which is $1.079 \cdot 10^9$ km/h. The energy of a photon can be related to its frequency v, wavelength λ ($v = 1/\lambda$), the PLANCK constant h, angular frequency ω, the reduced PLANCK constant $\hbar = h/2\pi$ and the speed c of photons ($v = c/\lambda$) via eq. (11.5).

$$E = h v = h \frac{c}{\lambda} = \hbar \omega \tag{11.5}$$

The spectrum of electromagnetic radiation is huge, as shown in Fig. 11.7. In terms of wavelength (λ in m) or frequency (v in s⁻¹, which is Hz) and in terms of energy (in eV), it covers many orders of magnitude: from λ ca. 10 pm to 1 km, from v ca. 10^{20} to 10^5 Hz, and E from keV to MeV. Only a limited part of the spectrum is visible to the human eye: about 400 to 700 nm in wavelength. This is neighbored by the ultraviolet and infrared regions, i.e. down to about 10 nm and up to about 10 µm, respectively. Beyond and towards longer wavelengths, there is, for example, the domain of radio waves covering an area of about 1000 MHz (UHF) to about 50 MHz (VHF). In the context of radioactive secondary transitions and post-processes, the high frequency and energy (i.e., short wavelength) branch of the electromagnetic spectrum becomes relevant. It is divided into two sub-parts: γ-radiation and X-rays. A good reason to make a distinction here is not the value of frequency, wavelength and energy, but rather their origin.

11.2.3.1 γ-rays
Gamma radiation[6] is understood to originate from the nucleus of a nuclide *via* de-excitation of an excited level of the nucleus. In terms of wavelength, it is approximately $<10^{-11}$ m (which is <10 pm or <0.1 Å, i.e. shorter than the diameter of a nucleus); in terms of energy, it is >0.1 MeV, as summarized in Tab. 11.2.

11.2.3.2 X-rays
In the framework of radioactive processes, electromagnetic radiation emitted from the shell of the atom (i.e., not from the nucleus) is signified as X-radiation.[7] Compared to γ-rays, the wavelengths are longer (ca. 0.01–10 nm), and energies are lower (ca. 0.1 keV to ca. 0.1 MeV). Thus, the domain of X-rays is between that of γ-rays and

Tab. 11.2: Regions of wavelength, frequency and energy of γ- and X-rays.

	γ-Rays	X-rays		
		General	Hard	Soft
Wavelength	<10 pm (10^{-11} m)	$\approx$0.01 – 10 nm (10^{-11} m – 10^{-8} m)		
Frequency	$>10^{19}$ Hz $<2\cdot10^{21}$ Hz	$>10^{16}$ Hz $<2\cdot10^{19}$ Hz		
Energy	>100 keV <10 MeV	>0.1 keV <0.12 MeV	>10 keV	<10 keV

6 Synonyms: γ-rays and γ-radiation.
7 When RÖNTGEN discovered this new phenomenon in 1896, he called it "X" because of its unknown nature at that time. The Nobel Prize in Physics, 1901 (the very first Nobel Prize), was awarded to W C RÖNTGEN "in recognition of the extraordinary services he has rendered by the discovery of the remarkable rays subsequently named after him".

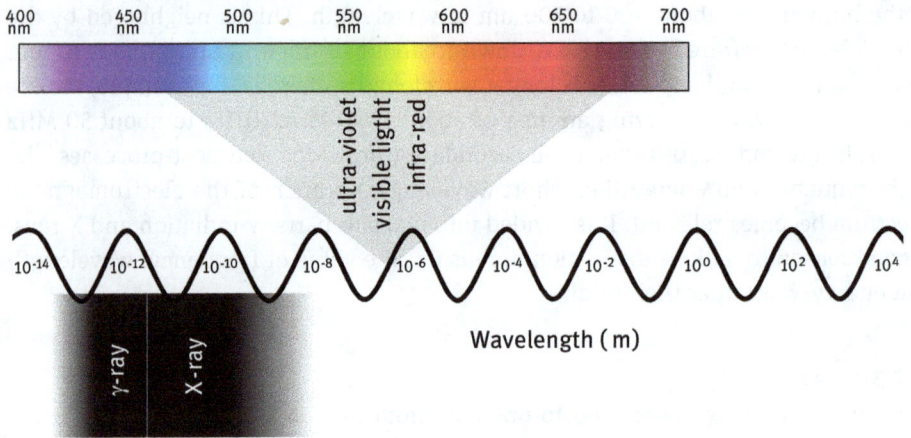

Fig. 11.7: Positioning of X- and γ-rays within the electromagnetic spectrum.

UV light. In terms of energy, X-rays are also discriminated by their penetration power into "soft" X-rays (energies < 10 keV) and "hard" X-rays (of about 10 to 120 keV).

11.2.4 Recoil energy

An ejected γ-photon contains almost all the energy released in a transition between two nuclear states. Similar to the small fragment x released in a primary transformation (x = α- and β-particles), the nucleus also experiences a relatively low recoil energy during secondary processes. The overall energy available within the transition is Q_γ in the present case, and is between ca. 0.1 and 10 MeV. The value for $^{\text{RECOIL}}E_{K2}$ is derived from the conservation of the two momenta as $p_\gamma = p_{\text{recoil}}$. With $E = h\nu$ and $p_\gamma^2 = h\nu/c$, eq. (11.6) gives a relationship between the mass number A of the recoil nucleus and the energy of the photon emitted:

$$^{\text{RECOIL}}E_{K2} = \frac{p_\gamma^2}{2m_{K2}} = \frac{E_\gamma^2}{2cm_{K2}} \approx \frac{E_\gamma^2}{2\text{GeV } A_{K2}} \tag{11.6}$$

The correlation is easy to understand through classical mechanics. If the nucleus is large and the energy of the emitted photon is small, the recoil of the nucleus is almost negligible on an eV scale. In contrast, for light nuclei and large photon energies, the recoil energy may approach keV scale.

11.2.5 Selection rules for photon emission

Figure 11.5 shows a hypothetical cascade of secondary transitions involving three states of K2: two excited states ($^{\odot 2}$K2 and $^{\odot 1}$K2) and the ground state (^{g}K2). Starting from *K1, there are six possible transitions,[8] namely:

1. *K1 → $^{\odot 2}$K2,
2. *K1 → $^{\odot 1}$K2,
3. *K1 → ^{g}K2,
4. $^{\odot 2}$K2 → $^{\odot 1}$K2,
5. $^{\odot 2}$K2 → $^{\odot 1}$K2,
6. $^{\odot 1}$K2 → ^{g}K2.

The question is: do these six options proceed with identical probabilities, or are some transition steps preferred over others? If the latter is true, then why? Indeed, all the potential steps have an individual branching. The reason for this defined protocol takes us back to the nuclear structure of the individual excited states, back to the shell model of the nucleus, i.e. back to the intrinsic quantum physical parameters of the nucleons and the resulting overall parameters of the nuclear state: overall orbital momentum (spin) and parity. Similar to β- and α-transformations, the process in secondary photon transitions is also defined by initial and final wave functions and a transition matrix element. The quantum physical parameters needed for each initial and final state and for the differences ΔE between the two states are:

– the orbital quantum number l,
– the magnetic orbital quantum number,
– the spin quantum number s and the resulting overall orbital momentum L,
– the overall angular momentum S,
– the overall momentum J obtained from orbital-spin coupling.

11.2.5.1 Overall momentum
Figure 11.8 identifies the differences in J between the initial and final nuclear states and the parities Π of the two states. Differences in overall momentum J between initial and final nuclear states are given according to eq. (11.7). Consequently, the photon's task is to carry away the equivalent of the quantum parameters described by ΔJ.

$$\left|\left(J_i - J_f\right)\right| \leq nJ \leq \left(J_i - J_f\right) \tag{11.7}$$

---

8 The higher the mass number of an unstable nucleus (which reflects high Z and N numbers), the higher the number of potential excited levels, and the more possible secondary transition steps.

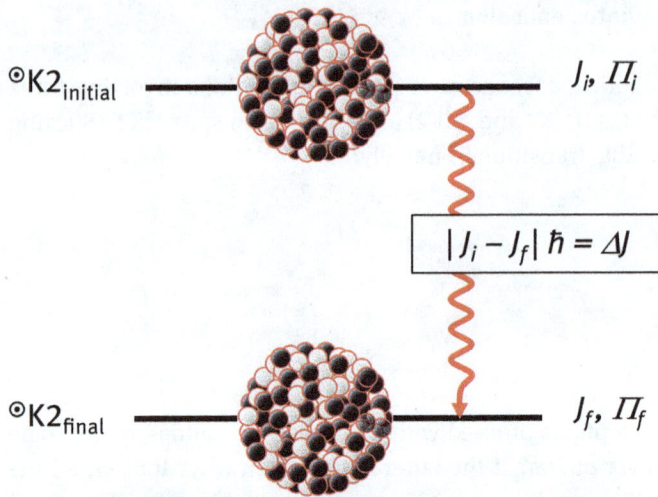

Fig. 11.8: De-excitation between two excited nuclear states defined by quantum physical parameters J and Π for the initial and final states. The transition *via* photon emission proceeds for specific values of ΔJ, and the photon must carry away this difference in overall momentum.

11.2.5.2 Electromagnetic multipolarity

The nucleus of an atom represents an ensemble of neutral and positively charged particles, the latter inducing electric fluxes. Within the dense matter of a nucleus, magnetic fields occur. The orbital momentum L and the overall momentum J determine the character of the corresponding field. Transitions between two excited nuclear states are always transitions between different electromagnetic states, and the difference is handled by the photon emitted according to multipole electromagnetic orders. These are derived as 2^l with $l > 0$ and are multipoles (MP), e.g. dipoles ($l = 1$), quadrupoles ($l = 2$) etc.[9]

Whenever *electromagnetic* radiation (i.e. the photon) is emitted from a multipole, it manifests as either an *electric* (E) or *magnetic* (M) variant. Electric dipole radiation (E_1) for example may be understood as caused by elliptic vibration of a homogeneous ensemble of nucleons. Magnetic quadrupole radiation (M_2) originates from deviations between the directions of magnetic moments and the overall momentum.[10]

---

9 Nuclei with overall orbital momentum 0 show a monopole electromagnetic moment, and because the angular momentum of a photon is 1, these transitions are impossible.
10 This defines a gyromagnetic parameter, which is the ratio (or proportionality factor) between a magnetic dipole moment and an angular momentum of a system (given by units of C / kg).

11.2.5.3 Conservation of orbital momentum

Whenever a secondary transition is accompanied by photon emission, i.e. as $^{\odot i}K2$
$\rightarrow$ $^{\odot\,(i-1)}K2 + \gamma$, the process must conserve quantum physical parameters. For the
orbital momentum, the spin is $J(^{\odot i}K2) = J(^{\odot\,(i-1)}K2) + l_\gamma$. The photon thus takes away
a spin of $l = 1, 2, 3$ etc.

There are two consequences:
1. Such a transition is impossible in cases where $J(^{\odot i}K2) = 0$ and $J(^{\odot\,(i-1)}K2) = 0$. In
   other words, ΔJ must be > 0.
2. The lowest multipole order of a photon is $l = 1$, which determines a dipole radiation according to $2^1 = 2$.

11.2.5.4 Parity *Π*

The parity of the photon is negative, while the parities of the electron, neutron and
proton are positive. For each secondary transition, the parity of the total system
must be conserved. However, changes in overall orbital momentum between initial
and final states act in an alternating way for electric and magnetic multipole radiation, according to the photon's parity as either $(-1)^l$ or $(-1)^{l+1}$, respectively. This is
shown in eqs. (11.8) and (11.9).

$$E_l\ \Pi_i = (-1)^l\ \Pi_f \tag{11.8}$$

$$M_l\ \Pi_i = (-1)^{l+1}\ \Pi_f \tag{11.9}$$

Thus, for the same order of the photon's multipole, the parity of electric multipole radiation is opposite to the parity of magnetic multipole radiation. Consequently, multipole radiations of type E_1, M_2, E_3, etc. stand for transitions changing the parity, while
in the case of M_1, E_2, M_3, etc. the parity is not changed between initial and final nuclear states. Combinations of $\Delta\Pi =$ "no" and $\Delta J = 2, 4$, etc. determine electric MP radiation, while combinations of $\Delta\Pi =$ "no" and $\Delta J = 1, 3$, etc. determine magnetic MP
radiation, shown in Tab. 11.3. If changes in ΔJ between initial and final nuclear states
are integer or noninteger, there are two options depending on whether the overall par-
ity remains ($\Delta\Pi =$ "no") or changes ($\Delta\Pi =$ "yes"). Values of ΔJ define dipole and quadrupole etc. monopoles of the corresponding type. For example, $\Delta J = 1$ gives either E_1
or M_1, depending on whether the parity is changed ($\Delta\Pi =$ "yes") or not ($\Delta\Pi =$ "no").
Figure 11.9 illustrates three examples of photon transitions.

11.2.5.5 Selection rules

With overall orbital momentum, photon parity and either electric or magnetic radiation
types explained, specific selection rules can finally be applied. They result in "allowed" and "forbidden" transitions, summarized in Tab. 11.4. Electromagnetic emissions structured between electric and magnetic photon radiation and in particular
along 2^l-multipoles are of very different abundances. This is the case for $l = 1$ (dipole

Tab. 11.3: Types of photon multipole (MP) radiation in secondary transitions. $|\Delta J|$ indicates the absolute value of the difference.

l	2^l	Type of MP	Electric			Magnetic		
			E_l	$|\Delta J|$	$\Delta\Pi$	M_l	$|\Delta J|$	$\Delta\Pi$
1	2	di-	E_1	1	yes	M_1	1	no
2	4	quadru-	E_2	2	no	M_2	2	yes
3	8	octu-	E_3	3	yes	M_3	3	no
4	16	hexadeca-	E_4	4	no	M_4	4	yes

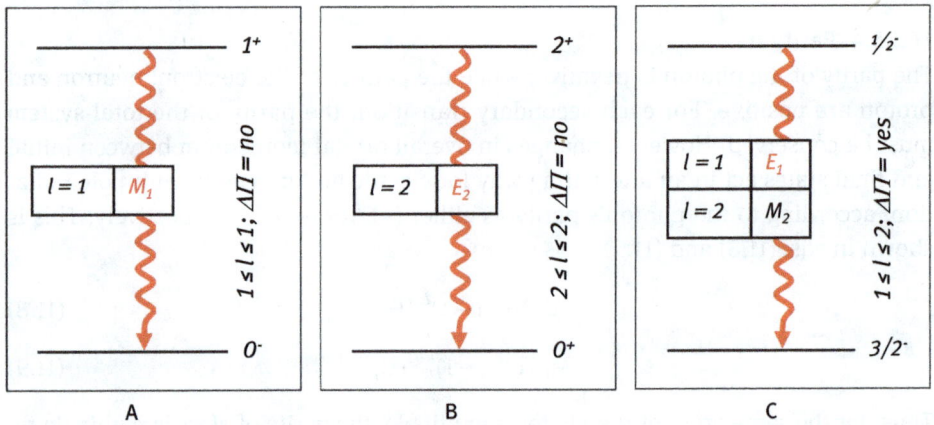

Fig. 11.9: Examples of photon transitions with different combinations of ΔJ and $\Delta\Pi$. For (A) $l=1$, $\Delta\Pi$ = yes and for (B) are $l=2$, $\Delta\Pi$ = no. For each of (A) and (B), there is only one option, namely M_1 and E_2, respectively. For the case (C) with $l=1$ and 2, and with $\Delta\Pi$ = yes, there are two transitions, namely E_1 and M_2; yet with a very different probability in favor of E_1.[11]

radiation) and $\Delta\Pi$ = "no". Yet, all of the transitions are allowed to occur, though with a very different branching. With increasing order 2^l, the probability of these photon emissions decreases successively.

[11] Example (A) could represent the transition of an excited level of ^{26}Mg (formed by primary processes of β^+- and ε-transformation from excited ^{26}Al, $t\frac{1}{2}=7.2\cdot10^5$ years) to its ground state with $t\frac{1}{2}=0.49$ ps. It emits a photon of $E_\gamma=1.809$ MeV, which is a sensitive measure of the astrophysical processes occurring in supernovae. Example (B) shows transitions which occur for instance for ^{58}Co $\rightarrow\beta^-\rightarrow{}^{58}$Fe. The two excited states are $^{\odot 2}(^{58}$Fe$)=2^+$ and $^{\odot 1}(^{58}$Fe$)=2^+$, while the ground state is of 0^+. The primary transformation populates $^{\odot 1}(^{58}$Fe$)$ with 99.48% efficiency. The main MP radiation is E_2. Example (C) is for two options in ΔJ, namely 1 and 2, and therefore two different transition options arise.

Tab. 11.4: Electric or magnetic MP radiation of the photon depends on changes in ΔJ and $\Delta \Pi$ between the initial and final nuclear states.

Electric MP radiation	Magnetic MP radiation	Allowed if	$\Delta \Pi$
$\Delta J = 2, 4, \ldots$	$\Delta J = 1, 3, \ldots$	$\Pi_i = \Pi_f$	"no" or " –"
$\Delta J = 1, 3, \ldots$	$\Delta J = 2, 4, \ldots$	$\Pi_i \neq \Pi_f$	"yes" or " +"

11.2.5.6 Transition probabilities for multipoles of different order

The selection rules refer to the important impact of the orbital moments. However, both the value of ΔE and the mass number A of the corresponding nuclide (*via* its radius $r = r_o A^{1/3}$) are involved. These parameters are semi-empirically linked in eqs. (11.10) and (11.11) through a constant S (eq. (11.12)),[12] yielding the transformation velocities of secondary γ-transitions. Individual velocities λ are a measure of transition probabilities: large values of λ (= short half-lives $t\frac{1}{2}$) would indicate high probabilities and *vice versa*. Typically, these half-lives are extremely short and cover nanoseconds and picoseconds.

$$\lambda_{el.MP} = \frac{2.4 S r^2}{(E/197)^{2l+1}} 10^{21} s^{-1} \tag{11.10}$$

$$\lambda_{magn.MP} = \frac{0.55 S A^{-2/3} r^2}{(E/197)^{2l+1}} 10^{21} s^{-1} \tag{11.11}$$

$$S = \frac{2(l+1)}{l\{135 \ldots (2l+1)\}^2} \left(\frac{3}{l+3}\right)^2 \tag{11.12}$$

Table 11.5 summarizes the resulting $t\frac{1}{2}$ values, depending on parameters of the multipoles and the energy of the γ-quanta emitted. The general trends are:

1. For same photon multipoles: The higher E_γ (which is $\approx \Delta E$) [1] the more abundant the transition. One order in ΔE makes a difference of many orders of magnitude in $t\frac{1}{2}$.[13]
2. For the same multipole order and same E_γ [2]: Electric transitions are faster than magnetic ones.
3. For the same E_γ but MP or EP increasing from e.g. dipole to quadrupole etc. [3] and [4], respectively: Transitions are slower, i.e. occur very seldom, by many orders of magnitude per each additional change in angular momentum.

---

12 Values of S are thus only a function of l. For $l = 1, 2, 3$ and 4 the corresponding values are $2.5 \cdot 10^{-1}$, $4.8 \cdot 10^{-3}$, $6.25 \cdot 10^{-5}$ and $5.3 \cdot 10^{-7}$, respectively.

13 . . . which is similar to those of primary transformations; see eq. SARGENT-type and GEIGER/NUT-TAL-type correlations.

Thus, the fastest transitions at high E_γ are those for E_1 (ca. 10^{-16} s); conversely, those for M_3 are very slow (ca. 10^{-4} s). For higher MP, transition velocities are not of seconds, but years and even thousands of years, i.e. proceed only with extremely low probability.

Tab. 11.5: Probabilities of individual γ-transitions, depending on parameters of photon multipoles and the energy of the γ-photons emitted. Half-life values are given for a nucleus of A = 100. See text above for [1], [2] and [3].

MP	ΔI	ΔΠ	t½ (s) at A = 100 and γ-energies of:		
			1.0 MeV	0.2 MeV	0.05 MeV
E_1	1	yes	$2 \cdot 10^{-16}$	$3 \cdot 10^{-14}$	$2 \cdot 10^{-14}$ [1]
M_1	1	no	$2 \cdot 10^{-14}$	$2 \cdot 10^{-12}$	$2 \cdot 10^{-10}$
E_2	2	yes	$1 \cdot 10^{-11}$ [2]	$3 \cdot 10^{-8}$	$3 \cdot 10^{-5}$ [4]
M_2	2	no	$9 \cdot 10^{-10}$	$3 \cdot 10^{-6}$ [3]	$3 \cdot 10^{-3}$
E_3	3	yes	$7 \cdot 10^{-7}$	$6 \cdot 10^{-2}$	$9 \cdot 10^{+2}$
M_3	3	no	$7 \cdot 10^{-5}$	$5 \cdot 10^{0}$	$8 \cdot 10^{+4}$
E_4	4	yes	$8 \cdot 10^{-2}$	$2 \cdot 10^{+5}$	$4 \cdot 10^{+10}$
M_4	4	no	$7 \cdot 10^{0}$	$3 \cdot 10^{+07}$	$4 \cdot 10^{+12}$

11.2.6 γ-spectra

Individual transitions between discrete excited states or from one excited state to the ground state show γ-emissions of individual and discrete energy. In a γ-spectrum, these are recorded according to their energy and relative abundance. Figure 11.10 illustrates the concept for some select transitions. Whereas the spectrum arranges the transitions according to increasing γ-energy (i.e. $^i\Delta E$ values), their intensities may be very different.

Let us discuss the secondary transformations following the primary β^--transformation of ^{131}I. The most frequently occurring primary transformation (89.4%) is the one populating the excited state ⊙4 (among 8 different nuclear levels of ^{131}Xe) – see Table 8.7. The de-excitations originating from this excited state will predominantly determine the γ-spectrum. The question is, which lower levels are addressed most often? The corresponding values of J^Π for levels 4, 3, 2, 1 and the ground state are $5/2^+$, $9/2^-$, $11/2^-$, $1/2^+$ and $3/2^+$, respectively (see Table 8.7).

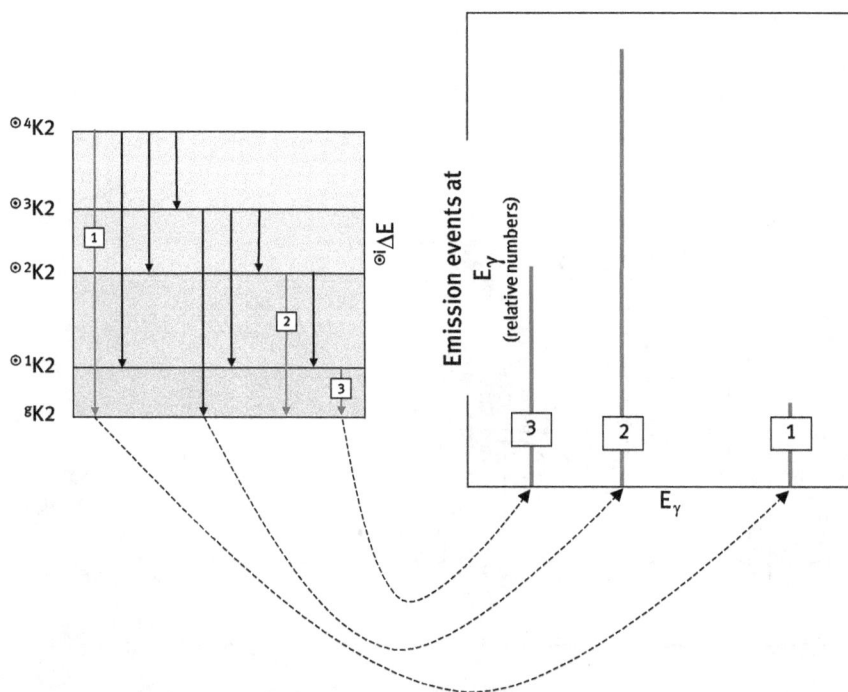

Fig. 11.10: The origins of a γ-spectrum. Transitions between discrete excited states or to the ground state show the hypothetical γ-emissions of individual discrete energy (left). In a γ-spectrum (right), the photons are recorded according to their energy (x-axis) and relative abundance or intensity (y-axis). In the artificial example given, three photon emissions are selected. The most energetic is (1) with $\Delta E = {}^{\odot 4}E_{K2} - {}^{g}E_{K2}$, but it proceeds at a relatively low probability. Transition (2) has a lower value of $\Delta E = {}^{\odot 2}E_{K2} - {}^{g}E_{K2}$, thus the photon emitted is of lower energy. It is, however, the most probable transition of the three. Transition (3) stands for $\Delta E = {}^{\odot 2}E_{K2} - {}^{\odot 1}E_{K2}$, thus the photon emitted is of lowest energy. This is obvious from the point of view of energy. But what about the intensity? There are two factors to consider, namely: (i) how populated the initial excited state was, and (ii) what its probability of relaxing to a certain lower-energy nuclear state was, according to selection rules. Note that later, the efficacy of the radiation detector will also become a factor, as it is not identical for all photon energies.

The transitions are:

$\gamma_{[\odot 4, \odot 3]}$ for $(2 \le \Delta J \le 7; \Delta\Pi = \text{yes})$,

$\gamma_{[\odot 4, \odot 2]}$ for $(2 \le \Delta J \le 7; \Delta\Pi = \text{no})$,

$\gamma_{[\odot 4, \odot 1]}$ for $(2 \le \Delta J \le 8; \Delta\Pi = \text{no})$,

$\gamma_{[\odot 4, g]}$ for $(1 \le \Delta J \le 4; \Delta\Pi = \text{no})$.

The most relevant ones are $M_1 + E_2$ for $\gamma_{[\odot 4, g]}$ (making up 83.1% of the photon emissions from all excited levels) and E_2 for $\gamma_{[\odot 4, \odot 1]}$ (making up 6.36% of the photon emissions from all excited levels). Subsequent to the photon emission $\gamma_{[\odot 4, \odot 1]}$,

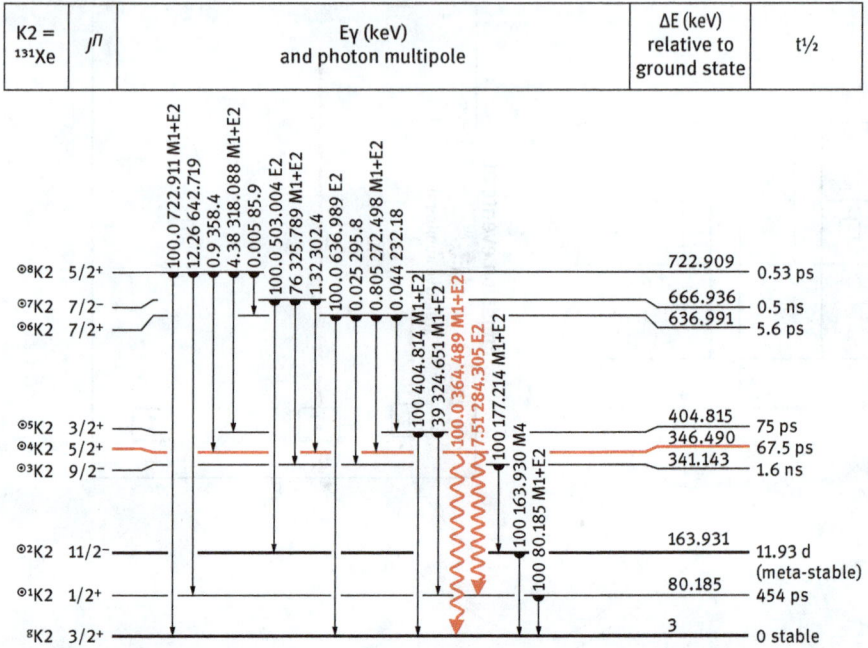

Fig. 11.11: Individual secondary photon transitions between discrete excited states of ^{131}Xe as populated from β^--transformation of ^{131}I. Excited nuclear states and electric and magnetic multipoles are given for each transition, together with the energy of the corresponding photon. The most relevant excited nuclear state is $^{\odot}4(^{131}$Xe) populated by 89.4% of all β^--transformations (red line). From each excited nuclear state, secondary transitions through emission of photons proceed simultaneously, though with specific half-lives, indicated in the right-hand column. Starting from $^{\odot}4(^{131}$Xe), the photon emissions would be $\gamma_{[\odot4,\odot3]}$, (5/2+ → 9/2−; 2 ≤ ΔJ ≤ 7; ΔΠ = yes), $\gamma_{[\odot4,\odot2]}$ (5/2+ → 11/2−; 3 ≤ ΔJ ≤ 8; ΔΠ = yes), $\gamma_{[\odot4,\odot1]}$ (5/2+ → ½+; 2 ≤ ΔJ ≤ 3; ΔΠ = no) and $\gamma_{[\odot4,g]}$ (5/2+ → 3/2+ ground state; 1 ≤ ΔJ ≤ 4; ΔΠ = no), respectively. Here, $\gamma_{[\odot4,g]}$ and $\gamma_{[\odot4,\odot1]}$ (wavy red lines) are relevant because of relatively low changes in overall momentum and conservation of parity. From RB Firestone, VS Shiley (eds.), Table of Isotopes, 8th edition, John Wiley & Sons, New York/Chichester/Brisbane/Toronto/Singapore, 1996.

there is the one of $\gamma_{[\odot1,g]}$ of type M1 + M2 and with 6.72% abundance. The individual half-lives of excited levels 4 and 1, according to eqs. (11.10) and (11.11), are 67.5 ps and 454 ps.

The corresponding energies of the different photons are 364.49 keV and 284.31 keV, respectively. Individual probabilities are relative to the dominant transition of $\gamma_{[\odot4,g]}$ (=100%) and $\gamma_{[\odot4,\odot1]}$ (= 7.51%), respectively. Referring back to the 89.4% initial population of the $^{\odot}4(^{131}$Xe) excited level *via* primary β^--transformation of ^{131}I, this gives overall values of photon emissions for $\gamma_{[\odot4,g]}$ = 83.1% and $\gamma_{[\odot4,\odot1]}$ = 6.36%, respectively. Usually, these individual percentages are given in terms of photon emission coefficients

$\varepsilon_{emission}$, which in this case are $\varepsilon_\gamma = 0.831$ and $\varepsilon_\gamma = 0.0636$.[14] This finally is reflected in the γ-spectrum of ^{131}I, Fig. 11.12. Note that the count rates measured here originate not from ^{131}I directly, but from individual excited nuclear states of ^{131}Xe.

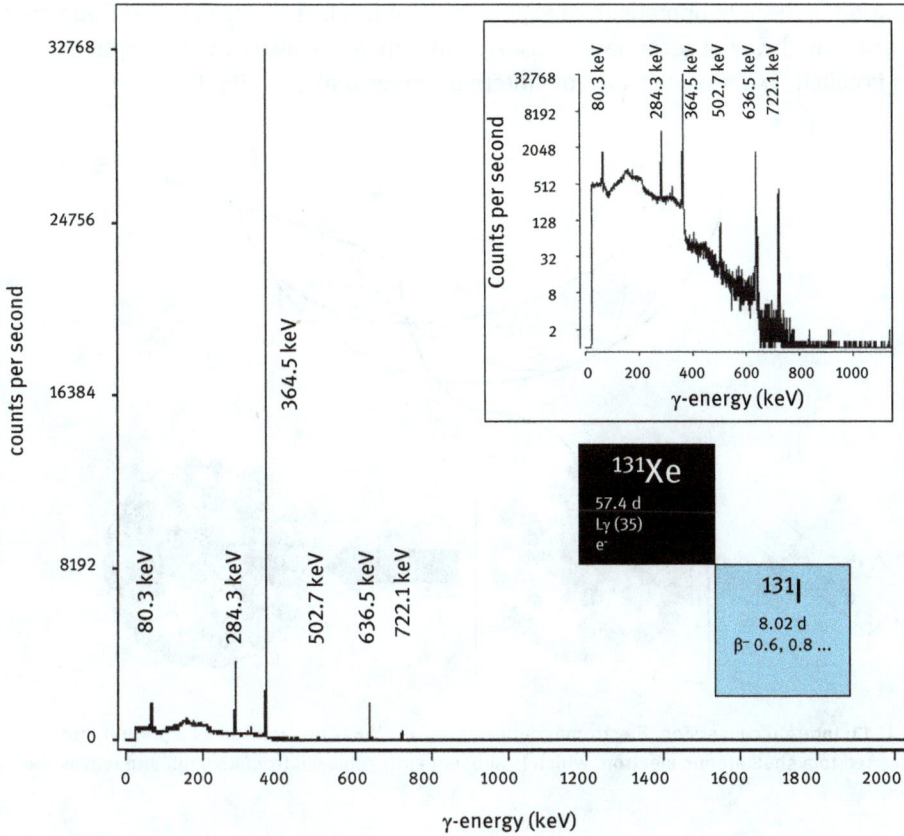

Fig. 11.12: The γ-spectrum of ^{131}I. Note the linear scale and the logarithmic scale (insert) of the y-axis. The two most abundant photon emissions are 364.489 keV and 284.305 keV. According to Fig. 11.11, other γ-emission energies are less probable, yet well identified when visualized *via* the semilog plot.

---

14 . . . and are needed to transfer experimentally measured count rates into absolute radioactivity of a nuclide sample; see eq. (5.11) and Chapter 1 in Volume II.

11.3 Conversion electrons

11.3.1 Internal conversion

Emission of discrete photons from excited nuclear levels is by far the most frequently occurring pathway of secondary transitions. But there are also two more options. The first is called "inner conversion" or "internal conversion" (IC), Fig. 11.13.

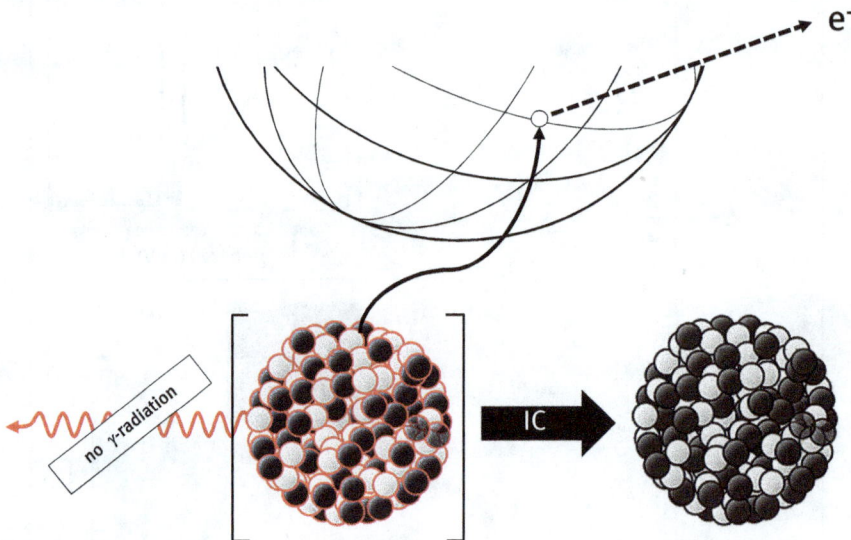

Fig. 11.13: Internal conversion. Electromagnetic energy of $^{\ominus i}\Delta E$ is not emitted as a photon, but converted to a shell atomic electron, which is subsequently released from its shell and leaves the atom.

It represents a situation in which the difference in energy $^{\ominus i}\Delta E$ between two nuclear levels is transferred to an atomic shell electron.[15] Similar to the electron capture process in primary β-transformations, this can only be understood in the context of quantum physics: s-electron orbital wave functions show a certain probability of existing close to and even inside the nucleus of an atom. Here it directly gains the equivalent of $^{\ominus i}\Delta E$, provided this amount of energy is larger than the electron binding energy of that electron shell $E_{B(e)}$. Consequently, all internal conversion electrons are former s-orbital electrons, and most of these originate from an s-orbital as "close" to the nucleus as possible. The kinetic energy E_{IC} of the ejected electron is equivalent

---

15 Internal conversion is well known in conventional chemistry, when high-energy states of a molecule or an atom may transit to lower energy states with or without emitting photons. In the latter case, this energy is absorbed internally, e.g. affecting vibrational systems.

to $^{\odot i}\Delta E$ minus the energy needed to overcome the electron binding energy, as per Fig. 11.14 and eq. (11.13). Internal conversion electrons thus have their own discrete energy – different from β^--particles.

$$E_{IC} = {}^{\odot i}\Delta E - E_{B(e)} \tag{11.13}$$

The IC electron is not excited to a higher energy electron shell – it is released from the atom.[16] Overall, no photon is emitted, because one was never created.[17] *Internal conversion refers to the origin of the energy that is created inside the nucleus, rather than deposited from external sources (such as in Roentgen spectroscopy).*

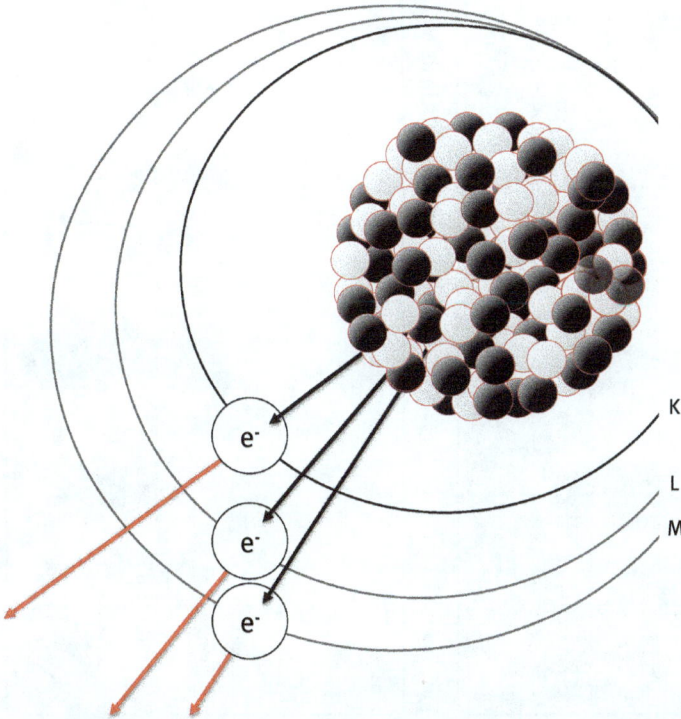

Fig. 11.14: Origin of internal conversion electrons from individual s-shell orbitals of individual main quantum number n, i.e. K-shell, L-shell etc.

---

16 Note that this is the emission of a "real" atomic electron, i.e. one which already existed in a lower electron shell of the atom. This is in contrast to the emission of electrons as β-particles, created instantaneously in the nucleus of that atom over the course of the β-transformation.
17 Note that this internal conversion should not be confused with the "photoelectric effect". In the latter case, an *existing* photon "hits" a shell electron and ejects it.

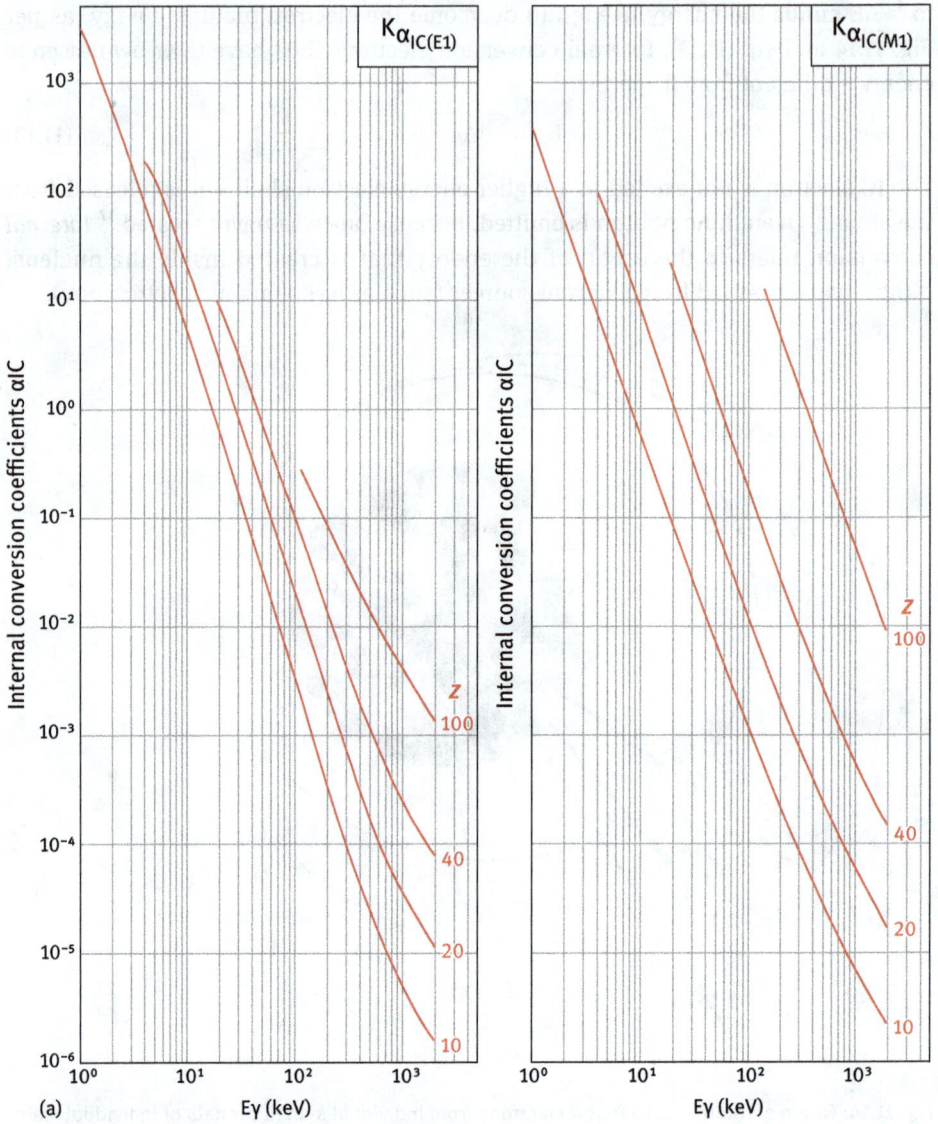

Fig. 11.15: Internal conversion coefficients α_{IC} depending on parameters of an excited nuclide: Z, Eγ, multipolarity and shell number. ICC values are higher for low γ-energies, for electric (*vs.* magnetic) multipoles, for higher multipolarities and for $^K\alpha_{IC}$ relative to $^L\alpha_{IC}$. From: IM Band, MB Trzhhaskovskaya, CW Nestor, PO Tikkanen, S Raman, Dirac–Fock Internal Conversion Coefficients, Atomic Data and Nuclear Data Tables 81 (2002) 1–334.

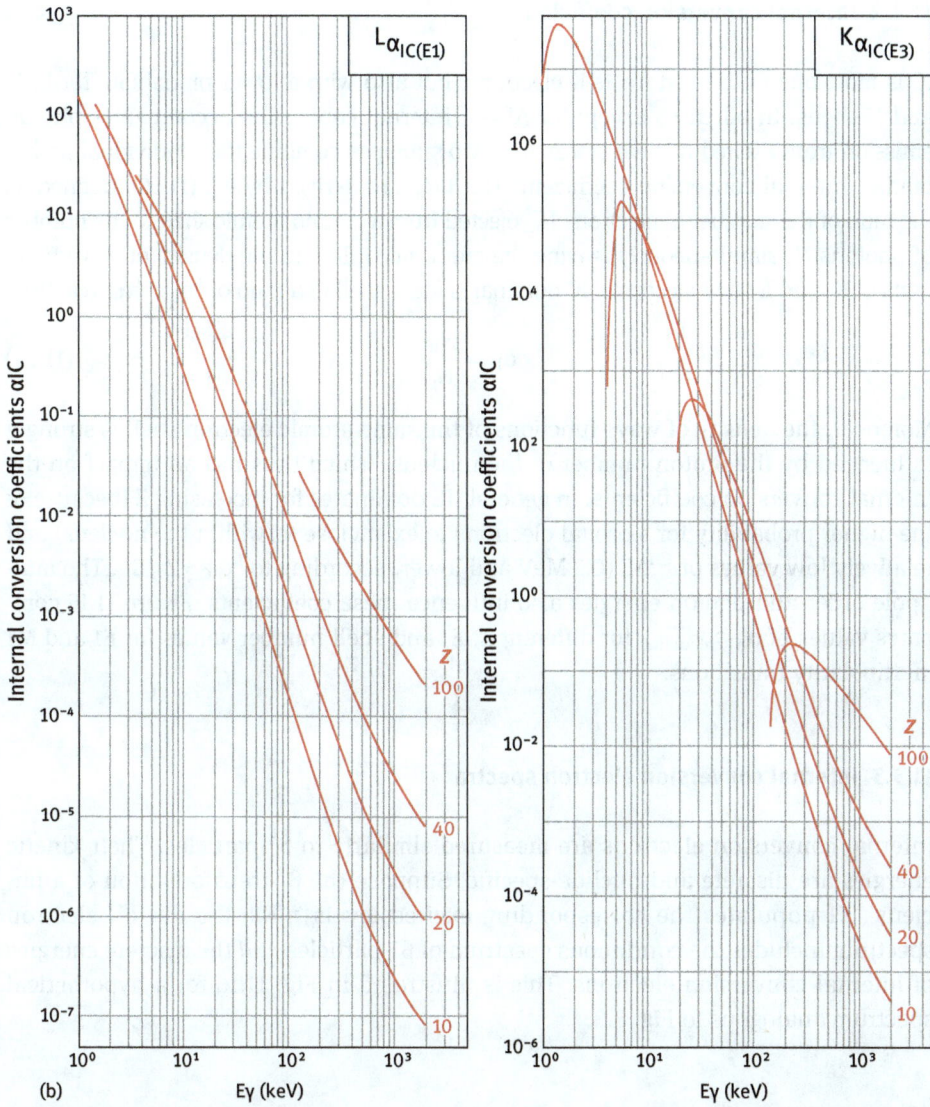

Fig. 11.15 (continued)

Internal conversion is a domain for low energy $^{\odot i}\Delta E$ transitions, particularly for monopole modes of $0^+ \rightarrow 0^+$ where γ-emission is impossible. Nevertheless, internal conversion is also possible for multipole transitions.

11.3.2 Internal conversion coefficients

IC is most often observed for 1s-K electrons, but also with a lower probability for L, M and N-shells. In all cases except for $\Delta J = 0$, internal conversion accompanies photon emission from excited nuclear states. The two subtypes combine their individual probabilities. Internal conversion coefficients (ICC) α_{IC} are derived by eq. (11.14), defined as the ratio of the number of electrons N_{IC} ejected from a certain atomic shell to the number of photons N_γ simultaneously leaving the atom. According to the density of wave functions of K- and L-shell electrons, IC dominates for K-shell at a ratio of $^K\alpha_{IC} > {}^L\alpha_{IC}$ (ca. 8:1).

$$\alpha_{IC} = \frac{N_{IC}}{N_\gamma} \tag{11.14}$$

Moreover, the density of wave functions of the same atomic electron shell is strongly influenced by the proton number of the nucleus, which thus has an impact on the internal conversion coefficients. In general, IC dominates for increasing Z (because of the higher probability for s-orbital electrons to exist close to/within the nucleus) and relatively low values of $^{\ominus i}\Delta E$ (0.2 MeV and lower, according to $^K\alpha_{IC} \approx Z^3/E_\gamma$). The multipole order and photon energies also influence these coefficients. Figure 11.15 compares values of $\alpha_{IC\ calculated}$ for different Z, E_γ and shell number values for E1 and M1 multipolarity transitions.

11.3.3 Internal conversion electron spectra

Internal conversion electrons are measured similarly to β^--particles. Their kinetic energies are discrete and nuclide-specific. Suppose the β^--transformation of a nucleus $^*K1$ populates the corresponding excited levels $^{\ominus i}K2$; the overall electron spectrum includes the continuous spectrum of β^--particles *and* the discrete energies of internal conversion electrons. This is illustrated in Fig. 11.16 for a hypothetical spectrum analogous to Fig. 8.8.

11.4 Pair formation

The third pathway of secondary transitions is called "pair formation", as shown in Fig. 11.17.

Here, the electromagnetic energy released from the nucleus is converted into real matter. This matter takes the form of an electron and its antiparticle, the positron, i.e. a pair of two equivalent particles. This means that basic parameters are conserved:

-   spin = 1 for the photon *vs.* spin = ½ for each of the two fermions,
-   lepton number (matter vs. antimatter) and
-   charge.

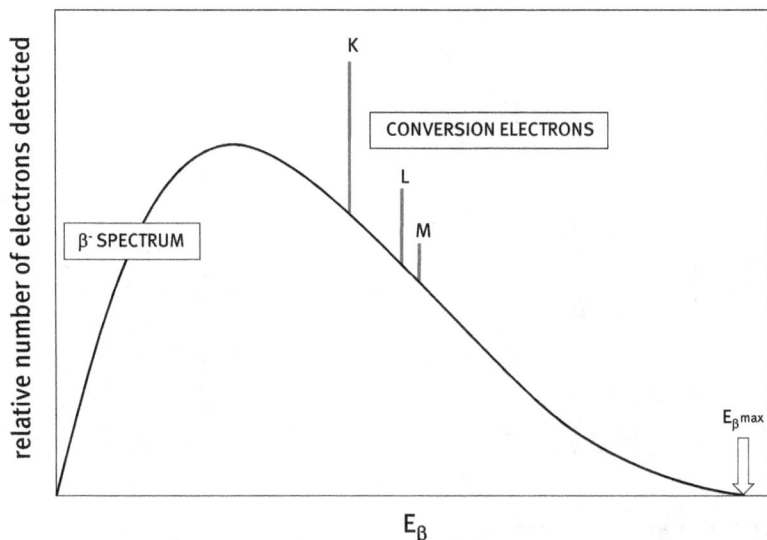

Fig. 11.16: (Hypothetical) continuous spectrum of β^--particles overlaid by discrete energies of IC electrons. The three discrete kinetic energies indicated by the red lines correspond to IC electrons emitted from K, L and M-shells of $^\odot$K2. For kinetic energy: $^K E_{IC} < {}^L E_{IC} < {}^M E_{IC}$, for intensity I: $^K I_{IC} > {}^L I_{IC} > {}^M I_{IC}$.

In order to conserve energy, the value of $^{\odot i}\Delta E$ must at least equal the mass of the two electrons. Electron masses are 0.511 MeV (or 0.00055 u) each, so the overall energy follows $^{\odot i}\Delta E \geq 1.022$ MeV.

Pair formation thus exclusively emerges at $^{\odot i}\Delta E > 1.022$ MeV, i.e. for a relatively high difference between the two nuclear energy levels involved. It is thus a relatively rare effect. Internal conversion, in contrast, arises at low values of $^{\odot i}\Delta E$, and a large number of radionuclides emit IC electrons. The emission of a γ-photon is, however, most frequent. It takes place over the whole range of $^i\Delta E$ values, and represents several thousand cases. Figure 11.18 illustrates the range of the three secondary transition options depending on the value of $^{\odot i}\Delta E$.

11.5 Energy and half life

The transformation of a nuclide existing in an energetically excited nuclear state proceeds exothermically, and the subsequent state is of lower energy. Because there is a defined energy for the ground and excited nuclear states of a given nuclide, the difference between any two levels reflects a specific value of ΔE. These differences cover a broad range from about 10 keV for individual levels energetically very close to each other, up to (in a few cases) about 10 MeV, as per Tab. 11.6. The same is true for the corresponding half-lives of secondary transformations. De-

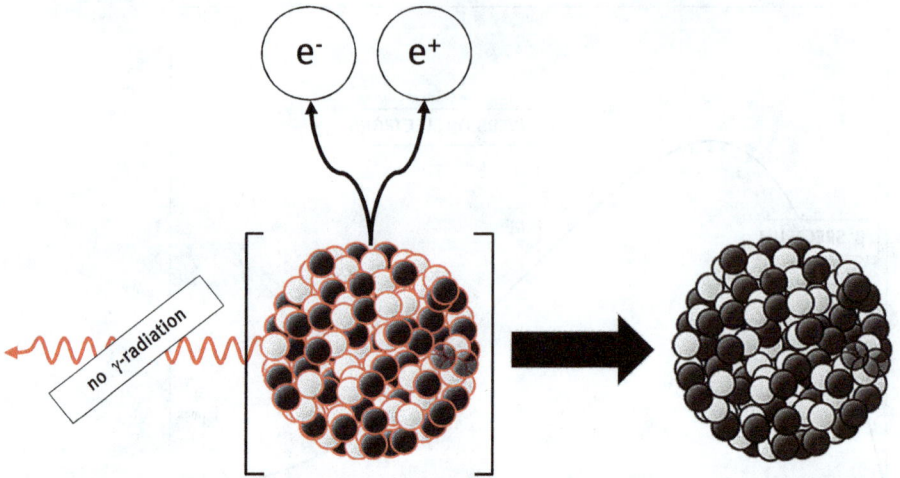

Fig. 11.17: Formation of an electron + positron pair.

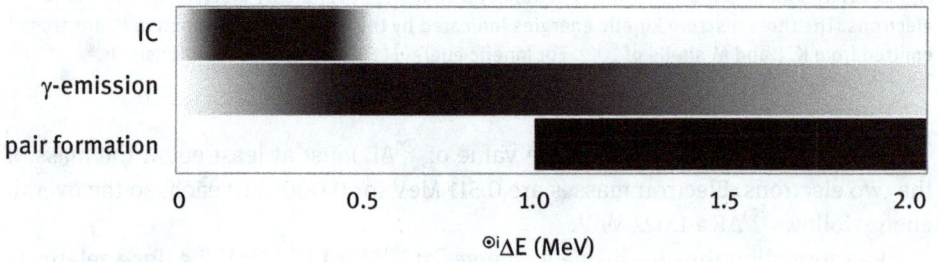

Fig. 11.18: Appearances of the three secondary transitions depending on the values of $^i\Delta E$.

excitation between individual excited states or the final de-excitation from one excited state to the ground state is very fast, typically lasting only 10^{-16} to 10^{-13} s. The overall secondary transformation is thus extremely fast, even when cascades of several transitions are involved.

Tab. 11.6: Ranges in energy and half-life of secondary transformations.

ΔE	Whole range	10 keV	–	10 MeV
	Typical	50 keV	–	2 MeV
Half-life	Typical	10^{-16} s	–	10^{-13} s
	Metastable states	10^{-03} s	–	10^{15} years

11.5.1 Metastable nuclear states and nuclear isomers

Suppose for the reasons introduced in Figs. 6.22 and 6.23, one (or more) of the individual excited nuclear levels of K2 "remain" for longer periods of time. Under these circumstances, this excited state $^{\odot i}$K2 will have a much longer half-life than the other excited levels of $^{\odot}$K2 – about five orders of magnitude longer than the next longest-lived transition.[18] These half-lives represent a particular state in a cascade of secondary transitions where time seems to stand still for a while (see Fig. 11.19).

This happens in the context of the selection rules, i.e. whenever the differences in overall angular momentum are large (octa-, hexadeca-, or higher multipole orders) or when parity is violated. The phenomenon is often observed for very small values of $^{\odot i}\Delta E$ between the initial and final states. This excited state is not really stable like the final, ground state of ^{g}K2, but remains intact for a significant period of time. Because it lies between unstable and stable (ground state) levels, it is called "metastable", ^{m}K2. Compared to the ground state, it reflects a nucleus of identical nucleon composition in terms of A, Z, and N, and is therefore also referred to as "nuclear isomer".[18] There are many metastable isomers with half-lives in the range of minutes, hours, days and years.[19]

Figure 11.11 showed the individual half-lives of the various excited levels in $^{\odot}(^{131}Xe)$. One of them, $^{\odot 2}(^{131}Xe)$, was of $t\frac{1}{2} = 12.93$ days, i.e. much longer than the other (extremely fast) transitions with nano- and picosecond half-lives. Indeed, the second excited level of ^{131}Xe represents such an isomer. Figure 11.20 shows the two pairs within the primary transformations of $^{131}I \rightarrow \beta^- \rightarrow {}^{131}Xe$ and $^{81g}Rb \rightarrow \varepsilon, \beta^+ \rightarrow {}^{81m}Kr/^{81g}Kr$. There are about 700 of these metastable isomers. Some are of interest in fundamental research; others are relevant to important applications.

For nuclides of Z > 92, a metastable nuclear isomer may transform *via* spontaneous fission, then called a "fission isomer".[20] The first sf isomers appear as ^{236m}U ($t\frac{1}{2} = 120$ ns) and ^{238m}U ($t\frac{1}{2} = 298$ ns). ^{242}Am has a fission isomer ^{242m}Am with a long half-life of 141 a. Interestingly, this half-life is longer lived than the ground state of ^{242g}Am of $t\frac{1}{2} = 16$ h.

---

18 Nuclear isomers are thus just another version of isomerism, in addition to the various isomer types known in classical chemistry. Stereoisomers, for example, are identical in the number of, identity of and chemical bonding between atoms, but are different in the orientation of those atoms in space. Organic compounds that contain a chiral carbon usually have two non-superimposable structures forming enantiomers. Just like our left and right hands, these two structures are mirror images of each other and are otherwise structurally the same – but may differ in some chemical properties. For example, carvone forms two enantiomers, namely R-(–)-carvone and S-(+)-carvone. The first one smells like spearmint, the other like caraway.

19 The ground state ^{180g}Ta has a half-life of 8.125 h. One excited nuclear state ^{180m}Ta remains at its excited level with $t\frac{1}{2} > 10^{15}$ years because of the high $\Delta J = 8$.

20 First discovered for ^{234m}Pa ($t\frac{1}{2} = 1.17$ m) by OTTO HAHN in 1921 when studying the β^--processes of ^{234g}Pa ($t\frac{1}{2} = 6.70$ h), the second product of the naturally occurring ^{238}U chain.

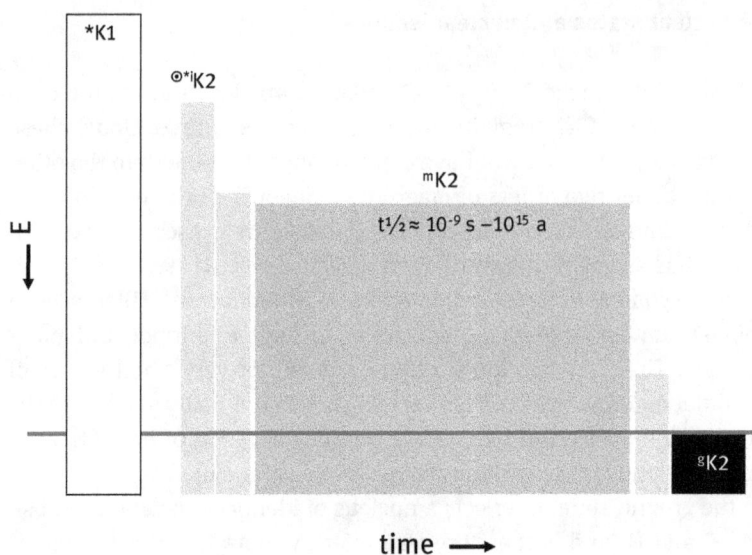

Fig. 11.19: Metastable excited nuclear states ^{m}K2. One (or more) individual levels $^{\odot i}$K2 shows a significantly prolonged half-life relative to the longest-lived of the other, very short-lived secondary transitions. The simplified representation assumes that velocities of a transforming nucleus follow the idea of a ball bouncing downstairs. Whenever the width of a step is small and the next stair is much deeper, the ball moves very fast (like the hillside model of the valley of β-stability concept, as in Fig. 6.13). If the ball meets a step which is only a bit deeper than the initial one, and much broader, its speed decreases.

Also interestingly, nuclear metastable isomers not only undergo secondary transitions as a form of de-excitation – they may also transform by a primary transformation pathway, as in Fig. 11.21. This may happen if the ground state ^{g}K2 does not represent the most stable nuclide of that particular isobar. Consequently, a metastable nuclide may undergo a separate primary β- or α- or sf-transformation approaching the next, more stable nuclide directly. Figure 11.22 shows such an example for a secondary β$^-$-transformation, namely ^{99}Mo → β$^-$ → ^{99m}Tc → β$^-$ → ^{99}Ru.

The appearance of the metastable ^{99m}Tc represents a transient radionuclide generator (see Chapter 5), because of the half-lives of 66.0 h for ^{99g}Mo and 6.0 h for ^{99m}Tc. The secondary transition releases a 140.511 keV photon of high branching: in ca. 89% of all the ^{99}Mo transformation events, this photon is emitted.

11.6 Example applications of photon emission

Among the three subtypes of secondary transitions, photon emission is the most relevant mode. The photons emitted are relatively easy to register quantitatively by adequate detectors (see Chapter 1 of Volume II). Compared to the emissions of the

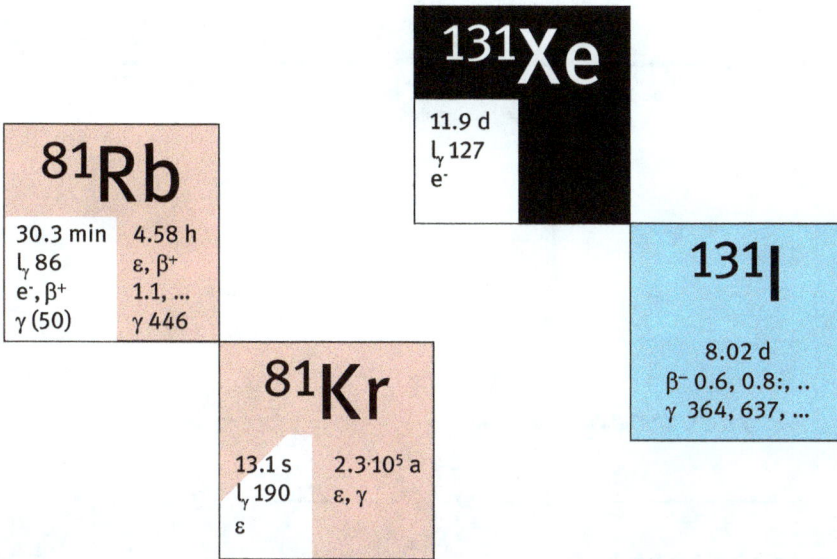

Fig. 11.20: Description of metastable isomers for secondary transitions in $^{131}I \rightarrow \beta^- \rightarrow {}^{131}Xe$ and $^{81g}Rb \rightarrow \epsilon,\beta^+ \rightarrow {}^{81m}Kr/{}^{81g}Kr$. For the primary transformations, the most relevant data are in the boxes of the initial nuclide *K1, such as half-life and main β^--energies, which also includes the main γ-energies of the subsequent secondary electromagnetic transitions. Within the Karlsruhe Chart of Nuclides the isomers are small white fields within the boxes of the stable product nuclides. They show the half-life of the nuclear isomer and parameters of transformation including the γ-energy (in keV). $^{131}I/{}^{131}Xe$ represents a secondary process subsequent to a β^--primary process.[21]

primary transformations, which (except electron neutrinos) all are particles, i.e. α-particles, β-particles and neutrons, the photons interact with matter only to a very small degree. They penetrate glassware and metallic containers and are only gradually affected in intensity, but not in energy. Consequently, photons are a noninvasive measure of the identity of the radionuclide and represent the most valuable diagnostic tool in nuclear research, nuclear technology and nuclear medicine. In the following, some directions of applications of γ-emitting radionuclides are indicated. Most of them will be discussed in more detail in Volume II.

In addition, photons can be used to noninvasively determine the absolute radioactivity of a sample within glass or metal vessels or containers. Photons emitted in secondary transitions of unstable nuclides are thus the most accurate tool to quantify nuclear transformation processes.

---

21 The $^{81g}Rb/{}^{81m}Kr$ process starts from the ground state of unstable ^{81g}Rb and populates the metastable isomer ^{81m}Kr. Due to the half-lives of 4.58 h relative to 13.1 s, this is a radionuclide generator system, used for diagnostic imaging in nuclear medicine.

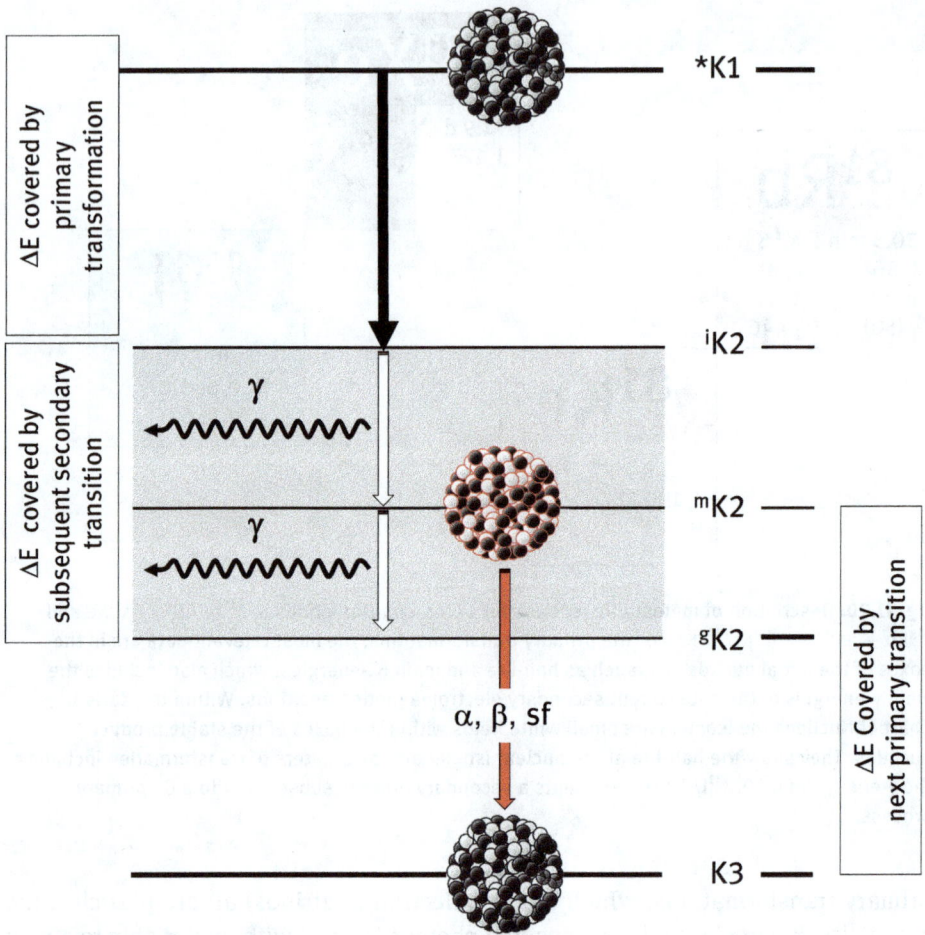

Fig. 11.21: Options for metastable isomers. Secondary transformation to lower energy nuclear states of $^{\odot}$K2 *via* e.g. photon emission *vs.* primary transformation to nuclide K3 (through α- or β-transformation or by spontaneous fission).

11.6.1 Astrophysics

High-energy photons traverse the universe and the atmosphere of the earth, making them useful in measuring astrophysical processes. ^{26}Al for example has a half-life of $7.16 \cdot 10^5$ years. When formed in supernovae, it transforms to stable ^{24}Mg *via* an excited nuclear level of ^{24}Mg of 6.35 s half-life. From that excited nuclear state, the subsequent secondary transition into the ground state ^{24g}Mg emits a photon of 1.809 MeV energy. Satellites register this γ-ray, emitted from the center of our

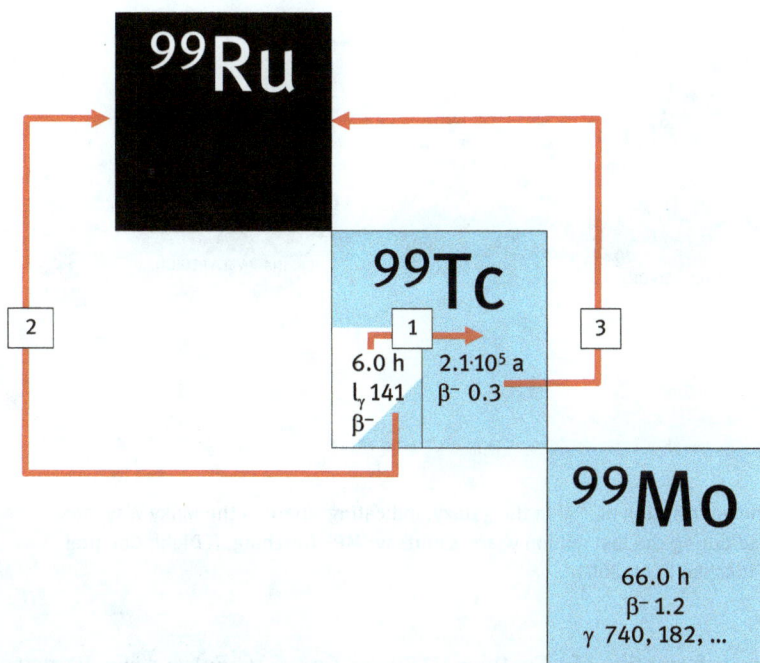

Fig. 11.22: Origin and fate of the nuclear isomer [99m]Tc. The isomer ^{99m}Tc is a small and basically white box within the blue box (β⁻-transformation) of the unstable [99]Tc. [99m]Tc transits to the ground state [99g]Tc *via* secondary processes (1), mainly through emission of a 141 keV photon. The quantum physical parameters belong to the $\gamma_{[\odot 1,g]}$ transition. The main photon multipole is M1 + 3.3% of E2 with 1/2⁻ and 9/2⁺ for ^{99m}Tc and [99g]Tc. Simultaneously, 99mTc transforms *via* β⁻-emission (blue triangle in the white field) to the more stable nuclide of the A = 99 isobar, which is stable [99]Ru (2). [99]Ru is thus predominantly obtained through the direct, primary β⁻-transformation of [99g]Tc (3), but additionally, even if to a very small percentage, by primary β⁻-transformation from [99m]Tc.

galaxy, as shown in Fig. 11.23. These data can contribute to the understanding of rotational movements of the galaxy.[22]

11.6.2 MÖßBAUER spectroscopy

MÖßBAUER spectroscopy is an important analytical technology based on the photons emitted in secondary transitions, and is applied in speciation and materials sciences. It utilizes γ-emission from an excited nuclear level to the ground state. The nuclides all are obtained *via* the primary transformation of an (ideally) relatively long-lived unstable nuclide *K1. The most prominent is [57]Fe and its excited nuclear state populated

---

22 For further applications of radionuclides in the context of dating, see Volume II, Chapter 5.

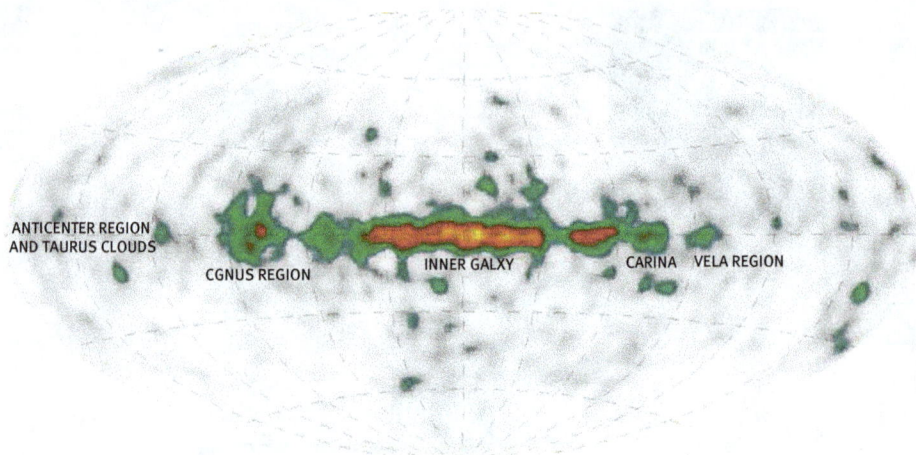

Fig. 11.23: Map of the distribution of ^{26}Al in the galaxy, indicating where in the Milky Way most supernovae exploded during the last million years. Courtesy: MPE Garching, R Diehl, Comptel 1991–2000, ME 7, Plüschke et al. 2001.

in the primary ε-transformation of ^{57}Co, t½ = 271.79 d: ^{57}Co → $^{\odot}(^{57}$Fe) → ^{57g}Fe. Particular effects are analyzed through moving the photon source and the material (with velocities <200 mm/s, for ^{57}Co ca. 15 mm/s) and resonances are registered[23] according to the DOPPLER effect.

11.6.3 Industry

Because γ-radiation is relatively easy to register, γ-emitting radionuclides are extensively applied to monitor processes in chemical and industrial technology. For example, γ-emitters are used to precisely determine the thickness of materials because the degree of adsorption of the γ-radiation serves as a measure of the thickness of the walls of any container.

11.6.4 Neutron activation analysis

The unique γ-energies, intensities and half-lives of photons serve as a fingerprint of the isotope they were emitted from. They thus represent a valuable tool for quantitative

---

23 Nobel Prize in Physics, 1961, to RUDOLF MÖSSBAUER for " . . . his researches concerning the resonance absorption of gamma radiation and his discovery in this connection of the effect which bears his name".

and noninvasive determination of trace amounts of chemical elements in materials analytics. In neutron activation analysis (NAA), for example, stable isotopes of an element located inside the irradiated sample are "activated" by neutron-capture nuclear reactions, i.e. generate an unstable isotope of the same element. In the course of its subsequent primary and secondary transformations and the characteristic γ-radiation emitted, the kind of chemical element and its concentration within the sample are quantified. Chapter 3 in Volume II systematically discusses the features of NAA.

11.6.5 Medical applications

Noninvasive medical imaging: Because photons are not completely adsorbed in aqueous media and biological tissue, they are also a sensitive tool for molecular imaging in medicine. The basic concept is:
(a) to label a medically relevant molecule with a photon-emitting radionuclide (ideally without changing the molecule's physico-chemical parameters and its pharmacology and pharmacokinetics, or at least with minor or "controlled" changes),
(b) to inject it into a patient, and
(c) to monitor how the labeled molecule is distributed within the human body (in terms of uptake by organs of interest, uptake kinetics, excretion route and velocity etc.).

For photons created in the course of secondary transformations, the medical technologies are scintigraphy and single photon emission computed tomography (SPECT). Requirements are thus:
− a physical half-life adequate to the medical purpose,
− a photon energy in the range of ca. 100 to 400 keV adequate to the optimum detection efficacy of the scintillation detectors used,
− the absence of further nuclear transitions (i.e. transformations should ideally lead to a stable transformation product) and
− robust availability of the nuclide and its labeled compounds.

The most common examples are listed in Tab. 11.7. Clearly, the metastable, generator-derived ^{99m}Tc is the most important radioisotope in nuclear medicine and SPECT imaging (Fig. 11.22). The interesting point here (different from the processes illustrated for ^{131}I, discussed below) is that this particular excited nuclear level has an extended period of existence, cf. Fig. 11.19. It is a metastable nuclide of ^{99}Tc and is therefore denoted by ^{99m}Tc. Its half-life is 6.0 h. This metastable ^{99m}Tc undergoes a secondary transition process to ^{99m}Tc, accompanied by the intense emission of low energy photons of 140.5 keV (see Chapter 11). Thus, ^{99m}Tc is available from a convenient radionuclide generator system, provides a daughter nuclide every day (according to the mathematics of the transient generator equilibrium, see chapter

5.4.3.4), and represents a source of photon radiation very useful for medical diagnostic imaging *via* SPECT (single photon emission computed tomography). Today, ^{99m}Tc is the number one medical isotope, with more than ten thousand applications for patient diagnoses worldwide every day.[24]

^{131}I is also applied in "diagnostic" activities, and is another key radioisotope in nuclear medicine. It is often used on an industrial scale as a "medical" radionuclide in the treatment of malignant tumors, or to treat diseases of the thyroid and other glands. As the excited states of ^{131}Xe de-excite, they emit electromagnetic radiation. According to the dominant population of one state, the photons emitted from this level approaching the ground state of ^{131}Xe either directly or *via* another excited state create intense photon emissions at 364 keV and 284 keV energy, respectively, as shown in Figs. 11.11 and 11.12.[25]

Tumor treatment from external radiation source: In contrast to the low energy photons of ^{99m}Tc and ^{131}I, which for diagnostic purposes pass through the human body without causing critical radiation doses, intense and high energy photons ($E_\gamma >$ 1 MeV) of photon-emitting radionuclides are applied in cancer treatment by means of external irradiation ("γ-knife"). Two examples are ^{60}Co ($t^{1/2} = 5.272$ a, $E\gamma = 1332$ and 1173 keV, shown in Fig. 8.21) and ^{137}Cs. Clinically, enormous activities (several thousand GBq) of ^{60}Co are shielded in a medical device, designed to lead the photons in sophisticated arrangements to well-localized brain tumors for noninvasive therapy.

Tumor treatment from internal radiation source: In contrast to the high energy photons of e.g. ^{60}Co, very low energy photons can be clinically generated from a sealed radiation source, which is deposited in the tissue of interest. It is referred to as "brachytherapy", from the Greek word for "soft". The most relevant radioisotopes to this type of therapy are ^{131}Cs ($t^{1/2} = 9.7$ d), ^{103}Pd ($t^{1/2} = 17.0$ d), ^{125}I ($t^{1/2} = 59.6$ d), and ^{192}Ir ($t^{1/2} = 73.8$ d).

---

24 For ^{99m}Tc to realize its full therapeutic potential still requires, in addition to the nuclear data, the contribution of radiopharmaceutical chemistry. The challenge is to turn this radionuclide, which does not exist naturally on earth, into biologically relevant molecules, thereby allowing for the measurement of a large variety of physiological processes in the living human body.

25 Due to its availability from a ^{99}Mo/^{99m}Tc radionuclide generator and the success of radiopharmaceutical chemistry to develop various biologically relevant ^{99m}Tc-labeled molecules, ^{99m}Tc-SPECT imaging and scintigraphy represent about 90% of all diagnostic studies in nuclear medicine and account for over 80% of the overall worldwide use of radionuclides in medicine. These procedures are performed on more than 300,000 patients worldwide every day (see Chapter 13 of Volume II.)

11.7 Outlook

Photon emission and pair formation finally de-excite the intermediate of a primary transformation to its stable ground state. In doing so, these two secondary transitions terminate the nuclear transformation processes initiated by the primary transformation. The unstable nuclide *K1 has been transformed into a new nuclide K2, either stable or at its ground state of an unstable nuclide.

However, this is not the case for IC. Why? Internal conversion is the ejection of an s-orbital electron, but leaves a vacancy (hole) in the electron shell of K2. The remaining atom is thus not yet "intact". The same is true for one subtype of the primary β-transformation, namely electron capture, which also created a hole in an atom's inner s-orbital electron shell. In both cases, the atom must now handle this situation in another subsequent step, somehow "filling" the vacancy in its own electron shell! What happens here is not part of a primary or secondary transformation, but a post-effect; it is definitely part of the whole story of a radionuclide transformation.

Furthermore – there is a second group of post-effects. Recall that primary β$^+$-transformations create a positron. While the nucleus has achieved its new nucleon configuration, the destiny of the positron still remains part of the overall transformation process. It is ready to annihilate with an electron, orbiting the shell of some far-away atom. Consequently, annihilation radiation is added to the other emissions characteristic for the primary transformation that created the positron, and the secondary emissions as consequences of the primary one.

Similarly, electron + positron pair formation was one of the outcomes of a secondary pathway within the $^{\ominus i}$K2 → $^{\ominus}$K2 or ^{g}K2 nuclear transformation. The pair of electron + positron is released from the nucleus, and immediately reacts further, inducing a characteristic annihilation radiation.

These processes all occur after (but due to the nano- and picosecond timescales, effectively parallel to) the primary and secondary processes and are denoted here as post-effects. They are needed to finally understand the various characteristic emissions, all induced by *K1, but caused by different pathways. This is discussed in the following Chapter 12.

Tab. 11.7: Important γ-emitting[a] radionuclides originating from primary β⁻-, EC and α-transformations and routinely applied for human diagnosis in nuclear medicine used for scintigraphy or SPECT and for therapy, respectively.

Radionuclide		Half-life	Main primary transformation	Photon emission E_γ (keV) and branching (%)
^{67}Cu		2.58 d	β⁻	861 (100), 833 (48)
^{67}Ga		3.26 d	EC	93 (38.8), 185 (21.4)
^{81m}Kr	from ^{81}Rb generator	13 s	IC	190 (67.7)
^{99m}Tc	from ^{99}Mo generator	6.0 h	IC	141 (89)
^{111}In		2.81 d	EC	167 (10.0), 235 (94.1)
^{123}I		13.2 h	EC	159 (83)
^{131}I		8.02 d	β⁻	364 (81.4), 637 (7.2)
^{153}Sm		46.27 h	β⁻	103 (29.3)
^{177}Lu		6.71 d	β⁻	208 (10.49), 103 (6.2)
^{186}Re		3.72 d	β⁻	137 (9.5)
^{188}Re		17.0 h	β⁻	106 (10.8)
^{201}Tl		3.05 d	EC	167 (10.0)
^{213}Bi		1.90 d	β⁻ + α	440 (25.9)
^{211}At		7.21 h	EC + α	687 (0.3)
^{223}Ra		11.43 d	α	269 (13.9), 154 (5.7)
^{224}Ra		3.66 d	α	241 (4.1)

[a]Photon emission data from www.nndc.bnl.gov/chart/chartNuc.jsp, the National Nuclear Data Center, Brookhaven National Laboratory. Most relevant photon emissions selected, energies in keV (rounded).

12 Post-processes of primary and secondary transformations

Aim: β-processes, α-emission and spontaneous fission are primary transformations leading either to the ground state of the new nuclide K2 or populate excited levels *K2. Excited levels de-excite *via* three subtypes of secondary transitions. Some of the primary and secondary transformations, however, induce additional inherent processes, which either continue as the final step of the nuclide's transformation or occur parallel to it. These post-effects are responsible for further radiative emissions beyond those constituting the basic features of primary and secondary transformations.

The first group of post-processes appears in nuclear processes that left a hole in the s-shell of the nuclide after (a) the primary electron capture or (b) a secondary internal conversion process. Logically, this hole must be filled immediately. This proceeds under emission of new kinds of radiation: both electromagnetic radiation, called X-rays, and ejection of electrons from all shells of the atom beyond the initial one, called AUGER and COSTER–KRONIG electrons.

Another, second source of a different kind of radiation is directly linked to the preceding primary β⁺-process: (c) the creation and emission of a positron. While the transformations continue within the nucleus, the positron leaves the atom to immediately combine with an electron of other atoms surrounding it. Similarly, the secondary process of (d) pair formation directly produces a pair consisting of a positron and an electron. In both cases (c) and (d), this pair of matter and antimatter particles annihilates and electromagnetic radiation of exactly 511 keV is created.

These post-transformation radiation emissions inextricably accompany the corresponding primary and secondary transformations. This chapter introduces the origin and the various manifestations of X-rays, AUGER and COSTER–KRONIG electrons and 511 keV annihilation photons.

Radionuclides that cause radiative post-process emissions find particular applications in science, technology and medicine.

12.1 Post-processes paralleling primary and secondary nuclear transformations

Processes of stabilization of an unstable nuclide *K1 result in a new nuclide of type K2, either directly or following secondary transitions. Simultaneously, small components x are released. These can be beta particles, electron neutrinos, α-particles or neutrons for primary processes. In many cases, small components y are released during secondary processes. These include photons, IC electrons or electron + positron

https://doi.org/10.1515/9783110742725-012

pairs. However, there are some additional emissions inherently[1] accompanying some of the primary and secondary processes. These represent either particle radiation (various kinds of electrons) or electromagnetic radiation (X-rays and bremsstrahlung), both of origin and characteristics not identical to the electron and photon emissions released in primary and secondary processes. In principle, there are two classes of post-effects occurring in condensed matter. Their origins are listed in Tab. 12.1.

Tab. 12.1: Types and origins of individual post-effects. Subtypes of primary and secondary transformation processes yield either a positron, electron + positron pair or create a vacancy in the electron shell of K2.

Transformation	Subtype	Result	Post-effect: emission of
Primary	β^+	Ejection of a positron from the atom	Annihilation radiation (after thermalization)
Secondary	Pair formation	Creation of a positron + electron pair	Annihilation radiation (immediate)
Primary	Electron capture	Vacancy in electron shell is immediately filled	Characteristic X-rays and cascade of AUGER and COSTER–KRONIG electrons
Secondary	Internal conversion		

The first type of post-effect relates to the appearance of a positron, which itself is an antimatter particle and thus destined to annihilate its matter counterpart, the electron (see Fig. 12.1). This necessarily happens for a primary β^+-transformation of a proton-rich nuclide, discussed in Chapter 8. Similarly, pair formation is a secondary transition in which a positron is created together with an electron. This occurs *via* the conversion of electromagnetic energy $^i\Delta E > 1.022$ MeV released between two nuclear states (see Chapter 11).

In the second type of post-process, a vacancy (hole) in the electron shell of an unstable nuclide is involved. The same hole can arise during a primary electron capture transformation of a proton-rich nuclide, or during a secondary internal conversion, as shown in Fig. 12.2.

These two principal effects stimulate two questions: What happens to the positron, and what happens to the electron vacancy?

---

1 These post-processes are inherently linked to the primary and secondary transformations of the unstable nuclide *K1. The corresponding radiation emitted thus indicates the preceding nuclear pathways. However, post-effects that address a hole in an atom's inner shell are also induced in nonradioactive processes. Whenever an electron vacancy is achieved by external action (e.g. in X-ray spectroscopy), the follow-up routes to fill these vacancies are the same.

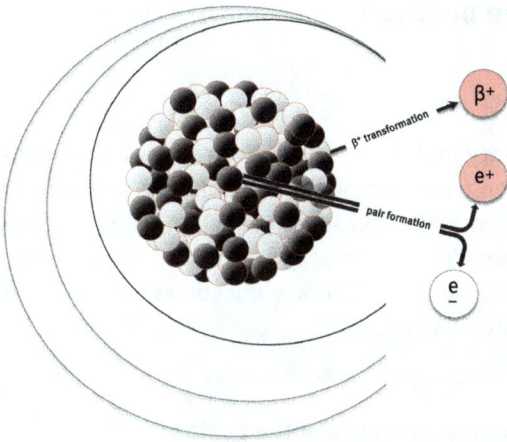

Fig. 12.1: Post-effects related to the appearance of a positron. This happens for a primary β^+-transformation and similarly for a secondary pair formation.

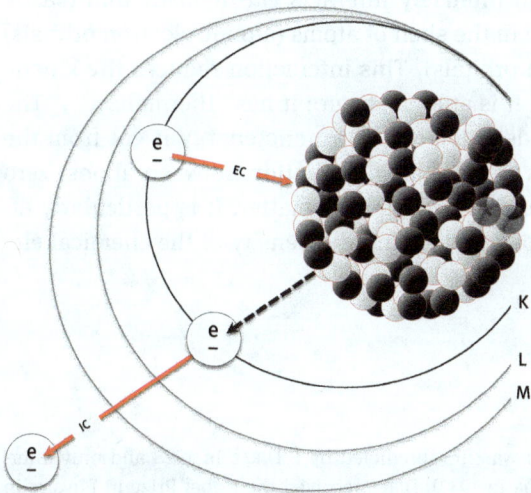

Fig. 12.2: Post-effects motivated by a vacancy in an electron shell of the atom. The vacancy arises after a primary electron capture transformation of a proton-rich nuclide, or after a secondary internal conversion.

12.2 Post-processes due to the positron

12.2.1 The positron and its formation

The positron is the antimatter elementary particle of the electron with all the quantum physical parameters of the electron.[2] The exception is the charge, which is +1 instead of –1. Since antimatter vanished about $13.75 \cdot 10^9$ years ago,[3] there are no positrons "naturally" left in the universe. Yet, positrons appear in radioactive transformations of many man-made unstable nuclides. The key question in the context of post-processes is: What *happens* with the positron?

12.2.2 Correlations between kinetic energy and range of the positron

What happens with the positron? In a vacuum, the answer is: nothing! Its lifetime is $\tau > 10^{20}$ a. This is different in (condensed) matter of course. Here (in contrast to the electron neutrino), the positron intensely interacts *via* inelastic and elastic scattering with the electrons located in the shell of atoms (atomic electron orbitals) or molecules M (molecular electron orbitals). This interaction reduces the kinetic energy of the moving positron until it is at almost zero: it has "thermalized".[4] The total distance the positron travels depends on the kinetic energy it got from the β^+-event (remember this includes a continuous distribution between almost zero and a maximum energy) and the kind of surrounding matter. It is particularly affected by the Q_{β^+} value, the values of Z and A, and the density of the chemical elements or compounds.

---

2 Like the electron neutrino, the positron was first predicted by P DIRAC in 1928 and only afterwards experimentally verified (CD ANDERSON in 1933). DIRAC received the Nobel Prize in Physics in 1936 "for his discovery of the positron".

3 The moment the universe was created within the Big Bang, matter and antimatter existed simultaneously in an (almost) 1:1 proportion. Within fractions of a second, however, the corresponding matter and antimatter elementary particles disappeared by fusing with each other. This process is called annihilation. There is no evidence of any antimatter in the universe, which would look like an atom made of negatively charged protons in the nucleus and positively charged electrons (positrons) in the shell.

4 "Thermalization" refers to the process in which a high energy object (such as a particle) interacts with objects (atoms and molecules) of the surrounding environment. If those objects show (vibration) energy equivalent to room temperature, the higher-energy particles lose energy through interaction, and finally obtain the same energy as the surrounding particles. For particles at room temperature, kinetic energy (of electrons) is $E_e = 3/2\ k_B T$ (k^B = BOLTZMANN constant, T = absolute temperature) and follows a MAXWELL distribution. "Thermal energy" is equivalent to 0.025 eV.

12.2.2.1 Kinetic energies and range

For most unstable nuclides undergoing β^+-transformation, maximum kinetic energies of the positron emitted (similar to β^--particles) are in the range of about 1 to 2.5 MeV. Mean kinetic energies are about 1/3 of these maximum values. The beta particles travel different distances in gaseous, liquid and solid phases and are affected by the specific chemical elements or compounds constituting those phases.[5] Typical correlations between β-particle energies and distances are shown in Fig. 12.3. In air, β-particles of 1 MeV kinetic energy travel a ca. 3 m distance, while their range in water and aluminum is 4–5 mm and ca. 1.5 mm only.

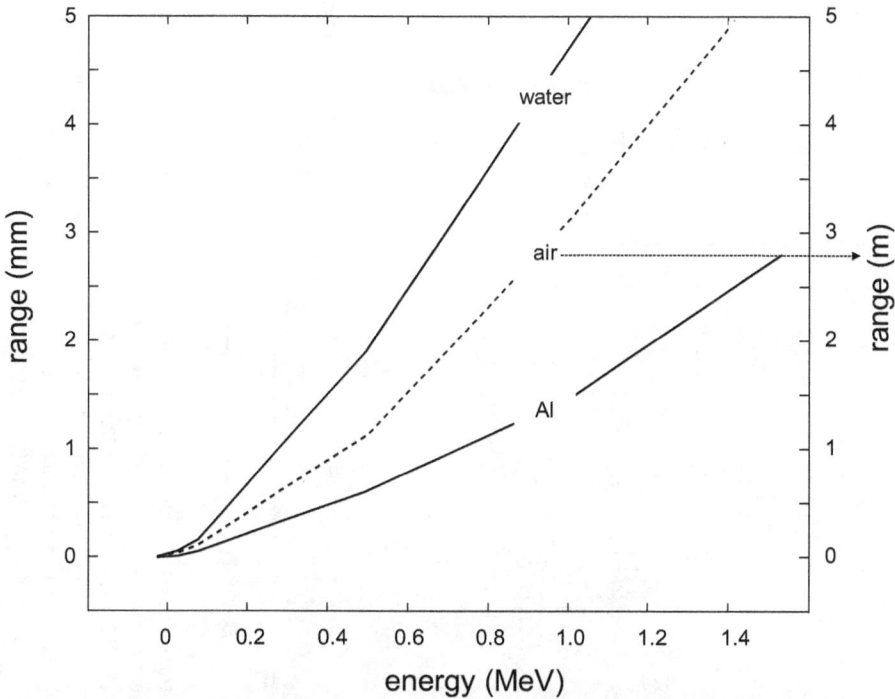

Fig. 12.3: Typical correlations between beta particle energy and "travel" distance in air, water and aluminum.

12.2.2.2 Continuous spectra of ranges of emitted positrons

Because of the continuous distribution of β-particle energies, there is a distribution in ranges of positrons for any given emitter. Figure 12.4 shows a simulation of how

---

5 Such correlations in terms of range of the beta particles in air have been used to correlate properties of nuclides undergoing beta-transformations such as beta-particle energy or Q_β-values; see. eq. (8.17).

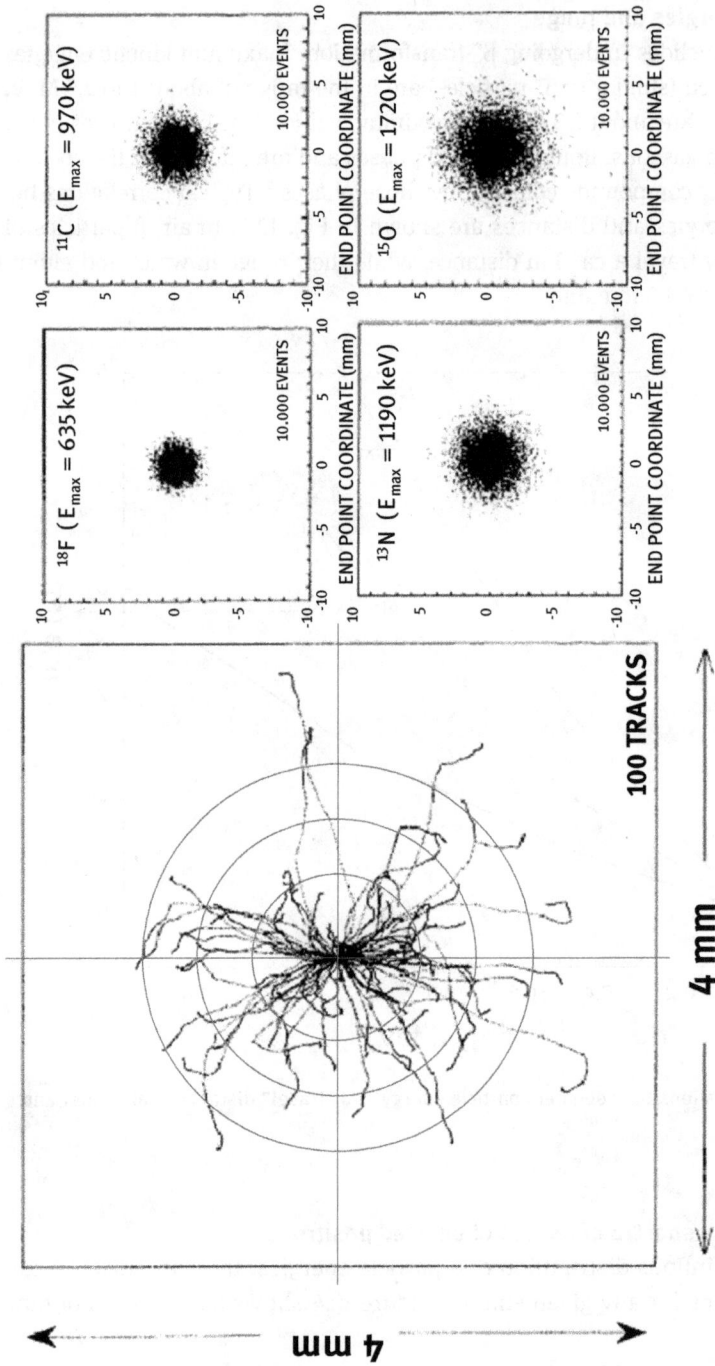

Fig. 12.4: Simulated tracks of positrons ejected from the decay of a point source of a sample of 100 nuclides of ^{18}F (top) and spherical distributions of ranges for prominent positron emitters (bottom). ^{18}F ($t_{1/2}$ = 110 min, E_{β^+} = 635 keV), ^{11}C ($t_{1/2}$ = 20 min, E_{β^+} = 970 keV), ^{13}N ($t_{1/2}$ = 10 min, E_{β^+} = 1190 keV) and ^{15}O ($t_{1/2}$ = 2 min, E_{β^+} = 1720 keV). Average range distributions in water are mainly determined by the mean kinetic energies of the positrons. They are in the order of ^{18}F < ^{11}C < ^{13}N < ^{15}O. See Fig. 8.7 for the distribution of kinetic energies of these nuclides. With permission: CS Levin, EJ Hoffman, Calculation of positron range and its effect on the fundamental limit of positron emission tomography system spatial resolution, Phys. Med. Biol. 44 (1999) 781–799.

positrons ejected from ^{18}F ($t\frac{1}{2}$ = 110 min, $E_{\beta^+}^{max}$ = 0.635 MeV) travel through water. The positrons show an average distance paralleling the mean kinetic energy (less than ca. 1 mm), while a relatively small number of β^+-particles reaches distances of about 1.5 to 2.5 mm. The maximum range is 3.8 mm, per Fig. 8.7.

12.2.3 The fate of the positron: Annihilation

In the case of pair formation due to secondary processes, the positron and the electron created annihilate immediately. In contrast, for positrons released in β^+-transformations, annihilation cannot occur directly due to the high kinetic energy of the positron emitted. Here, this positron first interacts by scattering with surrounding matter. Whenever inelastic scattering dominates, the positron stays alive and travels further. Finally, the kinetic energy of the positron approaches its zero value. Now, interaction of the positron with a shell electron becomes elastic and finally the positron may combine with an electron. This represents the formation of a pair of matter + antimatter particles at about zero kinetic energy and results in the annihilation of the two particles.

The mass of the two particles converts into energy according to $E = m_e c^2$. With the (rest) mass of the electron of m_e = 9.109383 · 10^{-28} g = 0.00054858 u and the energy equivalent of E_e = 510.9989 keV (see Table 2.1) the overall energy is 2 × 510.9989 keV = 1.0219978 MeV. This energy is emitted as electromagnetic radiation composed of two photons of 510.9989 keV each, emitted in opposite directions, shown in Fig. 12.5. These photons are measured γ-spectroscopically and appear in all the γ-spectra of unstable nuclides undergoing primary β^+-transformation or the secondary processes of pair formation. Figure 12.6 shows a γ-spectrum of ^{18}F.

12.2.4 More fates of the positron

Once inelastic scattering has reduced the kinetic energy of the positron to less than 1 keV, scattering becomes more and more elastic. With kinetic energies of about <100 eV, the positron has – in addition to immediate annihilation – two more interaction options. It can (1) form positronium Ps, and/or (2) ionize surrounding atoms and molecules M by combining with the atom. These competing processes of molecular ionization and/or positronium formation are schematically described by eqs. (12.2) and (12.3). The other pathway of the positron is not to hit a shell electron, but to combine with it. The Q-value for the formation of positronium depends on the ionization potential U_M of the counterpart electron of that molecule, eq. (12.4). This creates positively charged molecules, which remain for a very short period of time. Figure 12.7 summarizes the various process of a positron released through β^+-transformation.

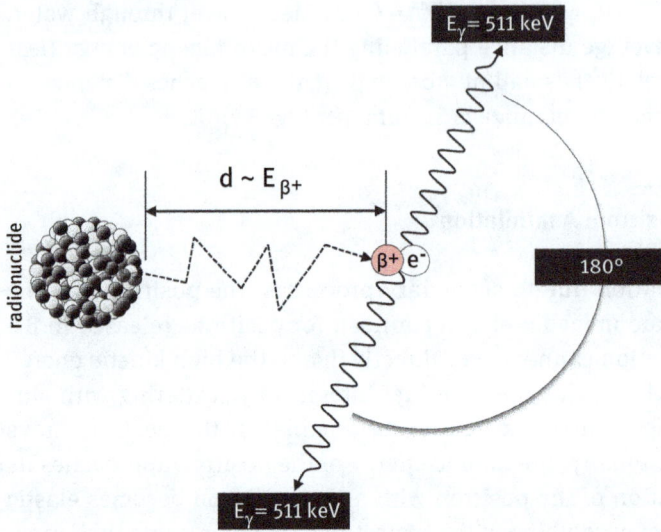

Fig. 12.5: The positron close to rest energy combines with an electron to annihilate. The overall mass of the positron and the electron is converted into two photons of 511 keV, which are emitted in opposite directions (shown here two-dimensionally).

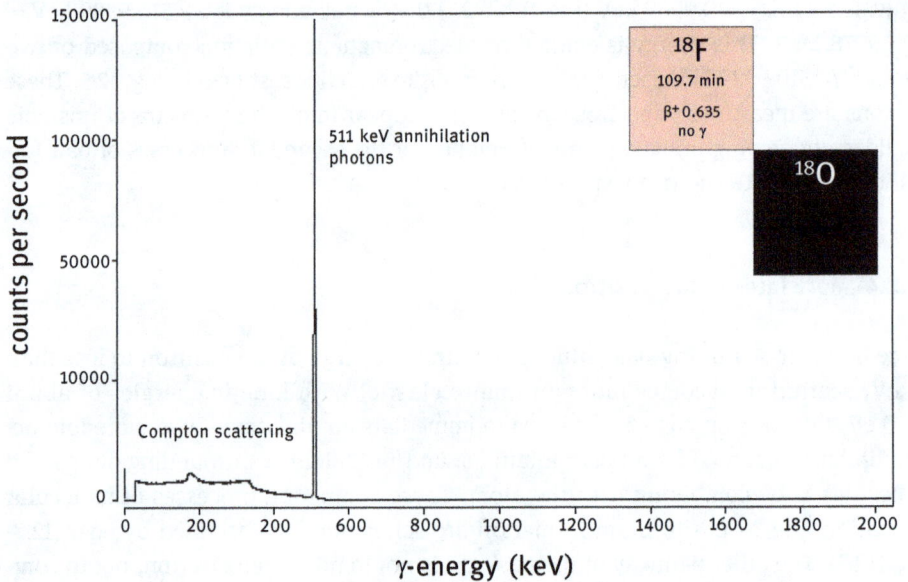

Fig. 12.6: γ-spectrum of a positron emitter (^{18}F). 511 keV annihilation photon emission is the dominant peak due to ca. 96.9% β$^+$-branching of the primary transformation ^{18}F → ^{18}O (3.1% ε). Other signals are low energy photons due to Compton scattering (see Chapter 1 of Volume II for detection of radioactivity). Note the linear scale of the y-axis.

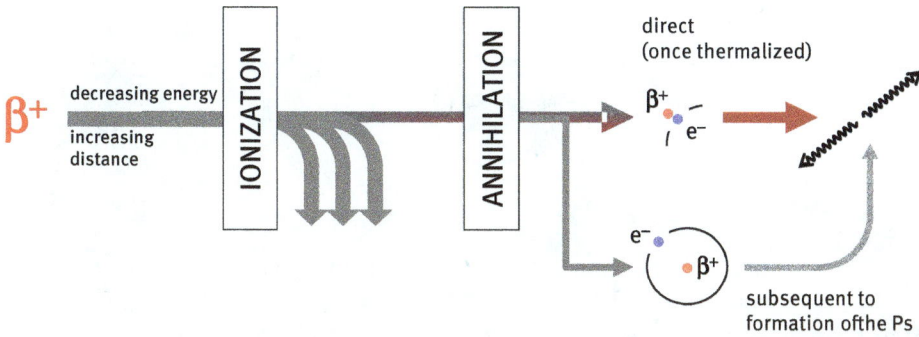

Fig. 12.7: Schematic overview of the late stage processes of positrons close to zero kinetic energy interacting with electrons, atoms or molecules. Direct annihilation is the most straightforward, but the formation of positronium may occur as an intermediate. Close to thermalization, the positron may also stick to the atom or molecule, forming positively charged intermediates e^+M (M^*).

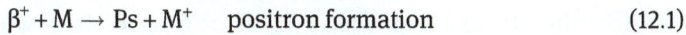

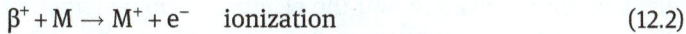

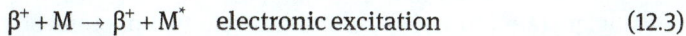

$$\beta^+ + M \rightarrow Ps + M^+ \quad \text{positron formation} \tag{12.1}$$

$$\beta^+ + M \rightarrow M^+ + e^- \quad \text{ionization} \tag{12.2}$$

$$\beta^+ + M \rightarrow \beta^+ + M^* \quad \text{electronic excitation} \tag{12.3}$$

$$Q_{Ps} = (U_M - 6.8)eV \tag{12.4}$$

12.2.5 Positronium

The formation and fate of positronium needs a more detailed discussion. First, what is positronium? The combination of the positron and electron may rest for a (very) short while, before it continues to annihilate. This intermediate status is (like a chemical element) called positronium, because it's understood to be atom-like and similar to hydrogen. It even has its own symbol: Ps. Figure 12.8 represents the structures of a hydrogen atom, anti-hydrogen atom, and positronium. The negatively charged shell electron is the same for H and Ps; the difference is the nucleus, with a positron for Ps instead a proton for H. Of course, the masses of Ps and H are very different. And because of the much lower mass of the positron, the electron is less attracted in Ps. This yields a larger radius for Ps and reduces its ionization potential.

Formation of positronium requires some kinetic energy of the positron to compensate for the binding energy of Ps, which is $E_{Ps} = 6.8$ eV. In addition, one must consider the ionization energy U_M of the counterpart electron in the shell of the atom or molecule. Thus, formation of Ps occurs at very low kinetic energy of the β^+-particle (a few eV, i.e. very close to rest energy).

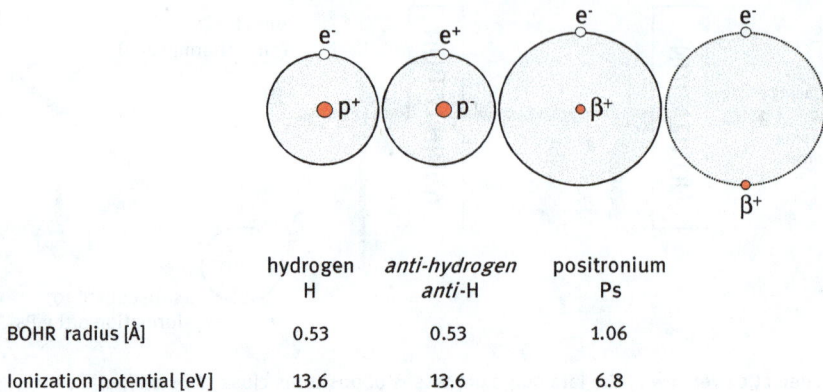

	hydrogen H	anti-hydrogen anti-H	positronium Ps
BOHR radius [Å]	0.53	0.53	1.06
Ionization potential [eV]	13.6	13.6	6.8

Fig. 12.8: The structure of positronium compared to the atoms of hydrogen and anti-hydrogen. Bohr radius and ionization potential are identical for H and anti-H, but different for Ps.

According to an ORE model, Ps may be formed either with anti-parallel or parallel spin of the positron and the electron, forming (similar to the hydrogen atom) either a singlet or a triplet quantum state (see Section 1.6.3 on multiplicity). Only the singlet state (ortho-Ps) annihilates into two photons of exactly 0.511 MeV. The triplet constellation (para-Ps) annihilates into three photons, still with an overall energy of 1.022 MeV, yet with individual energies of the three quanta distributed stochastically (see Tab. 12.2).

The formation of positronium is described through a formation probability (or reaction cross-section) σ_D in eq. (12.5).[6] It contains the dimension of the atom in terms of its area, $A = \pi r_o^2$. The velocity v of the positron is also included, relative to the speed of the light c. If this velocity approaches zero, eq. (12.5) simplifies to eq. (12.6). The direct formation of Ps prefers the ortho-Ps over para-Ps in a ratio of about 1:3. This would minimize the yield of the 2-photon annihilation. Yet in reality, the para-Ps transfers into the ortho-Ps under the influence of magnetic fields in the environment of positronium.[7] Following its formation, Ps annihilates with a characteristic speed. In theory, Ps has a mean lifetime of about 10^{-7} s. According to the different quantum states, annihilation velocities (and lifetimes) differ.[8] Annihilation constants of the singlet state are larger by a factor of 1115 compared to the triplet state. Interactions in

---

6 by DIRAC; therefore σ_D.

7 2-photon annihilation is also preferred because of the balance in angular momentum of the particles involved. Positrons and electrons are fermions of s = ½, while the photon is a boson. The two photons thus carry away either a total spin of 0 or 1, which matches the sum of the spins of the two fermions.

8 Both in part depend on the chemical environment the positronium is embedded in, i.e. the electron shell and nuclear values of the host atom, turning the concept into a spectroscopic method for materials analysis.

Tab. 12.2: Overview of the properties of ortho – and para-Ps. For its quantum mechanics, physico-chemical and nuclear parameters, see below.

Spin systems		ortho-Ps	para-Ps
	β^+ and e^- spin	Anti-parallel	Parallel
		Singlet $^1S_0$	Triplet $^3S_0$
	Quantum states	$L = 2J_s + 1 = 1$	$L = 2J_s + 1 = 3$
Energy [MeV]		$E_{\gamma 1} + E_{\gamma 2} = 1.022$ $E_{\gamma 1} = E_{\gamma 2} = 0.511$	$E_{\gamma 1} + E_{\gamma 2} + E_{\gamma 3} = 1.022$ continuous (stochastic distribution)
Annihilation constant λ [s^{-1}]		$\lambda^0{}_S = Z\,N\,\pi\,r_o{}^2\,c$	$\lambda^0{}_T = \lambda^0{}_S / 1115$
Life time τ^0_s[s]	Gas phase	$\approx 10^{-7}$	$\approx 10^{-7}$
	Condensed phase	$\approx 10^{-10}$	$\approx 10^{-10}$
Distribution of angular photon emission	$E_{kin} = 0$	180°	Stochastic
	$E_{kin} > 0$	180° ± x mradians	
Direction of transformation			

condensed phase, however, produce a kind of quenching, and the mean lifetime is much shorter, i.e. about 10^{-10} s. The basic equations are listed in Fig. 12.9.

Whereas the solid angle emission is stochastic for 3-photon annihilation, it is almost exactly 180° for 2-photon annihilation.[9] The effect of the simultaneous emission of two identical 511 keV photons in opposite directions is the basis of the so-called coincidence measurement in positron annihilation processes and represents one of the unique physical principles of Positron Emission Tomography (PET),[10] shown in Fig. 12.20.

---

9 This angle may only differ from the ideally opposite direction if the positron deposited a small amount of kinetic energy into the positronium being formed. In this case, relatively small deviations of a few milliradians occur.
10 See Chapter 13 in Volume II.

$$\sigma_D = \sigma_{2\gamma} = \frac{\pi r_0^2}{\Gamma + 1} \left[\frac{\Gamma^2 + 4\Gamma + 1}{\sqrt{\Gamma^2} - 1} \ln(\Gamma + \sqrt{\Gamma^2} - 1) - \frac{\Gamma + 3}{\sqrt{\Gamma^2} - 1} \right] \quad (12.5)$$

$$v \ll c$$

$$\Gamma = 1/\sqrt{1 - (v/c)^2}$$
$$r_0 = e^2/mc^2 = 2.8 \cdot 10^{-13}\,cm$$

$$\sigma_D = \sigma_{2\gamma} = \pi r_0^2 \frac{c}{v} \quad (12.6)$$

$$\frac{\lambda_{3\gamma}}{\lambda_{2\gamma}} = \frac{1}{1115} \quad (12.7)$$

$$\frac{\sigma_{3\gamma}}{\sigma_{2\gamma}} = \frac{\lambda_{3\gamma}\,(2/_T +1)}{\lambda_{2\gamma}\,(2/_s +1)} = \frac{1}{1155} \cdot \frac{1}{3} = \frac{1}{372} \quad (12.8)$$

Fig. 12.9: Formation probability σ_D and annihilation velocity λ of Ps, depending on singlet or triplet quantum states. Γ is the velocity of the electron relative to the speed of the light; r_0 is the radius of the electron. Transformation = annihilation constants are shown for 2-photon ($\lambda_{2\gamma}$) and 3-photon ($\lambda_{3\gamma}$) emissions.

12.3 Vacancies of shell electrons

Primary electron capture transformation of a proton-rich nuclide and a secondary internal conversion both release an electron from the electron shell of the unstable nucleus – leaving an empty space in that shell. Electron vacancies typically appear in electron shells closest to the nucleus. The K-shell is preferred, the L-shell is less affected, etc. Two different pathways handle this vacancy promptly. Both induce radiative emissions. The two options are:

(i) emission of electromagnetic radiation called X-rays and
(ii) ejection of electrons from shells of the transforming atom, called AUGER electrons (nowadays also called AUGER–MEITNER electrons) and COSTER–KRONIG electrons.

12.3.1 Electron vacancies filled *via* electromagnetic radiation emission

The first option to refill an electron vacancy proceeds through the import of electrons populating higher shells of that atom. Figure 12.10 illustrates the pathways for a vacancy in the K-shell or L-shell. Any electron hole is typically filled from an electron in the shell closest to it. Typically, an L-shell electron fills a K-shell electron vacancy. In parallel, an electron from the N-shell may transit, but this process is less abundant. Suppose the initial vacancy appeared in the L-shell; analogously, electrons from the M – or N-shell may fill that hole. The transitions are named K_α or K_β for L → K or M → K, and L_α or L_β for M → L or N → L routes, respectively. Here, K or L indicate the position of the hole to be filled, and the Greek index indicates

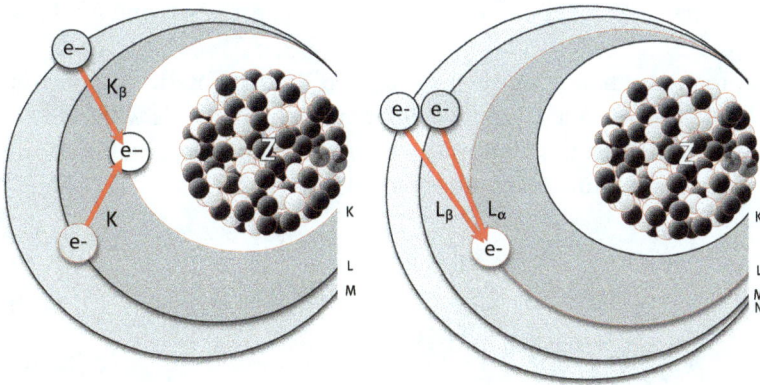

Fig. 12.10: Transition of electrons from higher shells to a vacancy in a K-shell (a) or L-shell (b) and their notations.

whether the arriving electron descended from the next proximate (α) or the next but one (β) main shell.[11]

12.3.1.1 Characteristic X-rays

Electron energies are a function of the main quantum number, as per eq. (1.1): $E_{B(e)}$ (n) $= - R_H Z^2 \cdot 1/n^2$. For a given atom (Z = constant), binding energies are $- E_{B(e)}(K) >$ $- E_{B(e)}(L) > - E_{B(e)}(M) > - E_{B(e)}(N)$ etc. The difference in binding energy of an electron between the initial state and final state (the original hole) is obtained through eq. (12.9). Consequently, there is a characteristic difference in energy ΔE, which is released in terms of electromagnetic radiation, depending on Z and Δn. As introduced in Chapter 11, this energy is called X-ray (rather than γ-ray because of its different origin).

For a given nucleus (Z), X-rays have characteristic values, depending on the shells involved. For transitions of type K_α and K_β, for example, energies (in eV) are calculated as $K_\alpha = f(Z)$, etc. With $R_H = 13.6$ eV $= 3.29 \cdot 10^{15}$ s^{-1}, $n_i = 1$ and $n_f = 2$ for e.g. K_α, direct proportionalities are obtained for the characteristic X-ray energies of each transition; see eqs. (12.10a, b) and (12.11). For a K_α transition, $1/n_i^2 - 1/n_f^2$ becomes $1/1^2 - 1/2^2 = 1 - \frac{1}{4} = \frac{3}{4}$ and eq. (12.10a) turns into $E_{K\alpha} = 10.2$ eV $\cdot Z^2$. The frequency of the X-ray emitted is found in accordance with the RYDBERG constant using eq. (12.10b).

---

[11] This is analogous to the hydrogen emission spectrum. For the BALMER series, transition for M $\rightarrow$ L or N $\rightarrow$ L would correspond to the notation of $L_\alpha = 656.3$ nm or $L_\beta = 486.1$ nm wavelength. Because much heavier atoms are concerned for the post-effects, the ΔE values between the shells are much lower and the electromagnetic emissions are not within the segment of visible light, as per Fig. 11.5.

The relationship of type $v \approx Z^2$ turns into $v^{1/2} \approx Z$ with $(K_\alpha) = \text{const} (Z - 1)$ and $v^{1/2} (K_\beta) = \text{const} (Z - 7.4)$.

Furthermore, there are notations such as $K_{\alpha 1}$ and $K_{\alpha 2}$. This indicates electron transitions into the K-shell vacancy from two different energetic levels within the L-shell (due to the different quantum numbers $l = 0$ and 1). Their differences in ΔE are very low, and their relative abundances are about 2:1.

$$\Delta E = h v = \frac{h c}{\lambda} = E_i - E_f = R_H \left(\frac{1}{n_i^2} - \frac{1}{n_f^2} \right) Z^2 \tag{12.9}$$

$$E_{K\alpha} = 13.6 \left(\frac{1}{1_i^2} - \frac{1}{2_f^2} \right) Z^2 \ (\text{eV}) \tag{12.10a}$$

$$v_{K\alpha} = 3.29 \cdot 10^{15} \left(\frac{1}{1_i^2} - \frac{1}{2_f^2} \right) Z^2 \ (\text{Hz}) \tag{12.10b}$$

$$K_\beta = 13.6 \left(\frac{1}{2_i^2} - \frac{1}{3_f^2} \right) Z^2 \ (\text{eV}) \tag{12.11}$$

12.3.1.2 MOSELEY's law

In 1913, Henry MOSELEY experimentally analyzed X-ray (emission) spectral lines of many chemical elements following external excitation with electrons. In doing so, he discovered a proportionality of type $v^{1/2} \approx Z$ for all the chemical elements. His expression of $(v_{K\alpha})^{1/2} = k_1 (Z - k_2)$ used two empirical constants, with $k_2 = 1$ for K_α transitions and $k_2 = 7.4$ for L_α transitions. With k_1 reflecting the term $(1/1_i^2 - 1/2_f^2)$ in eq. (12.10a), it yields eq. (12.13). Similarly, eq. (12.14) gives the correlation for L_α transitions. The term $(Z - 1)^2$ instead of Z^2 in eq. (12.9) is interpreted as a semi-empirical fit of experimental results, but also has a physical meaning, which is due to the different electron-electron repulsion of the transit electron for initial and final shell levels. This is sometimes denoted as "effective charge" Z_{eff} of the nucleus in eq. (12.9).

$$(v_{K\alpha})^{1/2} = k_1 (Z - k_2) \tag{12.12}$$

$$(v_{K\alpha})^{1/2} = 3.29 \cdot 10^{15} \cdot \frac{3}{4} (Z - 1)^2 = 2.47 \cdot 10^{15} (Z - 1)^2 \ (\text{Hz}) \tag{12.13}$$

$$(v_{L\alpha})^{1/2} = 3.29 \cdot 10^{15} \cdot \frac{5}{36} (Z - 7.4)^2 = 4.57 \cdot 10^{14} (Z - 7.4)^2 (\text{Hz}) \tag{12.14}$$

With the systematic correlation of X-ray energy to proton number Z of the nucleus, the discrete energies of either K or L emissions can be calculated. Furthermore, any

chemical element can be unambiguously analyzed from the experimentally obtained values for K_α or K_β (see Tab. 12.3).[12]

Tab. 12.3: X-ray emission energies of selected chemical elements. K_α, K_β, L_α, L_β, etc. values
increase with increasing Z. Because ΔE is smaller for L → K instead of M → K transitions, K_β-values
are larger by ca. 10% than the corresponding K_α-values of the same element [1]. Similarly, because
ΔE is smaller for M → L compared to L → K transitions, L_α-values are lower by a factor of ca. 10
than the corresponding K_α-values of the same element [2].

Element		X-ray energy (keV)			
Z		K_α	K_β	L_α	L_β
12	Mg	1.253	1.295		
14	Si	1.740	1.829		
16	S	2.307	2.464		
18	Ar	2.957	3.190		
20	Ca	3.691	4.012	0.341	0.345
22	Ti	4.510	4.931	0.452	0.458
24	Cr	5.414	5.946	0.573	0.583
26	Fe	6.403	7.057	0.705	0.718
28	Ni	7.477	8.263	0.851	0.869
30	Zn	8.637	9.570	1.012	1.034
38	Sr	14.163	15.833	1.806	1.871

---

12 This contributed to the better arrangement of the chemical elements within the PTE not just
according to mass, which sometimes gives an irregular positioning, but clearly by Z. In initial structures of the PTE with elements arranged according to increasing mass, the sequence for the last
elements of the fifth period according to their mass would have been Sb (1 mol = 118.71 g) < I
(126.90 g) < Te (127.60 g) < Xe (131.29 g). This contradicted the idea of groups in the PTE containing
elements of similar chemistry, thus the correct order here should be Sb < Te < I < Xe. MOSELEY's contribution could ultimately clarify the situation, arranging the elements by proton number, i.e. $_{51}$Sb
< $_{52}$Te < $_{53}$I < $_{54}$Xe. Moreover, the correlations of type $v^{1/2} \approx Z$ also confirmed the BOHR model of the
structure of electron shells. It was even used as an approach to search for chemical elements miss-
ing from the early PTE, such as 43 (technetium), 61 (promethium), 72 (hafnium), and 75 (rhenium).
In 1914 MOSELEY could *predict* their X-ray spectral lines. Accordingly, the two stable elements Hf
and Re were both discovered through X-ray spectroscopy. Both metals were chemically isolated
and their existence verified. In 1925, NODDACK, TACKE and BERG reported that they had detect element 43 in platinum ore, in the mineral columbite, a niobium ore [(Fe,Mn)Nb$_{O6}$], in gadolinite and
in molybdenite through the X-ray signal of a wavelength characteristic for element 43. Because the
element itself could not be isolated chemically, the final proof remained open. Today the latter appears to be logical, as this (radio)element does not exist in measurable amounts – there is no stable
isotope of technetium!

Tab. 12.3 (continued)

Element		X-ray energy (keV)			
40	Zr	15.772	17.665	2.042	2.124
56	Ba	32.188	36.372	4.465	4.827
58	Ce	34.714	39.251	4.839	5.261
60	Nd	37.355	42.264	5.229	5.721
72	Hf	55.781	63.222	7.898	9.021
82	Pb	74.965	84.922	10.550	12.612
90	Th	93.334	102.591	12.967	16.199
92	U	98.422	111.281	13.612	17.217

12.3.1.3 X-ray spectra

When low-energy electromagnetic radiation is measured from a radionuclide undergoing EC or IC transformations, spectra are recorded as illustrated in Fig. 12.11. For light elements of about Z < 30, K_α X-ray energies are <10 keV, while heavier elements cover the range of 10 to 100 keV. Absolute values of L_α cover a range of about 0.34 ($_{40}$Ca, i.e. with the beginning of the L-shell) to 14 keV for $_{92}$U. Electron transitions from, for instance, an L-shell into a K-shell vacancy (K_α) turn into a cascade process, because the L-shell electron that refilled the hole in the K-shell left a new hole in its original L-shell. This will be refilled by e.g. an M-shell electron, inducing an L_α X-ray emission, etc. These cascade X-ray emissions show decreasing energies.

12.3.2 Electron vacancies filled *via* electron emissions

Electron vacancies can also be refilled by a *radiation-free* process. The basic step remains the same: transit of an electron from a nearby higher shell. However, ΔE is not released as an X-ray, but is used to release another electron from a higher shell. The process is referred to as a radiationless "reorganization" due to a "direct" interaction of two electrons. This particular electron is immediately ejected from the atom; in this case, no electromagnetic radiation is emitted. If electrons are emitted in the course of processes between different main shells (interstate transitions), they are called AUGER electrons (A). If the pathway involves subshells within the same main shell, they are called intrastate transitions or COSTER–KRONIG electrons (CK).

12.3.2.1 Mechanism of (A) and (CK) electron emission

Figure 12.12(a) illustrates schematically how this looks for a K-shell vacancy, refilled by an L-shell electron under simultaneous release of an M-shell (A) electron, for

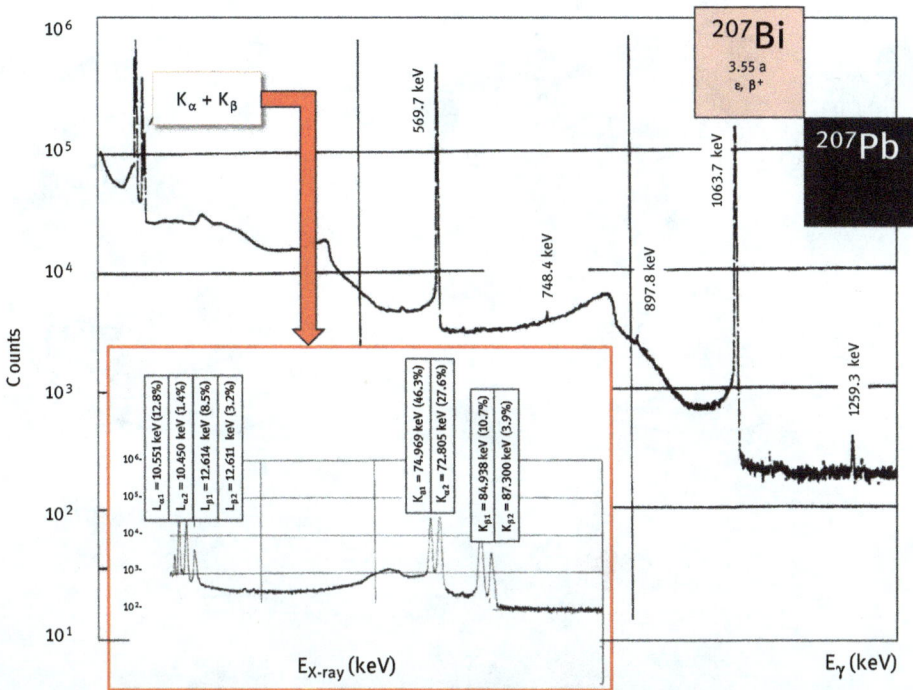

Fig. 12.11: Electromagnetic emissions of ^{207}Bi: X-ray spectra and γ-photons. Primary EC and β$^+$-transformations of ^{207}Bi (t$\frac{1}{2}$ = 31.55 a) induce secondary photon emissions (Chapter 11) and post-effect characteristic K$_\alpha$ and K$_\beta$ lines. These are due to the transformation product nucleus: stable ^{207}Pb. The individual X-rays are indicated with their corresponding energy and intensity. Courtesy: TS Vylov et al., Spektren der Strahlung radioaktiver Nuklide, gemessen mit Halbleiterdetektoren, ZfK-399, Rossendorf, 1980.

example. The terminology is KLM. This fills the initial hole, but leaves *two*[13] new holes: one in the L-shell and one in the M-shell. The subsequent vacancies may then be handled the same way, for example by using an M-shell electron to "refill" the L-hole, thereby (for example) releasing an N-shell (A) electron (denoted as LMN, Fig. 12.12(b)). (CK) electron pathways would be LLM, for example.

A complementary model of origin of (A) and (CK) electrons to explain their huge number is drafted in Fig. 12.12(c). It gives a somewhat oversimplified representation of cascade processes resulting in the release of a high number of (A) and (CK) electrons. Suppose not just one, but two L-shell electrons start to simultaneously transit to the initial K-shell vacancy; they would leave two new holes in their L-shell. Only one is "accepted" in the initial K-shell vacancy, while the other is ejected from the

---

13 Remember that the X-ray cascade mechanism only induced one new electron vacancy for each initial hole refilled.

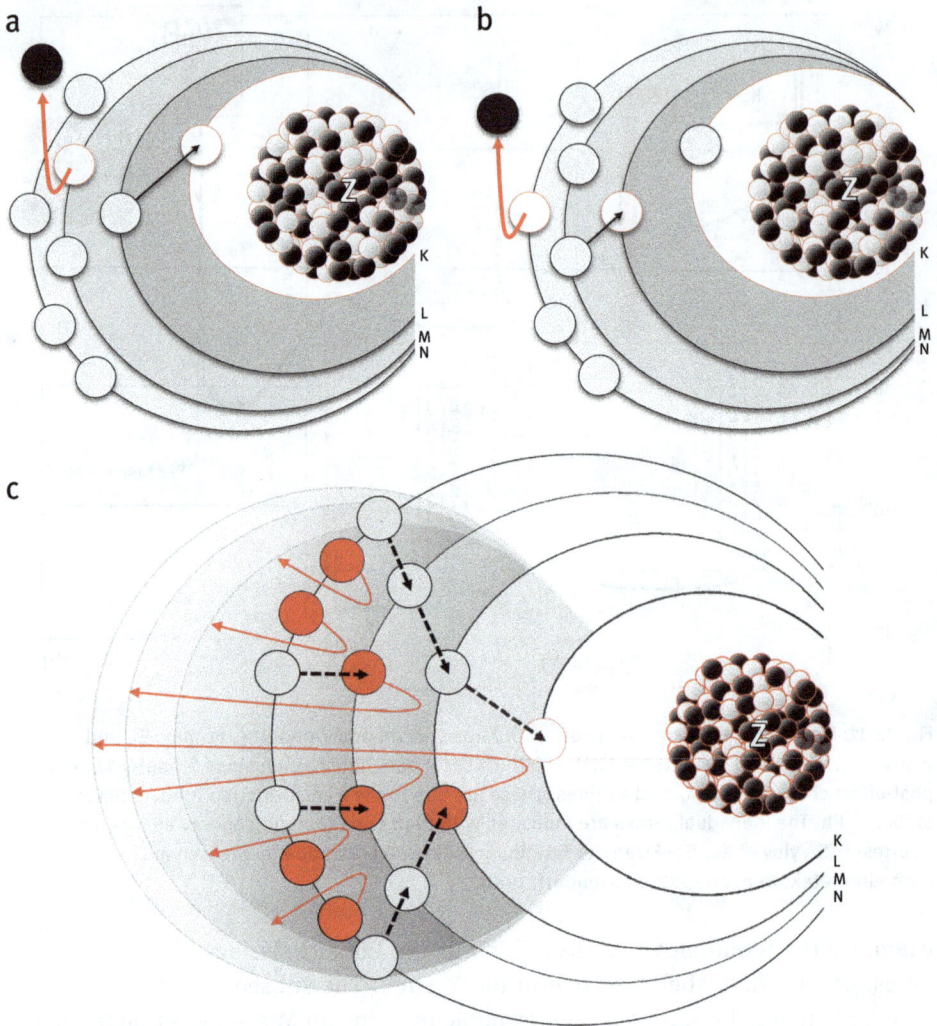

Fig. 12.12: Model of successive Auger electron emissions. (a) An initial vacancy in the K-shell is filled by an L-shell electron, thereby causing a new vacancy in the L-shell. While the initial hole is being filled, one M-shell electron leaves the atom (red line). The nomenclature of these emitted Auger electrons is KLM. (b) The vacancy created in the L-shell is refilled analogously by e.g. an LMN approach. Another Auger electron is emitted, this time from the N-shell. (c) Let us assume that not only one L-shell electron identified the "energetically promising" vacancy in the K-shell, but two. Only one would be successful in occupying the hole in the K-shell. The other, once departed from its former L-shell, cannot return. At the same time, the two holes the two L-electrons left when transiting to the K-shell vacancy are immediately identified by e.g. M-shell electrons. Let us assume that now four of the M-shell electrons start to improve their binding energy, but again: just two will be successful. All the electrons once released from their shells cannot return, because electrons that transited from the next highest energy shell have now occupied their earlier positions. They thus leave the atom. For the hypothetical model given, seven electrons have already left the atom. They represent three individual shells energies (gray areas).

atom. At the same time, the two new holes created in the L-shell are identified by four electrons of the M-shell. Similarly, all 4 start to transit, but again only two find their destiny in the L-shell. The two others are released as (A) electrons. At this stage, 7 (A) electrons have already been ejected. Next, 8 N-shell electrons address the four new holes in the M-shell (and so on). The process continues in a kind of cascade or avalanche. Figure 12.13 shows the frequency of the overall (A) + (CK) electrons released per primary transformation step of ^{125}I and their individual energies. Although each electron has an individual discrete energy, the various electrons emitted within several shell cascades represent a mixture of several characteristic energies, shown in Fig. 12.13.

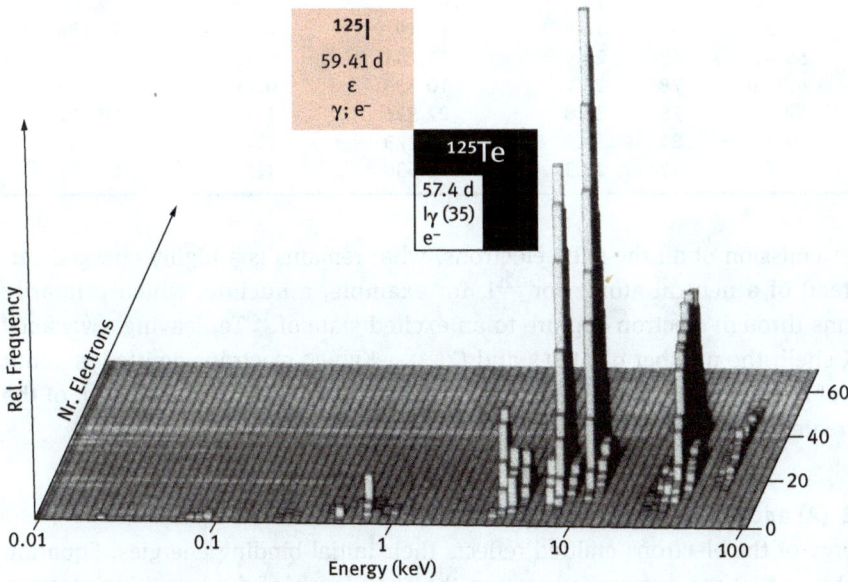

Fig. 12.13: Auger and Coster–Kronig electron emission from 125I. Frequency of the overall (A) + (CK) electrons released per primary transformation step of 125I and their individual energies. With permission: DE Charlton, J Booz, KFA J 1979.

The overall number of electrons ejected is thus larger than the number of all the shells (and subshells) of a given atom. Obviously, the number is larger in the case of chemical elements of high main quantum number relative to those located in lower quantum number periods. Table 12.4 summarizes the total number of AUGER and COSTER–KRONIG electrons ejected for representative nuclides, as well as their overall energy. It also compares this with the total number (and energy) of X-rays simultaneously emitted within a single nuclide transformation. Indeed, the overall number of (A) and (CK) electrons ejected increases with Z, but does not directly mirror the number of atomic electron shells in the corresponding elements.

Tab. 12.4: Total number of Auger and Coster–Kronig electrons ejected and their overall energy compared to the number and energy of X-rays simultaneously emitted for representative nuclides (calculated). From: W Howell, Radiation spectra of Auger-electron emitting radionuclides: Report No. 2 of AAPM Nuclear Medicine Task Group No.6. Medical Physics, 19 (1992) 1371.

Nuclide	t½	Z	(A) and (CK) electrons		X-ray fluorescence	
			Total yield (number)	Total energy (keV)	Total yield (number)	Total energy (keV)
			Per single transformation		Per single transformation	
^{67}Ga	78.3 h	31	4.7	6.264	0.57	4.936
^{99m}Tc	6.0 h	43	4.0	0.899	0.079	1.367
^{111}In	2.81 d	49	14.7	6.750	0.89	19.966
^{123}I	13.2 h	53	14.9	7.419	0.93	24.134
^{125}I	59.41 d	53	24.9	12.241	1.53	39.661
^{193m}Pt	4.33 d	78	26.4	10.353	0.43	12.345
^{195m}Pt	4.02 d	78	32.8	22.526	1.5	60.112
^{201}Tl	73.1 h	81	36.9	15.273	1.4	72.947
^{203}Pb	51.9 h	82	23.3	11.630	1.3	68.897

After emission of all the shell electrons, what remains is a highly charged cation instead of a neutral atom. For ^{125}I, for example, a nuclide, which primarily transforms through electron capture to an excited state of ^{125}Te, leaving a vacancy in the K-shell, the number of Auger and Coster–Kronig electrons emitted is about 25. This of course must cause dramatic changes in the chemical environment of the newly created atom.[14]

12.3.2.2 (A) and (CK) electron energies

The energy of the electrons emitted reflects their initial binding energies. Equation (12.15) shows that the difference between the shell levels of the transiting electron from initial to final level is defined by the individual shell parameters n and l in terms of $\Delta E_{(n,l)}$. This amount of energy is sufficient to release another electron, the final (A) or (CK) electron, from its shell defined through that electron binding energy $E_{B(e)}$. Most of the electrons released originate from outer shell levels, and accordingly their energies are relatively low. Energies of (CK) electrons are low because of the small differences in electron energy of subshells (quantum numbers s, p, d, f) within a given main quantum number n; this is particularly true for higher values such as n = 3, 4, 5 and 6. For probabilities between individual transitions, selection rules apply.

---

14 Chemical consequences of processes like this are discussed in Chapter 7 in Volume II in the context of radiochemical separations.

$$E_{(A), (CK)} = E_{B(e)} - \Delta E_{(n, l)} \tag{12.15}$$

Typically, (A) and (CK) electron energies are approximately 20 to 500 eV. Their range in aqueous phase (i.e. biological systems), for example, is short, about 1 to 10 nm.[15]

12.3.3 X-ray emission *vs.* electron emission

Once the vacancy of an s-shell electron has been handled, the radiative and radiationless post-effects proceed in parallel, yet to different extents. The fluorescence yield $\omega_{X\text{-ray}}$ (ω_K, ω_L, ...) gives the percentage for the filling of an electron vacancy through radiative processes. The total process of addressing the hole created by primary or secondary transformations of an unstable nuclide is then $\omega_{X\text{-ray}} + \omega_{(A)+(CK)} = 1$. The ratio between X-ray and AUGER and COSTER–KRONIG electron emission depends on the proton number of the nucleus and thus the absolute energies of the electrons and the relative differences between individual electron shell levels. Fluorescence dominates at higher Z, which is for nuclides of the heavier elements. Figure 12.14 illustrates this correlation in terms of the fluorescence yield for K-shell vacancies.

12.4 Example applications of post-effects

12.4.1 Medical applications: Positron and molecular imaging

Positrons are antimatter particles, derived from artificially produced radionuclides. It is fascinating that these elementary particles constitute the physical basis of a very important medical application: Positron Emission (sic!) Tomography (PET). The coincident mode of registering the two 511 keV photons created from the annihilation event is used to determine the distribution of a radionuclide or its labeled molecule within the human body. This is done with high spatial and temporal precision, as shown in Fig. 12.15. This technology represents one of the most exciting directions of *in vivo* molecular imaging.[16]

---

15 Yet the biological efficacy of (A) and (CK) electrons is high, because their kinetic energy is transferred to ionization processes over a very short distance; the Linear Energy Transfer is large (LET). This becomes relevant in terms of endogenous radiation therapy, discussed in Volume II, Chapter 14.
16 The state-of-the-art in clinical applications of the corresponding radiopharmaceuticals for PET is presented in Volume II: Modern Applications of Nuclear and Radiochemistry; specifically in Chapter 12, entitled "Life Sciences: Nuclear Medicine Diagnosis".

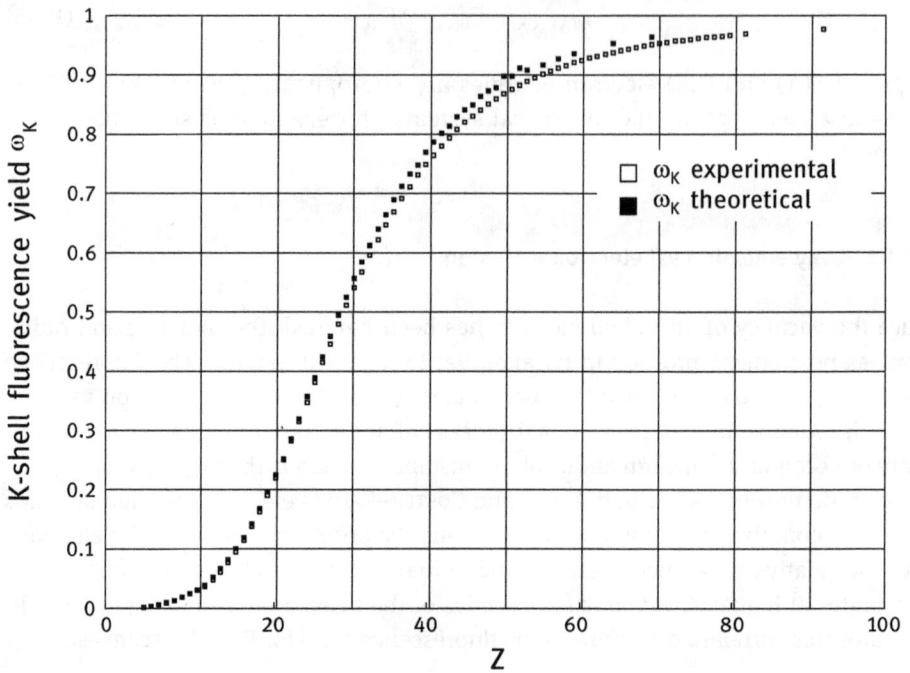

Fig. 12.14: Fluorescence yield for K-shell vacancies.

A variety of positron-emitting nuclides are utilized, with ^{18}F representing the most frequently applied one in clinical routines. Following a sophisticated radio-pharmaceutical chemistry, thousands of patients are diagnosed every day using biologically and medically relevant molecules with a high sensitivity for a given disease or physiological process. The most commonly applied PET radiopharmaceutical is a derivative of glucose,[17] namely 2-[^{18}F]fluoro-2-deoxy-glucose ([^{18}F]FDG). The radioactivity distribution measured in absolute units of Bq/cm^3 reflects the biochemistry *in vivo* and may be correlated with glucose consumption in mol/ml·min or g/ml·min or. This is mainly used to detect tumors and metastases because of their enhanced glycolysis rate, but is also very useful to analyze processes in the human brain. The brain utilizes about 90% of all glucose in a human body to satisfy its demand for energy. Figure 12.16 shows [^{18}F]FDG distribution in the brain of a healthy volunteer.

---

17 FDG is accumulated in cells of high glycolysis rate similar to glucose. Once it has entered the cell, the first enzymatic degradation forms the 2-[^{18}F]fluoro-2-deoxy-6-phosphate *via* the hexokinase enzyme similar to endogenous glucose-6-phosphate. However, in contrast to glucose-6-phosphate, the fluorinated 2-deoxy-6-phosphate of glucose is not further metabolized and remains trapped in the cell. This results in an enhanced uptake of ^{18}F in all cells of enhanced glucose turnover.

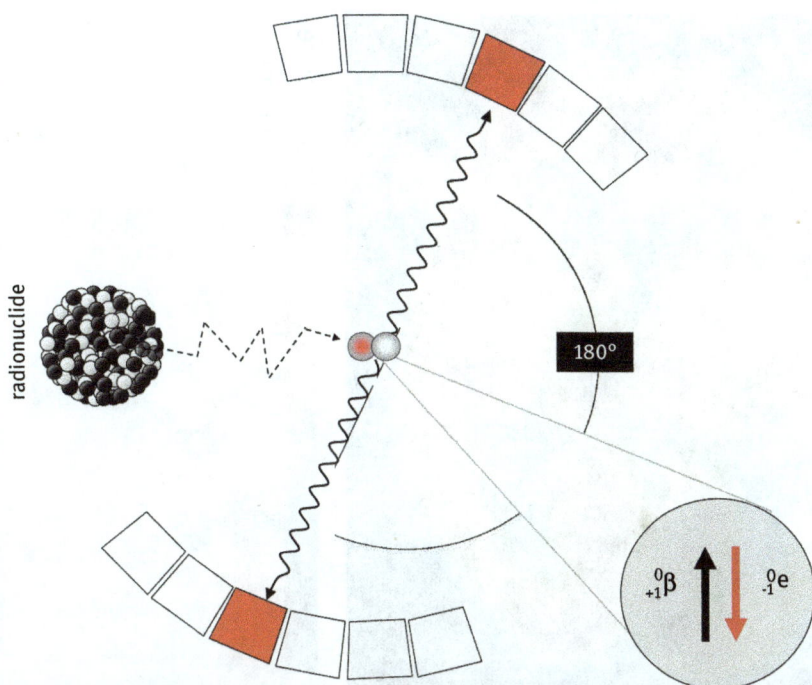

Fig. 12.15: The principle of PET detection. Coincident registration of two 511 keV photons originating from the annihilation of a positron + electron pair. Only those events are used for tomographic measurements that are registered simultaneously ("co-incident") in two opposite detectors (red) within ca. 10 ns.

Although [^{18}F]FDG is by far the most important radiopharmaceutical in nuclear medicine molecular imaging *via* PET and PET/CT, there are many other ^{18}F-labelled molecules which allow for the diagnosis of both normal and malignant physiological processes. In addition to ^{18}F, a number of other positron emitting radionuclides are applied clinically, such as ^{11}C, ^{13}N, and ^{15}O (see Tab. 12.5).

Practically relevant positron emitters of larger mass number A are rather rare (remember that electron capture is favored over β^+-processes with increasing A). Exceptions are ^{68}Ga and ^{82}Rb, which are both relatively short-lived (67.7 and 1.27 min) and available from radionuclide generators (^{68}Ge/^{68}Ga and ^{82}Sr/^{82}Rb with parent half-lives of 270.8 d and 25.34 d, respectively). Their positron emission branching is 88.0% and 96.0%, respectively. Other candidates between 40 < A < 90 are ^{44}Sc (t½ = 3.97 h, 94.3% β^+), ^{72}As (t½ = 26 h, 16.3% β^+), ^{74}As (t½ = 17.77 d, 26.0% β^+), ^{90}Nb (t½ = 14.6 h, 51.1% β^+) and ^{89}Zr (t½ = 3.27 d, 23.0% β^+). The only relevant positron emitter of A > 100 is ^{124}I (t½ = 4.15 d, 24.0% β^+). Both this positron-emitting halogen isotope and the metallic isotope ^{89}Zr are systematically used for imaging slower biological processes by means of PET.

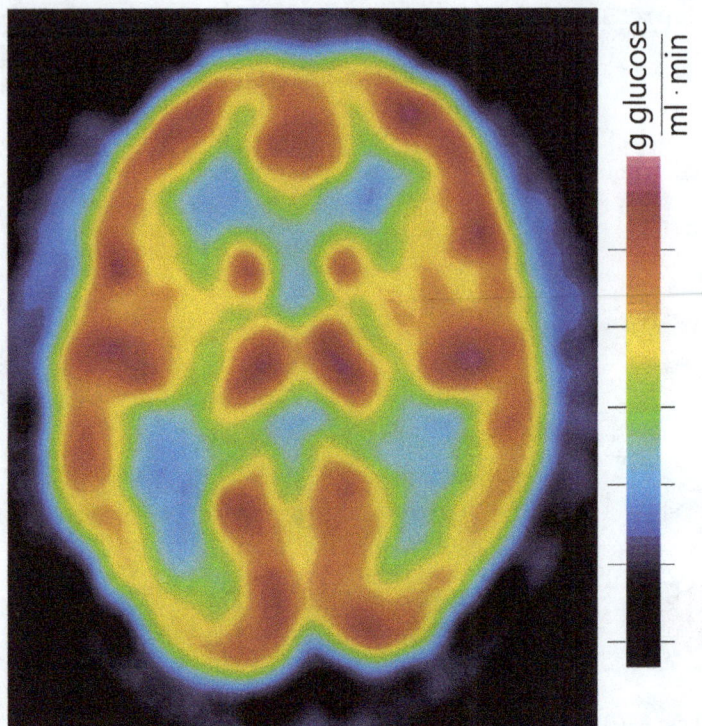

Fig. 12.16: [^{18}F]FDG PET scan illustrating glucose metabolism in the brain of a healthy living human. Normal glucose metabolism was seen in the 42-year-old volunteer (the author himself).

12.4.2 Medical applications: Single photon emitters

Single photon (or commonly γ-) emitters have been discussed in Chapter 11. Some of these radionuclides are characterized by only one (or mainly one) γ-emission between different excited nuclear states. Generally, this transition is in high abundance and it emits a relatively low energy photon (ca. 100–300 keV, ideal to be registered by dedicated detectors). Such radionuclides should also have a clinically and commercially convenient half-life, as well as an efficient production route. This profile qualifies a few radioisotopes to function as radioisotope for SPECT, which is single photon emission (sic!) computed tomography. Table 12.5 list the most prominent candidates.[18]

---

18 The state-of-the-art in clinical applications of the corresponding radiopharmaceuticals is presented in Volume II: Modern Applications of Nuclear and Radiochemistry; specifically in Chapter 12, entitled "Life Sciences: Nuclear Medicine Diagnosis". This chapter also includes important radiopharmaceuticals for nuclear medicine molecular imaging *via* SPECT and SPECT/CT.

Tab. 12.5: Important β⁺- and single photon (i.e. γ)-emitting radionuclides applied in nuclear medicine used for PET and for SPECT, respectively. (*) BNM-LNHB/CEA-Table de Radionucéides.

Positron emitters	Half-life	Z	Primary transformation	Branching (%)
^{11}C	78.3 h	6	β^+	99.75*
^{13}N	6.0 h	7		99.8*
^{15}O	2.81 d	8		99.9*
^{18}F	13.2 h	9		96.86*
^{62}Cu	9.74 min	29		97.6
^{64}Cu	12.7 h	31		17.9
^{68}Ga	1.13 h	31		88.88*
^{82}Rb	1.27 min	37		95
^{86}Y	14.74 h	39		34
^{89}Zr	3.27 d	40		22.8
^{124}I	4.15 d	53		22

Single photon emitters	Half-life	Z	Main secondary transformation	Branching (%)
^{67}Ga	3.26 d	31	EC	100*
^{99m}Tc	6.0 h	43	IC	100
^{111}In	2.81 d	49	EC	100
^{123}I	13.2 h	53	β^-	97*
^{125}I	59.41 d	53	EC	100
^{161}Tb	6.90 d	65	β^-	100
^{193m}Pt	4.33 d	78	IC	100
^{195m}Pt	4.02 d	78	IC	100
^{201}Tl	3.05 d	81	EC	100

12.4.3 X-rays and AUGER electrons induced through external excitation

The secondary and post-processes described above occur in the transformation of unstable nuclides. However, emission of X-rays and AUGER electrons can also be induced artificially. Whenever a vacancy in an inner shell of a stable atom is created, for example by ejecting it *via* the supply of electromagnetic energy, the processes of refilling that vacancy are exactly identical to those occurring in radionuclide transitions. For example, X-rays have been used for medical diagnosis since the early days of KONRAD RÖNTGEN. In order to induce a high flux of characteristic X-rays, heavy elements such as tungsten ($Z = 74$) (remember: fluorescence yield is high for high Z) are irradiated by energetic (up to 150 kV) and high-flux (1 mA to 1 A) electrons (at MOSELEY's time called an "electron gun"). This is organized within an "X-ray" vacuum tube with a cathode to release the electrons, a high voltage to accelerate them, and an anode which becomes the source of the X-rays, such as tungsten W. This yields a permanent emission of characteristic X-rays, though most of the energy (ca. 99%) of the electrons reaching the anode is converted to heat (which is the reason to prefer high-melting metals such as W of 3422 °C melting point).

In addition to the medical use of characteristic X-rays, there is an impressive number of physico-chemical analytical technologies based on characteristic X-rays such as X-ray emission spectroscopy (XES) and X-ray fluorescence spectroscopy (XRF).[19] The same is true for positron annihilation. There are also several analytical techniques based on the annihilation process of the positron, applied to physico-chemical and materials analytics, such as AUGER electron spectroscopy.

12.5 Outlook

12.5.1 Inner bremsstrahlung as another post-effect

In addition to the two post-effects discussed, there are some further effects that cause radiative emissions. They are basically discussed in terms of the interaction of radiation with condensed matter. One of these interactions should be mentioned here, because it induces a kind of electromagnetic radiation that is observed in the emission spectra recorded for a certain unstable nuclide. When an electron (liberated through internal conversion, for example) passes the proximity of its atomic nucleus, i.e. its strong electric field, it loses energy: The interaction leads to a loss in kinetic energy ("to brake"). Due to the law of conservation of energy the energy lost is converted into electromagnetic energy (in this case photons); referred to as "bremsstrahlung",[20] as per Fig. 12.17. It is also called internal or "inner (or internal) bremsstrahlung".[21]

Its branching, however, is relatively low. The number of bremsstrahlung photons per transformation is $<10^{-3}$ $(\Delta E)^2$. Their energy is lower than the energy of X-rays emitted for the same nuclide. A complete low-energy electromagnetic spectrum thus may reveal both the characteristic X-rays and, at lower energy and with lower intensity, continuous bremsstrahlung.

In addition, further effects arise from the interaction of the initial radiation caused by primary, secondary and post-effect, i.e. both particulate and electromagnetic, with condensed matter. However, these kinds of radiation interact with surrounding matter in very different ways. Because this interaction is (in part) an inherent effect in radiation measurement, these effects are discussed in Chapter 1 of Volume II in more detail. It also becomes relevant in the context of how radiation interacts with

---

19 See Chapter 6 of Volume II on "Chemical speciation of radionuclides in solution".

20 From the German wording *"bremsen"* for "to brake" and *"Strahlung"* for "radiation". Bremsstrahlung is induced whenever a charged particle transits with a given kinetic energy along other charged particles, i.e. deceleration of a charged particle.

21 Consequently, the interactions of electrons emitted in radioactive transformations with "foreign" atoms representing the surrounding condensed matter is the "outer bremsstrahlung".

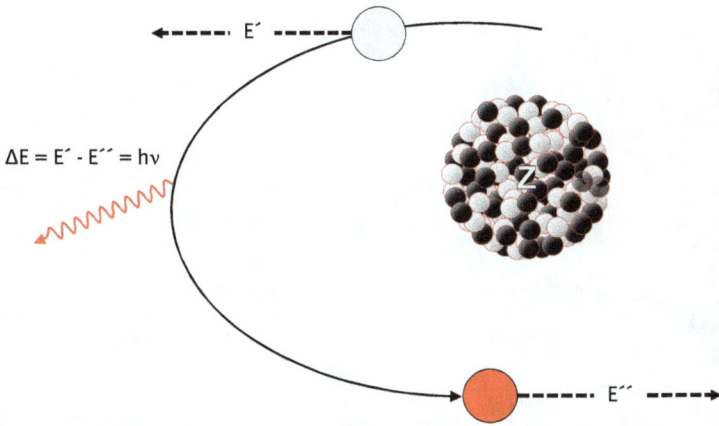

Fig. 12.17: The origin of bremsstrahlung. An electron originating from nuclear transformation processes is attenuated by the nucleus of an atom. The kinetic energy the electron loses (from E′ to E″) is emitted as electromagnetic radiation.[22]

matter, especially its effects on biological material. The latter aspect relates to "radiation dosimetry", which is discussed separately in Chapter 2 of Volume II.

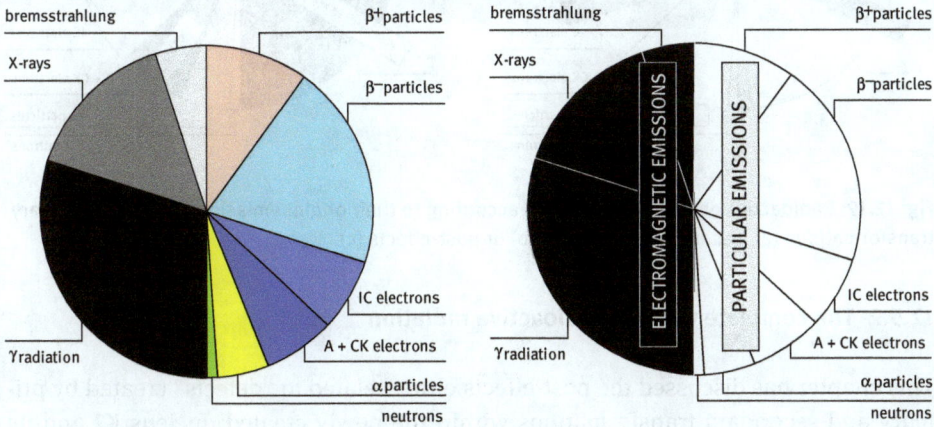

Fig. 12.18: Radioactive emissions structured according to their character, i.e. particulate vs. electromagnetic (a) and according to the individual types (b).

---

22 This, actually, was the reason RUTHERFORD's model of the atom was replaced by BOHR's: the latter model allows electrons to "move" around the nucleus of an atom in quantum orbits exclusively; see Section 1.3.3 (page 10).

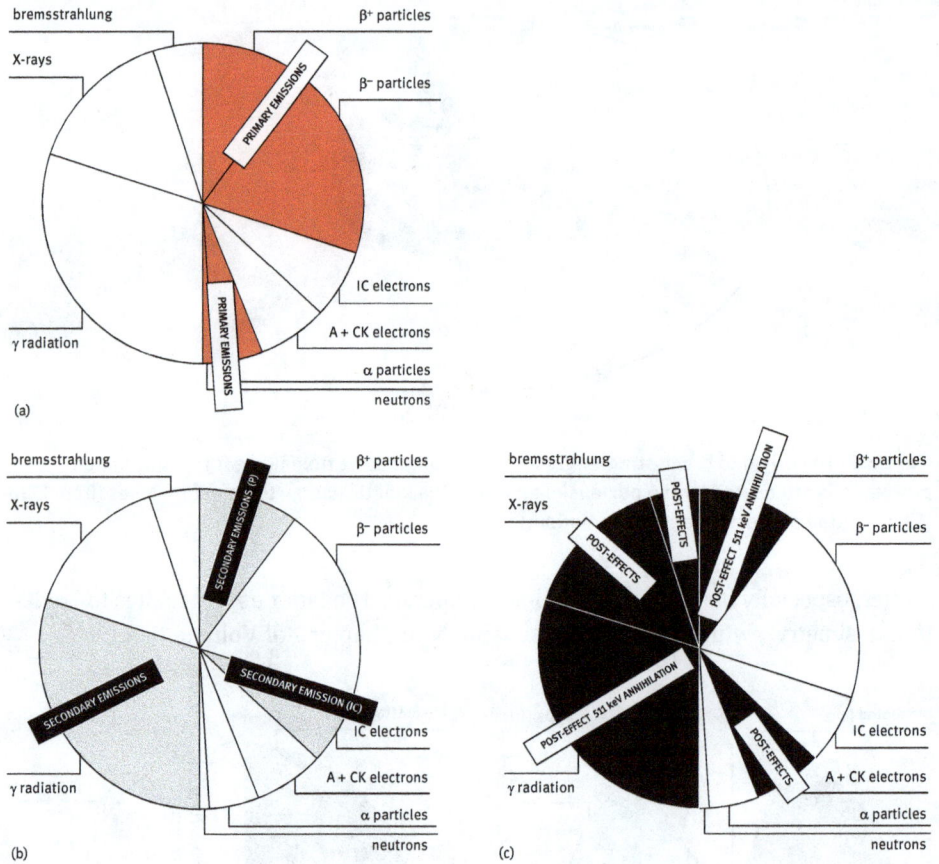

Fig. 12.19: Radioactive emissions structured according to their origin: emissions created in primary transformations (a), secondary transitions (b) or post-effects (c).

12.5.2 The complete world of radioactive radiation

This chapter has discussed the post-effects either related to "defects" created by primary and secondary transformations within the newly created nucleus K2 and its atom, or to the formation of an antimatter particle within that nuclide, the positron. The corresponding radiation generated by post-effects is either electromagnetic (X-rays, annihilation photons and/or low energy inner bremsstrahlung) or particulate (cascades of low-energetic AUGER and COSTER–KRONIG electrons). These radiations complete the spectrum of primary (β⁻- and β⁺-particles, α-particles, neutrons) and secondary (photons, IC electrons, electron-positron pair) emissions.

Only together do all of these effects reflect the essence of radioactive transformation processes.

Figures 12.18 and 12.19 group the various radiations by type, i.e. particulate *vs.* electromagnetic (Fig. 12.18(a)), according to the specific transformations (Fig. 12.18(b)), and according to the origin of the emissions created in primary transformations, secondary transformations and post-effects (Fig. 12.19). These radiative emissions are all part of nuclear and radiochemistry – they all are constituting what is referred to as RADIOACTIVITY.

13 Nuclear reactions

Aim: Unstable nuclei convert into stable ones by minimizing the absolute mass of the nuclide, which is typically expressed in terms of nucleon binding energy. The common feature of all these transformations is their exothermic character, with transformations occuring spontaneously. In contrast, nuclear reactions start *via* external "activation", delivered by a projectile. Balances in the mass of reactants and products may differ between exothermic and endothermic routes, and in the latter case the projectile delivers activation in the form of kinetic energy.

The general strategy of artificial nuclear transformation is to induce a nuclear reaction by delivering energy (either in the form of a particle or electromagnetic) of adequate kinetic energy into a (mainly) stable target atom. Thereby, a compound nucleus is formed, which instantaneously transforms into the nuclear reaction product(s).

There are several pathways for nuclear reactions and several classes may be defined. Some occur naturally in the universe, particularly in stars or interstellar space, and some occur in the earth's atmosphere. Another group is comprised of artificial, man-made nuclear reactions.

This chapter focuses on the latter class, and on the radionuclide product of the reaction. It discusses the principal parameters influencing the differential and integral yields of radionuclides produced in certain nuclear reactions. These parameters are: the cross-section and excitation function, particle energy and flux, stopping power, irradiation period, and target composition.

13.1 Artificially produced radionuclides

Radionuclides have been and are of great value to basic research, for analytical studies in astrophysics and environmental research, in chemistry and materials sciences, in industry, and for medical diagnoses and therapy. However, access to radionuclides occurring naturally on earth is limited to the members of the transformation families starting from ^{232}Th, ^{238}U, and ^{235}U contained in ores in the earth's crust, and, at least in principle, to a very limited number and in a very limited way to radionuclides permanently created in the atmosphere. Consequently, it is the task of man-made nuclear reactions to produce "artificial" unstable nuclides, i.e. those not usually found on earth.

A nuclear reaction is, in analogy the chemical reaction, defined as the interaction of (in almost all cases) two reactants to yield the desired reaction product(s). If a radioactive nuclide is to be produced, one reactant is typically a nucleus of a stable atom (the "target"). The other is either a subatomic particle (mainly neutrons and small charged particles like protons, deuterons, α-particles, etc.) or the nucleus of a medium-large atom (electromagnetic radiation can also be used.). Thus, the initial nucleus is "bombarded" or "irradiated" by the small "projectile" of characteristic kinetic

https://doi.org/10.1515/9783110742725-013

energy and is transformed into a new nucleus. Efficient production, in terms of high yields and high purity of the desired radionuclide, requires the following:
- adequate choice of the type of the nuclear reaction,
- availability of the corresponding projectiles and targets,
- handling the irradiation process, and
- (in many cases) the (radio)chemical separation of the radionuclide produced from the irradiated target.

This chapter focuses on aspects of nuclear reactions dedicated to the production of a specific radionuclide.

13.1.1 Historical reference

The use of a neutron as a projectile basically produces another, unstable isotope of the initially irradiated chemical element. Because this product isotope is neutron-rich, it may subsequently transform *via* β^--transformation to create an isotope of another chemical element $(Z + 1)$. In contrast, a nuclear reaction induced by charged particles immediately yields a radioactive isotope of a chemical element different from the initial one. In both cases, this is, metaphorically speaking, transmuting one chemical element into another.

It takes one back to the heroic attempts of alchemists to make gold out of other elements.[1] While they were destined to fail because the "chemically organized" exchange of shell electrons of an atom will never change one element into another (see DALTON'S principles, Chapter 1), nuclear reactions are finally able to achieve that goal! Stable gold, which is represented by the mono-isotope ^{197}Au, can be artificially produced through several nuclear reactions from other chemical elements.

The first man-made nuclear reactions were already introduced in preceding chapters, namely when RUTHERFORD used a natural source of α-particles emitted from ^{226}Ra or ^{214}Po to irradiate naturally existing nitrogen as a mixture of two stable isotopes ^{14}N (99.636%) + ^{15}N (0.364). For ^{14}N, the man-made nuclear reaction with an α-particle created the stable product nucleus ^{17}O together with a flux of protons. Similarly, when CHADWICK discovered the neutron, it was through a man-made nuclear reaction directing α-particles at beryllium composed of the mono-isotope ^{9}Be (100%). Here, in addition to the ejected neutron, ^{12}C was formed as stable product.

---

1 This is the search for the "philosopher's stone" (lat.: *lapis philosophorum*; arab.: *el Iksir*). Its discovery was said to be the *magnum opus*. Despite laborious efforts over centuries, it was never found, although chemistry is thankful for the discovery of many initially unrevealed properties of chemical substances. A fortunate spin-off effect was the contribution of JF BÖTTGER in 1707 on how to prepare European porcelain. At the time, he had been hired by the monarch of Saxony in Dresden to convert less noble metals into gold.

In both cases, the nuclear reaction product formed was <u>stable</u> and thus "invisible" in those reactions – but even then, these products were already known, stable isotopes (see Tab. 13.1).

In contrast, when I CURIE and F JOLIOT in 1934 used the same α-particles to irradiate natural elements such as boron, magnesium or aluminum,[2] the irradiated samples became radioactive (see Tab. 13.1). As chemists, they were able to chemically separate the nuclear reaction product (e.g. ^{13}N as ammonia from the irradiated boron nitride) and thereby could identify the elemental identity of the radioisotope. This was the first artificial creation of <u>unstable</u>, "man-made" radionuclides, deemed "artificial" because these nuclides do not exist naturally on earth![3]

Tab. 13.1: First man-made nuclear reactions. Based on α-projectiles obtained from, for example, the primary transformation of natural ^{226}Ra yielded either stable or unstable (i.e. radioactive) product nuclei. Half-lives for the latter were t½ = 9.96 min for ^{13}N and t½ = 2.50 min for ^{30}P.

stable → stable		stable → unstable	
RUTHERFORD 1919, CHADWICK 1932		I. CURIE + F. JOLIOT 1934	
$^{14}_{7}$N $+ \alpha =$	$^{17}_{8}$O$_{9}$ $+ p$	$^{10}_{5}$B$_{5}$ $+ \alpha =$	$^{13}_{7}$N$_{6}$ $+ n$
$^{9}_{4}$Be$_{5}$ $+ \alpha =$	$^{12}_{6}$C$_{6}$ $+ n$	$^{27}_{13}$Al$_{14}$ $+ \alpha =$	$^{30}_{15}$P$_{15}$ $+ n$

---

2 The natural isotopic compositions of the irradiated stable isotopes (in today's nomenclature, "targets", see below) are: ^{10}B (19.9%) + ^{11}B (80.1%) for boron, ^{24}Mg (78.99%) + ^{25}Mg (10.00%) + ^{26}Mg (11.01%) for magnesium, ^{27}Al (100%). FRÉDÉRIC JOLIOT reported: "We have proposed that these new radio-elements (isotopes, not found in nature, of known elements) be called radio-nitrogen, radio-phosphorus, radio-aluminum (in the case of magnesium irradiated by alpha rays) and designated by the symbols: R^{13}N, R^{30}P, R^{28}Al". Note that an (α,n) reaction on one of the magnesium isotopes would not yield an isotope of aluminum, but of silicon. The isotope ^{28}Al (t½ = 2.246 min) could be formed via a ^{25}Mg(α,p) process.

3 The Nobel Prize in Chemistry, 1935, was awarded jointly to FRÉDÉRIC JOLIOT and IRÈNE JOLIOT-CURIE in recognition " . . . of their synthesis of new radioactive elements . . . ". The Nobel lecture was entitled: "Chemical Evidence of the Transmutation of Elements"; note *Transmutation* of *Elements*. This was organized through α-particles of relatively low energy, which originated from naturally occurring radioactive sources. Another Nobel Prize was awarded for "the *transmutation of atomic - nuclei*" using artificially created charged particles, as per footnote 37.

13.2 General mechanism of nuclear reactions

13.2.1 Nomenclature

The course of a nuclear reaction is typically written as if a small component "a" delivers mass and/or energy to a larger nucleus "A", forming a product nucleus "B" and eventually a small component "b", as in eq. (13.1): $a + A = B + b$. The short notation is $A(a,b)B$ given in eq. (13.2). The nucleus A is called the "target". The small component a is the "projectile", and b is the "ejectile". The nuclear reaction procedure is often said to "irradiate" a target, and the target is "irradiated" or "bombarded".

$$a + A = B + b \qquad (13.1)$$

$$A(a, b)B \qquad (13.2)$$

Whenever the projectile consists of a nucleon (n or p) or cluster of nucleons (such as the nucleus of deuterium (d), He (e.g. the α-particle) or larger nuclei), and if the kinetic energy of these projectiles is below a limit which would split the target nucleus into many pieces (a type of nuclear reaction called "spallation", see below), nucleons are "transferred" in the course of the reaction to/from the target nucleus. This type of reaction follows a "nucleon transfer mechanism". It will be discussed in this chapter in detail, because it delivers artificial radionuclides relevant to radio- and nuclear chemistry. The focus is on the radioactive product of the reaction, rather than the smaller particles emitted.

Nuclear reactions between two nuclei of chemical elements of masses covering almost the complete range of A, which are often pertinent to heavy element research, are discussed in detail in Chapter 10 of Volume II. Nuclear reactions occurring between two subatomic particles, which is mainly the situation in nuclear physics and particle physics research, are only briefly mentioned.

13.2.2 Classification of nuclear reactions

Nuclear reactions may be categorized by many different criteria. One criterion is whether the reaction is exothermic vs. endothermic. If the projectile does not fuse with the target nucleus, but scatters in an elastic or inelastic way, the processes are called "elastic scattering" and "inelastic scattering" respectively, as per eqs. (13.3) and (13.4). The initial nucleus thus remains unaltered in terms of its nucleon composition, but either becomes energetically excited (via inelastic scattering, expressed by *A) or gains no energy at all (elastic scattering). Another point of view considers the change or exchange of nucleons or nucleon clusters, yielding nuclear reaction products of nucleon composition different from that of the target nucleus.

$$A(a, a)A \qquad (13.3)$$

$$A(a, a')^* A \qquad \cdot \qquad (13.4)$$

One may further distinguish between naturally occurring and man-made nuclear reactions. Some nuclear reactions take place naturally in the universe, in stars (such as the fusion of light elements and genesis of heavier elements) or in supernovae (such as the permanent nuclear synthesis of ^{26}Al in supernovae, as mentioned in Chapter 11). Some naturally occurring nuclear reactions take place in the earth's atmosphere (such as formation of ^{14}C).

Artificial, man-made nuclear reactions govern the direct interaction of projectiles, either naturally available or artificially produced, with dedicated target nuclei.

In the context of "particle research", nuclear reactions aim at infinitesimal elementary particles and their properties[4] and at the origin of the infinite universe and astrophysics in general. Another group may be identified revealing the mechanism of nuclear reactions itself, which is basic physics again.

In the framework of nuclear and radiochemistry, however, there is a specific goal: The production of artificial radionuclides for
-   chemical and physical research and development,
-   research on nuclear materials,
-   industrial analytics, and
-   the synthesis and application of radiopharmaceuticals in medicine.

13.2.3 Mechanism of nucleon transfer reactions

For nucleon transfer reactions, mechanistically, an intermediate state C^* is formed before the two components B and b appear. It represents the sum of the mass and energy of components A and a. If the projectile did enter the target nucleus with a given kinetic energy, this amount of energy is deposited homogeneously within the nucleus C^* formed. C^* thus denotes a collectively excited state.[5] It is called a "compound" nucleus in the case of low excitation and a "composite" nucleus in the case of high excitation, cf. Figure 13.1.

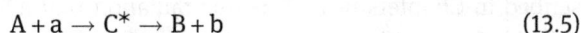

$$A + a \rightarrow C^* \rightarrow B + b \qquad (13.5)$$

---

4 For example, in 2012, CERN, Geneva (Conseil Européen pour la Recherche Nucléaire) confirmed the existence of the HIGGS boson – see Fig. 7.1.

5 The total excitation energy may be expressed in temperature scale: C^* is a "hot" nucleus. Parts of this excitation are released in a prompt way, typically in terms of electromagnetic radiation. In parallel, nucleons or nucleon clusters may "evaporate" or "distill off" – like molecules above their boiling point in "hot" solvents.

In nuclear reactions, fundamental parameters such as the sum of reactants and the sum of products must remain the same. The intermediate compound nucleus C* thus obtains a nucleon composition that is exactly the sum of the nucleon numbers of a + A. For the total mass number A, this gives $A_A + A_a = A_{C*}$ and $A_{C*} = A_B + A_b$. This is the same for each type of nucleon; for example, the number of neutrons is $n_A + n_a = n_{C*}$ and $n_{C*} = n_B + n_b$. This holds true for mass m of the participating species, and thus also for energy as $E = mc^2$. The same conservation holds true for other parameters such as the electric charge, momentum (linear and angular) and parity.

$$A_A + A_a = A_{C*} = A_B + A_b \qquad (13.6a)$$

$$m_A + m_a = m_{C*} = m_B + m_b \qquad (13.6b)$$

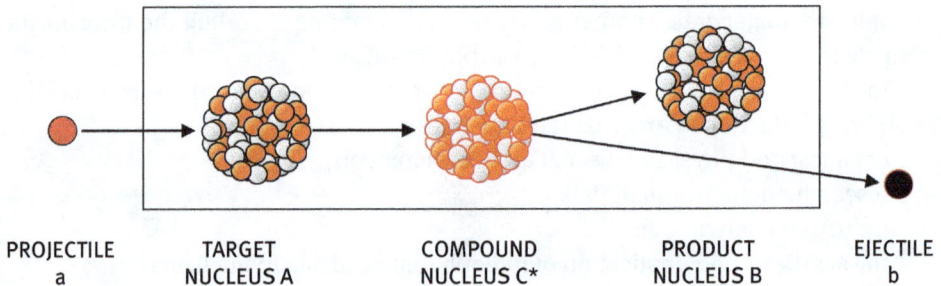

Fig. 13.1: Mechanism of nuclear reactions, including an intermediate compound nucleus C*.

This composite state C* splits instantaneously (typically with a lifetime of about 10^{-13} s) into B and b. The mass of component b ejected from C* is typically much less than that of the radionuclide B formed. It thus may leave the process at high velocity. In many cases, the ejectile b is a photon γ, but in other cases can be a neutron, proton, deuteron, triton or α-particle. Whenever, for example, a neutron is released from the compound nucleus, it removes about 8 MeV from the excitation energy of C* – which is the mean binding energy of a nucleon. The second species formed is the nucleus B. If unstable, i.e. radioactive, it immediately starts to transform according to the processes described in Chapters 8–12. It is this radiation that makes the product radionuclide B valuable for specific applications!

13.2.4 Reaction channels

Any given composite C* (as defined by Z, N, and A) may be obtained through different combinations of type A + a, all having the same sum of protons and sum of neutrons. The processes of forming a given C* are called "in-channels", shown in Fig. 13.2.

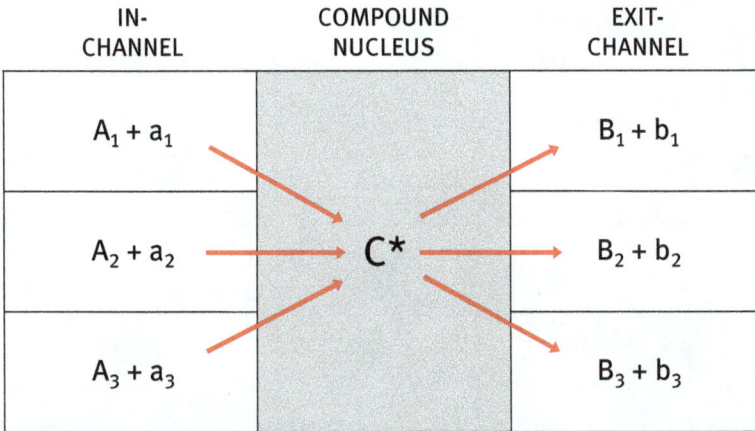

IN-CHANNEL	COMPOUND NUCLEUS	EXIT-CHANNEL
$A_1 + a_1$		$B_1 + b_1$
$A_2 + a_2$	C*	$B_2 + b_2$
$A_3 + a_3$		$B_3 + b_3$

Fig. 13.2: Mechanism of nuclear reactions including nucleon transfer processes. A compound nucleus C* defined by a given (Z, N, and A) may be obtained through various combinations of type A + a, all representing the same sum of protons and the same sum of neutrons. These processes of forming C* are called "in-channels". From C*, various "exit-channels" may open, depending on the energy deposited into C*.

The same is true for the fate of compound nucleus C*. This energetically rich intermediate state is transformed into the products B and b. Again, different combinations of B + b may exist, reflecting the same sum of proton number, neutron number, and mass number as C*. These are the "exit-channels" (or "out-channels").

A variety of in- and exit-channels are conceivable for a given compound nucleus – all candidate channels must, however, be permitted by energy and conservation laws. Why one particular in- or exit-channel is favored over others is decided quantum mechanically, in terms of the degree of overlap between nuclear wave functions of initial and final states.

13.2.5 Pathways of nucleon transfer reactions

Figure 13.4 summarizes the most commonly used nuclear reactions in producing artificial radionuclides with the projectiles γ, n, p, d, ^{3}He, and α. The Chart of Nuclides indicates the stable target nuclide and the various product nuclei according to the (a,b) nuclear reaction.

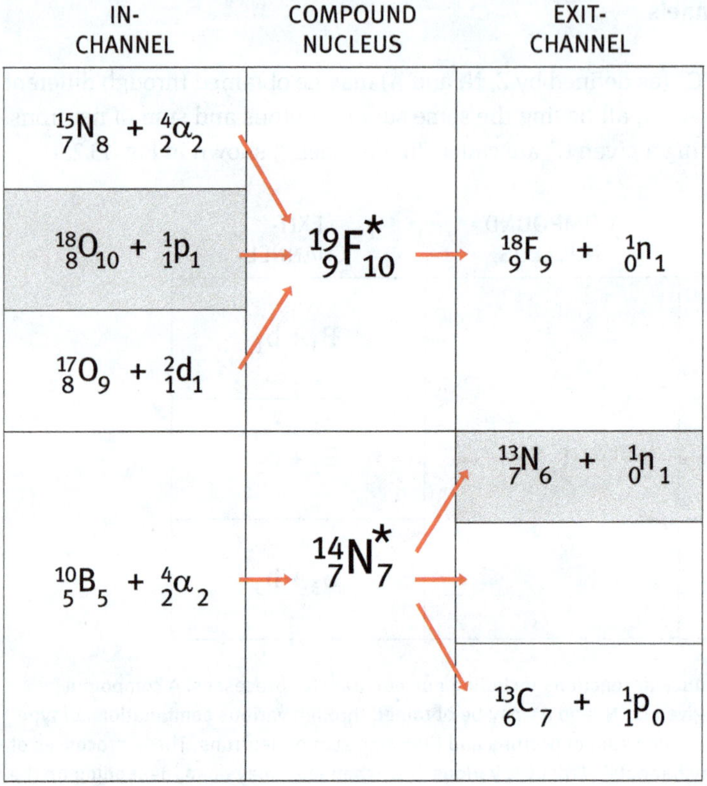

IN-CHANNEL	COMPOUND NUCLEUS	EXIT-CHANNEL

Fig. 13.3: Examples of in-channels and exit-channels. The most relevant channels are highlighted. The example for in-channel options is on the composite nucleus $^{19}F^*$, which leads to the positron emitter ^{18}F through an $^{18}O(p,n)$ ^{18}F reaction. For the exit-channels, an example for one of the first man-made nuclear reactions is selected, namely $^{10}B(\alpha,n)$ ^{13}N *via* the $^{14}N^*$ composite nucleus formation. The in-channel was the one initiated by α-particles of relatively low kinetic energy (offered by α-emitting members of naturally occurring chains), the exit-channel creating ^{13}N was the one energetically favored under these conditions.[6]

13.2.6 A short note on stoichiometry

The energy balance in nuclear reactions based on nucleon transfer parallels a fundamental rule known in chemical reactions: stoichiometry. Yet, there is a very important

---

6 For the discussion of which in-channel works best and which exit-channel is preferred, see below (energetics of nuclear reactions). The other exit-channels may appear with the increasing kinetic energy of the α-projectile. Possible exit-channels are $[^{14}N^*] = {}^{13}N + {}^{1}n$ and $[^{14}N^*] = {}^{13}C + {}^{1}p$. The latter exit-channel "opens" through a higher kinetic energy E of the incoming α-particle. The higher energy deposited in the composite nucleus may also allow for the separation of not just one, but two nucleons.

	(α,3n) (³He,2n)	(α,2n) (³He,n)	(α,n)	
(p,2n) (d,3n)	(p,n) (d,2n)	(p,γ) (d,n) (α,t)	(t,n)	(α,p)
(d,t) (n,2n) (γ,n) (p,d)	**TARGET NUCLEUS**	(n,γ) (d,p)	(t,p)	
(p,α)	(d,α)	(n,d) (γ,p)	(n,p) (d,2p)	
	(n,α)			

PROJECTILES:
γ,
n,
p, d, ³He, α

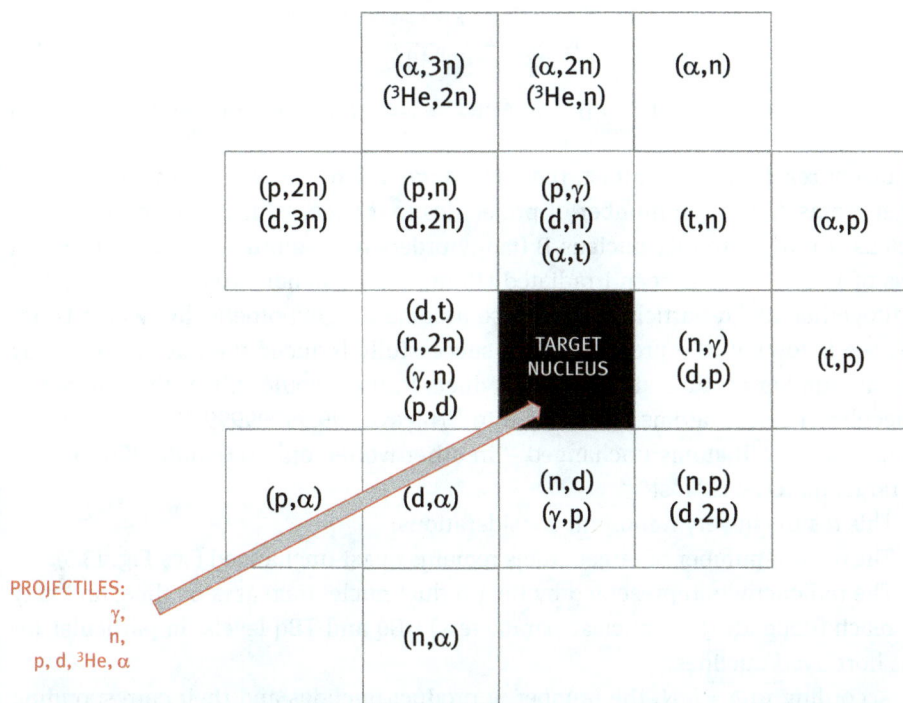

Fig. 13.4: Typical nuclear reactions used to produce radionuclides with the projectiles γ, n, p, d ³He, and α within the Chart of Nuclides. The stable target nuclide is black, and the various product nuclei are denoted with the balance of (a,b)-type nuclear reaction. For example, the neutron capture reaction (n,γ) yields a nuclide just to the right of the stable target nuclide.

difference between the chemical and nuclear reactions. In chemistry, depending on the value of the chemical reaction (or equilibrium) constant, the concentration of reactants decreases proportional to the increase in concentration of the products. For a given chemical reaction of $\alpha_A A + \alpha_a a \rightarrow \alpha_B B + \alpha_b b$, with a forward reaction rate of $k^+ = [A]^{\alpha A} [a]^{\alpha a}$ and a backward reaction rate of $k^- = [B]^{\alpha B} [b]^{\alpha b}$, the law of mass action yields a reaction equilibrium constant K^{EQ} with [A], [a], [B], and [b] reflecting the concentration of the species in mol/L and α_i being the stoichiometric factors or better: activity coefficients. For example, syngas is a mixture of carbon monoxide and water. It can be converted into carbon dioxide according to eq. (13.8), reaching a state of equilibrium.[7]

---

7 The latter reaction has an equilibrium constant K^{EQU} = 4 at 800 °C, which corresponds to 80% reaction yield of CO_2 (and still containing 20% CO). By starting from 100% CO, this concentration decreases while CO_2 is simultaneously formed. This equilibrium is reached at a certain time, as shown in Fig. 13.5, left.

$$K = \frac{k^+}{k^-} = \frac{[A]^{\alpha_A}[a]^{\alpha_a}}{[B]^{\alpha_B}[b]^{\alpha_b}} \tag{13.7}$$

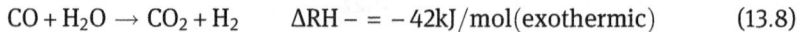

$$CO + H_2O \rightarrow CO_2 + H_2 \qquad \Delta RH- = -42kJ/mol(exothermic) \tag{13.8}$$

In nuclear reactions, in contrast, there is a huge difference between the numbers of target atoms A and the number of projectiles that successfully transform a target nucleus A into a product nucleus B (many orders of magnitude less). Let us say a mass of 10 mg ^{10}B had been irradiated (10 mg ^{10}B = 1 mmol ≈ $6 \cdot 10^{20}$ atoms of ^{10}B) by altogether 10^{10} α-particles. Let us also assume that hypothetically (which is not true, see below) all the projectiles had successfully induced the nuclear reactions and the number of nuclear reaction products formed would mirror the number of projectiles. The 10^{10} atoms converted into ^{13}N leaves 99.999999999999% of the initial number of ^{10}B atoms unchanged – in other words, only 0.000000000001% of the target material is "lost".[8]

This results in very important considerations:
1. The overall number of target atoms remains quasi-unchanged (see Fig. 13.5).
2. The radioactivity represented by the product nuclei increases by time, and may reach (depending on nuclear parameters) GBq and TBq levels, in particular for short-lived nuclides.
3. According to Á = λ Ń, the number of product nuclides and their corresponding mass available will be quite limited.[9]

13.3 Energetics

13.3.1 Nuclear reaction and nuclear transformation

Once the unstable radionuclide B is created, it will transform to stabilize following one of the primary transformation routes, i.e. β-processes, α-emission or fission, as per Fig. 13.6. Formation of unstable product nuclei in the course of a nuclear reaction thus must always consider the correlation of creation and decay.

---

8 Then each α-particle had reacted a ^{10}B atom into ^{13}N, cumulatively yielding 10^{10} atoms of radioactive ^{13}N. With a half-life of $t_{1/2}$ = 9.96 min (and supposing that the irradiation was arranged within a very short period of time relative to this half-life in order to avoid the need to correct the radioactivity lost by simultaneous transformation), the radioactivity of 10^{10} atoms of ^{13}N would be 11.6 MBq. (Á = λ Ń = (ln2 / $t_{1/2}$) Ń = (0.693 / 9.96 × 60 s)·10^{10} = 0.00116 s^{-1}·10^{10} = 11.6·10^6 s^{-1} = 11.6·10^6 Bq.) This activity is significant, yet the changes in atoms of the target atom ^{10}B are negligible.
9 F JOLIOT in his Nobel lecture, entitled "Chemical Evidence of the Transmutation of Elements": "The yield of these transmutations is very small, and the weights of elements formed by using the most intense sources of projectiles which we are able to produce at the present time are less than 10^{-15} g, representing at the most a few million atoms."

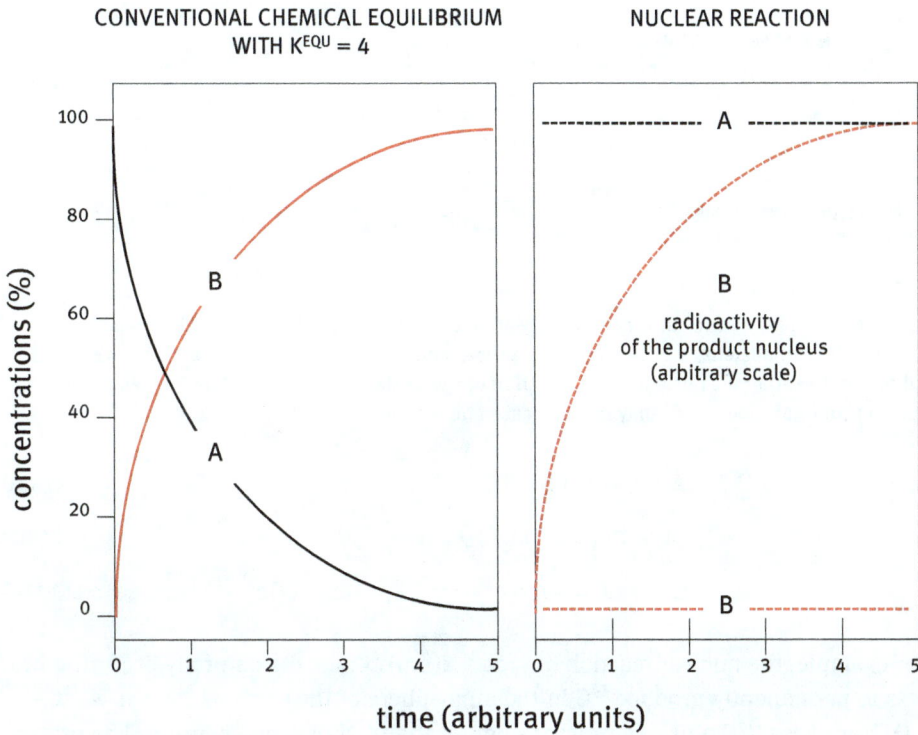

Fig. 13.5: Comparison of chemical and nuclear reactions. Conventional chemical equilibrium of $CO + H_2O \rightarrow CO_2 + H_2$ with $CO = A$ and $CO_2 = B$, defined by the simultaneous decrease in A and increase in B (left). In contrast, nuclear reactions do not result in a significant decrease in the concentration of A, because the number of product nuclei B is comparably small (right). However, this number reflects a significant radioactivity of B (dotted line).

13.3.2 Q-values

When nuclear reactions are expressed in analogy to nuclear transformations (and to chemical reactions), the reaction energy ΔE appears on the side of the products, as in eq. (13.9) and Fig. 13.6. Nuclear reaction energies ΔE may either be positive (exothermic reaction) or negative (endothermic). This means that the sum of the reactant masses is either larger or smaller than the sum of the product masses, respectively.

For nuclear reactions, as for spontaneous radioactive nuclear transformations, the amount of ΔE is expressed as the Q-value of the process. In nuclear reactions, the corresponding Q-value is given by $Q = -\Delta E$, eq. (13.11).

endothermic or exothermic

| NUCLEAR REACTION formation of a radionuclide | $K1 + a \longrightarrow {}^*K2 + b + \Delta E$ |

| NUCLEAR TRANSFORMATION formation of (more) stable nuclide | ${}^*K2 \longrightarrow K3 + x + \Delta E$ |

exothermic

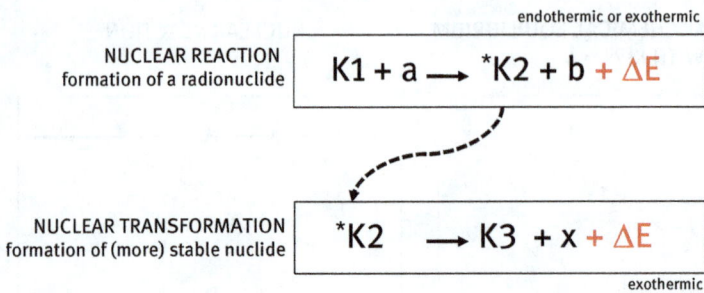

Fig. 13.6: Interplay of nuclear processes. Creation of an artificial radioactive nuclear reaction product B (here: nucleus *K2) by inducing a process on a stable nuclide A (here: nucleus K1) and subsequent exothermic nuclear transformation of the unstable product nuclide into a stable nucleon configuration K3 (K3 may in some cases be identical to K1).[10]

$$A + a \rightarrow B + b + \Delta E \tag{13.9}$$

$$\Delta m = m_{(B+b)} - m_{(A+a)} \tag{13.10}$$

$$Q = -\Delta E = -\Delta m\, c^2 = \left(m_{(A+a)} - m_{(B+b)}\right)c^2 \tag{13.11a}$$

For example, the nuclear reaction of type $^{14}N(n,p)^{14}C$, i.e. the naturally occurring process to permanently produce ^{14}C in the atmosphere of the earth, is $^{14}N + n \rightarrow {}^{14}C + p + Q$. Mass-wise, ^{14}C plus one proton is lighter than ^{14}N plus one neutron. The process is exothermic, with a Q-value of $-(-0.000672$ u$)^{11}$ = 626 keV.[12] Actually, all (n,γ) and (p,γ) type nuclear reactions are exothermic. Nuclear reactions induced by a charged projectile larger than the proton can be either exothermic or endothermic.

While eqs. (13.10) and (13.11) use absolute mass, the mass excess values Δ^{excess} (or simply Δ, see Chapter 2) can also be applied as tabulated in the atomic mass evaluation database (see also Table 14.4). The analogous equation to (13.11a) is:

$$Q = -\Delta E = \Delta_A + \Delta_a - \Delta_B - \Delta_b \tag{13.11b}$$

---

10 See for example the process of producing ^{18}F from Fig. 13.3. The nuclear reaction is $^{18}O(p,n)^{18}F$, and the primary transformation of the proton-rich ^{18}F is β^+-emission to stable ^{18}O.

11 $\Delta m = [m(^{14}N) + m(^1n)] - [m(^{14}C) + m(^1p)] = (14.003074$ u $+ 1.008665$ u$) - (14.003242$ u $+ 1.007825$ u$) = (15.011\,739 - 15.011\,067)$ u $= -0.000672$ u. (1 u $c^2 = 931.49$ MeV).

12 Similar to nuclear transformation processes, this energy is mainly "located within" the new nucleus in terms of nucleon binding energy.

13.3.3 Endothermic reactions and projectile energies

The first man-made synthesis of an artificial unstable nuclide was $^{27}Al(\alpha,n)^{30}P$. The atomic masses are 26.981538 u for ^{27}Al and 29.978313 u for ^{30}P. With atomic mass of the α-projectile of 4.002603 u and 1.008665 u for the neutron ejected, the balance for the A(a,b)B nuclear reaction according to eq. (13.10) is Δm = 30.986978 u - 30.984141 u = 0.002837 u. Though this is a small change in atomic mass units relative to the masses of (A + a), just 0.00915%, it is endothermic: the sum of the masses of the two products is larger than the sum of the two reactants. When expressed as energy, it gives a Q-value of −2.644 MeV.[13] The reaction may thus only work if the α-particle delivers a kinetic energy above that level.[14] For projectiles with a certain kinetic energy, this kinetic energy ($E^{kin} = \frac{1}{2}mv^2$) must be added to eq. (13.11). If this referred exclusively to the incoming projectile, this would result in eq. (13.12). In fact, this is the main feature of most of the man-made nuclear reactions to produce a desired artificial nuclear reaction product: to induce the reaction by delivering energy[15] (either in the form of a particle or electromagnetic radiation) to a target nucleus. In most cases of radionuclide production, the target is at rest energy.[16]

$$E(A+a) = m_A c^2 + \left(m_a c^2 + \frac{1}{2}m_a v^2 \right) \qquad (13.12)$$

In addition, the product species also receive an impulse, and thus the complete balance in energy is even more complex. Both nuclear reaction products are thus distributed along a virtual axis between projectile and target in an angular distribution. The experimental measurement of these angles represents a valuable approach to determine the characteristics of a nuclear reaction process.

---

13 ^{27}Al: E_B = 224.952 MeV, $\bar{E}_B$ = 8.332 MeV; ^{30}P: E_B = 250.605 MeV, $\bar{E}_B$ = 8.354 MeV.
14 Because this was in fact the case with the kinetic energies of α-particles emitted from the long-lived α-sources of ^{226}Ra and its daughter nuclides (4.784 MeV and 4.601 MeV for ^{226}Ra, 5.489 MeV for ^{222}Rn, 6.002 MeV for ^{218}Po etc.), this endothermic process took place.
15 Here is another difference between chemical and nuclear reactions – the absolute scale of the energy involved. Exchanges between electrons of different species to form a molecule shift atomic electrons to molecular electron levels. This proceeds on the scale of electron binding energies, i.e. in the order of eV to keV. In nuclear reactions, nucleon binding energies are concerned, and "activation" energies may reach MeV scales. In modern particle physics, nuclear reactions are induced on the level of GeV or even TeV.
16 For most of the commonly organized nuclear reactions, the target nucleus represents a stable nuclide and is mechanically fixed. All the stable isotopes of the chemical elements can be used as targets, either in the chemical form of the atom or in chemical compounds including that atom. In some cases, such as the synthesis of super-heavy elements (see Volume II, Chapter 10), radioactive target nuclei are also used.

13.3.4 Nuclear reaction threshold energy

Exothermic nuclear reactions may occur at any kinetic energy of the projectile, even (hypothetically) at its rest mass energy (i.e. projectile kinetic energy = 0). This is different in endothermic nuclear reactions. The corresponding Q-value represents a kind of energy that must be overcome by adding kinetic energy to the rest mass of the projectile – it serves as a kind of activation or excitation energy E^* for the reaction itself: $E^* = -Q$. The amount of E^* is the excess energy deposited within the compound nucleus C^*.

However, the kinetic energy of a projectile is not 100% used. Some part of this energy is not available for inducing the nuclear reaction, because the projectile and the target nucleus also obey conservation of impulse. The product nucleus gets a recoil energy $^{RECOIL}E_B$, which depends on the masses of both a and A, as per eq. (13.13) (see also Fig. 9.2). Consequently, the kinetic energy of the projectile must be "a bit" larger than the value of –Q in order to compensate for the recoil energy. To be precise, it must be larger by $-Q(m_a/m_A)$. The kinetic energy derived represents the "threshold energy" $\check{E}_a$ of a nuclear reaction, shown in eq. (13.14): ·

$$^{RECOIL}E_B = E^* \frac{m_a}{m_a + m_A} \tag{13.13}$$

$$\check{E}_a = -Q\left(\frac{m_A + m_a}{m_A}\right) = -Q\left(1 + \frac{m_a}{m_A}\right) \tag{13.14}$$

RUTHERFORD's nuclear reaction $^{14}N(\alpha,p)^{17}O$, for example, has a Q-value of –1.19 MeV. The threshold energy is deduced *via* the ratio of the mass numbers of the α-particle and the ^{14}N nucleus: $\check{E}_\alpha = -(-1.19\ \text{MeV})\ (4 + 14)\ /\ 14 = 1.19\ \text{MeV} \cdot 1.2857 = 1.53\ \text{MeV}$. The threshold energy in this case is larger than the Q-value by 0.34 MeV, i.e. about 28.5%. Nevertheless, the reactions could proceed because the α-particles kinetic energies were >4 MeV. For nuclear reactions of type (p, n) on target nuclei of mass number 100, for example, the additional amount of kinetic energy of the protons needed is only (1 + 100) / 100 = 1% of the Q-value.

13.3.5 Coulomb barrier

Both the activation energy E^* and the threshold energy $\check{E}_a$ discussed above exclusively referred to the mass of the participating nuclear reaction components.

The nucleons of the projectile will benefit from the strong force if they are close enough to the central attractive well of the target nucleus, which is about the dimension of a nucleon. However, target nuclei do have a positive charge. Thus, projectiles of positive charge like protons and α-particles (actually, all projectiles except for photons and neutrons) must overcome a repulsive coulomb force – much longer ranging than the attractive strong force. Figure 13.7 compares the difference between a neutral projectile (a neutron), and a positively charged projectile (a proton), in terms of approaching the target nucleus.

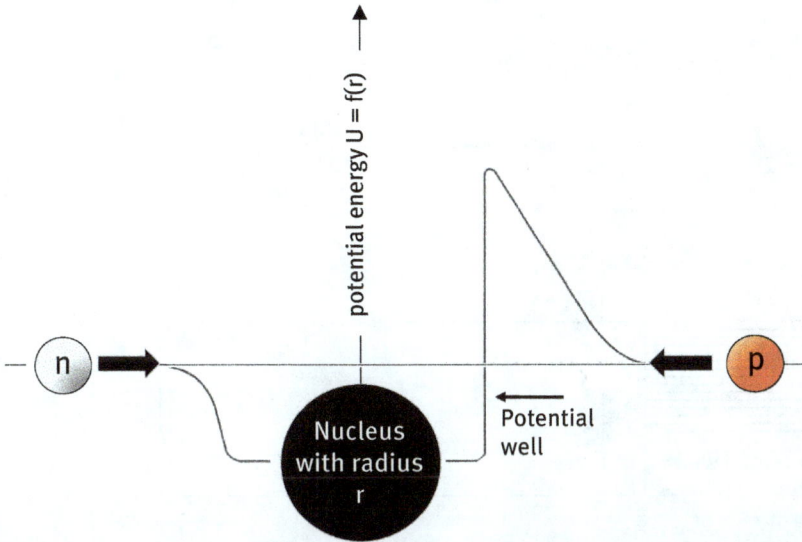

Fig. 13.7: Projectiles entering the potential well of a target nucleus. Coulomb barriers apply to charged projectiles, such as the proton, but not to uncharged projectiles, such as the neutron. Note: Coulomb barriers must be taken into account both when entering the target nucleus and when leaving it, such as discussed for primary transformations of unstable nuclei through α-particle emission.

The coulomb barrier applies only to charged projectiles, not to noncharged projectiles. The former need to have sufficient kinetic energy in addition to the threshold energy, as in Fig. 13.8. The potential is $U = f(r)$, with r^C being the separation distance between the (centers of charge of the) two components A and a. A maximum value is obtained for the closest distance r^C between them, which is given by eq. (13.15); r_o is the radius parameter, discussed in Chapter 2. It yields the value of the coulomb barrier E^C for the corresponding components, as in eq. (13.15) or (13.16). Suppose the kinetic energy of a charged projectile is below this level; it will not be able to reach the attractive strong forces – it just may scatter elastically from the target nucleus.

$$E^C = \frac{Z_A \cdot Z_a \cdot e^2}{r^C} \tag{13.15}$$

$$E^C \approx 1.44 \, \text{MeV fm} \, \frac{Z_A \cdot Z_a}{r^C} \tag{13.16}$$

$$r^C = r_0 \left(A_A^{1/3} + A_a^{1/3} \right) \tag{13.17}$$

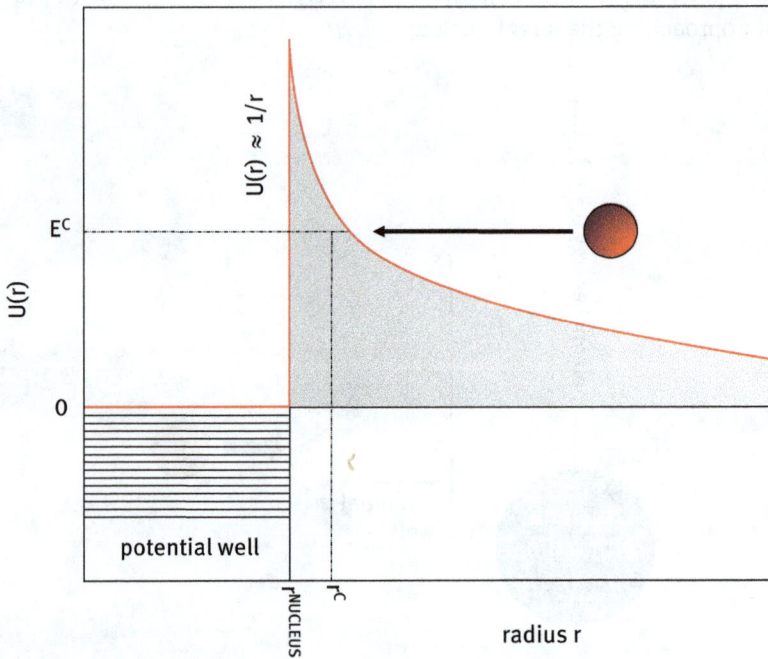

Fig. 13.8: The coulomb wall as seen by an approaching charged projectile.

13.4 Yields of nuclear reaction products

Following the intention to artificially produce radioactive nuclides, either for basic research or for industrial and medical applications, nuclear reactions should be experimentally designed in a way to achieve maximum product nuclide yields (activities) and highest product purities (similar to conventional preparative chemistry). Again, similar to conventional chemistry, fundamental correlations exist, defining nuclear reaction yields. These parameters are discussed below.

13.4.1 Nuclear reaction cross-section

Suppose a projectile has sufficient kinetic energy to enter the target nucleus; by far, not every hit represents a nuclear reaction. Usually, many "hits" are needed to induce just one nuclear reaction. The ratio of the number of hits to the number of nuclear reactions induced represents a reaction probability. In the terminology of nuclear reactions, this probability is called the "cross-section" σ.

Let us consider the spherical target nucleus as the real target for a virtual arrow (or bullet), representing the projectile a. If the target has a very small area ($A = 2\pi r^2$), then given many arrow shots some may hit the target, others may miss. Assuming for identical arrows that the area of the target was now doubled, more arrows would hit the target. This is schematically illustrated in Fig. 13.9. What is the probability for an arrow to meet the target? The larger the target area, the easier it is to hit.

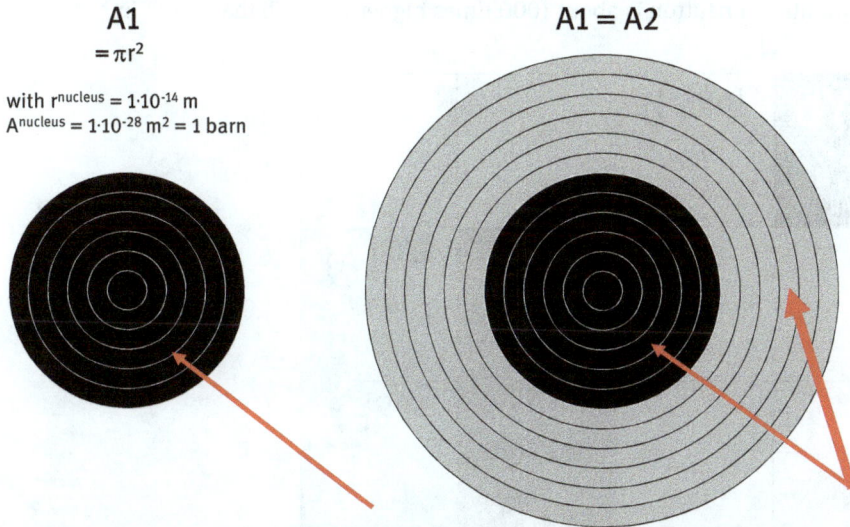

A1
$= \pi r^2$

with $r^{nucleus} = 1 \cdot 10^{-14}$ m
$A^{nucleus} = 1 \cdot 10^{-28}$ m^2 = 1 barn

A1 = A2

Fig. 13.9: Intention of the "nuclear reaction cross-section". Schematic representation of the different probabilities for a projectile to hit a small and a large target. For the nucleus on the right, the probability of a bullet hitting the target is twice as high as for nucleus on the left: two out of two projectiles reached the target as opposed to just one. For the same bullet, the nuclear reaction "cross-section" is small for the target nucleus on the left and large for the one on the right. Note that the gray area on the right is virtual – the real nucleus (black) does not change its dimension.

How does this apply to nuclear reactions? The target nucleus is taken with its real dimension, which is on average $r^{nucleus} \approx 10^{-14}$ m. Its diameter represents a plate or a circle, whose area is about the square of $r^{nucleus}$, i.e. of $A^{nucleus} \approx 10^{-28}$ m^2. This unit is called 1 barn (b) $= 1 = 1 \cdot 10^{-28}$ m^2, which is extremely small.[17] This area stands perpendicular to the arriving projectile.

---

[17] This is a joke; "barn" in English refers to a barn door, which is big enough to let horses and farm tractors pass.

From a simple geometric point of view, the same number of identical projectiles should then yield the same nuclear reaction rate for target nuclides of the same (or very similar) mass number. However, this is not even remotely true. Figure 13.10 gives three stable target nuclides of same (isobar A = 168) or similar mass (A = 169). However, the probabilities for a neutron to induce a nuclear reaction (neutron capture, (n, γ)) differ by three orders of magnitude. For the incoming neutron, the ^{169}Er nucleus "looks" like a "normal" nucleus with ca. $2 \cdot 10^{-28}$ m² area (σ = 2.3 b), but the ^{169}Tm nucleus appears to be about 50 times larger (σ = 108 b), and the ^{168}Yb nucleus about 1000 time larger (σ = 2400 b)! All three nuclei are almost identical in size, but the probability of incorporating a neutron is about 1000 times higher for ^{168}Yb than for ^{169}Er.

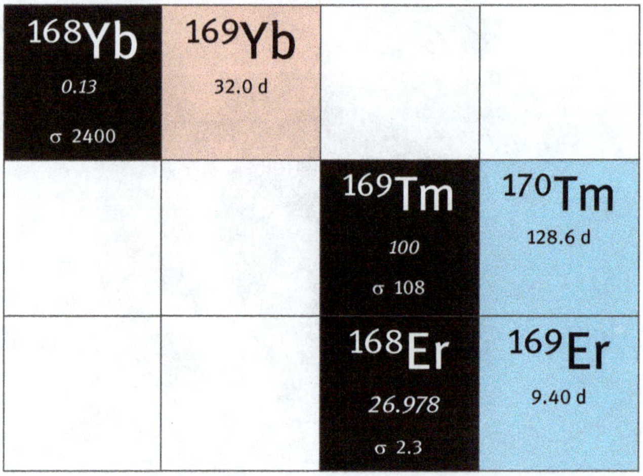

^{168}Yb 0.13 σ 2400	^{169}Yb 32.0 d		
		^{169}Tm 100 σ 108	^{170}Tm 128.6 d
		^{168}Er 26.978 σ 2.3	^{169}Er 9.40 d

Fig. 13.10: Neutron capture cross-sections σ for ^{A}A(n, γ)$^{A+1}$B reactions on lanthanide nuclides of identical or similar mass number A. Values are quite different, i.e. 2.3, 108 and 2400 barn for the ^{168}Er(n, γ)^{169}Er, ^{169}Tm(n, γ)^{170}Tm and ^{168}Er(n, γ)^{169}Er, respectively. (Italic numbers give the abundance H in % of the stable isotope for the natural isotope composition of the element.).

This is because the very simplified geometric explanation has a quantum mechanical background. The nuclear reaction probability W_{if} is bridged with the overlap of initial and final wave functions,[18] and the cross-section finally mirrors the transition matrix element (see Tab. 13.2). The probability of a certain nuclear reaction is high or low because of quantum mechanical considerations. The cross-section serves as a proportionality constant. If – for the same number of projectiles, given in units of 1/m² – the probability of a certain nuclear reaction is high, then the

---

18 . . . analog to the probability P_{fi} of transition (per unit time) between initial (i) and final (f) states of unstable nuclides undergoing radioactive transformation, in particular for primary β-transformation, as shown in eq. (7.2). For radioactive transformation, quantum mechanics defined a transition rate that parallels the transformation constant or half-life of the processes.

cross-section value is high and *vice versa*. In this respect, the cross-section is similar to the equilibrium constant used in conventional chemistry. There are even extremely low probabilities, expressed as $\sigma < 1$ (see Fig. 13.11) indicating a cross section much smaller than the area of the target atom![19]

Tab. 13.2: Quantum mechanical background of the cross-section. M_{if} = transition matrix element, E_{if} = density of energy states, ρ^{psf} = phase space factor.

Quantum mechanics	Wave functions
Probability W_{if} = reaction rate (events/s)	$W_{if} = 2\pi/\hbar \; r(E_{if}) \{M_{if}\}^2 = \rho^{psf} \{M_{if}\}^2$
Cross-section ≈ transition matrix element	$W_{if} \approx \sigma$ $\sigma \approx \{M_{if}\}^2$

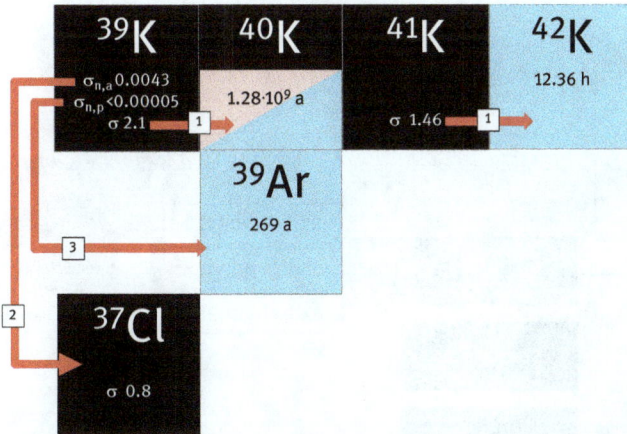

Fig. 13.11: Cross-section values σ (in barn) for different neutron-induced reactions on stable potassium target nuclei. The ^{39}K(n, γ)^{40}K process has a cross-section of $\sigma = 2.1$ b, that of the ^{41}K(n, γ)^{42}K process is $\sigma = 1.46$ b [1]. For ^{39}K, the ^{39}K(n,α)^{42}Sc ($\sigma = 0.0043$ b) [2] and the ^{39}K(n,p)^{39}Ca ($\sigma < 0.0005$ b) [3] processes are much less probable.

Cross-section values cover a very broad range, from a few mb (1 mb = 10^{-27} cm^2) to 10^5 b. Values (in barn) for neutron capture processes are indicated in the Chart of Nuclides, shown in Figs. 13.10 and 13.11. If not indicated in detail, the values stand for (n,γ) processes induced by neutrons of thermal energy, which is 0.025 eV (up to 0.1 eV). For some special cases, cross-section values of (n,γ) processes induced

---

[19] The σ-value of the ^{4}He (n,γ) reaction is practically zero, although the area of the ^{4}He nucleus is not, of course.

by fast neutrons (10 keV to 20 MeV kinetic energy) and for (n,α) and (n,p) reactions are listed, such as ^{39}K in Fig. 13.11.

13.4.2 Individual *vs.* total cross-sections

The interaction of a projectile with a target may be described by a total cross-section, including different types of scattering (elastic, inelastic) and projectile adsorption in the target nucleus (causing various reactions). While in the production of many radionuclides the nucleon transfer type dominates, spallation (at very high projectile energies) and fission (for fissile target nuclei) are also important. Suppose the desired product radionuclide is obtained through a nucleon transfer mode; its exit-channel may compete with other exit-channel options, shown in Fig. 13.12. For each single channel, there is an individual cross-section.

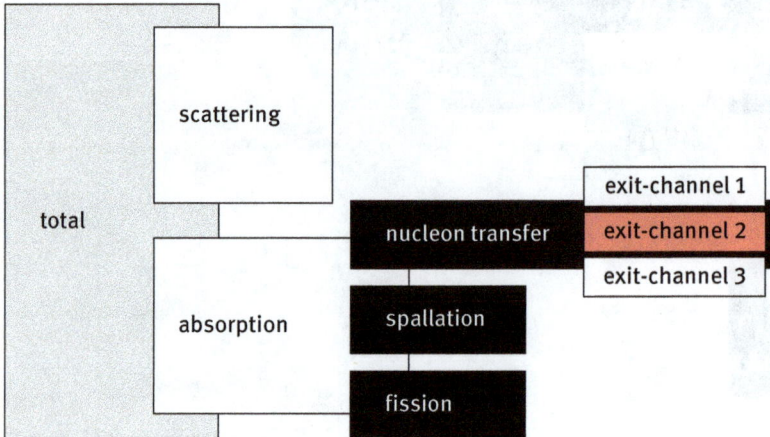

Fig. 13.12: Individual *vs.* total cross-section. While the total cross-section covers all the possible interactions of the projectile with the target nucleus, the specific route yielding the desired nuclear reaction product (here exit-channel no. 2 of a nucleon transfer reaction) just represents a part of the total cross-section, an individual one.

At a given excitation energy, an exit-channel opens for the compound nucleus. The "easiest" way to de-excite is to release electromagnetic radiation: in doing so, the (n,γ) or (p,γ) nuclear reaction occurs. At very low excitation energy of C*, the release of one single nucleon or cluster (e.g. the α-particle) is possible. The release of several individual nucleons may only occur at higher excitation energies.

If the excitation energy is high enough, several exit-channels are possible from the point of view of energy. For a proton-induced process, for example, these are (p,γ), (p,n), (p,2n), etc., (p,pn) etc., (p,d), (p,α), etc. Figure 13.13 illustrates some of

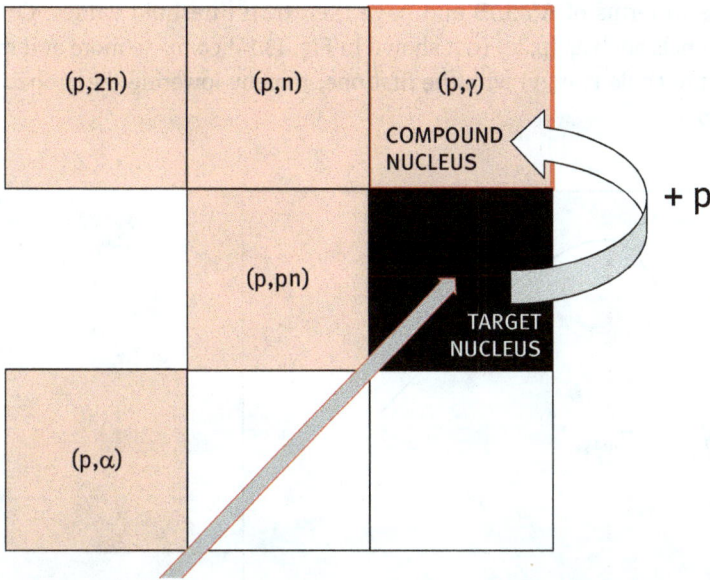

Fig. 13.13: Some exit-channel pathways for a proton-induced nuclear reaction within the Chart of Nuclides. Target nucleus A, compound nucleus C*, and projectile a according to the A(a,b)B, i.e. A(p,b)B processes.

these exit-channel pathways and the position of the product nuclei relative to the target nuclide and the composite nuclide.

13.4.3 Excitation function

The question is, which exit-channel works best? In fact, a certain individual exit-channel may dominate over others. For a given combination of target nucleus and projectile, each exit-channel is individually populated depending on the kinetic energy of the projectile; that is, on the excitation energy of the composite nucleus. This correlation of type $\sigma = f(E^{projectile})$ is called the "excitation function". A typical excitation function for a nuclear reaction induced by charged projectiles is given in Fig. 13.14, showing a (p,n) process on target nuclei of masses of about 50–150 and for proton energies below 25 MeV. To illustrate the general principle, this graph only shows the (p, n) exit-channel. Figure 13.14 reveals that after a maximum cross-section value at a certain energy, the cross-section decreases with further increasing projectile energy. Why is this? With increasing energy of the projectile deposited in the compound nucleus, another exit-channel "opens" because the higher excitation energy of the compound nucleus allows more nucleons to "evaporate". These alternative channels define new

nuclear reactions in terms of A(a,b)B and have their own threshold values. Consequently, exit-channels such as (p,2n) (not shown in Fig. 13.14) compete more and more
(with increasing projectile energy) with the first one, thereby lowering the probability
of the (p,n) channel.

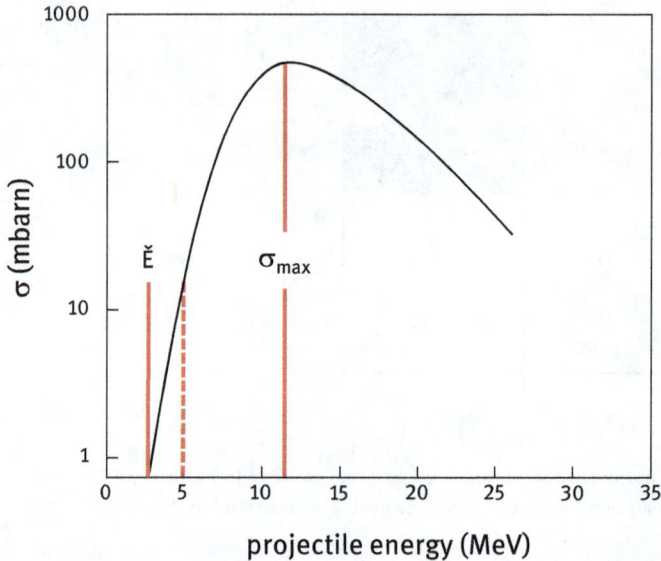

Fig. 13.14: Typical excitation function for a (p,n) nuclear reaction. Note the logarithmic scale of the
cross-section. The function indicates the threshold energy $\check{E}$ of about 2.5 MeV. The nuclear reaction
starts at $E^{projectile} > \check{E}$. From $\check{E}$ onwards, the cross-section values increase with increasing projectile
energy. Here, the value of σ is about 10 mb at about 5 MeV. It further increases, but tends to
increase more slowly at about $E^{projectile} > 10$ MeV. It then reaches a maximum value, here of about
500 mb at about 12 MeV. From now on, cross-section values decrease with increasing projectile
energy due to the onset of a "competing" exit-channel.

Figures 13.15 and 13.16 give concrete results for medically relevant iodine radioisotopes in comparison with other general diagrams. They show the experimentally
determined cross-section values for nuclear reactions leading to [124]I, a positron
emitter used in molecular imaging. Figure 13.15 gives the [125]Te(p,xn) reactions with
x = 1, 2 and 3, corresponding to product nuclei ^{125}I, [124]I and [123]I. Figure 13.16 illustrates α-induced processes on [123]Sb in terms of (α,xn) reactions with x = 1, 2 and 3,
corresponding to [126]I, ^{125}I and [124]I, respectively.

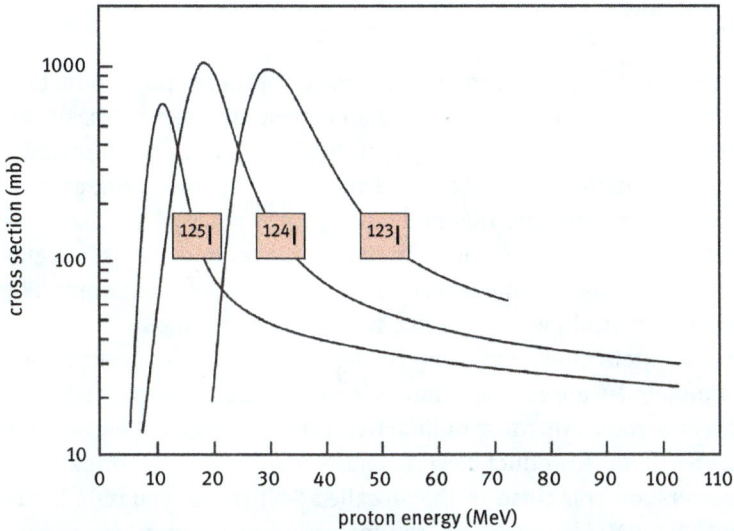

Fig. 13.15: Excitation function of proton-induced nuclear reactions leading to iodine radioisotopes. ^{125}Te(p,xn) reactions with x = 1, 2 and 3 correspond to product iodine isotopes ^{125}I, ^{124}I and ^{123}I. With permission: A Hohn, FM Nortier, B Scholten, TN van der Walt, HH Coenen, SM Qaim, Excitation functions of ^{125}Te(p,xn)-reactions from their respective thresholds up to 100 MeV with special reference to the production of ^{124}I, Appl. Radiat. Isot. 55 (2001) 149.

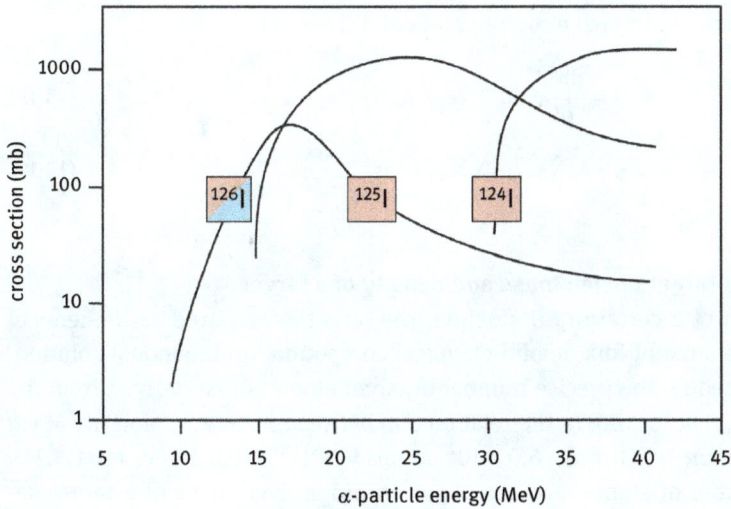

Fig. 13.16: Excitation function of α-induced nuclear reactions on ^{123}Sb leading to iodine radioisotopes. ^{123}Sb (α,xn) reactions with x = 1, 2 and 3 correspond to ^{126}I, ^{125}I and ^{124}I, respectively. With permission: MS Uddin, A Hermanne, S Sudár, MN Aslam, B Scholten, HH Coenen, SM Qaim, Excitation functions of α-particle induced reactions on enriched ^{123}Sb and ^{nat}Sb for production of ^{124}I, Appl. Radiat Isot 69 (2001) 699.

13.4.4 Activation equation

In conventional synthetic chemistry, the chemist has to optimize the production yield in the chemical reaction process. The tools a chemist may use are basically the amounts of the reactants (the more one puts into the reaction, the larger the product yield), reaction temperature, pressure and volume (which will work according to rules of chemical equilibria and thermodynamics), the reaction process time (usually the longer a reaction lasts, the more product is made) in some cases catalysts to facilitate the process, etc. In nuclear reactions, the tool set is different. For example, conventional chemical parameters like temperature and pressure applied on the scale of nucleon interactions and catalysts are meaningless. Concentrations of target atoms and projectiles are relevant, but in a more complex sense. The reaction period also plays a role, but for a radioactive product nuclide the nuclear chemist has to consider that the product decays simultaneously with its formation.

All these parameters are combined in the so-called "activation equation". The nuclear reaction rate $R = d\hat{N}_B/dt$ represents the formation of the product nuclei, which is the product of cross-section, projectile flux and number of target nuclei $\hat{N}_A$. Simultaneously, R depends on the transformation of the unstable product nucleus, which depends on its decay constant λ, as shown in eq. (13.18). If the reaction rate is expressed as product radioactivity (i.e. the nuclear reaction yield) according to $\acute{A}_B = \lambda_B \cdot \hat{N}_B$, then eq. (13.19) gives the individual radioactivity of each irradiation period t^{irr}. However, this is a rather "abstract" mathematical expression, and needs some more discussion when real experimental conditions are concerned.

$$R = \frac{d\hat{N}_B}{dt} = \sigma\Phi_a\hat{N}_A - \lambda_B\hat{N}_B \qquad (13.18)$$

$$\acute{A}_B = \lambda_B\hat{N}_B = \sigma\Phi_a\hat{N}_A\left(1 - e^{-\lambda_B t^{irr}}\right) \qquad (13.19)$$

13.4.4.1 Number of target nuclei: mass and density of a target

For the production of a certain radionuclide, the target is prepared as a chemical sample. It could be a metal foil, a solid chemical compound, an (aqueous) solution, or a gas.[20] Consequently, the precise number of target atoms is first derived from the mass of the target (g) according to the relationship between mass and molarity M *via* the Avogadro constant N^A (1 mol = $6.02 \cdot 10^{23}$ atoms = $N^{AVOGADRO}$), as per eq. (13.24). Secondly, the number of stable target isotopes within a given mass of a target depends on the natural isotopic composition of the given element. In the ^{209}Bi$(\alpha,2n)^{211}$At

---

20 For example, ^{211}At is produced *via* the ^{209}Bi $(\alpha,2n)$ reaction irradiating metallic bismuth (see Volume II, Chapter 8), ^{124}I through (p,n) or (d,2n) reaction on ^{124}Te-TeO$_2$, ^{18}F through ^{18}O(p,n) reaction irradiating ^{18}O-water, and ^{123}I through p-induced reactions on ^{124}Xe gas.

production, the target represents mono-isotopic ^{209}Bi, so its abundance is H = 1. In contrast, if ^{124}I were produced through ^{124}Te(d,2n) reaction by irradiation of natural tellurium, the abundance of ^{124}Te among all the stable isotopes of tellurium is 4.74% and H = 0.0474. For the ^{18}O(p,n)^{18}F reaction irradiating "normal" water, the abundance of ^{18}O would be only H = 0.00205.[21] Thus, the yield of the nuclear reactions is directly proportional to H. For situations like these, the nuclear chemist would prefer to use isotopically enriched target materials, such as water enriched in ^{18}O.

13.4.4.2 Target mass *vs.* energy

Interaction of the projectiles with the atoms of the target material (i.e. electron shell and nucleus) causes a stepwise loss in kinetic energy in terms of $\Delta E \approx \Delta x$. This is schematically illustrated in Fig. 13.17. The projectile enters the first slice of the target (i.e. its surface) at its initial energy E_0.

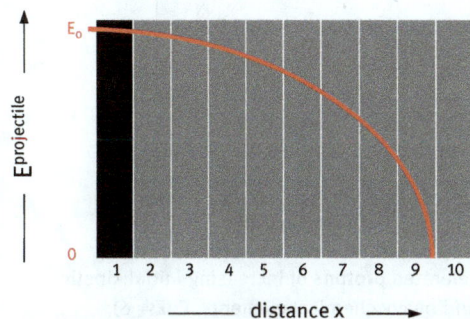

Fig. 13.17: Loss of kinetic energy of an incoming projectile when penetrating a target material along a distance x. The real target dimension here is divided into 10 slices. These may be virtual, but may also be realized practically, for example in metal targets by combining 10 thin metal foils. For an overall 0.10 mm thick metal target, this would correspond to 10 separate foils of 10 μm thickness. The projectile enters the first slice at its initial energy E_0. It then starts to lose kinetic energy in terms of E = f(x). Within the first slices, the gradual loss of ΔE is low, but increases with decreasing kinetic energy. It completely loses its kinetic energy at slice 9, and the final slice of the complete target does not even "see" any projectile – and thus does not contribute to the nuclear reaction. The same may be true for slice no. 9, if the kinetic energy of the projectile is below the nuclear reaction threshold.

For a given target material, projectile penetration depends on the properties of the projectile (what kind of projectile, how much kinetic energy). At the same time, a given projectile shows very different penetration performances for different kinds of

---

21 Indeed, the production of ^{18}F utilizes water enriched with >98% ^{18}O. Now, H is 0.98 instead of 0.00205 and the yield with identical irradiations is about 500 times higher compared to irradiating the same amount of natural water.

target materials. The incoming projectile "feels" a certain "resistance", originating from the physico-chemical properties of the target material. For one and the same projectile (of identical kind and kinetic energy) and a given size of the target (in particular the length x), the penetration length basically relates to the chemical element (namely its proton number Z) and the density ρ of a particular chemical species of that element.

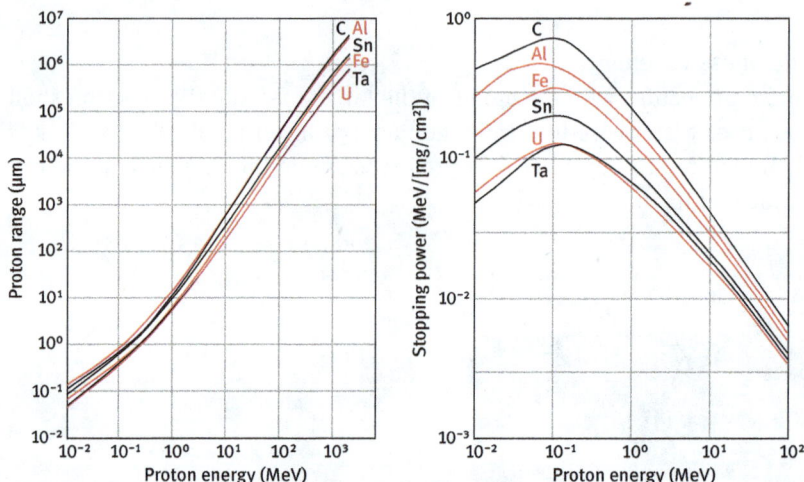

Fig. 13.18: Range (a) and stopping power (b) of accelerated protons of increasing initial kinetic energies for target nuclei covering light, medium and heavy chemical elements. C (Z = 6), Al (Z = 13), Fe (Z = 26), Sn (Z = 50), Ta (Z = 73) and U (Z = 92). With permission: JF Ziegler, Transport of ions in matter, International Business Machines, 1992.

This kind of "resistance" is expressed as "stopping power" $S = -dE/dx$ of a target material.[22] It depends on the nuclear (mainly Z) and chemical (chemical species, material density) parameters of the target, the nuclear parameters of the projectile (mass and charge Z), and yields a characteristic retardation (Bremszahl) B, as per eq. (13.20a). The loss of energy a projectile undergoes with increasing distance is described by the BETHE–BLOCH equation, eq. (13.20b). This equation describes the loss of kinetic energy of a charged projectile (charge z) within a target material defined by proton number Z, density ρ and the molar mass M (in g/mol). Here, U^{pot} is the ionization energy of the target material, c is the speed of light, v the velocity of

---

22 Stopping power is given in units of MeV/cm. A more appropriate unit is the mass stopping power, which is $-dE / dx / \rho$, in units of MeV g/cm^2. In more detail, the stopping power includes the effect of the projectiles interacting with both the electron shell of the target atoms ($S^{electric}$) and with the nuclei of the target atoms ($S^{nuclear}$).

the projectile, ε_o the vacuum permittivity, e and m_e the charge and rest mass of the electron.

$$-\frac{dE}{dx} = f\left(\frac{Z^2}{v^2}\rho\,B\right) \tag{13.20a}$$

$$-\frac{dE}{dx} = Z\,\rho\,\frac{\dot{N}_A}{M}\frac{z^2}{\left(\frac{v}{c}\right)^2}\frac{e^2}{4\pi\varepsilon_0}\left[\ln\frac{2m_e c^2\left(\frac{v}{c}\right)^2}{U^{pot}\left(1-\left(\frac{v}{c}\right)^2\right)} - \left(\frac{v}{c}\right)^2\right] \tag{13.20b}$$

Figure 13.18 shows the range of accelerated protons of initial kinetic energies of 10 keV to 2 GeV for 6 target species covering light, medium and heavy chemical elements: carbon (Z = 6), aluminum (Z = 13), iron (Z = 26), tin (Z = 50), tantalum (Z = 73) and uranium (Z = 92). Ranges continuously increase with increasing projectile energy and are larger for low-Z targets relative to high-Z targets. The figure also translates these ranges and energies into the stopping power for protons of kinetic energies up to 100 MeV. Here, stopping powers increase from 10 keV to 100 keV, reaching maximum values. With further increasing kinetic energy of the protons, the values start to decrease continuously.

13.4.4.3 Target mass *vs.* cross-section
The cross-section for a charged particle induced reaction depends on the energy of the projectile. Figure 13.19 combines the course of the energy of the charged projectile in the target stack (Fig. 13.17) with the excitation function shown in Fig. 13.14. Values of the cross-section change dramatically within the target dimension.

Consequently, there is no single value of σ for a real nuclear reaction to produce a radionuclide. Instead, the cross-section is integrated over the adequate value of energy, i.e. distance. This turns σ into an integral covering the corresponding changes in projectile energy and cross-section. The term now includes the values of the two kinetic energies E_1 and E_2 of interest, the distance x covered by projectile a, and the density ρ of the target material, eq. (13.21).

When starting from relatively large projectile energies, several different targets may be installed within the projectile beam along the decreasing energy path of the projectile. Whenever a certain region of energy fits with a relevant cross-section of a nuclear reaction, the corresponding target material may be inserted at this particular position, as per Fig. 13.20.

$$\sigma\,(E_x) \rightarrow \int_{E_1}^{E_2}\left[\frac{dE_a}{d(\rho x)}\right]^{-1}\sigma\,(E_a)dE_a \tag{13.21}$$

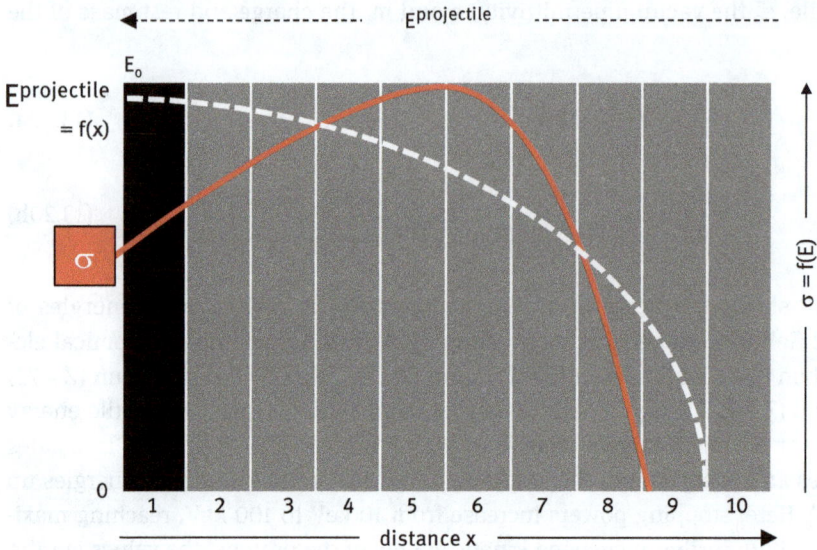

Fig. 13.19: Excitation function $\sigma = f(E^{\text{projectile}})$ with $E^{\text{projectile}} = f(x)$ of Fig. 13.17. Note that the excitation function is left/right image reversed: the cross-section is zero within slice no. 8, i.e. below the threshold energy, increases with increasing energy towards the surface of the target, reaches a maximum value between slices 5 to 6 and decreases with further increasing energy towards slice no. 1.

13.4.4.4 Target mass *vs.* projectile flux

Finally, the flux density of the projectile also decreases when it penetrates a target of a certain mass, as per Fig. 13.21. Similar to projectile energy, the loss in projectile flux density is an exponential function and mathematically expressed by eqs. (13.22a) and (13.22b).

$$-\frac{d\Phi}{dx} = \sigma\,\Phi\,\acute{N}_A \tag{13.22a}$$

$$\Phi = \Phi_0 e^{-\sigma \acute{N}_A x} \tag{13.22b}$$

For neutrons, the interaction is much lower than in the case of charged projectiles. For charged particles the so-called BRAGG effect can occur. This refers to an extremely increased intensity of interactions at close to zero kinetic energy of the projectile. The yield is an increased deposition of ionization within a very short distance, referred to as the BRAGG peak.

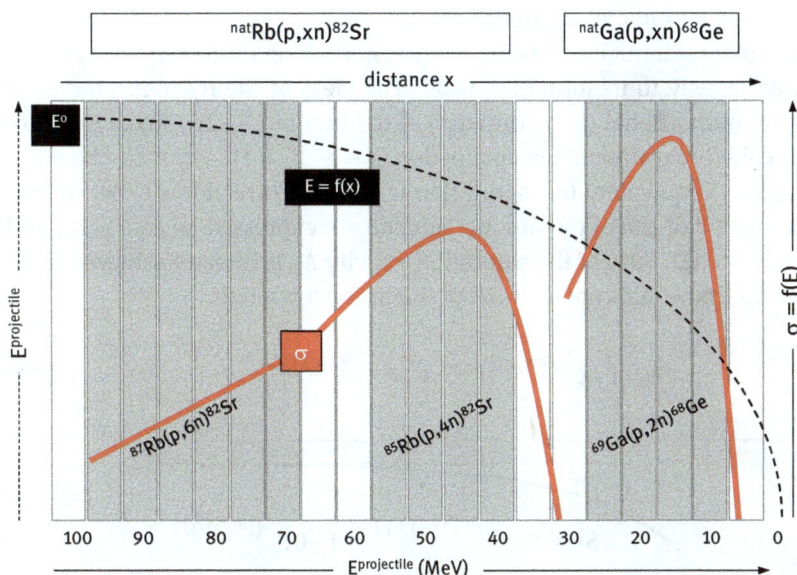

Fig. 13.20: Schematic illustration of radionuclide production at Los Alamos National Laboratory for (p,xn) reactions and proton energies of $E_0 = 100$ MeV. The high proton flux density of 250 mA makes the linear accelerator-based Isotope Production Facility a powerful radionuclide production unit. Three different targets may be inserted in a row (stack), namely two separate natural rubidium chloride targets and one natural gallium target. The protons first traverse the two Rb targets and then arrive at the Ga target with less kinetic energy; i.e. ca. 100–70 MeV for the first ("high-energy") RbCl target, 60–40 MeV for the next ("low-energy") RbCl target, and ca. 30–10 MeV for the Ga target. These intervals each correspond to the maximum cross-sections of the three different nuclear reaction routes, which are $^{87}Rb(p,6n)^{82}Sr$, $^{85}Rb(p,4n)^{82}Sr$ and $^{69}Ga(p,2n)^{68}Ge$. After irradiation, the target stack is dismantled mechanically and each target is further treated radiochemically to isolate the corresponding product nuclide, ultimately to extract ^{82}Sr and ^{68}Ge. Adapted from: M Nortier, Recent advances in large scale isotope production at LANL, NRC8, Como, September 2012. With permission.

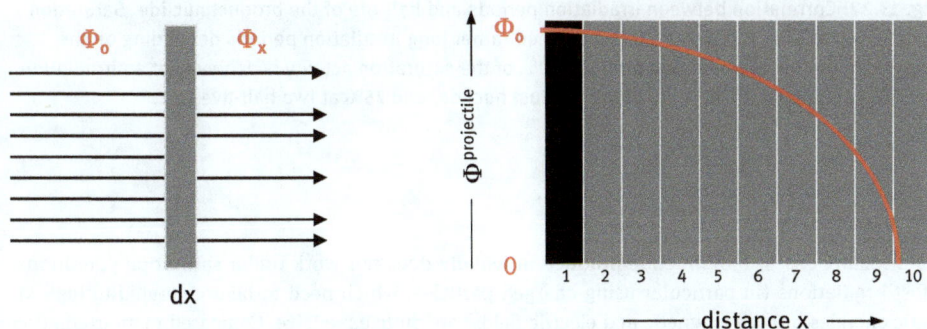

Fig. 13.21: Decrease in projectile flux when penetrating a target of dimension dx.

13.4.4.5 Irradiation period and saturation factor

The production of a radionuclide reflects the transformation of the nuclear reaction product simultaneously to its formation yield. The yield of the reaction product is thus not directly proportional to the duration of the irradiation. The maximum production yield, called "saturation", is obtained at $(1 - e^{-\lambda t^{irr}} = 1)$. Any range between $0 \leq (1 - e^{-\lambda t^{irr}}) \leq 1$ is equivalent to a saturation factor SF. Correlations between irradiation periods and half-life of the product nuclide are expressed by eq. (13.23) and graphically in Fig. 13.22. 50% of the saturation activity $\acute{A}_B$ is already achieved at an irradiation period that equals one half-life of the product nuclide.[23]

$$SF = 1 - e^{-\left(t^{irr}/t^{1/2}\right)\ln 2} = 1 - 0.5^{-\left(t^{irr}/t^{1/2}\right)} \tag{13.23}$$

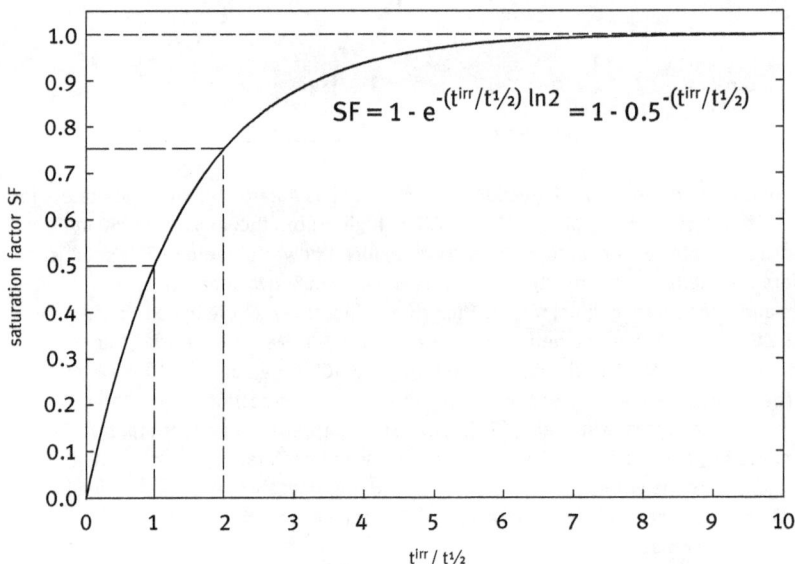

Fig. 13.22: Correlation between irradiation periods and half-life of the product nuclide. Saturation is achieved at SF = 1; however, this may need rather long irradiation periods depending on the half-life of the nuclear reaction product. 50% of the saturation activity is achieved at an irradiation period that equals one half-life of the product nuclide, and 75% at two half-lives etc.

---

23 In many cases, radionuclide production usually does not work under saturation conditions. Most irradiations (in particular using charged particles, which need to be accelerated to high kinetic energies in high magnetic and electric fields) are quite expensive. Compared to an irradiation lasting, for example, for two half-lives of the product nuclide (representing 75% of the maximum achievable product activity) a further irradiation by one or two periods of half-life would add just 12.5% or 18.75% to the 75% value.

13.4.5 Final activation equation

When all the above-mentioned aspects are taken into consideration, the basic equation (13.18) turns into integral versions, shown in Fig. 13.23. For thermal neutrons as projectiles eq. (13.19) applies, while for charged particles as projectiles eq. (13.25) applies. The latter integral over the excitation function considers effects of thick targets, causing degradation in projectile energy and projectile flux.

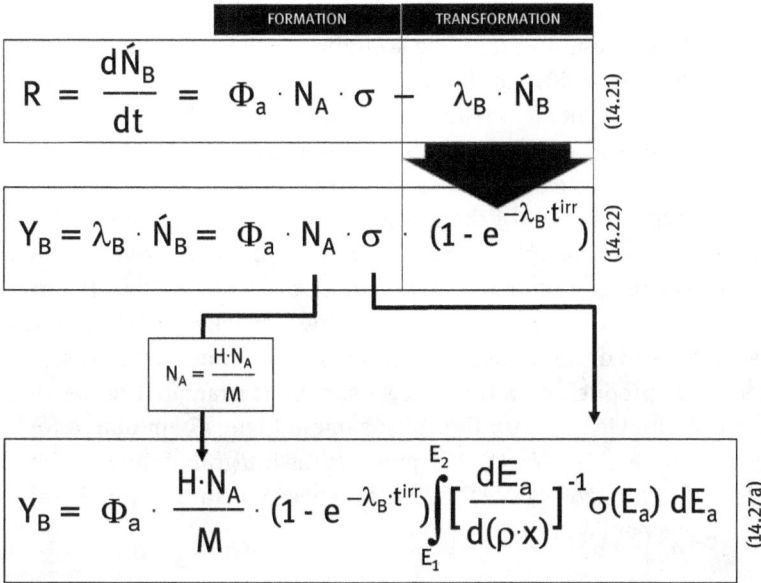

Fig. 13.23: Activation equations for neutrons (13.19) and charged projectiles (13.25). R is the nuclear reaction rate, σ the cross-section for a certain exit-channel in barn (b), i.e. 10^{-24} cm^2. $N^A = N^{AVOGADRO}$. The projectile flux ϕ_a is cm^{-2} s^{-1} for neutrons and s^{-1} for charged projectiles. In the latter case, it is typically expressed in units of Ampere (such as, e.g., μA).

For neutrons, the flux density is given in cm^{-2} s^{-1}, which finally yields the unit for R and $Á_B$ (s^{-1} = Bq). For charged projectiles, the flux is given in projectiles per second and, as for the unit known for electrons (the electricity), the Ampere is preferred.[24] M is the target atomic mass (g), $\sigma(E_a)$ is the excitation function with cross-sections (mb), and $S(E_a)$ is the stopping power in g/cm^2 for 1 MeV. For protons, and for saturated reaction yield, a specific expression is given in eq. (13.26). The product radionuclide yield $Á_B$ then has the unit Bq / μA.

---

24 The flux density of neutron is given in this unit. For charged particles (in analogy to electric current) it is given in Ampere, typically μA or mA (1 A = $6.24 \cdot 10^{18}$ electrons per second, 1 A = 1 C/s or 1 A = 1 W/V).

$$\acute{A}_B = 6.24 \cdot 10^{12} m \frac{H\,N^{AVOGADRO}}{M} \int_{E_1}^{E_2} \left[\frac{\sigma\,(E_a)}{S(E_a)} \right] dE_a \qquad (13.26)$$

13.4.6 Differential *vs.* integral radionuclide production yields

Suppose a real target is (either virtually or physically) divided into many extremely thin slices (dx → 0). The nuclear reaction product B may be different and its product radioactivity may be different in each slice according to the individual values of the thickness of the slice, according to the projectile energy (a higher neutron energy may result e.g. in a (n,2n) nuclear reaction instead of (n,γ)), the corresponding value of the cross-section, and the projectile flux. If the production yield is thus a function of $\acute{A}_B$ = f(dx) ≈ f(E), this can be measured experimentally by e.g. irradiating a stack of several very thin samples of the target material. Suppose the individual projectile energy within each sample is known, and the product activity is measured for each slice. The differential radionuclide production rate follows the excitation function, i.e. it is low at low projectile energies, reaches a maximum at higher projectile energy and drops off as projectile energy continues to increase.

In real radionuclide production, a thick target sample is irradiated rather than several thin samples. The yield within the thick sample is thus cumulative (over many virtual infinitesimally thin targets). It represents the integral of the individual differential yields within a region of $\Delta E^{projectile}$, i.e. between a starting (high) value E_1 and a lower one E_2, as per Fig. 13.24.

13.5 Radionuclide production using neutrons

Neutrons are very commonly used projectiles for two reasons: Neutrons do not need to overcome the coulomb barriers, and they are available from various sources.

13.5.1 Sources of neutrons

First, let us explore where the neutrons we need may origin. Table 13.3 lists various sources of neutrons based on permanent facilities and laboratory samples. This represents a quite versatile and well-accessible spectrum of neutrons for many kinds of applications. It includes nuclear reactions induced by charged particles of type (projectile,xn) to deliver high-energy neutrons, which in turn can be used to induce new neutron-induced reactions. The neutrons "generated" show different parameters in terms of energy and flux density. High initial kinetic energies can be moderated down to lower values.

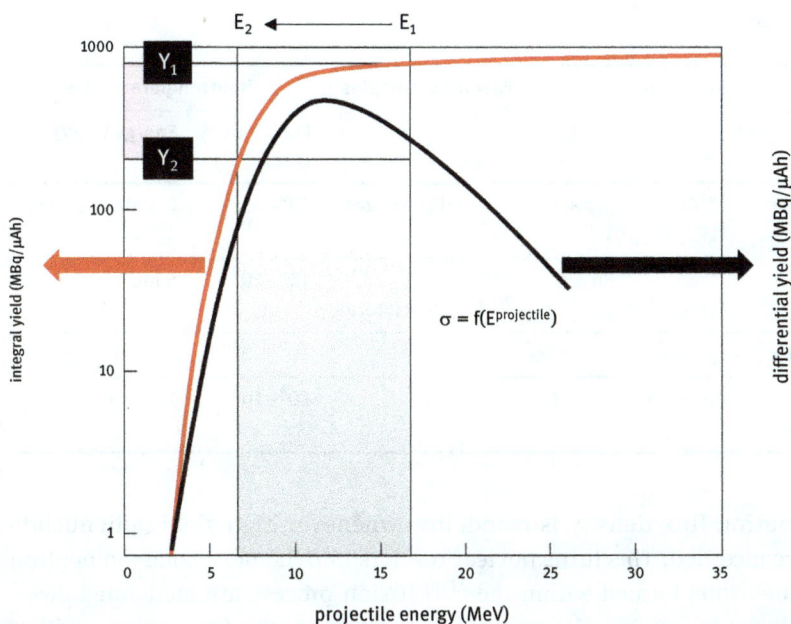

Fig. 13.24: Differential vs. integral yields of a charged projectile-induced nuclear reaction. For a given cross-section value at a specific projectile energy, *differential* nuclear reaction yields are obtained according to eq. (13.18) if infinitesimally thin targets are irradiated. If a thick target is irradiated, production yields are cumulative from $E^{threshold}$ to increasing $E^{projectile}$. Suppose the initial energy E_1 of the projectile is 17 MeV and covers the whole thick target until the threshold value, an integral yield $Á_{B(1)}$ is obtained (here ca. 800 MBq/µAh) for 17 MeV → threshold. Suppose the target is not as thick, and the projectile beam leaves the target at 7 MeV, the projectile energy acting in the target is 17 → 7 MeV. In this case, the theoretical production yield $Á_{B(2)}$ between 7 MeV → threshold (which is ca. 200 MBq/µAh in this example) is not included. The overall production yield is $Á_B = Á_{B(1)} - Á_{B(2)} = 800-200 = 600$ MBq/µAh.

Tab. 13.3: Neutrons available at nuclear reactors and from other sources used in neutron-induced nuclear reactions.

Source	Reaction	Facilities/samples	Neutron parameters	
			Flux density (1/cm² s¹)	Energy (MeV)
Nuclear reactors (induced fission)	$^{235}U(n,f)$	Research reactors	4.10^{12}	Up to 10 MeV down to thermal after moderation
		Production/military reactors	$>10^{15}$	
Spontaneous fission	$^{252}Cf(sf,x)$	^{252}Cf	$10^4–10^{10}$	

Tab. 13.3 (continued)

Source	Reaction	Facilities/samples	Neutron parameters	
			Flux density $(1/cm^2\ s^1)$	Energy (MeV)
α-Induced nuclear reactions	$^{241}Am \rightarrow\ ^{237}Np + α$ $^7Li(α,n)$	$^{241}Am/Li$ mixtures	10^4–10^5	0.5 MeV (average)
	$^{226}Ra \rightarrow\ ^{222}Rn + α$ $^7Li(α,n)^{10}B$	$^{226}Ra/Be$ or $^{241}Am/Be$ mixtures	10^4–10^5	5 MeV (average)
Neutron generator	$t(d,n)α$		10^8–10^{10}	14 MeV
Proton-induced nuclear reactions	$A(p,xn)B$		10^4–10^5	1 –8 MeV

A high neutron flux density is mandatory whenever high-yield radionuclide production is concerned. This turns nuclear reactors into the most common neutron sources. The neutrons formed within the ^{235}U fission process are moderated down to thermal energy (ca. 0.025 eV) and used to irradiate massive targets – either within the reactor core or at outside target positions.

13.5.2 Neutron capture reactions

Figure 13.25 shows the typical options of neutron-induced processes, namely (n,γ), 2x(n,γ) and (n,2n) (both along the line of isotopes), (n,p) (along the line of isobars), and (n,d) and (n,γ) + β⁻.

The most frequently used pathway is the (n,γ) route, which produces a neutron-rich radioisotope K2 of the same chemical element compared to the stable target isotope K1.[25] At very high neutron flux densities, a subsequent neutron capture takes place (double neutron capture), producing the mass (A + 2) isotope of the target nucleus. In both cases, the radionuclide cannot be separated chemically from the target isotope.[26] The radioisotope remains incorporated within the excess of mass of the stable target isotope.

If the (n,γ) nuclear reaction product K2 is not of practical interest as directly formed via eq. (13.27), but rather its radioactive subsequent (n,γ) → β⁻-transformation product, then the process in eq. (13.28) applies instead. The (n,γ) produced nucleus is neutron-rich and thus transforms *via* a β⁻-process into a nuclide K3 of (Z + 1) proton

---

25 This also represents the most commonly applied process in Neutron Activation Analysis – see Chapter 3 of Volume II.

26 . . . unless SZILLARD–CHALMER type processes are utilized, see Chapter 7 in Volume II.

PROJECTILE = NEUTRON

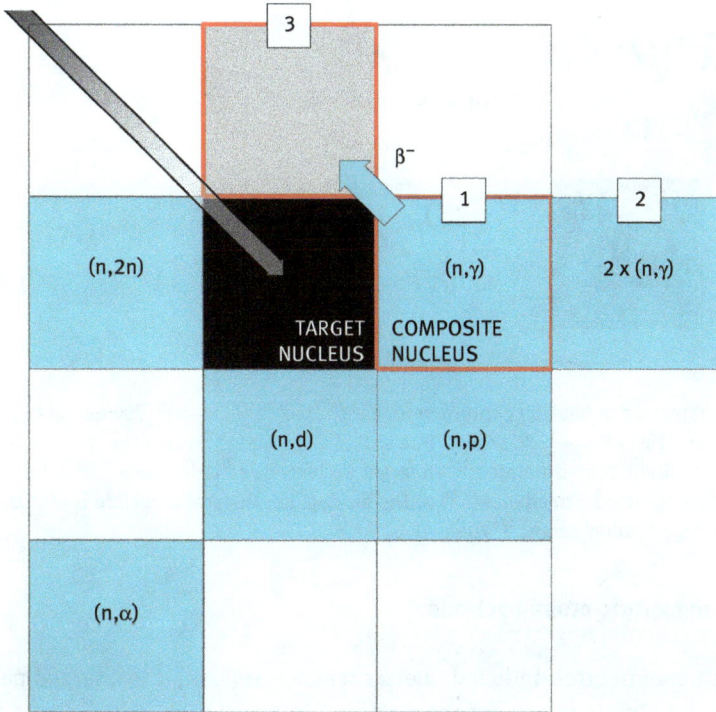

Fig. 13.25: Potential neutron-induced nuclear reactions and the subsequent β^--transformation processes of the reaction products within the Chart of Nuclides. Color-coding in blue, because the composite nucleus is neutron-rich. For these products in particular, β^--transformation is an inherent strategy to obtain the $(n,\gamma) \rightarrow \beta^- \rightarrow$ product [1].

number relative to the target nuclide K1. K3 represents a different chemical element relative to the irradiated target, and can be separated radiochemically. The product is thus free of the initial target nuclide.

$$K1(n,\gamma)_Z^{A+1}(^*K2)_{N+1} \tag{13.27}$$

$$K1(n,\gamma)_Z^{A+1}(^*K2)_{N+1} \xrightarrow{\beta^-} {}_{Z+1}^{A+1}K3_{N+1} \tag{13.28}$$

Figure 13.26 illustrates the two pathways of producing ^{177g}Lu, a low-energy β^--emitter systematically used in nuclear medicine therapy. For more information, see also Chapter 14 of Volume II.

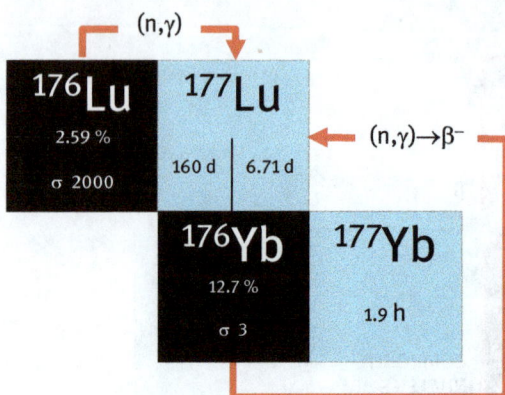

Fig. 13.26: Two pathways of neutron capture producing ^{177g}Lu.[27] ^{176}Lu(n,γ)^{177g}Lu as direct process and the ^{176}Yb(n,γ)^{177}Yb → β$^-$ → ^{177g}Lu route. Due to t½(^{177}Yb) < t½(^{177g}Lu), already about one day after the end of irradiation the irradiated ^{176}Yb target exclusively contains pure ^{177g}Lu, i.e. all the intermediate ^{177}Yb has transformed into ^{177g}Lu. The product lutetium radionuclide is effectively separated from the ytterbium target ^{176}Yb.

13.5.3 Neutron capture cross-sections

The cross-section of neutron-induced nuclear reaction is (similar to charged particles) a function of the kinetic energy of the neutrons in terms of $\sigma \approx 1/v$.[28] Cross-section values at low "thermal" energy (in particular for room temperature with energy of ca. 0.025 eV and a speed of ca. 2 km/s) decrease with increasing velocity by a 1/v correlation.[29] At about 0.01 eV, cross-section values are in the range of several hundred barns, and decrease to less than 100 barns at 10^{-1} eV. If simply extrapolated, cross-

---

27 In this case of the (n,γ) process on isotopically enriched ^{176}Lu, the cross-section is very high (2000 barn), but the abundance of the isotope 176 in natural lutetium is low (2.59%). Consequently, isotopically enriched ^{176}Lu is used (96%). Irradiations use neutron flux densities of 10^{14} to 10^{15} neutron/cm^2 s, and irradiation periods last for a 1, 2 or weeks. Batch activities approach the TBq level. Specific activities after EOB (end of bombardment) reach values of about 20–30 Ci/mg, which is below the theoretical maximum value (100 Ci/mg). Simultaneously, the metastable isomer ^{177m}Lu is formed (t½ = 160 days). At EOB, it represents up to 0.05% radioactivity relative to ^{176}Lu. The other route starts from isotopically enriched ^{176}Yb. The primary reaction product is ^{177}Yb with a half-life of 1.9 h. Already within the course of the irradiation, it transforms to ^{177g}Lu. After irradiation, ^{177g}Lu is radiochemically separated from the ^{176}Yb target of about 100 mg mass. It thus principally exists at ultimate specific activity. The β$^-$-transformation of ^{176}Yb to ^{177g}Lu does not populate ^{177m}Lu. Practical specific activities reach 80% of the theoretical value.
28 The longer the "slow" neutron is in contact with the target nucleus, the higher the probability that "something" happens, i.e. a neutron gets trapped.
29 Neutron energies and velocities correlate by E ≈ 5.23 v^2 with E in meV units and v in km/s. At 0 °C, it is 0.0353 eV (ca. 2.6 km/s); at room temperature, 0.025 eV (ca. 2 km/s).

section values at energies of 0.1 to 1 eV (epithermal) and 0.5 eV to 10 keV (medium-fast)[30] should be as low as about 50 to 5 barn.

However, this is not necessarily the case. Instead, at those kinetic energies, cross-section values may peak at specific energies reaching values larger than expected by one or two orders of magnitude, as in Fig. 13.27. This effect is due to characteristic "resonances".[31] In nuclear reactions (and transitions), this relates to the overlap of the wave function of the projectile at specific energies with the wave function of the target nucleus at a specific nuclear state. It should be noted that the same feature applies to all kinds of projectiles.

With further increasing neutron velocities, reaching keV and MeV magnitudes for "fast" neutrons, these resonances disappear, and cross-section values more or less follow the 1/v correlation again, i.e. are within a range of about 5 to 2 b.

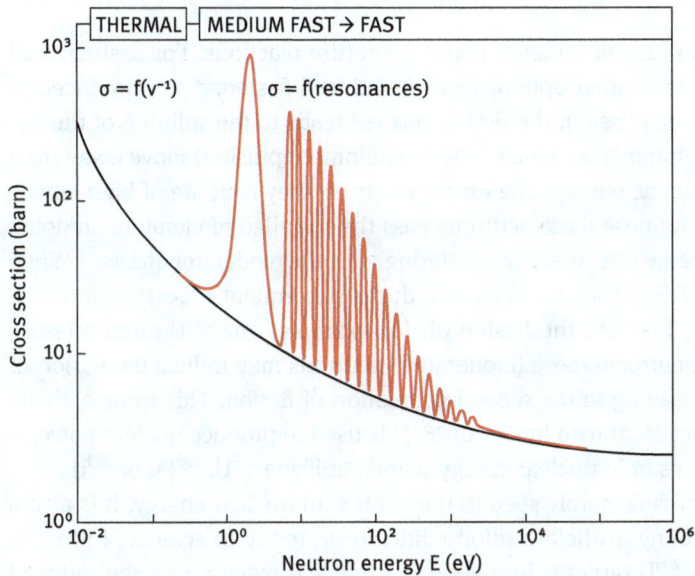

Fig. 13.27: Course of cross-sections of neutron-induced nuclear reactions over the range of kinetic energies of 1 meV to 1 MeV. Neutrons are grouped into thermal, epithermal and medium to fast neutrons.

---

30 According to increasing energy and velocity, neutrons are grouped into ultra-cold (<0.2 meV, <0.2 km/s), cold (<2 meV, <0.6 km/s), thermal (<0.1 eV, <4.4 km/s), epithermal (<1 eV, <14 km/s), medium-fast (0.5 eV–10 keV, 10–44 km/s), fast (10 keV–20 MeV, 44–62,000 km/s), and relativistic (>20 MeV, >62,000 km/s).

31 By definition, resonance is the specific response of a system (basically in natural sciences) to specific external activation. It is typically described in terms of the oscillation of the system with significantly enhanced amplitude in the case that the excitation occurred by a characteristic frequency of an electromagnetic activation – the system shows a resonant frequency.

There is a broad range of thermal neutron capture cross-sections among the chemical elements. Values range from practically zero (e.g. for ^{4}He, where the double magic nucleon configuration is not at all receptive to a further neutron), through values below 1 (i.e. from 0.0001 to 1 barn) up to several thousand barns. Cross-sections for the individual isotopes of a given element again cover a range of values.[32] The values also reflect a consequence of the shell model of the nucleus: neutron cross-sections are typically larger when a nucleus with odd neutron number captures a neutron, yielding a new nucleus of even neutron number. This alternation between the magnitude of cross-sections of unpaired and paired neutron nuclei nicely illustrates the impact of the shell model of the nucleus: paired nucleons provide a gain in binding energy.

13.5.4 Neutron-induced fission

Within nuclear reactors, nuclei undergo neutron capture reactions. For fissile nuclei (see Chapter 12), there is another option: neutron-induced fission.[33] As introduced in the context of spontaneous fission, the fission process leads to the splitting of the nucleus into two primary fragments, which (due to coulomb repulsion) move away from each other with high kinetic energy. The excess neutrons they emit are of high kinetic energy, up to 10 MeV. Suppose these neutrons pass the so-called moderators; in doing so, they lose kinetic energy due to elastic scattering with the moderator atoms.[34] When moderated down to 0.025 eV, such a neutron induces subsequent processes on fissile nuclei, especially ^{235}U. Because the fission of, for example, one ^{235}U atom releases more than two single neutrons, these (moderated) neutrons may induce the fission of two other ^{235}U nuclei, leading to the second generation of fission. This turns a single event into a fission cascade, shown in Fig. 13.28. It is used to produce nuclear power – either in nuclear weapons or in nuclear energy plants, utilizing ^{235}U, ^{239}Pu or ^{233}U.

Although this process is mainly seen in the context of nuclear energy, it is also a significant source of many artificial radionuclides used today in science, industry, and medicine. When a ^{235}U target is irradiated in a nuclear reactor, neutron-induced fission creates the typical profile of binary asymmetric fission, a characteristic distribution of primary and secondary fission fragments and final neutron-rich fission products. Some of these isobars include radionuclides of real practical value e.g. in nuclear medicine, especially ^{99}Mo, ^{90}Sr and others. The radiochemical separation of

---

32 For example, for the stable isotopes of gadolinium (Z = 64), the corresponding σ-values are ^{152}Gd = 700 b, ^{154}Gd = 60 b, ^{155}Gd = 61 000 b, ^{156}Gd = 2.0 b, ^{158}Gd = 2.3 b, ^{157}Gd = 254 000 b, ^{158}Gd = 2.3 b, ^{160}Gd = 1.5 b.

33 This deserves a dedicated explanation; see Chapter 11 in Volume II.

34 In order to maximize elastic scattering and minimize neutron absorption, low Z elements are preferred: hydrogen or deuterium as components of normal or "heavy" water, but also graphite (C), cadmium, boron and lithium.

such individual radionuclides is an extremely challenging task – both from a chemical and radiation point of view. An irradiated ^{235}U target contains an incredible radioactivity caused by radionuclides representing almost the entire PTE. Radiochemical separation first consists in the recovery of the ^{235}U target. Next, the various radionuclides are extracted in multistep separation protocols under remote-operation technologies. All the radionuclides finally obtained are at "no-carrier-added" state and thus superior to (n,γ) production routes. This is particularly relevant for ^{99}Mo, which is used to construct ^{99}Mo/^{99m}Tc radionuclide generators and where the "fission"-produced ^{99}Mo is carrier free compared to ^{98}Mo(n,γ)-produced ^{99}Mo (see Chapters 8 and 13 of Volume II).

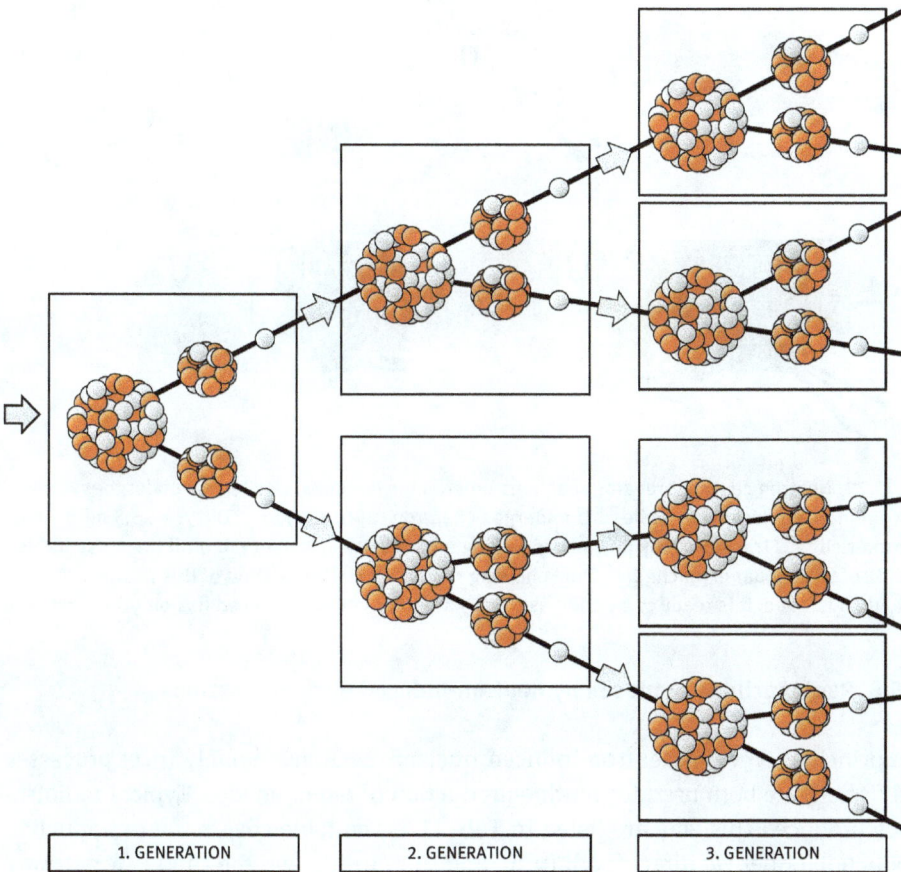

| 1. GENERATION | 2. GENERATION | 3. GENERATION |

Fig. 13.28: Cascade scheme of neutron-induced fission. The fission cascade is used to produce nuclear power. Today's nuclear power plants use ^{235}U.

13.5.5 Fission *vs.* breeding

Neutron capture on nonfissile nuclei may produce new generations of fissile nuclides, such as ^{239}Pu and ^{233}U. For ^{239}Pu, the pathway is illustrated in Fig. 13.29.[35] In terms of nuclear power, the so-called breeding process sets in at the neutron-rich reaction product ^{239}U. It transforms by the primary β^--process to ^{239}Np (t½ = 2.355 d), which continues the same way to ^{239}Pu (t½ = 2.4 · 10^4 a). This long-lived "bred" nuclide may be separated chemically from used ^{235}U fuel and represents another generation of a fissile nuclide.

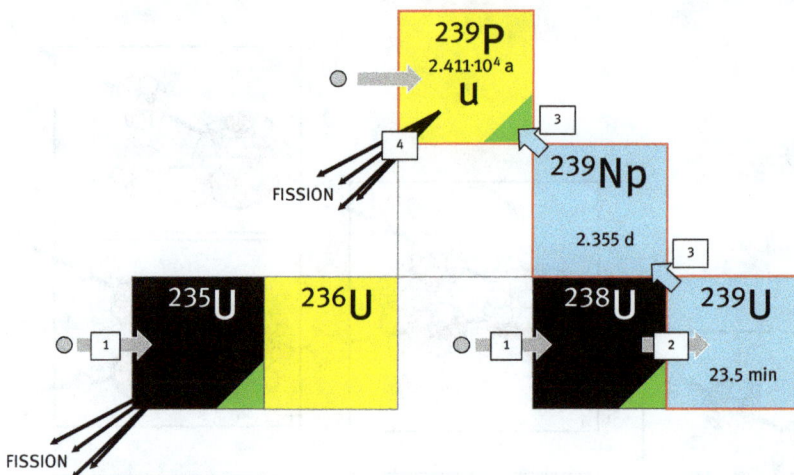

Fig. 13.29: Neutron capture breeding reactions on uranium isotopes. Fissile ^{235}U undergoes neutron-induced fission (1), while ^{238}U undergoes neutron capture (2) to ^{239}U (t½ = 23.5 min). The neutron-rich ^{239}U transforms by β^--emission along an A = 239 = constant (3), until the most stable nuclide of that isobar is reached, ^{239}Pu. Following radiochemical separation of this fissile plutonium isotope, it is ready to be used as a new target for neutron-induced fission (4).

13.5.6 Radionuclides produced by neutron-induced nuclear reactions

The principal types of neutron-induced nuclear reactions, namely (n,x) processes and fission, are both used for routine production of radionuclides. Typical radioisotopes produced this way are listed in Tab. 13.4. The table covers the radionuclide production either *via* direct (n,γ), (n,p), and (n,α) routes and the (n,γ) + β^- pathway or *via* neutron-induced fission. As the production is supposed to have high yields, the cross-sectional values and the natural abundance of the target isotope are

---

35 For ^{233}U, the analogue path starts from ^{232}Th: ^{232}Th (t½ = 1.405 · 10^{10} a) → (n,γ) → ^{233}Th (t½ = 22.3 min) → (β^-) → ^{233}Pa (t½ = 27.0 d) → (β^-) → ^{233}U (t½ = 1.592 · 10^5 a).

Tab. 13.4: Radionuclides produced *via* neutron-induced nuclear reactions and directions of their applications. (IVMI = *in vitro* molecular imaging, NAA = neutron activation analysis, CRA = chemical reaction analytics, RIA = radioimmunoassay assay, EBT = external beam therapy, ERT = endoradiotherapy, SPECT = single photon emission tomography).

Product nuclide	t½	Target isotope	Natural abundance (%)	Reaction, if not (n,γ)	σ (in b). (For fission, yields are given in (%) for the mass number.)	Application
^{3}H	12.323 a	^{6}Li	7.59	(n,t)	940	IVMI, CRA
^{14}C	5730 a	^{14}N	99.636	(n^{fast},p)	1.8	IVMI, CRA
^{24}Na	14.96 h	^{23}Na	100		0.53	NAA
^{42}K	12.36 h	^{41}K	6.7302		1.46	NAA
^{32}P	14.26 d	^{31}P	100		0.18	IVMI, RIA
		^{32}S	94.99	(n^{fast},p)	0.07	
^{35}S	87.5 d	^{35}Cl	75.76	(n^{fast},p)	≈0.08	IVMI, RIA
^{60}Co	5.272 a	^{59}Co	100		16.5	Sterilization, EBT
^{75}Se	119.64 d	^{74}Se	0.89		46	IVMI
^{89}Sr	50.5 d	^{88}Sr	82.58		0.0058	ERT
^{90}Sr	28.64a	^{235}U	–	(n,f)	5.89%	Generator for ERT: ^{90}Y/^{90}Sr
^{99}Mo	66.0 h	^{98}Mo	24.18		0.14	Generator for SPECT: ^{99}Mo/^{99m}Tc
		^{235}U	–	(n,f)	6.14%	
^{131}I	8.02 d	^{130}Te	34.08	(n,γ) + β^{-}	0.19	ERT, SPECT
		^{235}U	–	(n,f)	2.84%	
^{137}Cs	30.17 a	^{235}U	–	(n,f)	6.236%	Calibration
^{153}Sm	46.27 h	^{152}Sm	26.75		206	ERT
^{160}Tb	72.3 d	^{159}Tb	100		23.2	ERT
^{166}Dy	81.5 d	^{164}Dy	28.26	2x(n,γ)	2700 + 3500	Generator for ERT: ^{166}Dy/^{166}Ho
^{166}Ho	26.80 h	^{165}Ho	100		58	ERT
^{169}Er	9.40 d	^{168}Er	26.978		2.3	ERT
^{175}Yb	4.2 d	^{174}Yb	31.83		63	

Tab. 13.4 (continued)

Product nuclide	t½	Target isotope	Natural abundance (%)	Reaction, if not (n,γ)	σ (in b). (For fission, yields are given in (%) for the mass number.)	Application
^{177}Lu	6.71 d	^{176}Lu	2.59		2100	ERT
		^{176}Yb	12.76	(n,γ) + β⁻	3.1	
^{186}Re	89.25 h	^{185}Re	37.40		110	ERT
^{188}Re	16.98 h	^{187}Re	62.60		72	ERT
		^{186}W	28.43	2x(n,γ) + β⁻	36 + 70	Generator for ERT: ^{188}W/^{188}Re
^{197}Pt	16.3 h	^{196}Pt	25.242		0.55	ERT

listed. Here, target isotopes of high natural abundance are preferred. If this value is low but the cross-section high, target isotopes are isotopically enriched. The table also lists typical applications of the produced radionuclides.

13.6 Radionuclide production using charged particles

Neutron capture reactions cover a certain domain of product nuclides relative to the target nucleus and product nuclides are neutron-rich in almost all cases. Charged particles, in contrast, open access to a different neighborhood of the irradiated target nucleus, shown in Figs. 13.4 and 13.13. In most cases of large-scale radionuclide production, nuclei of hydrogen and helium are used as projectiles, namely the proton ^{1}H, the deuteron ^{2}H, ^{4}He and ^{3}He. Because these projectiles introduce one or two protons to the target nucleus, most of the product nuclei are proton rich. Chemically, they thus differ from the irradiated target, can be isolated and finally represent "no-carrier-added" states.

13.6.1 Sources of charged projectiles: Particle accelerators

Charged projectiles must overcome the coulomb barrier to reach the nucleus, and thus should have sufficient kinetic energy. This requires a technology that can increase the kinetic energy of charged projectiles.[36]

13.6.1.1 Ionization

The general approach is to first ionize hydrogen or helium gas to produce ions. This is done through dissociation of one (H) or two (He) electrons from the corresponding nucleus, creating positively charged ions. This results in either $^1H^+$ (the proton itself, p) or $^4He^{2+}$ (the alpha particle) – the nuclei of the ionized atoms. Optionally, H_2 can be ionized in such a way that negatively charged ions are obtained ($^1H^-$).

13.6.1.2 Acceleration

Subsequently, the ions are accelerated in electromagnetic fields. There are several technological concepts, such as electrostatic VAN DER GRAAF generators, linear accelerators and different versions of cyclotrons.

13.6.1.3 Linear accelerator

A straightforward approach to accelerating ions is to apply a potential inside a vacuum chamber. Within a given segment, the anode attracts the negative ions, which thereby gain kinetic energy. After passing the first segment, they enter a subsequent one. Between these segments, there is a charge converter, which changes an anode into a cathode and *vice versa*. Within a constant potential gradient, the ions are accelerated again within the second segment, and the process repeats. After a given number of accelerations, the ions reach several MeV of kinetic energy. They are finally deviated by external magnets and extracted from the vacuum chambers.[37]

Linear accelerators (LINAC) have been developed in order to obtain higher kinetic energy ion beams. Separating tube accelerators apply high-frequency potentials. Acceleration is achieved not within the segments, but between them by periodically switching the electrode charge between segments. Whenever the ion has left one segment, it is thus repulsed by a backward electrode and attracted to a forward electrode. The length L of the segments increases with increasing kinetic

---

36 Interestingly, in the case of neutrons, the opposite applies. The challenge was to moderate their kinetic energy.

37 The first artificial nuclear reactions were initiated by accelerated protons. Nobel Prize in Physics in 1951 to DOUGLAS COCKROFT and ERNEST SINTON WALTON " . . . for their pioneering work on the *transmutation* of atomic nuclei by artificially accelerated atomic particles".

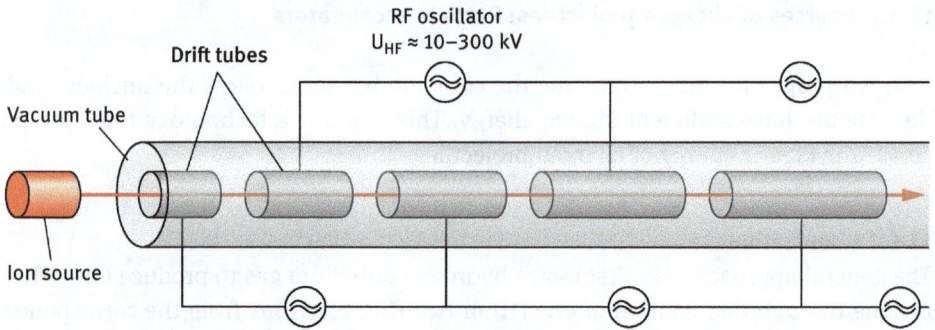

Fig. 13.30: Concept of linear acceleration of ions. Ions released from the ion source are accelerated in a long vacuum chamber within potential gradients between separate drift tube segments. Each time the charged projectile leaves one drift tube and is in-between two segments, the radiofrequency between these two segment switches.

energy of the ions, shown in Fig. 13.30. This concept may be extended significantly, from several meters up to a length of kilometers. Energies reach values of GeV.

13.6.1.4 Cyclotrons

To achieve high kinetic energies of charged projectiles without an endless extension of linear accelerator facilities, the idea of accelerating ions within cyclic paths was proposed; it was one of many brilliant ideas in nuclear sciences.[38] The concept of acceleration is similar to the LINAC, having an alternating radio HF-based electric field organized within a gap between two chamber segments. The two chambers look like a flat tin can cut into two halves. Because they have a shape like the letter D, they are called "dees", shown in Fig. 13.31. Ions are released from the ion source within the center of the magnetic field, and accelerating ions are forced to move along a spiral trajectory by a static magnetic field B provided by two electromagnets. Each time an ion of charge q passes the acceleration gap (and this appears always at the same frequency time t), the switch of the HF is according to its cyclotron resonance frequency f, as per eq. (13.29) with m being the (relativistic) mass of the ion.[39] At that frequency, the centripetal force and the magnetic LORENTZ force are equal. After each half cycle, the increased kinetic energy of the ion results in an

---

38 ERNEST LAWRENCE and STANLEY LIVINGSTON first realized a prototype of such a device (1930 at the now Lawrence Livermore National Laboratory). LAWRENCE received the 1939 Nobel Prize in Physics for " . . . for the invention and development of the cyclotron and for results obtained with it, especially with regard to artificial radioactive elements".

39 For nonrelativistic situations, the rest mass of the ions can be used, and cyclotrons of constant magnetic field work at constant frequencies. This changes if projectiles such as the proton reach energies of 100 MeV and above. 500 MeV protons already move at an incredible 75% of the velocity of light.

increased radius of its path.[40] Finally, the ions are extracted[41] and directed towards the target. Their final kinetic energy thus depends on the HF acceleration and the magnetic field, but also on the radius of the cyclotron. The projectile flux intensity ranges from a tenth of µA to several mA.

$$f = \frac{qB}{2\pi\,m}$$

(13.29)

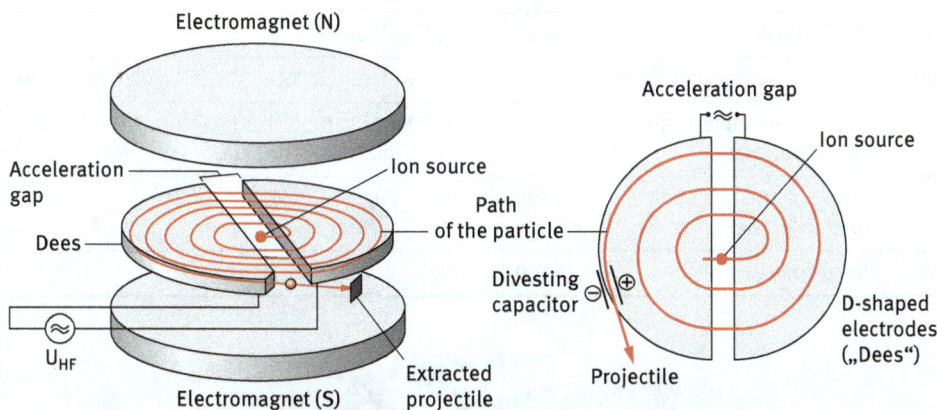

Fig. 13.31: Concepts of cyclotron acceleration of ions. (left) Assembly of the vacuum chambers (two dees with the acceleration gap in-between) embedded by the strong electromagnets. (right) Schematic drawing of the path of the ions from the ion source to the beam extraction position.

There are several categories of cyclotrons, summarized in Table 3.5. They range from huge facilities capable of accelerating all kinds of ions (from protons to large ions (such as Ca) at energies reaching GeV per mass unit) down to dedicated, often so-called "medical" cyclotrons. These units provide protons, deuterons, ^{4}He and ^{3}He projectiles only, with maximum kinetic energies of about 20 MeV for p and d and about 30 MeV for ^{4}He and ^{3}He, respectively. Recently, so-called Mini-Cyclotrons have become available. They are even smaller and provide protons only with maximum

---

40 The increasing mass of the projectile would "delay" the arrival time of the projectile at the acceleration gap between the dees. This requires technological solutions. One option is to change the frequency of the radio-HF accordingly, and these types of cyclotrons are called "synchro"-cyclotron. The other option is to adapt the magnetic field to the increasing velocity of the projectile. Cyclotrons with corresponding alternating field gradients may use a constant radio-HF and are therefore called "isochronous"-cyclotron.

41 For positive ions, this is accomplished by a deviating magnet. For accelerated negative ions, such as H⁻, this occurs by directing the ions through a thin carbon stripping foil. The foil catches the two electrons and thus converts an H⁻ ion into a proton, which immediately moves differently from the anions.

Tab. 13.5: Types of accelerators used for radionuclide production. Courtesy: M Qaim, Cyclotron production of medical radionuclides, in: Handbook of Nuclear Chemistry, volume 4, A Vèrtes, S Nagy, Z Klencsár, RG Lovas, F Rösch (eds.), second edition, Springer, Dordrecht, Heidelberg, London, New York, 2011.

Level	Characteristics	Energy [MeV]	Major radionuclides produced
I	Single particle (d)	<4	^{15}O
II	Single particle (p)	≤11	^{11}C, ^{13}N, ^{15}O, ^{18}F
III	Single or double particle (p, d)	≤20	^{11}C, ^{13}N, ^{15}O, ^{18}F, ^{64}Cu, ^{86}Y, ^{124}I (^{123}I, ^{67}Ga, ^{111}In)
IV	Single or multiple particle (p, d, ^3He, ^4He)	≤40	^{38}K, ^{73}Se, $^{75-77}Br$, ^{123}I, ^{81}Rb (^{81}Kr), ^{67}Ga, ^{111}In, ^{201}Tl, ^{22}Na, ^{57}Co, ^{44}Ti, ^{68}Ge, ^{72}As, ^{140}Nd
V	Single or multiple particle (p, d, ^3He, ^4He)	≤100	^{28}Mg, ^{67}Cu, ^{72}Se, ^{82}Sr, ^{117m}Sn, ^{123}I
VI	Single particle (p)	≥100	^{26}Al, ^{67}Cu, ^{68}Ge, ^{82}Sr, etc.

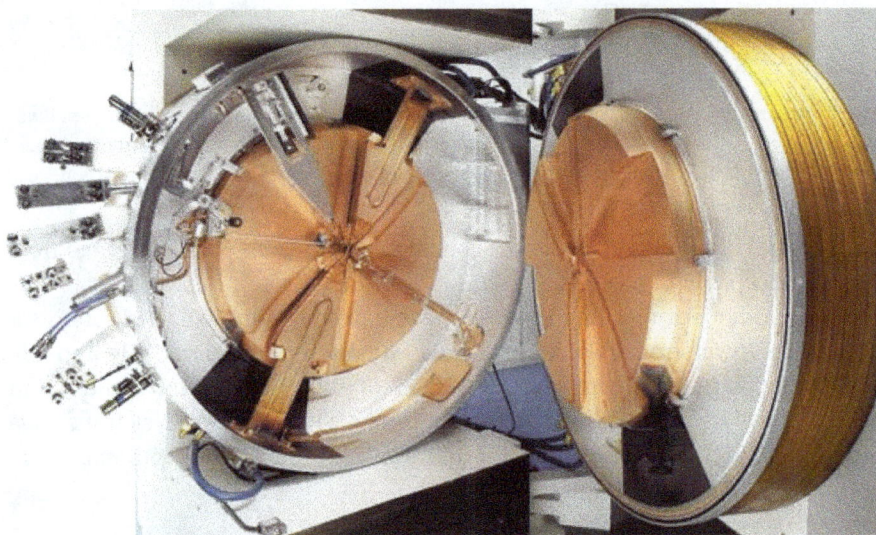

Fig. 13.32: Vertical components of a "medical" cyclotron. Photograph of the PETtrace 700 "medical" cyclotron (courtesy: GE, Uppsala, manufacturer). The cyclotron is arranged in a vertical position, allowing easy access by opening the door that represents one copper-based electromagnet. Acceleration of H- ions is organized by 4 dees. Maximum proton energies are 9.6 MeV and maximum proton flux is 100 μA. Proton extraction occurs at the 9 o'clock position on the left half, by mounting a carbon foil into the outer radius of the proton path. The protons created are directed towards target stations. There are several target stations mounted, left. Copyright General Electric Company. Used with permission.

energies below 10 MeV. Figure 13.32 shows a photograph of a medium-energy type "medical" cyclotron.

13.6.2 Target systems

The target atoms to be irradiated must be mechanically prepared in a sophisticated manner, which is commonly called "targetry". Targets should be able to withstand the heat created during the irradiation,[42] but also allow for effective subsequent radiochemical separation of target atoms and product nuclide. Targets are grouped according to their chemical state: solid, liquid and gas.

Solid target materials include the metallic form of the target element (mainly used as foils of various thickness), salts, oxides and alloys. Liquid targets are used much less frequently, though water enriched in ^{18}O is a very important example to produce ^{18}F. Several gaseous targets are applied in commercial radionuclide production, such as e.g. ^{123}I through a nuclear reaction process starting from enriched ^{124}Xe.

Solid targets are often prepared on (or connected to) a so-called target backing, which mechanically stabilizes the thin target layer and helps to externally cool the backside of the target. Salts, oxides or powder targets are principally prepared by pressing them into pellets, which are inserted in a correspondingly shaped backing. Because these compounds are mechanically less stable, thin metal foils may cover their frontside. The frontside of the target is typically cooled by gases (air, He etc.).

Liquids and gases are filled into (mainly cylindrical) containers through tube inlets, sometimes at elevated pressure. Again, the walls of the container are cooled with air, water, He, etc. In order to avoid degradation of the projectile's energy, the frontside of the cylinder is very thin – thin foils are especially common.

The complete targetry consists of the target itself, its container, and the cooling arrays at back and front, and is connected to the arriving projectile.

13.6.3 Optimum irradiation conditions

In order to produce the highest yield of a certain radionuclide, several aspects must be considered in detail.

1. The excitation function of the nuclear reaction, its defining threshold energies and the optimum range of the particle energy.

---

42 A target bombarded by charged projectiles of 10 MeV energy and 0.1 mA flux intensity accumulates a power of 10 MeV × 0.1 mA = 100 W. For an irradiation period of 10 h, this gives 100 W × 10 h × 3600 s/h = 36 MW s. Because projectile beams are only cm^2 in diameter, this corresponds to a high power density per target area.

2. The excitation functions of *competing* nuclear reactions, in order to avoid co-production of undesired radionuclides.
3. Targetry.
4. Radiochemical processing of the irradiated target.

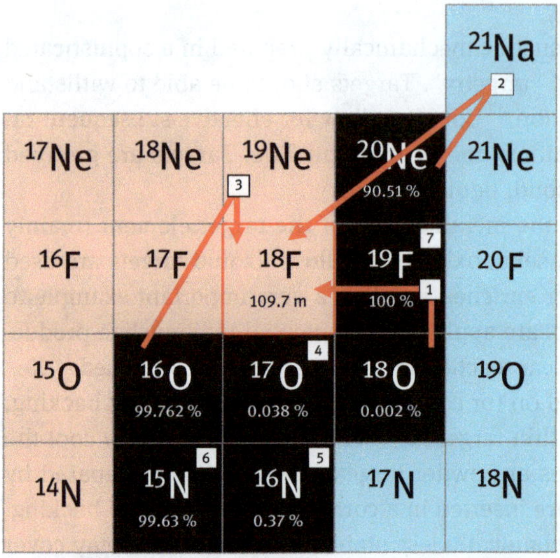

Fig. 13.33: Charged particle-induced nuclear reactions leading to ^{18}F. Isotope production starts from stable nuclei. Reactions 1–3 are used to generate ^{18}F, and the squares give the corresponding compound nuclei C*. (1) ^{18}O(p,n), (2) ^{20}Ne(d,α), (3) ^{16}O(^{3}He,p). Theoretically, there are many more options. The squares indicate the target isotopes: (4) ^{17}O(d,n), (5) ^{16}N(α,2n), (6) ^{15}N(α,n).

The optimum choice considers the appropriate combination of target A, projectile a (and its kinetic energy), the natural abundance of the target isotope and several practical aspects. For example, Fig. 13.33 illustrates the many routes leading to ^{18}F, the most frequently used positron emitter for molecular imaging and PET.[43] Figure 13.34 shows the integral (thick target) yields of the ^{125}Te(p,xn) reactions. If the positron emitter ^{124}I is the desired product nuclide, the energy of the protons should cover a range from ca. 81 to ca. 10 MeV. This would avoid substantial co-production of ^{123}I as a contaminant. Co-formation of ^{125}I cannot be avoided, but integral yields of ^{124}I are higher by almost two orders of magnitude – thus, percent contamination of ^{125}I would be low. Table 13.6 summarizes production parameters of several production routes leading to ^{124}I.

---

43 Details on the ^{18}F-production routes, the targets used and the implications for subsequent nucleophilic or electrophilic ^{18}F-labeling chemistry are discussed in Chapter 13 of Volume II.

Tab. 13.6: Experimental parameters of several production routes leading to ^{124}I. Courtesy: SM Qaim, Cyclotron production of medical radionuclides, in: Handbook of Nuclear Chemistry, volume 4, A Vèrtes, S Nagy, Z Klencsár, RG Lovas, F Rösch (eds.), second edition, Springer, Dordrecht, Heidelberg, London, New York, 2011.

Nuclear reaction	Suitable energy range [MeV]	Thick target yield of ^{124}I [MBq $\mu A^{-1}h^{-1}$]	Impurity [%] ^{123}I	^{125}I
^{124}Te(p,n)	12 → 8	16	1.0	<0.1
^{124}Te(d,2n)	14 → 10	17.5	–	1.7
^{125}Te(p,2n)	21 → 15	81	7.4	0.9

13.6.4 Radionuclides produced by charged particle-induced nuclear reactions

There are an impressive number of artificial radionuclides produced *via* charged particle-induced processes. Table 13.7 lists the most relevant ones.

Tab. 13.7: Radionuclides produced via charged particle-induced nuclear reactions and their main applications. Projectile energies give approximate ranges. PET = positron emission tomography, SPECT = single photon emission tomography, ERT = endoradiotherapy.

Nuclide	t½	Nuclear reaction	Projectile energy range (MeV)	Application
^{11}C	20.4 min	^{14}N(p,α)	13 → 3	PET
^{13}N	10.0 min	^{16}O(p,α)	16 → 7	PET
^{15}O	2.0 min	^{14}N(d,n) ^{15}N(p,n)	8 → 0 10 → 0	PET
^{18}F	109.6 min	^{18}O(p,n)	16 → 3	PET
		^{20}Ne(d,α)	14 → 0	
^{55}Co	17.6 h	^{58}Ni(p,α)	20 → 5	PET
		^{54}Fe(d,n)	15 → 3	
^{61}Cu	3.4 h	^{61}Ni(p,n)	20 → 5	PET
^{64}Cu	12.7 h	^{64}Ni(p,n)	20 → 5	PET
^{67}Cu	2.6 d	RbBr(p,spall)	800	ERT
		^{68}Zn(p,2p)	150 → 50	
		^{70}Zn(p,α)	25 → 10	

Tab. 13.7 (continued)

Nuclide	t½	Nuclear reaction	Projectile energy range (MeV)	Application
^{62}Zn	9.13 h	^{63}Cu(p,2n)	25 → 10	Parent of ^{62}Zn/^{62}Cu generator
^{66}Ga	9.4 h	^{66}Zn(p,n)	20 → 5	PET
^{67}Ga	3.26 d	^{68}Zn(p,2n)	26 → 18	SPECT
^{68}Ge	270 d	^{69}Ga(p,2n)	25 → 10	Parent of ^{68}Ge/^{68}Ga generator
^{73}Se	7.1 h	^{75}As(p,3n)	45 → 20	PET
^{75}Br	1.6 h	^{76}Se(p,2n)	30 → 10	PET
		^{75}As(^{3}He,3n)	20 → 40	
^{76}Br	16.0 h	^{76}Se(p,n)	20 → 5	PET
		^{76}As(^{3}He,2n)	130 → 10	
^{82}Sr	25.34 d	^{nat}Rb(p,xn)	100 → 70 + 70 → 40	Parent of ^{82}Sr/^{82}Rb generator
^{86}Y	14.7 h	^{86}Sr(p,n)	20 → 5	SPECT
^{89}Zr	78.4 h	^{89}Y(p,n)	15 → 5	PET
^{90}Nb	14.6 h	^{90}Zr(p,n)	15 → 5	PET
^{94m}Tc	52 min	^{94}Mo(p,n)	15 → 5	PET
^{99m}Tc	6.0 h	^{100}Mo(p,2n)	25 → 10	SPECT
^{103}Pd	17.0 d	^{103}Rh(p,n)	15 → 5	ERT
^{111}In	2.8 d	^{112}Cd(p,2n)	25 → 10	SPECT
^{123}I	13.2 h	^{123}Te(p,n)	14 → 10	SPECT
		^{124}Te(p,2n)	26 → 23	
		^{127}I(p,5n)^{123}Xe	65 → 45	
		^{124}Xe(p,x)^{123}Xe	29 → 23	
^{124}I	4.2 d	^{124}Te(p,n)	20 → 5	PET
		^{124}Te(d,2n)	25 → 10	
		^{125}Te(p,2n)	30 → 15	
^{186}Re	3.8 d	^{186}W(p,n)	20 → 10	ERT
		^{186}W(d,2n)	20 → 10	

Tab. 13.7 (continued)

Nuclide	t½	Nuclear reaction	Projectile energy range (MeV)	Application
^{201}Tl	3.06 d	^{201}Tl(p,3 n)^{201}Pb	28 → 20	SPECT
^{211}At	7.2 h	^{211}Bi(α,2n)	28 → 20	ERT
^{225}Ac	10.0 d	^{226}Ra(p,2n) (t½ = 1600 a)	30 → 15	ERT

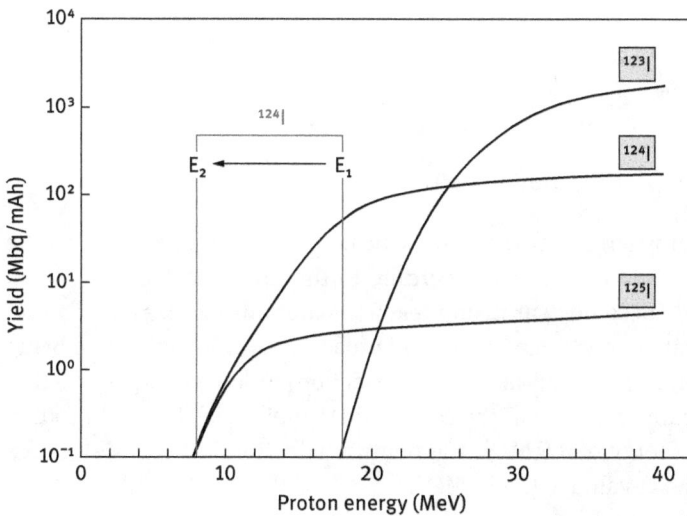

Fig. 13.34: Integral yield of iodine radioisotopes ^{125}I, ^{124}I, ^{123}I in MBq/μAh. This corresponds to the cross-section values for the nuclear reaction ^{125}Te(p,xn) reactions with x = 1, 2 and 3 shown in Fig. 13.15. The optimum proton energy covers the range from E_1 = 18 MeV to E_2 = 10 MeV. This achieves a compromise between production yield and product purity. With permission from: A Hohn, FM Nortier, B Scholten, TN van der Walt, HH Coenen, SM Qaim, Appl. Radiat. Isot. 55 (2001) 149.

13.7 Outlook

There are specific examples of nucleon transfer nuclear reactions that are not relevant for the man-made production of artificial radioisotopes, but are important from a scientific point of view. In addition to induced nuclear fission as a source of atomic energy, there are two extremely relevant nuclear processes: nuclear fusion and the genesis of the chemical elements in the stars.

13.7.1 Other classes of nuclear reactions: Nuclear fusion

Figure 13.35 illustrates the main parts of the nuclear fusion cycle. First, two protons within the sun plasma fuse together. This process is exclusively possible in the core[44] of the sun. This is not nucleon transfer, but nucleon fusion. The process does not simply fuse two protons, creating a helium nucleus without a neutron; instead, the heavy hydrogen isotope deuterium is formed, meaning one proton is converted to a neutron *via* simultaneous release of a positron and an electron neutrino. In a subsequent step, this deuterium nucleus fuses with another proton to create a stable isotope of the next chemical element: Helium-3. High energy photons are simultaneously released. When two ^{3}He nuclei fuse, ^{4}He is created and two excess protons are released, eqs. (13.30a–c).

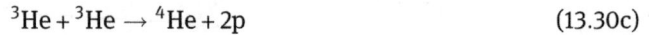

$$p + p \rightarrow {}^2H + \beta^+ + \nu_e \tag{13.30a}$$

$$^2H + p \rightarrow {}^3He + \gamma \tag{13.30b}$$

$$^3He + {}^3He \rightarrow {}^4He + 2p \tag{13.30c}$$

The energy released in the nuclear reactions can be calculated according to the gain in mean nucleon binding energies according to the LDM and the WEIZSÄCKER equation for each step. These nuclear fusion reactions are thus exothermic, and release energy. To generate energy using man-made fusion, several options are being discussed. The most promising one (largest cross-section) at low energy is reaction (13.31). It fuses deuterium and tritium, the so-called DT fuel cycle. The ^{4}He product nucleus has a kinetic energy of 3.5 MeV, the neutron is emitted at 14.1 MeV maximum kinetic energy, the Q-value is 17.588 MeV.

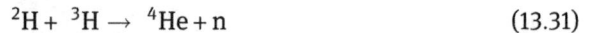

$$^2H + {}^3H \rightarrow {}^4He + n \tag{13.31}$$

13.7.2 Other classes of nuclear reactions: Element genesis

Stellar (and supernova) fusion reactions start with the proton plasma in the core of a star, slowly and steadily converting the initial reservoir of hydrogen nuclei into helium nuclei. In the later stages of a star's life, i.e. when the abundance of helium is large enough (and that of hydrogen nuclei small enough), fusion of helium nuclei begins.

---

44 The core of the sun is a sphere of <0.7 solar radius. The core of our sun exists at temperatures of about $15.6 \cdot 10^6$ °C. The temperature rapidly drops towards the outer regions, reaching $0.1 \cdot 10^6$ °C already in the connective zone, and "only" ca. 5500 °C at the photosphere. The huge density and temperature of the core allow the two protons to overcome the coulomb barrier.

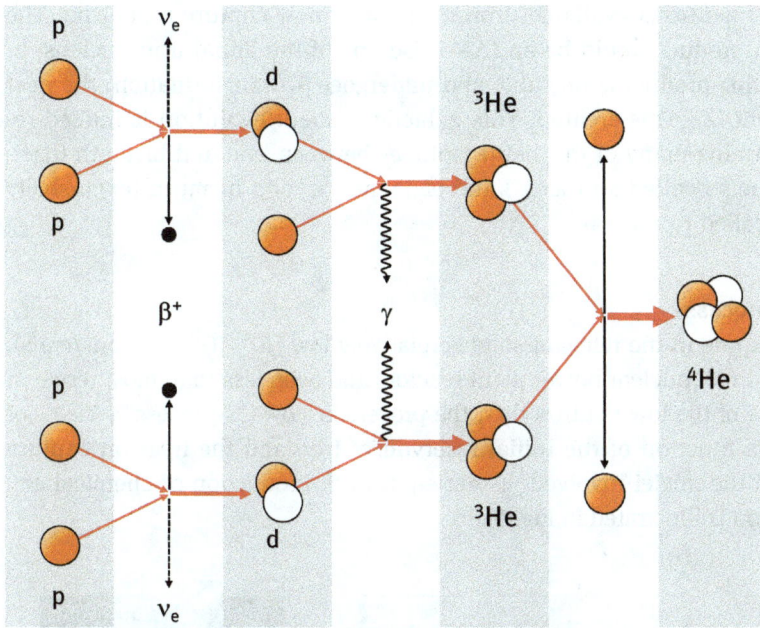

Fig. 13.35: Nuclear fusion in the core of stars. Two protons fuse to form deuterium. Another proton is added to the deuterium to form a Helium-3 nucleus. Finally, two ^{3}He nuclei fuse to form ^{4}He and two "free" protons (which become part of the next fusion cycle).

This process requires the star to have high temperature ($>1.4 \cdot 10^9$ °C), mass and density: such stars are several times larger than our sun (red giants). Here, three ^{4}He nuclei fuse to ^{12}C. The newly formed nuclei of increasing mass combine with each other (^{12}C + ^{12}C = ^{24}Mg and smaller product nuclei) or continue to fuse with ^{4}He nuclei or free nucleons to form nuclei of even more chemical elements; at temperatures $>6 \cdot 10^8$ °C, masses of >6 times the mass of our sun, and densities of $>2 \cdot 10^8$ kg/m^3. According to the LDM, this process remains exothermic until the maximum mean nucleon binding energy is achieved, which is at iron, $Z = 56$.

From nuclei of iron ($Z = 26$) onwards, fusion reactions are endothermic. Thus, these reactions cannot explain the genesis of heavier chemical elements. Additionally, such reactions do not contribute to the generation of energy in stars – in contrast, they need energy due to their endothermic character. Instead, nuclear element formation proceeds *via* neutron capture reactions. Neutrons appear in the stars mainly as a side product of two fusion reactions:

$$^{13}\text{C} + {}^4\text{He} \rightarrow {}^{16}\text{O} + \text{n} \tag{13.32a}$$

$$^{22}\text{Ne} + {}^4\text{He} \rightarrow {}^{25}\text{Mg} + \text{n} \tag{13.32b}$$

With a source of neutrons available, iron nuclei may now capture a neutron. The nuclear reaction product would be an (A + 1) isotope of the initial iron nucleus; in the case where this product is unstable and undergoes β^--transformation, the next chemical element (Z + 1) is created. This galactic element evolution is indeed responsible for forming many of the stable isotopes between iron and bismuth (basically through the so-called s-process) and elements beyond bismuth (exclusively through the so-called r-process).

13.7.2.1 The s-process

The neutron flux rate in the relevant stars is relatively low (10^5–10^{11} neutrons/cm^{2}s). This is less than in all nuclear power plant reactors and even less than most research reactors. Because of the low neutron flux, the process is slow ("s-process"). Yields of new nuclei are a function of the initial reservoir of iron and the neutron capture cross-sections of the nuclei involved, as per eq. (13.19). Production of chemical elements beyond iron is illustrated in Fig. 13.36.

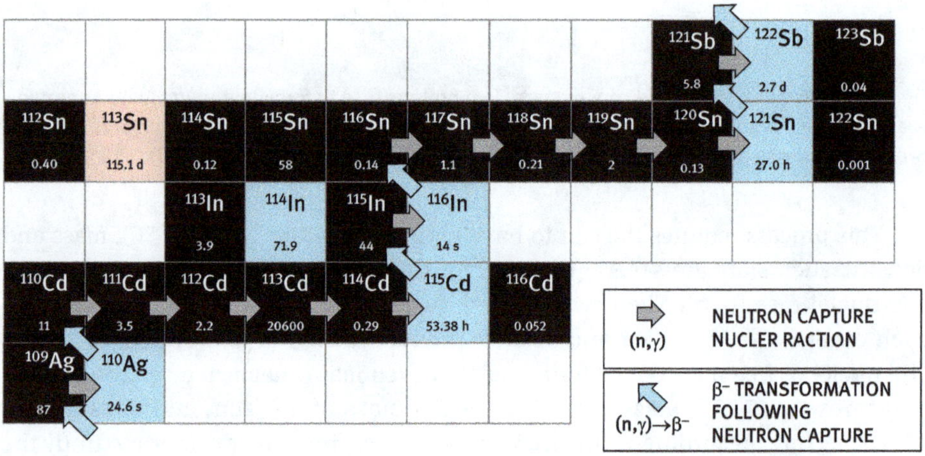

Fig. 13.36: Profile of the s-process of element genesis. ^{56}Fe is the starting nuclide, and (n,γ) reactions are followed by direct β^--transitions.

13.7.2.2 The r-process

At and above ^{209}Bi, s-type element genesis stops. The formation of heavier elements requires a different pathway. These pathways also starts with a neutron capture reaction forming the product nucleus (A + 1); however, before it can transform *via* the β^--process, it captures another neutron and forms another isotope of mass (A + 2).

This continues, creating an excess of many neutrons.[45] The competition between β⁻-transformation and neutron capture of an unstable neutron-rich nucleus is in favor of the (n,γ) process, but only in cases of extremely large neutron flux density. Only in supernovae do flux densities of ca. 10^{22} n/cm^2 s appear. Here, subsequent neutron capture processes occur rapidly, within fractions of a second, which is the "r-process".

Finally, the probability of multiple (n,γ) capture steps drops off. At a given nucleon configuration, the extremely neutron-rich product nucleus prefers the β⁻-process rather than the next neutron capture. Figure 13.37 illustrates the principal pathways of r-process driven nuclear reactions to generate nuclides of new elements.

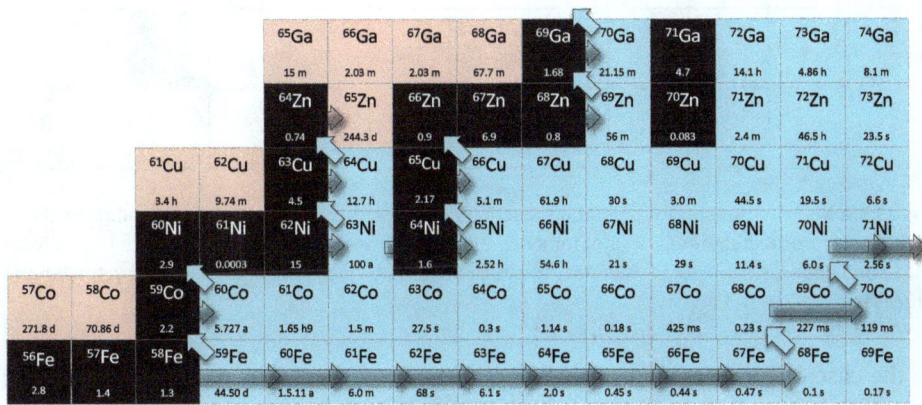

				^{65}Ga	^{66}Ga	^{67}Ga	^{68}Ga	^{69}Ga	^{70}Ga	^{71}Ga	^{72}Ga	^{73}Ga	^{74}Ga
				15 m	2.03 m	2.03 m	67.7 m	1.68	21.15 m	4.7	14.1 h	4.86 h	8.1 m
				^{64}Zn	^{65}Zn	^{66}Zn	^{67}Zn	^{68}Zn	^{69}Zn	^{70}Zn	^{71}Zn	^{72}Zn	^{73}Zn
				0.74	244.3 d	0.9	6.9	0.8	56 m	0.083	2.4 m	46.5 h	23.5 s
		^{61}Cu	^{62}Cu	^{63}Cu	^{64}Cu	^{65}Cu	^{66}Cu	^{67}Cu	^{68}Cu	^{69}Cu	^{70}Cu	^{71}Cu	^{72}Cu
		3.4 h	9.74 m	4.5	12.7 h	2.17	5.1 m	61.9 h	30 s	3.0 m	44.5 s	19.5 s	6.6 s
		^{60}Ni	^{61}Ni	^{62}Ni	^{63}Ni	^{64}Ni	^{65}Ni	^{66}Ni	^{67}Ni	^{68}Ni	^{69}Ni	^{70}Ni	^{71}Ni
		2.9	0.0003	15	100 a	1.6	2.52 h	54.6 h	21 s	29 s	11.4 s	6.0 s	2.56 s
^{57}Co	^{58}Co	^{59}Co	^{60}Co	^{61}Co	^{62}Co	^{63}Co	^{64}Co	^{65}Co	^{66}Co	^{67}Co	^{68}Co	^{69}Co	^{70}Co
271.8 d	70.86 d	2.2	5.727 a	1.65 h9	1.5 m	27.5 s	0.3 s	1.14 s	0.18 s	425 ms	0.23 s	227 ms	119 ms
^{56}Fe	^{57}Fe	^{58}Fe	^{59}Fe	^{60}Fe	^{61}Fe	^{62}Fe	^{63}Fe	^{64}Fe	^{65}Fe	^{66}Fe	^{67}Fe	^{68}Fe	^{69}Fe
2.8	1.4	1.3	44.50 d	1.5.11 a	6.0 m	68 s	6.1 s	2.0 s	0.45 s	0.44 s	0.47 s	0.1 s	0.17 s

Fig. 13.37: Profile of the r-process of element genesis. ^{56}Fe is shown here as the starting nuclide, but r-processes may start from almost any stable nuclide. Multiple (n,γ) reactions occur prior to systematic β⁻-transitions.

13.7.3 Other classes of nuclear reactions: Spallation

Nucleon absorption within the whole nucleus induces nuclear reactions according to nucleon transfer. This is the domain of relatively low kinetic energy projectiles (≤ 10 MeV per nucleon), resulting in so-called low energy nuclear reactions. For proton-induced nuclear reactions, for example, on average the mean nucleon binding energy of 8–9 MeV must be delivered to the compound nucleus in order to achieve a (p,n) nucleon transfer. To release two neutrons, approximately twice as much energy must be deposited by the incoming projectile to allow for a (p,2n) exit-channel.

---

[45] This is similar to the neutron-induced radioisotope production pathways of 2 × (n,γ) – see Fig. 13.25.

This strategy works for (p,xn) mechanisms of up to x = 8 or 9, shown in Fig. 13.20 for the production of ^{82}Sr.

However, this does not continue forever (see Fig. 13.38). The release of 8 neutrons, for example, would require up to 90 MeV kinetic energy of the proton. At this level of ca. $E_p > 100$ MeV, the conventional nucleon transfer mechanism starts to turn into a different process: spallation.

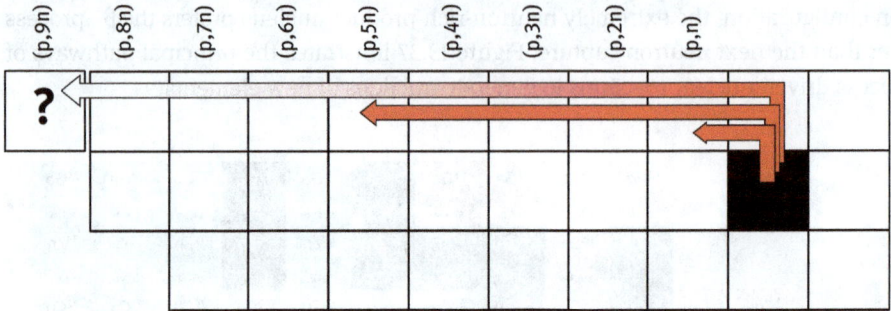

Fig. 13.38: Nucleon transfer *vs.* spallation reactions. Processes of type (p,xn) cover product nuclei of the same proton number Z for (p,n) for x = 2, 3, 4, etc. However, this cannot be extrapolated to x ca. 10 and above. At about $E_p > 100$ MeV, the nucleon transfer mechanism is replaced by spallation processes.

In the case of high energy charged projectiles (ca. 20–250 MeV per nucleon and kinetic energies >100 MeV, even reaching TeV levels), the projectile transfers its energy directly to the nucleons of the target atom. This induces an intra-nuclear cascade of energy transfer between individual nucleons. Finally, the very excited nucleus splits into several pieces of varying mass number, instead of reacting to form one particular product nucleus B. This is called spallation. Figure 13.39 schematically illustrates the process. It is accompanied by the release of many individual nucleons, protons and (in particular) neutrons of high kinetic energy. Three major steps can be distinguished, and protons are used as prototype of the projectile:

1. Initial proton (= projectile nucleon) interacts with target nucleon at very high kinetic energy. Because the cross-section is $\sigma = f(1/v)$, there is almost no reaction and the protons pass the target nucleus as if it were "transparent".
2. The incoming proton hits a target nucleon and eject it from the nucleus. This generates a highly excited large fragment of the former target nucleus.
3. The excited nucleus de-excites. This occurs either by dividing into two fragments similar to the fission process, i.e. with mass distributions around $A^{fragment} \approx 1/2 \, A \pm i$, or by evaporating a number of particles such as neutrons, protons, deuterons or α-particles. This leaves many fragments of larger mass, where $A^{fragment} > 2/3 \, A$.

Spallation is interesting from several practical points of view. First, it produces nuclear reaction products that are not accessible by the nucleon transfer reaction pathway. Second, the large number of neutrons emitted at high kinetic energy makes them a new projectile for subsequent nuclear reactions. In this context, spallation targets are also called "proton-neutron converters". Third, spallation offers hope of transmuting long-lived nuclear waste into short-lived, less ecologically toxic products.

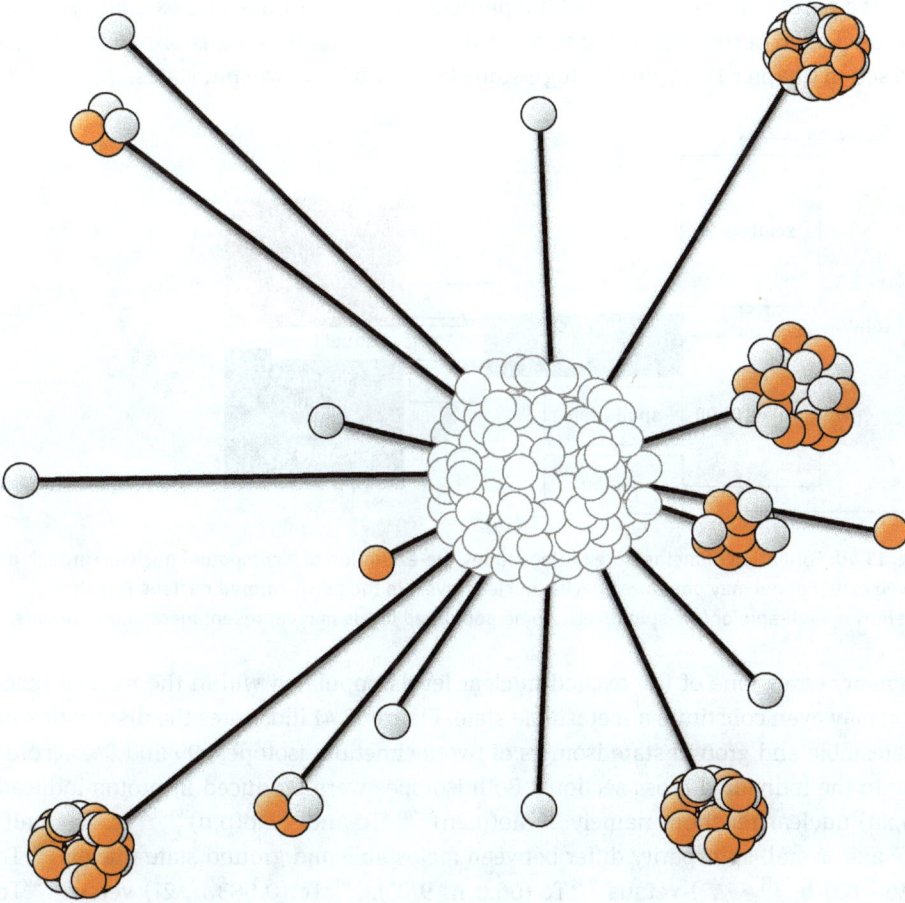

Fig. 13.39: Schematic illustration of spallation processes. Several product nuclei of varying mass are released, together with single nucleons.

13.7.4 Excited nuclear states

Like the processes of spontaneous nuclear transformation, which can promote electrons to higher energy levels in the formed nuclide, individual nuclear states may also be produced in nuclear reaction processes, as shown in Fig. 13.40. The corresponding nuclear states of a given product nucleus B are again defined by a characteristic set of quantum numbers, with an emphasis on the overall moment J. Either high-spin or low-spin levels may be preferred.

If a particular excited level of the product nucleus is formed, de-excitation within the nucleus B occurs according to the character of secondary transformations as discussed in Chapter 11. Typically, high-spin states turn into low-spin states.

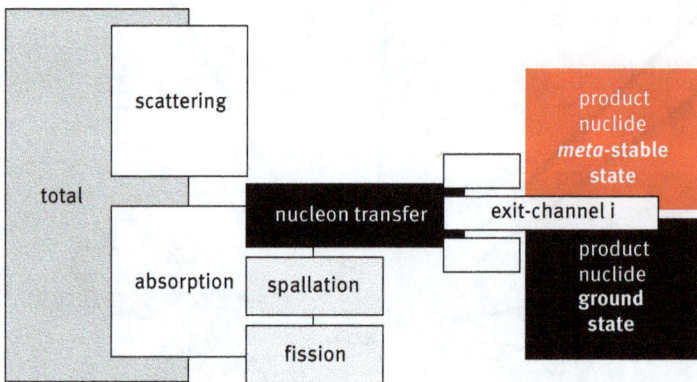

Fig. 13.40: Formation of metastable radionuclides. De-excitation of a compound nucleus through a given exit-channel may populate specific nuclear levels in the newly formed nucleus B, either preferring high-spin or low-spin levels. Some populated levels may represent metastable isomers.

In many cases, one of the excited nuclear levels populated within the nuclear reaction may even constitute a metastable state. Figure 13.41 illustrates the distribution of metastable and ground state isomers of two technetium isotopes, 99 and 94, according to the individual cross-sections. Both isotopes were produced in proton-induced (p,xn) nuclear reaction, namely ^{100}Mo(p,2n)^{99m,g}Tc and ^{94}Mo(p,n)^{94m,g}Tc. The half-life and overall spin/parity differ between metastable and ground state nuclei: ^{99m}Tc ($t\frac{1}{2} = 6.0$ h, $J^{\Pi} = \frac{1}{2}^{-}$) versus ^{99g}Tc (66.0 h, 9/2^{+}), ^{94m}Tc (0.883 h, 2^{+}) versus ^{94g}Tc (4.9 h, 7^{+}). The formation of the two different nuclear states delivers specific ratios between the metastable and the ground state of the product nucleus. It may depend on the kinetic energy of the projectile. For example, take Tc-94: the formation of ^{94m}Tc is favored at low proton energy (at $E_p = 7$ MeV, the ratio of ^{94m}Tc:^{94g}Tc is ≈10:1 in terms of differential production yield), while at higher proton energies (>$E_p = 18$ MeV) both states are populated in a 1:1 ratio, as shown in Fig. 13.42.

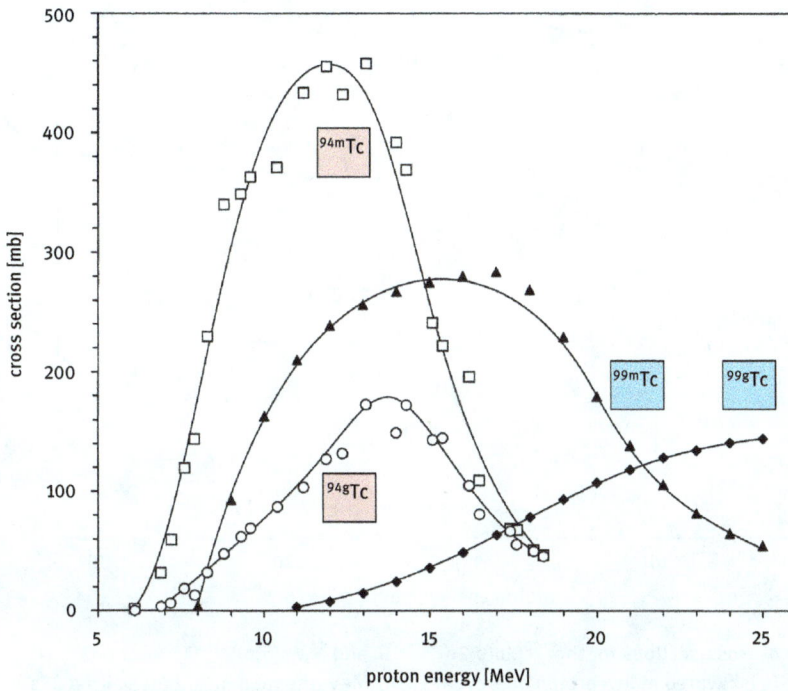

Fig. 13.41: Individual cross-sections of ^{100}Mo(p,2n)^{99m,g}Tc and ^{94}Mo(p,n)^{94m,g}Tc reactions. The metastable and the ground state isomer of the two technetium isotopes, 99 and 94, are formed in different abundances according to the individual cross-sections. Experimental data from: F Rösch and SM Qaim, Radiochim. Acta 62 (1993) 115, and SM Qaim et al., Appl. Radiat. Isot. 85 (2013) 101.

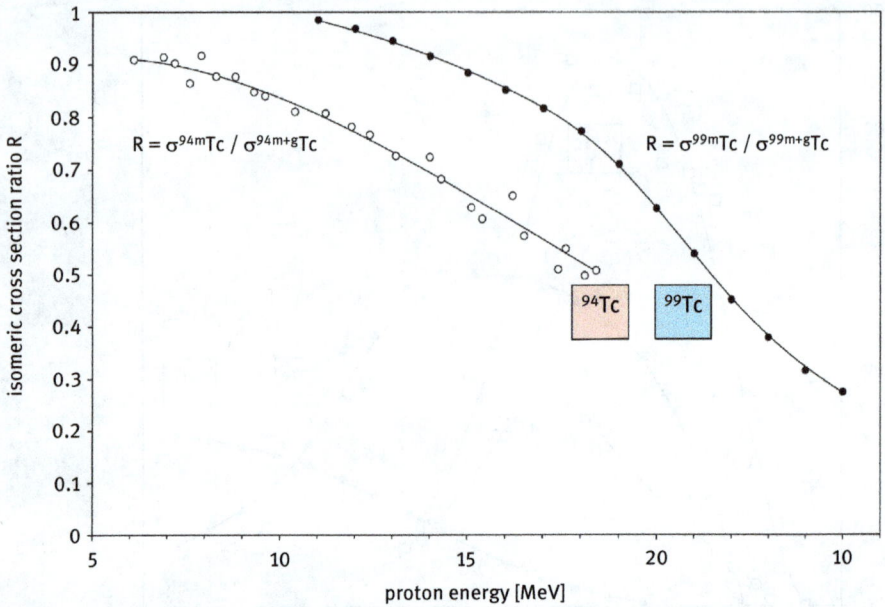

Fig. 13.42: Ratio of cross-sections for the $^{100}Mo(p,2n)^{99m,g}Tc$ and $^{94}Mo(p,n)^{94m,g}Tc$ reactions. Formation of ^{94m}Tc is favored at low proton energy. At $E_p = 7$ MeV, the ratio $\sigma_{94m} / (\sigma_{94m} + \sigma_{94g})$ is about 0.3, which corresponds to a $^{94m}Tc:^{94g}Tc$ ratio of $\approx$10:1 in terms of differential production yield. At higher proton energies ($>E_p = 18$ MeV) the value of $\sigma_{94m} / (\sigma_{94m} + \sigma_{94g})$ is ca. 0.5, i.e. both states are populated in a 1:1 ratio. Experimental data according to Fig. 13.41.

14 Appendix

Tab. 14.1: Atomic units and selected fundamental constants.

Quantity	Symbol	Value	Unit
Speed of light in vacuum	c, c_0	299 792 458	$m \cdot s^{-1}$
Energy	$m_e c^2$	$8.18710414 \times 10^{-14}$	$J \cdot s$
		0.510998902	MeV
Energy equivalent	$m_u c^2$	931.494013	MeV
Momentum	$m_e c$	$2.73092398 \times 10^{-22}$	$kg \cdot m \cdot s^{-1}$
		$0.510998902 \times 10^{-15}$	MeV/c
Elementary charge	e	$1.602176462 \times 10^{-19}$	C
BOHR radius (bohr)	$a_o = \alpha/4\pi R^H$	$0.5291772083 \times 10^{-10}$	m
Magnetic constant	μ_0	$4\pi \times 10^{-7} = 12.56637 \times 10^{-7}$	$N \cdot A^{-2}$
Electric constant $1/\mu_0 c^2$	ε_o	$8.854187817 \ldots \times 10^{-12}$	$F \cdot m^{-1}$
PLANCK constant	h	$6.626068762 \times 10^{-34}$	$J \cdot s$
		$4.13566727 \times 10^{-15}$	$eV \cdot s$
Reduced PLANCK constant	$\hbar = h/2\pi$	$1.054571596 \times 10^{-34}$	$J \cdot s$
		$6.58211889 \times 10^{-16}$	$eV \cdot s$
Atomic mass constant	m_u	$1.66053873 \times 10^{-27}$	kg
AVOGADRO constant	N^A, L	$6.02214199 \times 10^{23}$	
BOLTZMANN constant	k^B	$1.3806503 \times 10^{-23}$	$J K^{-1}$
		8.617342×10^{-5}	$eV K^{-1}$
FARADAY constant	$F = N^A e$	96 485.3365	s A / mol = (C / mol)
RYDBERG constant	R^H	10 973 731.569	m^{-1}

https://doi.org/10.1515/9783110742725-014

Tab. 14.2: Selected atomic and nuclear constants of electron, proton, neutron, deuteron and alpha particle. Atomic masses from: The 2020 Atomic mass evaluation. M Wang et al., Chinese Physics C Vol. 45, No. 3 (2021) 030003.

Quantity	Symbol	Value	Unit
Electron (e)			
Electron mass	m_e	$9.10938188 \times 10^{-31}$	kg
Relative mass	$A_r(e)$	$5.485799110 \times 10^{-4}$	u
Energy equivalent of mass	$m_e c^2$	$8.18710414 \times 10^{-14}$	J
Electron molar mass $N_A m_e$	m_e	$5.485799110 \times 10^{-7}$	kg mol^{-1}
		0.510998902	MeV
Proton (p)			
Proton mass	m_p	$1.67262158 \times 10^{-27}$	kg
		$1.0078250319 \times 10^{-6}$	u
Energy equivalent of mass	$m_p c^2$	$1.50327731 \times 10^{-10}$	J
		938.271998	MeV
Neutron (n)			
Neutron mass	m_n	$1.67492716 \times 10^{-27}$	kg
		$1.0086649159 \times 10^{-6}$	u
Energy equivalent of mass	$m_n c^2$	$1.50534946 \times 10^{-10}$	J
		939.595330	MeV
Neutron-proton mass ratio	m_n/m_p	1.001378418887	
Deuteron (d)			
Deuteron mass	m_d	$2.0141017778 \times 10^{-6}$	u
Alpha particle (α)			
Alpha particle mass	m_α	$4.00260325413 \times 10^{-6}$	u

Tab. 14.3: Energy equivalents. From: A Vértes, S Nagy, Z Klencsár, RG Lovas, F Rösch (eds.), Handbook of Nuclear Chemistry, second edition, Springer, Dordrecht, Heidelberg, London, New York, 2011. With permission.

1	J	kg	m^{-1}	Hz	K	eV	u
J		$(1\ \mathrm{J})/c^2 = 1.112650 \times 10^{-17}$	$(1\ \mathrm{J})/hc = 5.034117 \times 10^{24}$	$(1\ \mathrm{J})/h = 1.509191 \times 10^{33}$	$(1\ \mathrm{J})/k = 7.242964 \times 10^{22}$	$(1\ \mathrm{J}) = 6.241510 \times 10^{18}$	$(1\ \mathrm{J})/c^2 = 6.700537 \times 10^{9}$
kg	$(1\ \mathrm{kg})\,c^2 = 8.987552 \times 10^{16}$		$(1\ \mathrm{kg})\,c/h = 4.524439 \times 10^{41}$	$(1\ \mathrm{kg})\,c^2/h = 1.356393 \times 10^{50}$	$(1\ \mathrm{kg})\,c^2/k = 6.509651 \times 10^{39}$	$(1\ \mathrm{kg})\,c^2 = 5.609589 \times 10^{35}$	$(1\ \mathrm{kg}) = 6.022142 \times 10^{26}$
m^{-1}	$(1\ \mathrm{m}^{-1})\,hc = 1.986445 \times 10^{-25}$	$(1\ \mathrm{m}^{-1})\,h/c = 2.210219 \times 10^{-42}$		$(1\ \mathrm{m}^{-1})\,c = 299792458$	$(1\ \mathrm{m}^{-1})\,hc/k = 1.438775 \times 10^{-2}$	$(1\ \mathrm{m}^{-1})\,hc = 1.239842 \times 10^{-6}$	$(1\ \mathrm{m}^{-1})\,h/c = 1.331025 \times 10^{-15}$
Hz	$(1\ \mathrm{Hz})\,h = 6.626069 \times 10^{-34}$	$(1\ \mathrm{Hz})\,h/c^2 = 7.372496 \times 10^{-51}$	$(1\ \mathrm{Hz})/c = 3.335641 \times 10^{-9}$		$(1\ \mathrm{Hz})\,h/k = 4.799237 \times 10^{-11}$	$(1\ \mathrm{Hz})\,h = 4.135667 \times 10^{-15}$	$(1\ \mathrm{Hz})\,h/c^2 = 4.439852 \times 10^{-24}$
K	$(1\ \mathrm{K})\,k = 1.380650 \times 10^{-23}$	$(1\ \mathrm{K})\,k/c^2 = 1.536181 \times 10^{-40}$	$(1\ \mathrm{K})\,k/hc = 69.50356(12)$	$(1\ \mathrm{K})\,k/h = 2.083664 \times 10^{10}$		$(1\ \mathrm{K})\,k = 8.617342 \times 10^{-5}$	$(1\ \mathrm{K})\,k/c^2 = 9.251098 \times 10^{-14}$
eV	$(1\ \mathrm{eV}) = 1.602176 \times 10^{-19}$	$(1\ \mathrm{eV})/c^2 = 1.782662 \times 10^{-36}$	$(1\ \mathrm{eV})/hc = 8.065545 \times 10^{5}$	$(1\ \mathrm{eV})/h = 2.417989 \times 10^{14}$	$(1\ \mathrm{eV})/k = 1.160451 \times 10^{4}$		$(1\ \mathrm{eV})/c^2 = 1.073544 \times 10^{-9}$
u	$(1\ \mathrm{u})\,c^2 = 1.492418 \times 10^{-10}$	$(1\ \mathrm{u}) = 1.660539 \times 10^{-27}$	$(1\ \mathrm{u})\,c/h = 7.513007 \times 10^{14}$	$(1\ \mathrm{u})\,c^2/h = 2.252343 \times 10^{23}$	$(1\ \mathrm{u})\,c^2/k = 1.080953 \times 10^{13}$	$(1\ \mathrm{u})\,c^2 = 931\ 494013$	

Tab. 14.4: Masses of nuclides (atomic mass), mean nucleon binding energies $\bar{E}_B$, and mass excess values Δm^{excess} for selected nuclides. From: M WANG et al., The AME2020 atomic mass evaluation, Chinese Physics C Vol. 45, No. 3 (2021) 030003.

Z	Element	N	A	Atomic mass in atomic mass units (µu)	Mean nucleon binding energy $\bar{E}_B$ (MeV)	Δm^{excess} (MeV)
0	n	1	1	1.008665	0	8.071
1	H	0	1	1.007825	0	7.289
2	He	1	3	3.016049	2.573	14.931
2	He	2	4	4.002603	7.074	2.423
3	Li	3	6	6.015122	5.332	14.087
3	Li	4	7	7.016003	5.606	14.907
4	Be	5	9	9.012182	6.463	11.348
6	C	4	10	10.016853	6.032	15.699
6	C	5	11	11.011433	6.676	10.649
6	C	6	12	12.000000	7.680	0
6	C	7	13	13.003355	7.470	3.125
6	C	8	14	14.003242	7.520	3.020
6	C	9	15	15.010599	7.100	9.873
7	N	6	13	13.005739	7.239	5.345
7	N	7	14	14.003074	7.476	2.863
7	N	8	15	15.000109	7.699	101
7	N	11	18	18.014078	7.039	13.113
8	O	7	15	15.003066	7.464	2.856
8	O	8	16	15.994915	7.976	−4.737
8	O	9	17	16.999131	7.751	−809
8	O	10	18	17.999160	7.767	−783
8	O	11	19	19.003580	7.566	3.335
9	F	8	17	17.002095	7.542	1.952
9	F	9	18	18.000937	7.632	873
9	F	10	19	18.998403	7.779	−1.487
10	Ne	14	24	23.993611	7.993	−5.952

Tab. 14.4 (continued)

Z	Element	N	A	Atomic mass in atomic mass units (μu)	Mean nucleon binding energy $\bar{E}_B$ (MeV)	Δm^{excess} (MeV)
11	Na	13	24	23.990963	8.063	−8.418
12	Mg	13	25	24.985837	8.223	−13.193
13	Al	14	27	26.981538	8.332	−17.197
13	Al	15	28	27.981910	8.310	−16.851
14	Si	13	27	26.986705	8.123	−12.385
15	P	15	30	29.978313	8.354	−20.201
15	P	17	32	31.973908	8.464	−24.305
16	S	16	32	31.972071	8.493	−26.016
16	S	19	35	34.969032	8.538	−28.846
17	Cl	18	35	34.968853	8.520	−29.014
18	Ar	22	40	39.962383	8.595	−35.040
19	K	21	40	39.963998	8.538	−33.535
20	Ca	20	40	39.962591	8.551	−34.846
25	Mn	30	55	54.938045	8.765	−57.711
26	Fe	29	55	54.938293	8.747	−57.479
28	Ni	36	64	63.927966	8.777	−67.099
29	Cu	35	64	63.929764	8.739	−65.424
30	Zn	34	64	63.929142	8.736	−66.004
30	Zn	38	68	67.924844	8.756	−70.007
31	Ga	37	68	67.927980	8.701	−67.086
32	Ge	36	68	67.928094	8.688	−66.980
39	Y	51	90	89.907152	8.693	−86.488
39	Y	60	99	98.924636	8.472	−70.201
40	Zr	50	90	89.904704	8.710	−88.767
40	Zr	52	92	91.905041	8.693	−88.454
40	Zr	59	99	98.916512	8.541	−77.768
41	Nb	51	92	91.907194	8.662	−86.448
41	Nb	58	99	98.911618	8.579	−82.327

Tab. 14.4 (continued)

Z	Element	N	A	Atomic mass in atomic mass units (μu)	Mean nucleon binding energy $\bar{E}_B$ (MeV)	Δm^{excess} (MeV)
42	Mo	50	92	91.906811	8.658	−86.805
42	Mo	57	99	98.907712	8.608	−85.966
43	Tc	56	99	98.906255	8.614	−87.323
44	Ru	55	99	98.905939	8.609	−87.617
45	Rh	54	99	98.908132	8.580	−85.574
48	Cd	83	131	130.940670	8.207	−5.5270
49	In	82	131	130.926850	8.299	−68.137
50	Sn	81	131	130.917000	8.363	−77.314
50	Sn	87	137	136.945990	8.153	−50.310
51	Sb	80	131	130.911982	8.393	−81.988
51	Sb	86	137	136.935310	8.220	−60.260
52	Te	71	123	122.904270	8.466	−89.172
52	Te	72	124	123.902818	8.473	−90.525
52	Te	73	125	124.904431	8.458	−89.022
52	Te	79	131	130.908524	8.411	−85.210
52	Te	85	137	136.925320	8.282	−69.560
53	I	68	121	120.907367	8.442	−86.287
53	I	69	122	121.907589	8.437	−86.080
53	I	70	123	122.905589	8.449	−87.943
53	I	71	124	123.906210	8.441	−87.365
53	I	72	125	124.904630	8.450	−88.836
53	I	73	126	125.905624	8.440	−87.911
53	I	74	127	126.904473	8.445	−88.983
53	I	75	128	127.905809	8.433	−87.738
53	I	76	129	128.904988	8.436	−88.503
53	I	77	130	129.906674	8.421	−86.932
53	I	78	131	130.906125	8.422	−87.444
53	I	84	137	136.917871	8.327	−76.503

Tab. 14.4 (continued)

Z	Element	N	A	Atomic mass in atomic mass units (μu)	Mean nucleon binding energy $\bar{E}_B$ (MeV)	Δm^{excess} (MeV)
54	Xe	69	123	122.908482	8.421	−85.249
54	Xe	77	131	130.905082	8.424	−88.415
54	Xe	83	137	136.911562	8.364	−82.379
55	Cs	68	123	122.912996	8.380	−81.044
55	Cs	82	137	136.907090	8.389	−86.546
82	Pb	125	207	206.975897	7.870	−22.452
82	Pb	126	208	207.976652	7.867	−21.789
82	Pb	127	209	208.981090	7.849	−17.614
83	Bi	124	207	206.978471	7.855	−20.054
84	Po	127	211	210.986653	7.819	−12.433
84	Po	129	213	212.992857	7.794	−6.653
84	Po	132	216	216.001915	7.759	1.784
84	Po	134	218	218.008973	7.732	8.358
86	Rn	134	220	220.011394	7.717	10.613
86	Rn	136	222	222.017578	7.694	16.374
88	Ra	136	224	224.020212	7.680	18.827
88	Ra	138	226	226.025410	7.662	23.669
90	Th	138	228	228.028741	7.645	26.772
90	Th	144	234	234.043601	7.597	40.614
92	U	143	235	235.043930	7.591	40.921
92	U	146	238	238.050788	7.570	47.309

Tab. 14.5: Symbols and abbreviations.

Symbol/ abbreviation	Meaning	Chapter
A		
A	atom mass number (sum of Z + N)	2
A	area	13
A	Ampere	9
(A)	AUGER electron	12
Á	(radio)activity	5
Á$_S$	specific (radio)activity	5
^{THEORY}Á$_s$	theoretical (maximum) specific (radio)activity	5
A^{SARGENT}	coefficients	8
A$^{GEIGER/NUTTAL}$	coefficients	9
^{S}A$_V$	mass number of a sphere of volume V	10
^{S}A$_S$	mass number of a sphere of surface S	10
[A], [a]	concentration of chemical species (mol/l) in chemical reactions	13
a$^{D-S,R}$	coefficient in DASGUPTA–SCHUBERT, REYES equation	9
a^{SARGENT}	coefficients	8
a$^{GEIGER/NUTTAL}$	coefficients	9
a als ^{E}r$_a$	radius of an ellipsoid	10
Å	Angström (1 Å = 10^{-10} m = 100 pm)	1
$^{qs}\alpha$	quadrupole stretching factor	10
α^{LDM}	coefficient of the WEIZSÄCKER equation	3
α	alpha particles	
$\alpha_{A,a,B,b}$	stoichiometric coefficients	13
$\alpha^{COULOMB}$	coefficients	10
$\alpha^{SURFACE}$	coefficients	10
α^H	alpha-line of hydrogen emission series	1
α^{FS}	fine-structure constant	1
α_S, α_C	coefficients for surface and coulomb energy	10
α_{IC}	coefficients for internal conversion	12

Tab. 14.5 (continued)

Symbol/ abbreviation	Meaning	Chapter
B		
B	retardation (Bremszahl)	13
$B^{SARGENT}$	coefficients	8
$B^{GEIGER/NUTTAL}$	coefficients	9
[B], [b]	concentration of chemical species in mol/l in chemical reactions	13
$b^{SARGENT}$	coefficients	8
$b^{GEIGER/NUTTAL}$	coefficients	9
b	barn	13
$b^{D-S,R}$	coefficient in DASGUPTA-SCHUBERT, REYES equation	9
b as $^{E}r_b$	radius of an ellipsoid	10
β	beta particles, beta electron	6
β^{LDM}	coefficient of the WEIZSÄCKER equation	3
C		
C	coulomb	1
C als C^{FERMI}	constant in FERMI'S golden rule	7
C_1, C_2	constants in TAAGEPARA, NURMIA equations	9
$\acute{C}$	count rate	5
C^*	compound nucleus, composite nucleus	13
(CK)	COSTER–KRONIG electron	12
c	speed of light	1
$c^{D-S,R}$	coefficient in DASGUPTA-SCHUBERT, REYES equation	9
c as $^{E}r_b$	radius of an ellipsoid	10
D		
d	range	8
d^{max}	maximum range of β particles	8
d_{air}^{max}	maximum range of β particles in air	8
d	subshell of electrons or nucleons	1
d_V	specific volume elements of the volume V of the nucleus	2

Tab. 14.5 (continued)

Symbol/ abbreviation	Meaning	Chapter
$\varnothing$	diameter	2
Δ	difference	1
Δ	LAPLACE operator $\nabla \cdot \nabla$, ∇^2 or Δ	1
$\rightarrow \nabla$	LAPLACE operator $\rightarrow \nabla^2 = \Delta$	1
δ^{LDM}	coefficient of the WEIZSÄCKER equation	3
E		
E	symbol of the chemical element	2
E	energy	1
$E^{SURFACE}$	surface energy	10
$E^{COULOMB}$	coulomb energy	10
E^C	coulomb barrier	13
EC	electron capture	6
$^{RECOIL}E_{K2}$	recoil energy	8
$\bar{E}_\beta$	mean kinetic energy of β particles	2
$E_\beta{}^{max}$	maximum kinetic energy of β particles	7
$E_{B(e)}$	electron binding energy	1
$E_{I(e)}$	ionization energy	1
E^*	excitation energy	13
E_o	zero-point energy (ground state energy)	1
E_B	nucleon binding energy	2
$\bar{E}_B$	mean nucleon binding energy	2
E_s	nucleon separation energy	3
δE_s	partial nucleon separation energy	3
$\check{E}_a$	threshold energy	13
$\hat{E}$	energy operator	1
$\acute{E}_n$	eigenvalues	1
E_{if}	density of energy states	13
E_i	electric multipoles	11

Tab. 14.5 (continued)

Symbol/ abbreviation	Meaning	Chapter
E_{IC}	kinetic energy of an internal conversion electron	11
e	natural (EULER'S) constant (e = 2.71828183)	5
e	electron	1
e_ρ	electrical charge density	1
eV	electron volt	1
ε	electron capture	6
$\varepsilon^{DEFORMATION}$	degree of nuclei deformation, eccentricity	10
ε	coefficients: $\varepsilon_{emission}$, $\varepsilon_{detector}$, $\varepsilon_{geometry}$	5
F		
F	restoring force of an oscillator	1
F	FERMI correction function	8
f	subshell of electrons or nucleons	1
f	frequency factor, reduced transition probability	9
φ	angle in a polar coordinate system	1
G		
G	GAMOW factor	9
g	gram	1
Γ	velocity of the electron relative to the speed of the light	12
γ	photon	6
γ^{LDM}	coefficient of the WEIZSÄCKER equation	3
H		
H	(isotopic) abundance	13
H	HAMILTON function	1
$\hat{H}$	HAMILTON operator	1
H	hindrance factor	9
H_{fb}	fission barrier	10
H	eigenvalue	1
h	PLANCK constant	1

Tab. 14.5 (continued)

Symbol/ abbreviation	Meaning	Chapter
$\hbar$	"reduced" PLANCK constant $\hbar = h/2p$	1
$\Delta_R H^-$	chemical reaction enthalpy	13
I		1
I	intensity	11
IC	internal conversion	11
J		
J	joule	1
J	total angular momentum	1
j	total angular momentum quantum number	1
K		
K	electron main shell	1
K^{EQU}	chemical reaction equilibrium constant	13
$^{\odot}K$	excited nuclear state	6
$^m K$	metastable nucleus	6
$^g K$	ground state nucleus	6
k	spring constant of an oscillator	1
k^+, k^-	chemical reaction rates	13
K^B	BOLTZMANN constant	14
k^C	COULOMB'S constant	9
L		
L	electron main shell	1
L	length	1, 13
L	orbital quantum moment	1
LDM	liquid drop model	3
l	orbital quantum number	1
λ	wavelength	1
λ	transformation (or "decay") constant	5
λ_{total}	sum of transformation constants	11

Tab. 14.5 (continued)

Symbol/ abbreviation	Meaning	Chapter
M		
M	electron main shell	1
M	molarity	13
M	muliplicity	1, 2
M	molecules	12
M_{fi}	matrix element in FERMI'S golden rule	7
MP	multipole	11
M_i	magnetic multipoles	11
m	mass	1
m_o	rest mass	2
m	magnetic quantum number	1
m_s	magnetic quantum number referred to the angular spin momentum of the electrons	1
Δm^{defect}	mass defect	2
Δm^{excess}	mass excess	2
μ^-	muon	7
N		
N	neutron number	2
N	electron main shell	1
N^{magic}	magic proton number	3
$N^{AVOGADRO}$	AVOGADRO constant or LOSCHMIDT number ($1\ mol = 6.02 \cdot 10^{23}$ atoms)	13
Ń	number of unstable nuclides	5
NSM	shell model	3
n	main electron shell number	1
n	main (principal) quantum number	1
n	neutron	xxx
n	number of (nuclear) states	7
n	exponent	8
v_e	electron neutrino	7

Tab. 14.5 (continued)

Symbol/ abbreviation	Meaning	Chapter
$\bar{\nu}_e$	electron anti-neutrino	7
ν_μ	muon neutrino	7
$\bar{\nu}_\mu$	muon anti-neutriono	7
ν_τ	tau neutrino	7
$\bar{\nu}_\tau$	tau anti-neutriono	7
O		
O	electron main shell	1
Ω	solid angle in a polar coordinate system	1
ω	eigenfrequency of the harmonic oscillator	1
P		
P	electron main shell	1
P	parity operator	8
P	penetrability, penetration factor, transition factor	9
P_{fi}	probability of transition in FERMI's golden rule	7
PSE	Periodic System of (chemical) Elements	1
p	impulse	1
p	proton	1
p	subshell of electrons or nucleons	1
Π	parity	8
ψ	wave function	1
Ψ_i and Ψ_f	wave functions of initial and final nuclear states	7
Q		
Q	electron main shell	1
Q	energy in nuclear transformations	7
Q	energy in nuclear reactions	13
Q	quadrupole	10
$^S Q$	quadrupole of a sphere	10
$^E Q$	quadrupole of an ellipsoid of rotation	10

Tab. 14.5 (continued)

Symbol/ abbreviation	Meaning	Chapter
q	electric charge	1, 2
R		
$R_{n,l}$	radial distance in a polar coordinate system	1
R_H	RYDBERG constant	1
R	reaction rate (of nuclear reactions)	13
r	radius	1
r_0	radius parameter	3
r^{mass}	mass radius	2
r^{charge}	charge radius	2
r_0^{mass}	mass radius parameter	2
r_0^{charge}	charge radius parameter	2
$^s r$	radius of a sphere	10
r^C	radius at coulomb barrier	9
$r_{1/2}$	average radius value	
$\sqrt{<r^2>}$	effective radius: "root mean square" (r.m.s.)	2
r_{long}, r_{short}	radius of an ellipsoid of rotation	10
ρ^{psf}	phase space factor	13
ρ	density	2
ρ_0	ultimate density of matter	2
S		
S	factor in electromagnetic multipole transitions	11
S	stopping power	13
s	spin	1
$^s S$	surface of a sphere	10
$^E S$	surface of an ellipsoid of rotation	10
$S^{TENSION}$	surface tension factor	10
s	subshell of electrons or nucleons	1
sf	spontaneous fission	6

Tab. 14.5 (continued)

Symbol/ abbreviation	Meaning	Chapter
σ	cross-section	3
σ	nuclear reaction cross-section	13
σ	sigma molecule orbitals	1
T		
T	temperature (absolute) in Kelvin	13
t	time	1
$t^{1/2}$	half-life	5
τ	"exponential time constant" or "mean lifetime" or "lifetime"	5
τ^-	tau	7
θ	angle in a polar coordinate system	1
U		
U	potential energy	1, 9, 10
U	harmonic potential	1
U_{HF}	high-frequency potential	13
U_M	ionization potential of a molecule	12
U^{pot}	ionization energy of the target material	13
U^{pot}	ionization potential	12
u	atomic mass unit	2
V		
V	volume	2
V	volume segment $V \approx \Delta p_x \Delta p_y \Delta p_z$	7
$^S V$	volume of a sphere	10
$^E V$	volume of an ellipsoid of rotation	10
$\bar{v}$	wavenumber	1
W		
W_{if}	nuclear reaction rate or probability	13
X		
x	position along a coordination system	1

Tab. 14.5 (continued)

Symbol/ abbreviation	Meaning	Chapter
x	fissility parameter	10
Y		
$Y_{l,m}$	angular positions in a polar coordinate system	
y	position along a coordination system	1
u	frequency	1
Z		
Z	chemical element number	2
Z	proton number (also: charge number)	1
Z^{magic}	magic proton number	3
Z_A	proton number at optimum mean nucleon binding energy for A = constant	6
z	position along a coordination system	1
ζ^{LDM}	coefficient of the WEIZSÄCKER equation	3

Index

https://doi.org/10.1515/9783110742725-015

www.ingramcontent.com/pod-product-compliance
Lightning Source LLC
Chambersburg PA
CBHW080126220326
41598CB00032B/4973